Materials Science and Engineering: Concepts and Applications

Materials Science and Engineering: Concepts and Applications

Edited by
Heather Dale

MURPHY & MOORE
www.murphy-moorepublishing.com

Published by Murphy & Moore Publishing,
1 Rockefeller Plaza,
New York City, NY 10020, USA

ISBN: 978-1-63987-353-1

Cataloging-in-Publication Data

Materials science and engineering : concepts and applications / edited by Heather Dale.
p. cm.
Includes bibliographical references and index.
ISBN 978-1-63987-353-1
1. Materials. 2. Materials science. 3. Engineering. I. Dale, Heather.
TA403 .M38 2022
620.11--dc23

For information on all Murphy & Moore Publications
visit our website at www.murphy-moorepublishing.com

MURPHY & MOORE

Contents

Preface VII

Chapter 1 **Mechanical Properties of Scrap Tire Crumbs-Clayey Soil Mixtures Determined by Laboratory Tests** 1
ShanShan Li and Dayong Li

Chapter 2 **Study on the Permeability Characteristics of Polyurethane Soil Stabilizer Reinforced Sand** 11
Jin Liu, Xiaohui Qi, Da Zhang, Qiao Feng, Yong Wang and Debi Prasanna Kanungo

Chapter 3 **The Compressibility and Swell of Mixtures for Sand-Clay Liners** 25
Muawia A. Dafalla

Chapter 4 **In Situ Testing of Square Footing Resting on Geobelt-Reinforced Gravel Thin Cushion on Soft Silt** 34
Wei Zhang, David Du and Xiaohong Bai

Chapter 5 **Estimation of Sand Production Rate Using Geomechanical and Hydromechanical Models** 46
Son Tung Pham

Chapter 6 **Experimental Investigation on the Deformation, Strength and Acoustic Emission Characteristics of Sandstone under True Triaxial Compression** 56
Zhaolin Li, Lianguo Wang, Yinlong Lu, Wenshuai Li and Kai Wang

Chapter 7 **Electric Resistance Tests on Compacted Clay Material under Dynamic Load Coupled with Dry-Wet Cycling** 72
Zheng Lu, Xiaowen Wu, Zhi Hu, Shaohua Xian and Ran Fang

Chapter 8 **Intelligent Machine Vision Based Modeling and Positioning System in Sand Casting Process** 78
Shahid Ikramullah Butt, Umer Asgher, Umar Mushtaq, Riaz Ahmed, Faping Zhang, Yasar Ayaz, Mohsin Jamil and Muhammad Kamal Amjad

Chapter 9 **Investigation into the Heat of Hydration and Alkali Silica Reactivity of Sustainable Ultrahigh Strength Concrete with Foundry Sand** 89
Federico Aguayo, Anthony Torres, Tate Talamini and Kevin Whaley

Chapter 10 **Strength and Stiffness of Stabilized Alluvial Silt under Frost Actions** 100
Haibo Wang, An Deng and Ping Yang

Chapter 11 **Experimental Investigation on the Behavior of Iron Powder-Reinforced Sand under Electromagnetic Field** 113
Ying Lai, Bin Zhu and Xiangtian Xu

Chapter 12 **The Effect of Polymer-Fiber Stabilization on the Unconfined Compressive Strength and Shear Strength of Sand** 128
Jin Liu, Qiao Feng, Yong Wang, Yuxia Bai, Jihong Wei and Zezhuo Song

Chapter 13 **Discretization of the Water Uptake Process of Na-Montmorillonite Undergoing Atmospheric Stress: XRD Modeling Approach** 137
Walid Oueslati and Mahdi Meftah

Chapter 14 **Dynamic Behavior of Clay-Aggregate Mixtures** 154
Kang Fei and Jinxin Xu

Chapter 15 **Experimental Study on Sedimentary Rock's Dynamic Characteristics under Creep State Using a New Type of Testing Equipment** 164
Wanpeng Huang, Wenbin Xing, Shaojie Chen, Yang Liu and Kai Wu

Chapter 16 **Microstructures, Mineral Compositions and Mechanical Properties of Red-Layers** 177
Huizhi Zhang and Jifeng Liu

Chapter 17 **Chemical, Mineralogical and Refractory Characterization of Kaolin in the Regions of Huayacocotla-Alumbres** 186
L. M. Romero-Guerrero, R. Moreno-Tovar, A. Arenas-Flores, Y. Marmolejo Santillán and F. Pérez-Moreno

Chapter 18 **Experimental Research on Microstructure and Physical-Mechanical Properties of Expansive Soil Stabilized with Fly Ash, Sand and Basalt Fiber** 197
Qin-yong Ma, Zi-ming Cao and Pu Yuan

Chapter 19 **Aggregate Evolution Mechanism during Ion-Adsorption Rare Earth Ore Leaching** 210
Kui Zhao, Yulong Zhuo, Xiaojun Wang and Wen Zhong

Chapter 20 **A Nonlinear Strength Criterion for Frozen Sulfate Saline Silty Clay with Different Salt Contents** 220
Yanhu Zhao, Yuanming Lai, Jing Zhang and Chong Wang

Chapter 21 **A New Method to Simulate Free Surface Flows for Viscoelastic Fluid** 228
Yu Cao, Xiao-Guang Ren, Xiao-Wei Guo, Miao Wang, Qian Wang, Xin-Hai Xu and Xue-Jun Yang

Permissions

List of Contributors

Index

Preface

Every book is a source of knowledge and this one is no exception. The idea that led to the conceptualization of this book was the fact that the world is advancing rapidly; which makes it crucial to document the progress in every field. I am aware that a lot of data is already available, yet, there is a lot more to learn. Hence, I accepted the responsibility of editing this book and contributing my knowledge to the community.

The design and discovery of new materials falls under the domain of materials science and engineering. It primarily focuses on solids. It studies the factors which influence the structure, properties and performance of materials. The study of the relationship between the properties, structure and processing of materials is known as materials paradigm. The fields of nanotechnology, metallurgy and biomaterials have advanced with the understanding of this paradigm. The major categories of materials studied under this field include semiconductors, polymers, biomaterials, metals, nanomaterials and ceramics. The common methods of materials processing include welding, crystal growth, casting, sintering, glassblowing, etc. This book is a valuable compilation of topics, ranging from the basic to the most complex advancements in the field of materials science and engineering. There has been rapid progress in this field and its applications are finding their way across multiple industries. In this book, using case studies and examples, constant effort has been made to make the understanding of the difficult concepts of materials science as easy and informative as possible, for the readers.

While editing this book, I had multiple visions for it. Then I finally narrowed down to make every chapter a sole standing text explaining a particular topic, so that they can be used independently. However, the umbrella subject sinews them into a common theme. This makes the book a unique platform of knowledge.

I would like to give the major credit of this book to the experts from every corner of the world, who took the time to share their expertise with us. Also, I owe the completion of this book to the never-ending support of my family, who supported me throughout the project.

Editor

1

Mechanical Properties of Scrap Tire Crumbs-Clayey Soil Mixtures Determined by Laboratory Tests

ShanShan Li[1] and Dayong Li[1,2]

[1]College of Civil Engineering and Architecture, Shandong University of Science and Technology, Qingdao 266590, China
[2]College of Civil Engineering, Fuzhou University, Fuzhou 350116, China

Correspondence should be addressed to Dayong Li; ldydy@163.com

Academic Editor: Jun Liu

Some laboratory tests, such as Proctor compaction test, direct shear and cyclic direct shear tests, consolidation test, and unconfined compression test, were performed on scrap tire crumbs-clayey soil mixtures to study the mechanical properties of the mixtures. The results show that (1) the maximum dry unit weight and the corresponding optimum moisture content of the mixtures decrease rapidly with the increase of scrap tire crumbs content (C_{STC}), showing good potential for using the mixtures as lightweight fill material; (2) it is not possible to prepare the mixture when C_{STC} exceeds 30% due to the occurrence of cracks in the mixture after removing from a mould; and (3) the shear strength of mixtures approximately increases by 20% when C_{STC} is up to 30%, while the residual strength decreases by 15%, compared with that of pure clayey soil. During shearing, the dilation of the mixtures occurs, particularly under the condition of a high C_{STC} and a low vertical pressure. Besides, the compressive strength and consolidation settlement of the mixtures decrease with C_{STC} increasing. The results indicate that it is possible for scrap tire crumbs used to improve clayey soil, which is suitable to act as a fill material.

1. Introduction

Six hundred million scrap tires were generated in China by 2015, while two hundred million scrap tires still remained in stockpiles [1]. Huge quantities of scrap tires were generated in other countries as well. For example, nearly three hundred million scrap tires were generated in the European Union by 2015 [2]; two hundred million scrap tires were generated in the US [3]; and one hundred and fifty million scrap tires were estimated to be generated per year in India [4]. Land filling or stockpiling of scrap tires is prone to cause environmental problems, such as (a) largely occupied spaces, (b) health hazards (scrap tires will provide a natural breeding place for diseases caused by insects and rodents), and (c) air pollution (stockpiling of scrap tires could pose a major fire risk and then cause air pollution). Therefore, there exists an urgent need to explore new and beneficial ways to recycle or reuse scrap tires.

Scrap tire pieces have been applied in civil engineering, such as using as embankment fill material [5–8], drainage material to collect leachate landfill [9], retaining wall backfill material [10, 11], and bridge abutment fill material [12]. Laboratory testing and field demonstration of the application of sands mixed with scrap tire pieces with various shapes and sizes have shown that the ductility and shear strength of sands could be significantly enhanced [13]. Furthermore, the mixture has lower unit weight and higher compressibility, compared with clean sands [14].

It is concluded that scrap tire pieces with various shapes and sizes can be successfully used for modification of sand. However, it is necessary to study effects of amounts of scrap tire pieces on the mechanical performance for clayey soil. Hasan et al. [15] and Sellaf et al. [16] investigated the mechanical properties of scrap tire chips-cohesive clayey soil mixtures. The results revealed that it is possible to use clayey soil mixed with tire chips as a fill material. Attom et al. [17] found that increasing the content of scrap tire shreds can improve the shear strength and permeability of clayey soil but can cause a decrease in the plasticity index, the expansion pressure, and expansion potential, compared with that of pure

(a) (b)

FIGURE 1: Test materials. (a) Clayey soil. (b) Scrap tire crumbs.

TABLE 1: Components of clayey soil.

Sand content (%)	3.80
Silt content (%)	20.6
Clay content (%)	75.6
Soil classification	Silty clay

TABLE 2: Some properties of clayey soil.

Liquid limit (%)	37.2
Plastic limit (%)	17.1
Plastic index (%)	20.1
Nature void ratio	1.13
Specific gravity	2.72
Nature moisture content (%)	45.5

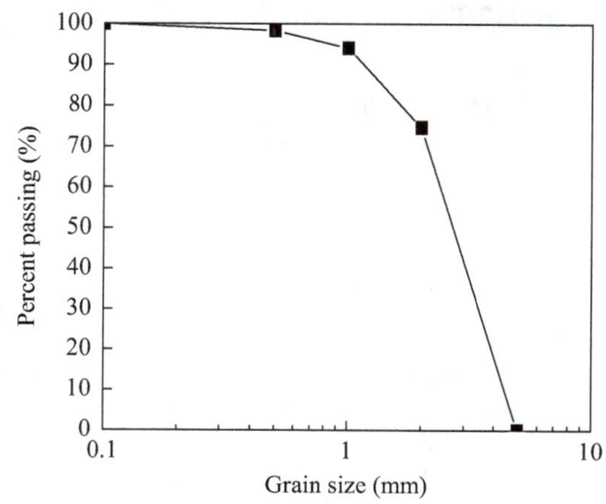

FIGURE 2: Particle-size distribution of scrap tire crumbs.

clayey soil. Kalkan [18] concluded that the silica fume-scrap tire rubber fiber mixture materials can be used to improve clayey soils.

The mechanical properties of scrap tire crumbs-clayey soil mixtures, such as shear strength, residual shear strength, unconfined compressive strength, and consolidation settlement, were explored by a series of laboratory tests. Three scrap tire crumbs contents (C_{STC}), 10, 20 and 30%, were used in this study. The effects of the C_{STC} on the mechanical properties of the mixtures were discussed.

2. Materials

2.1. Soil Specimens. Figure 1 shows the materials used in the experiments. The components of clayey soil are shown in Table 1. Besides, the engineering parameters of clayey soil were measured according to the Standard for Soil Test Method (GB/T50123) [19]. Such parameters are summarized in Table 2.

2.2. Scrap Tire Crumbs. Scrap tire crumbs were obtained from a local rubber processing factory. It should be noted that the steel and fluff in the scrap tire crumbs have been removed. The scrap tire crumbs vary from 0.5 to 4.5 mm in diameter. The specific gravity of scrap tire crumbs was tested to 1.15 according to ASTM C127 (2007) [20]. The sieve analysis of scrap tire crumbs was determined according to ASTM D422-63 (1998) [21]. The result of the sieve analysis is shown in Figure 2.

3. Experiment Descriptions

3.1. Sample Preparation. Hasan et al. [15] found that the shear strengths increase up to 30% for fine tire chip and 20% for coarse tire chip-cohesive clayey soil mixtures. The diameter of fine tire chips is less 0.425 mm, and the diameter of coarse tire chips is between 2 and 4.75 mm. Moreover, it is not possible to prepare the mixture when C_{STC} exceeds 30% due to the occurrence of cracks in the mixture after removing from a mould. Therefore, the content of scrap tire crumbs used in this study is less than 30%. The scrap tire crumbs content, C_{STC}, is defined as the ratio of weight of scrap tire crumbs to the total dry weight of the mixture [15].

The process of preparing the mixtures includes three steps:

Step 1: the clayey soil was oven-dried at approximately 65°C [19], and then the soil mass was ground to particles of less than 0.05 mm in diameter.

Step 2: the mixtures were compacted at the optimum moisture content to obtain the maximum dry unit

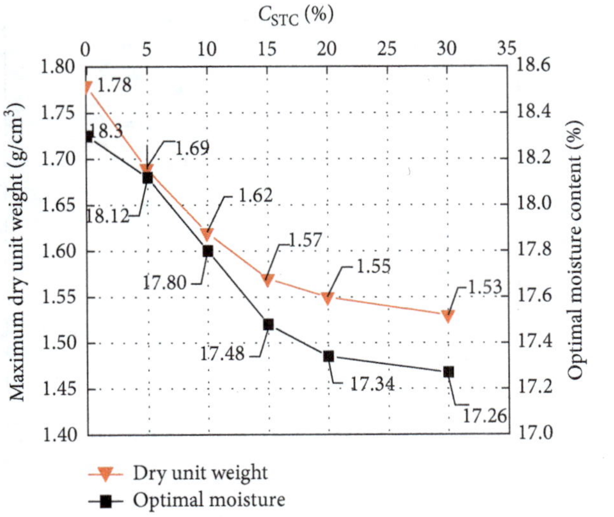

FIGURE 3: Variations of dry unit weight and the corresponding optimal moisture content for mixtures.

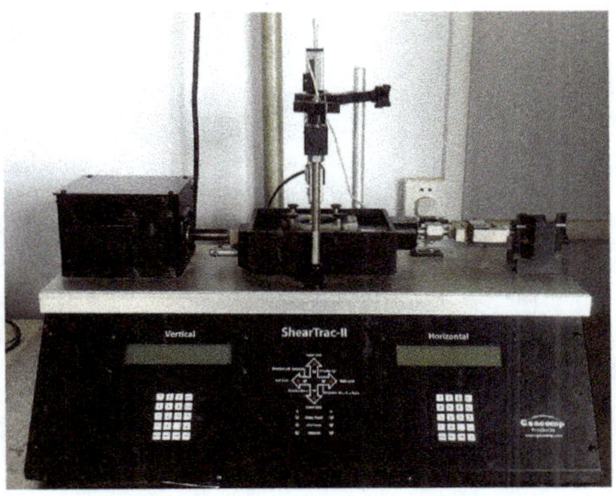

FIGURE 4: Direct/residual shear test apparatus (ShearTrac II).

weight. For the mixtures with various C_{STC}, weights of needed clayey soil, scrap tire crumbs, and water were calculated, respectively. The dry clayey soil particles and dry scrap tire crumbs were mixed uniformly in a laboratory mixer. Water needed was then added in the laboratory mixer until the optimum water content of the mixture is reached.

Step 3: the mixture was compacted in three equal-thickness layers with a hammer that delivers 25 blows to each layer. The hammer has a mass of 2.5 kg and has a drop of 30.5 mm. The mixture specimens were obtained using a standard ring sampler, and the specimens were saturated in a vacuum pump.

To determine the maximum dry unit weight and the corresponding optimum moisture of the mixtures, the Proctor compaction tests were carried out according to ASTM D698-78 (1989) [22]. The variations of maximum dry unit weight and the corresponding optimal moisture content for the mixtures are given in Figure 3. It is shown that the maximum dry unit weight and the corresponding optimum moisture of mixtures are less than that of pure clayey soil and decrease rapidly with C_{STC} increasing. It is similar to that

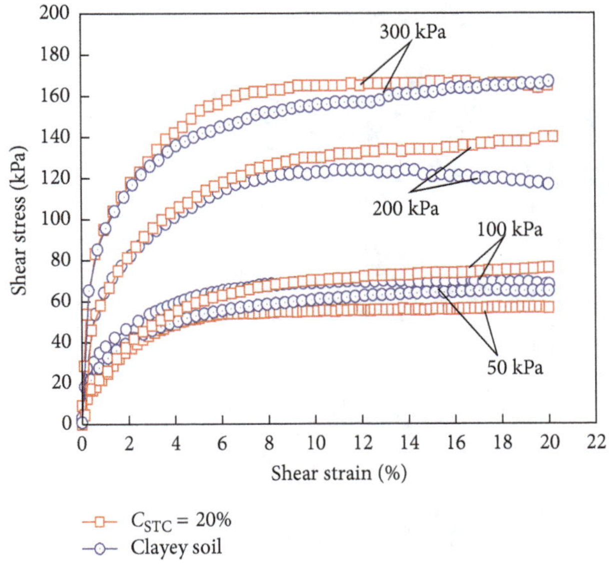

FIGURE 5: Variations of shear stress versus shear strain for mixtures.

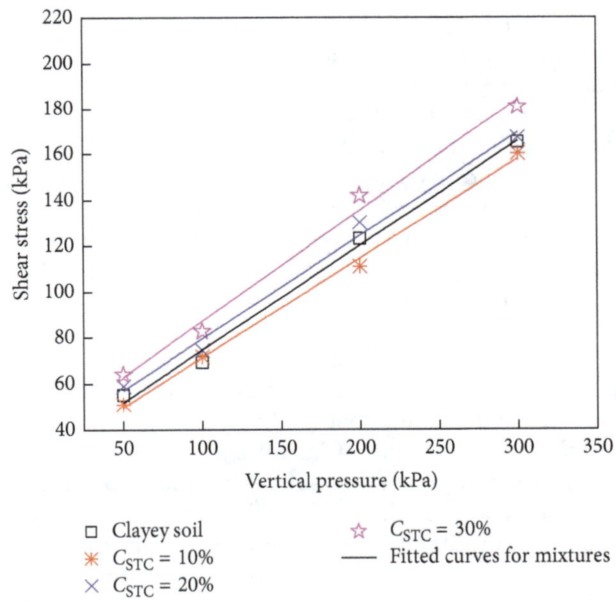

FIGURE 6: Mohr–Coulomb failure envelopes for mixtures.

found by Kalkan [18] and Akbulut et al. [23], who mixed clayey soil with scrap tire rubber fiber. The reason of the reduction in maximum dry unit weight of the mixtures is the low density of scrap tire crumbs [24]. In addition, bibulous rate of the scrap tire crumbs is very low compared with that of the clayey soil, resulting in the decrease of optimum moisture content for the mixtures.

3.2. Test Methods

3.2.1. Direct Shear Tests. Figure 4 shows the direct/residual shear test apparatus (ShearTrac II) used in this study. This apparatus is capable of performing the consolidation and shearing phases of a standard direct shear test and cyclic

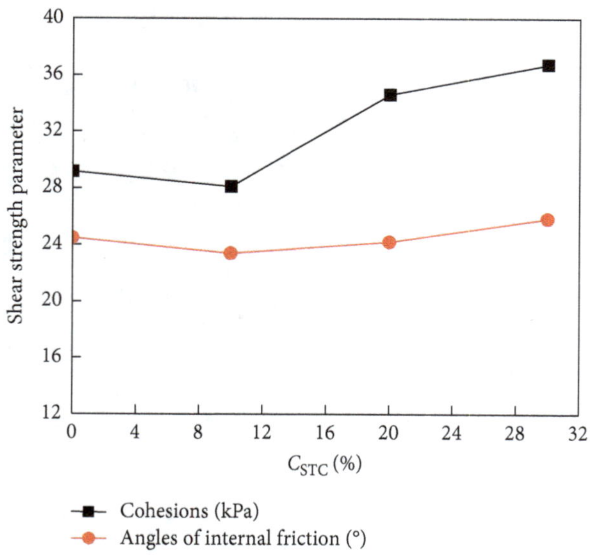

FIGURE 7: Variations of cohesion and internal friction angle for mixtures.

direct shear tests. Besides, it is capable of performing the consolidation tests under different stress paths. The specimens were 63.5 mm in diameter and 20 mm in height.

The direct shear tests were carried out according to the procedure described by ASTM D3080-98 (2003) [25]. The vertical pressures of 50, 100, 200, and 300 kPa were applied. During shearing, the shear rates were kept 0.1 and 0.8 mm/min. The shear strengths, shear parameters (cohesion and the angle of internal friction), and volume changes of the mixtures were derived from test results.

3.2.2. Cyclic Direct Shear Tests. The cyclic direct shear tests were carried out to elucidate the influence of C_{STC} on the residual shear strength of mixtures with the apparatus shown in Figure 4. The mixtures were placed in the shear box; after about 12 hours (to allow for consolidation), shearing was recommenced at a rate of 0.1 mm/min under the vertical pressure of 50, 100, 200, and 300 kPa, respectively.

3.2.3. Consolidation Tests. One-dimensional consolidation tests for the mixtures using incremental loading were conducted with the apparatus shown in Figure 4. The mixtures were consolidated at 50, 100, 200, and 300 kPa, respectively, and each vertical pressure increment is maintained until excessive pore water pressure completely dissipates.

3.2.4. Unconfined Compression Tests. The mixture specimens are 39.1 mm in diameter and are trimmed to 80 mm height. Compressive strengths of the mixtures with the C_{STC} of 10%, 20%, and 30% were obtained using a strain rate of approximately 3%/min under unconfined compaction conditions.

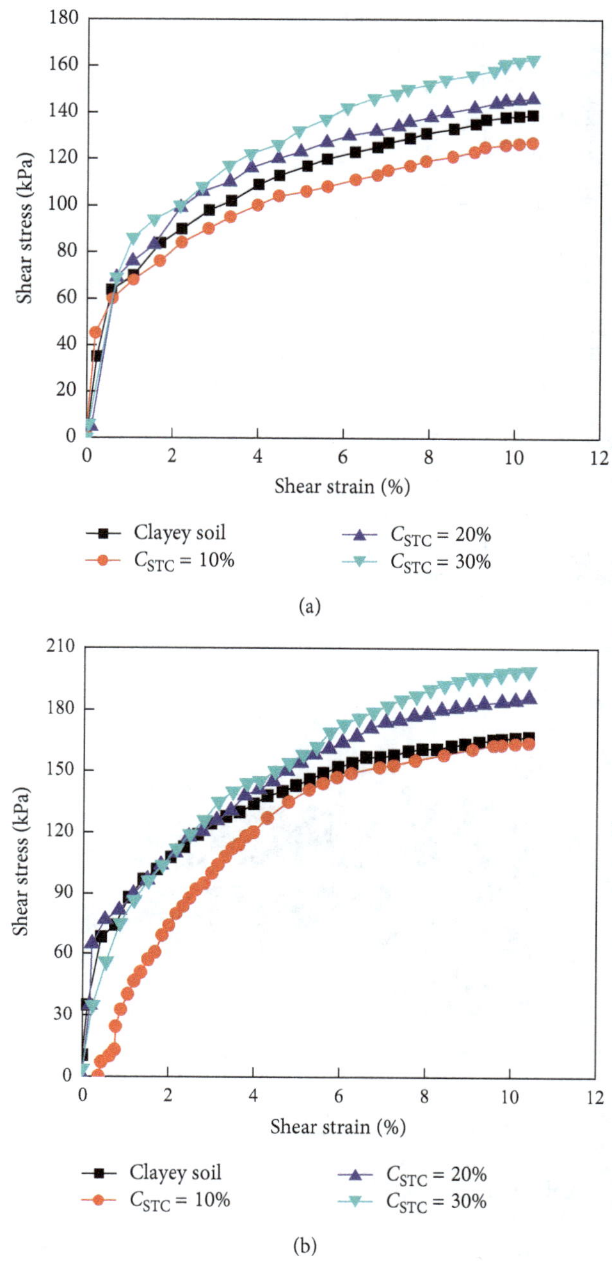

FIGURE 8: Shear stress-shear strain curves for mixtures under vertical pressure of 200 kPa. (a) Shear rate of 0.8 mm/min. (b) Shear rate of 0.1 mm/min.

4. Results and Discussion

4.1. Direct Shear Behavior of the Mixture

4.1.1. Effects of Vertical Pressure and C_{STC} on Shear Strength. Figures 5–7 compare the results of direct shear tests of pure clayey soil with those of the mixtures. Figure 5 shows some typical curves of the shear stress versus shear strain for the test specimens, where the shear stress has a general tendency to increase with the shear strain increasing. It is similar to the researches by Hasan et al. [15] and Tang et al. [26]. In addition, it should be noted that the shear strength of the mixture is equal to the shear stress at the shear strain of 10% [15].

Table 3: Effects of shear rates on shear strength of mixtures.

C_{STC} (%)	Shear rate (mm/min)	Vertical pressure (kPa)			
		50	100	200	300
0	0.1	56.6	83.5	157	215
	0.8	54.1	71.8	123	160
10	0.1	52.0	81.8	147	207
	0.8	44.0	71.6	115	156
20	0.1	65.0	92.5	166	219
	0.8	60.7	75.3	128	168
30	0.1	68.2	99.3	174	229

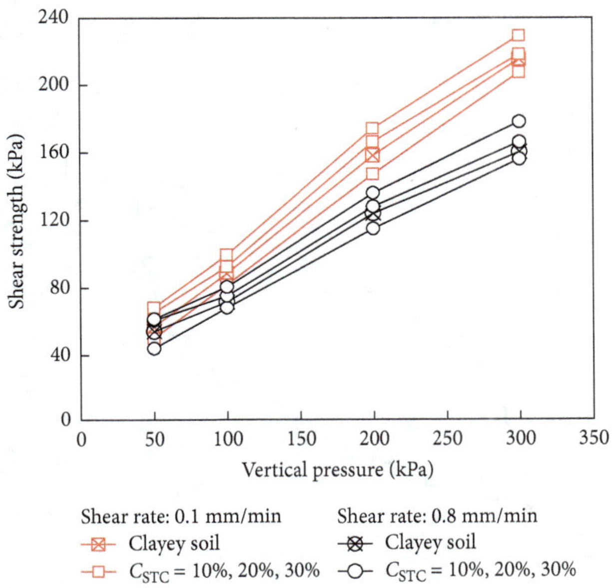

Figure 9: The shear strength of mixtures under different shear rates.

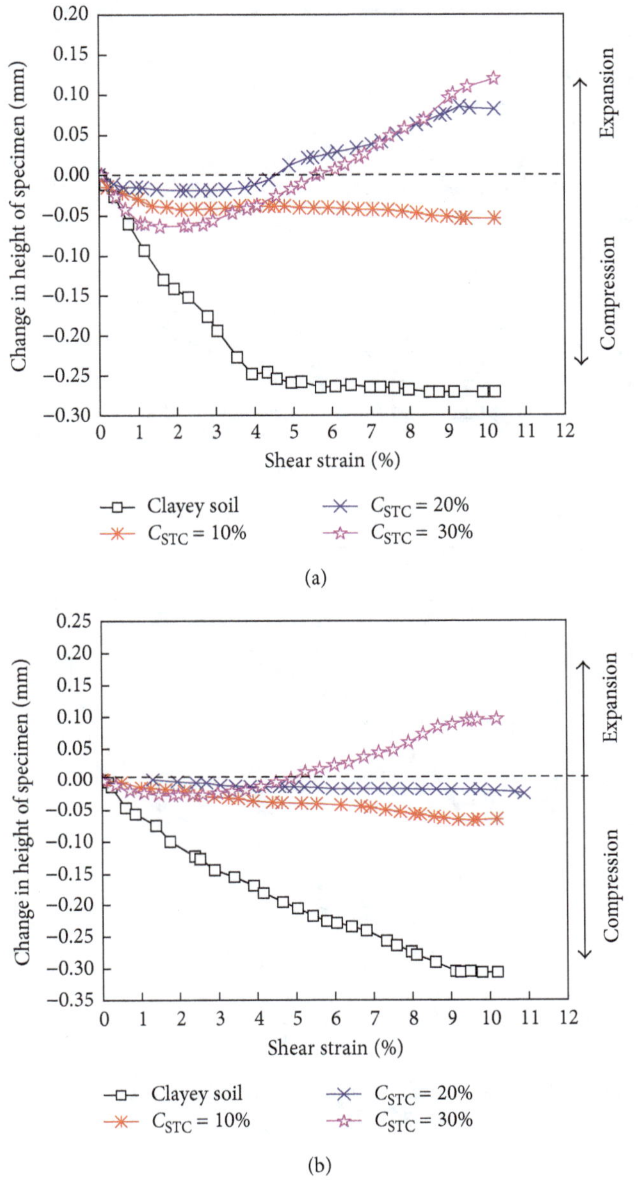

Figure 10: Variations of height of mixtures against shear displacement. (a) Under vertical pressure of 50 kPa. (b) Under vertical pressure of 300 kPa.

The Mohr–Coulomb failure envelopes and curves of the cohesion and angle of internal friction versus C_{STC} are shown in Figures 6 and 7. In Figure 6, each point represents the shear strength of mixtures under a certain vertical pressure, where the shear strengths of mixtures increase with the increase of vertical pressure. Besides, an approximately linear correlation between the shear strength and vertical stress was found in Figure 6. It is also found that C_{STC} has a strong influence on the shear strength of the mixture. The shear strength approximately enhances by 20% when C_{STC} is up to 30%. However, when C_{STC} is less than 10%, the shear strength of mixtures decreases compared with that of pure clayey soil.

The shear strength parameters of mixtures, such as the cohesion and the angles of internal friction, were determined to further analyze of the effect of C_{STC} on the shear strength of mixtures. The cohesion is calculated to be 29.15, 28.11, 34.61, and 36.74 kPa for pure clayey soil and the mixture with C_{STC} of 10, 20, and 30%, respectively. Moreover, the angle of internal friction is calculated to be 24.46°, 23.38°, 24.18°, and 25.83° for pure clayey soil and the mixture with C_{STC} of 10%, 20%, and 30%, respectively. The relationship between shear strength parameters and C_{STC} is illustrated in Figure 7, showing that C_{STC} has a significant influence on the cohesion of the mixture. However, the angle of internal friction of mixtures does not change significantly with C_{STC} increasing. With the increase of C_{STC}, there is an increase in the friction between clayey soil particles and scrap tire crumbs, as well as the friction between scrap tire crumbs and scrap tire crumbs. However, the value of C_{STC} is small ($\leq$30%) in this study. Furthermore, the dilation of mixtures during shearing will weaken the friction. Therefore, the angle of internal friction does not change significantly with the increase of C_{STC}.

4.1.2. Effects of Shear Rate on Shear Strength. Figure 8 shows the variations of shear stress with the increase of shear strain

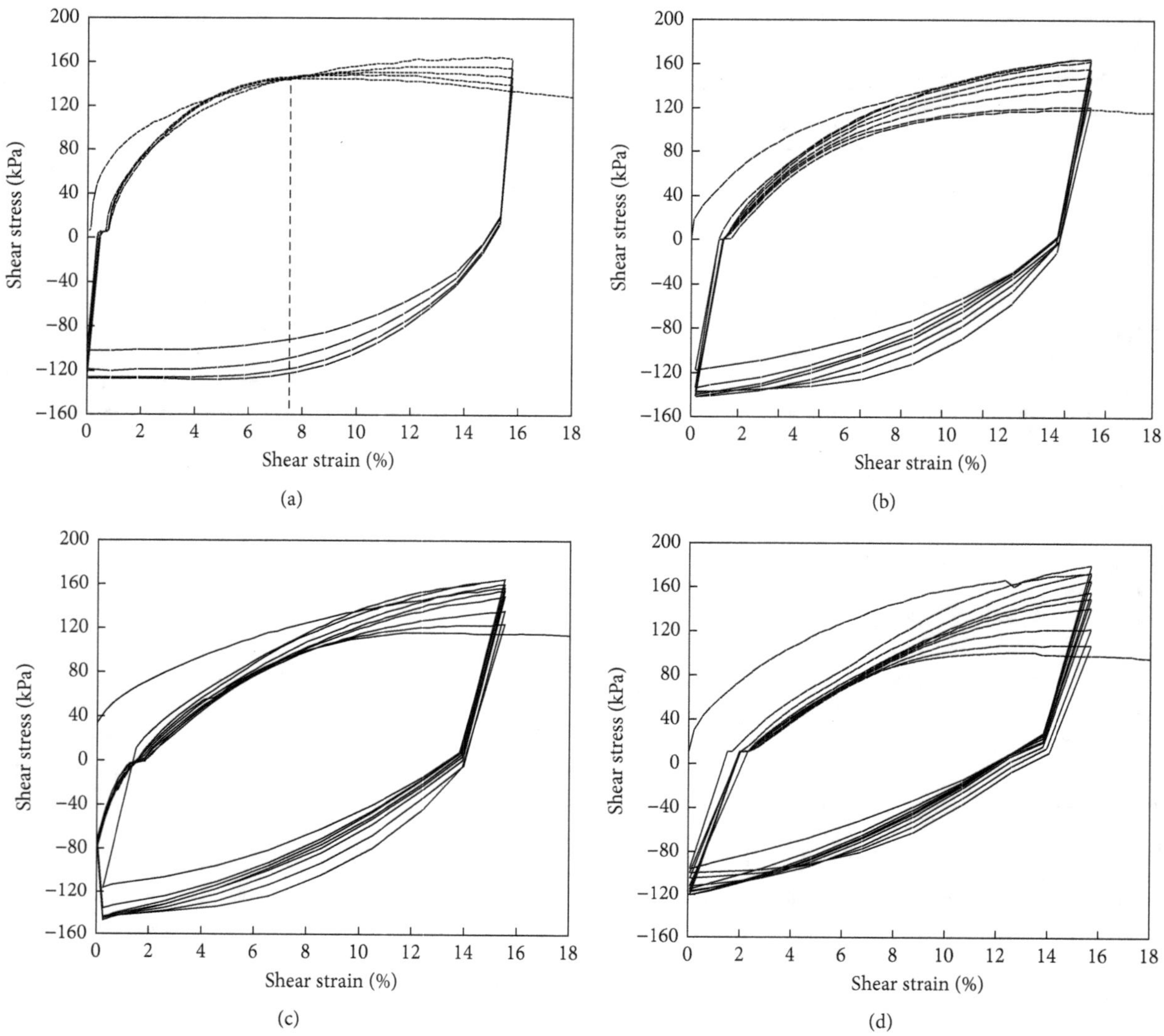

Figure 11: Variations of shear stress versus shear strain for mixtures under vertical pressure of 200 kPa. (a) Clayey soil. (b) $C_{STC} = 10\%$. (c) $C_{STC} = 20\%$. (d) $C_{STC} = 30\%$.

for mixtures under vertical pressure of 200 kPa. It indicates that there is no peak shear stress observed in the shear stress-shear strain curves. Therefore, the shear stress at the shear strain of 10% can be used as the shear strength of the mixtures. The shear strengths of mixtures with various C_{STC} are given in Table 3, and the Mohr–Coulomb failure envelops for mixtures are shown in Figure 9. The results show that when the vertical pressure is less than 100 kPa, the shear strength of mixtures does not vary significantly with the decrease of shear rate. However, when the vertical pressure exceeds 100 kPa, the shear strength significantly increases with the shear rate increasing.

4.1.3. Effects of Vertical Pressure and C_{STC} on Volumes of Mixtures. Figure 10 shows the volume changes of the mixtures (in height) versus shear strain for the mixtures and possible specimen volume compression (with a "−" sign) and dilation (with a "+" sign) shown in this figure. It was found that, when the vertical pressure is 50 kPa, the clayey soil and the mixture with C_{STC} of 10% compress during shearing. However, the mixture with C_{STC} of 20% initially compresses and then begins to dilate after the shear strain exceeds 4.5%. Furthermore, with C_{STC} of 30%, the mixture compressed at first and then began to dilate after the shear strain of 5% to 6%. This shearing dilation for scrap tire rubber shreds-soil mixture was also observed by Hasan et al. [15]. However, when the vertical pressure is 300 kPa, during shearing the clayey soil and the mixtures with C_{STC} of 10 and 20% are compressed. It can be concluded that during shearing, the dilation of mixtures occurs, particularly under the condition of a high C_{STC} and a low vertical pressure.

4.2. Evaluation of Cyclic Direct Shear Tests Results. The residual shear strengths of mixtures were obtained from cyclic direct shear tests. In addition, the volume changes of mixtures were measured during cyclic direct shearing. The test results for 48 specimens are presented in Figures 11–13 and Table 4. Figure 11 presents the variations of shear stress

TABLE 4: The results of repeated direct shear tests for mixtures.

Rubber content (%)	Normal stress (kPa)	Peak strength (kPa)	Residual strength (kPa)	Shear displacement at residual strength (mm)	Moisture content increase ratio (%)
Clayey soil	100	83.5	78.5	45~48	1.58
	200	157	135		
	300	221	202		
10	100	71.6	68.7	65~70	1.15
	200	140	125		
	300	215	180		
20	100	88.5	63.5	83~86	1.46
	200	137	116		
	300	219	175		
30	100	95.3	62.6	94~97	1.27
	200	155	109		
	300	210	170		

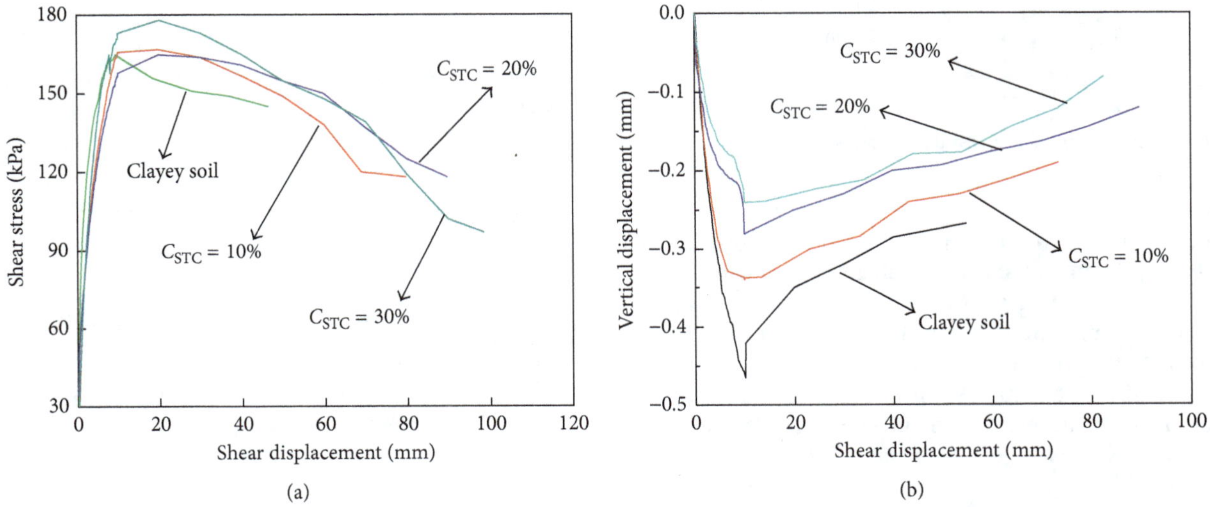

FIGURE 12: Plots of shear stress and vertical deformation of mixtures against shear displacement for mixtures under vertical pressure of 200 kPa. (a) Shear stress versus shear displacement. (b) Vertical displacement versus shear displacement.

versus shear strain for mixtures under vertical pressure of 200 kPa.

Figure 12(a) presents that the shear stress of mixtures increases initially, after reaching a maximum value, it decreases with the shear displacement increasing, indicating a strain-softening phenomenon for the mixtures. It also can be found that the peak strength of mixtures enhances with the increase of C_{STC}. Conversely, the residual shear strengths of mixtures rapidly decrease with C_{STC} increasing. The residual shear strength of the mixture with C_{STC} of 30% decreases by about 15%, compared with that of pure clayey soil. The reason is that the volume of void in the mixture is larger than that of the pure clayey soil.

Figure 12(b) shows that, during cyclic direct shearing, the vertical displacements decrease initially and then increase with increasing the shear displacement. It can be also observed that the mixtures can effectively limit the vertical deformation under the same shear rate and the same vertical pressure, compared with that of pure clayey soil.

The envelopes of the residual shear strength for the mixtures are shown in Figure 13. The results reveal that the residual shear strengths of mixtures gradually decrease with C_{STC} increasing. However, when the C_{STC} is larger than 20%, the effect of the C_{STC} on the residual strength of the mixture can be neglected.

It can be also concluded from Table 4 that the required shear displacement to obtain the residual shear strength of mixture increases with increasing the C_{STC} value. As the C_{STC} increases from 10 to 30%, the corresponding maximum shear displacements increase by 45.8%, 79.2%, and 102%, respectively, compared with that of pure clayey soil. Furthermore, changes in the moisture contents of the mixtures were measured (Table 4). It is concluded that the moisture content increases by an average of 1.3%, compared with the initial moisture content of the shear surface. The increase of moisture content is caused by the increase of the void ratio of mixture located in shear zone during cyclic shearing.

4.3. Evaluation of Consolidation Tests Results. The estimation of consolidation settlement is one of the most important

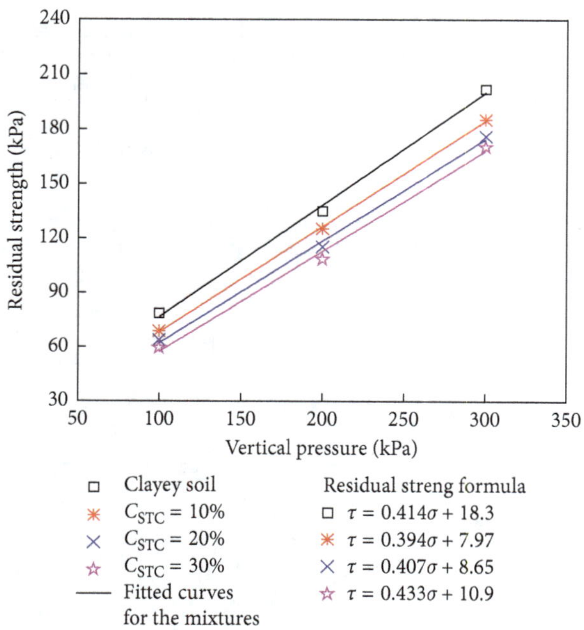

FIGURE 13: Envelopes of the residual shear strength for mixtures.

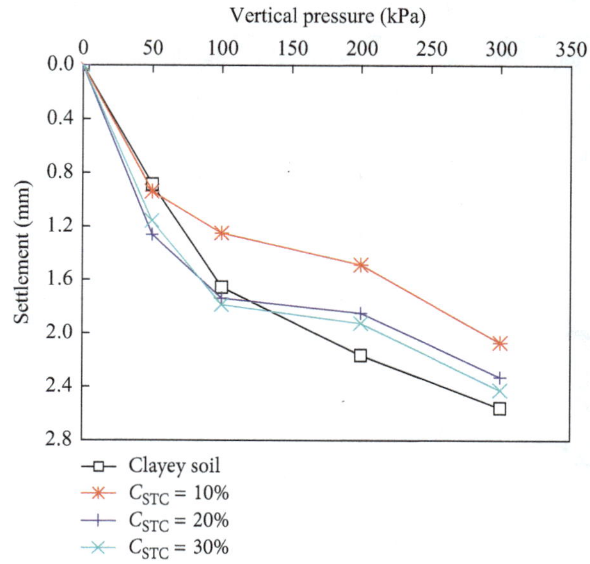

FIGURE 14: Settlement development with vertical pressure for mixtures.

procedures in the design of soft ground improvement projects [17]. Figure 14 shows the changes of consolidation settlement for pure clayey soil and the mixtures under various vertical pressures. When the vertical pressure is less than 100 kPa, the consolidation settlement of mixtures does not vary significantly with C_{STC} increasing. However, as the vertical pressure is greater than 100 kPa, the settlements of the mixtures with C_{STC} of 10, 20, and 30% decreased approximately 25%, 12%, and 8%, respectively, compared with that of pure clayey soil. The reason of the reduction in consolidation settlement for mixtures is that, with the increase of C_{STC}, the mixtures tend to be more elastic and resilience.

4.4. Compressive Properties

4.4.1. Effect of C_{STC} on Compressive Strength of Mixtures. For each compression test under the undisturbed, remolded, and compacted conditions using a strain-controlled axial load, the compressive strength of mixtures, which is taken as the load per unit area at 15% axial strain, can be obtained. The variations of axial stress for the mixtures with axial strain increasing are plotted in Figure 15. The results from Figure 15 present that the unconfined compressive strength of these specimens was decreased from 111 to 90 kPa, from 111 to 69 kPa, and from 111 to 66 kPa for pure clayey soil and mixtures with C_{STC} of 10%, 20%, and 30%, respectively. It is pointed out that, when C_{STC} is 20%, there is an approximately 40% decrease in compressive strength of the mixture compared with that of pure clayey soil. However, when C_{STC} exceeds 20%, the effect of C_{STC} on the compressive strength is weak for the mixtures. The decrease in the compressive strength of mixtures with C_{STC} increasing is attributed to the presence of scrap tire crumbs, which causes the increase of

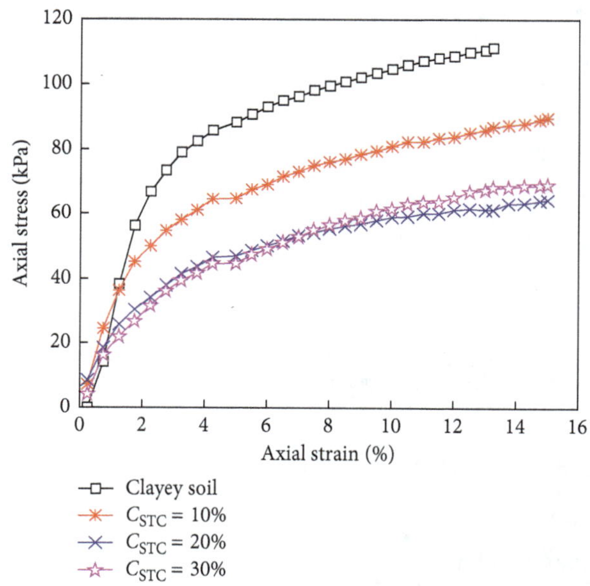

FIGURE 15: The axial stress-strain relationships for mixtures.

contact between scrap tire crumbs, resulting in higher resilience, higher deformation, and less strength.

4.4.2. Effects of C_{STC} on the Failure Modes of Mixtures. The failure modes of pure clayey soil and the mixture with C_{STC} of 10% by unconfined compression tests are presented in Figures 16(a) and 16(b), respectively. Figure 17 shows the failure modes of the mixture with C_{STC} of 20%; this failure mode is similar to that of the mixture with C_{STC} of 30%. As shown in Figures 16(a) and 16(b), the inclined plane failure occurs for pure clay and the mixture with C_{STC} of 10%. Instead, when the C_{STC} exceeds 20%, mixture specimens

(a) (b)

Figure 16: Inclined plane failure of (a) clayey soil and (b) mixture with C_{STC} of 10%.

Figure 17: Bulging failure of mixture with C_{STC} of 20%.

exhibit bulging failure. It is concluded that C_{STC} may have influence on the failure modes of the mixtures effectively.

5. Conclusions

A series of laboratory tests were performed on scrap tire crumbs-clayey soil to study the mechanical properties. The following conclusions can be obtained:

(1) Maximum dry unit weight and the corresponding optimum moisture content of the mixtures decrease with increasing the C_{STC} value.

(2) For a given C_{STC}, the direct shear strength of the mixtures increases with the increase in vertical pressure and decreases with the decrease in shear rate. Moreover, the dilation of the mixtures occurs during shearing, particularly under the condition of a high C_{STC} and a low vertical pressure.

(3) Residual shear strength, compressive strength, and consolidation settlement decrease markedly with C_{STC} increasing, while start to decrease slightly when the C_{STC} exceeds 20%.

Scrap tire crumbs-clayey soil mixture has the advantages of higher shear strength, and lower density and settlement compared with that of pure clayey soil. This mixture can be used in many geotechnical applications, such as backfills behind retaining structures and embankments on soft compressible soil.

Conflicts of Interest

The authors declare that there are no conflicts of interest regarding the publication of this paper.

Acknowledgments

This work is jointly supported by the SDUST Research Fund (Grant no. 2015TDJH104).

References

[1] http://news.blhbnews.com/2017/w_0323/109675.html.

[2] M. Miles, "RMA exec: more scrap tires likely," *Rubber and Plastics News*, vol. 46, pp. 1–12, 2016.

[3] D. Don, "Scrap tire management rules vary across world," *Rubber and Plastics News*, vol. 44, pp. 36–46, 2015.

[4] A. Al-tabbaa and T. Aravinthan, "Natural clay-shredded tire mixtures as landfill barrier materials," *Waste Management*, vol. 18, no. 1, pp. 9–16, 1998.

[5] D. N. Humphrey, "Effectiveness of design guidelines for use of tire derived aggregate as lightweight embankment fill," in *Proceedings of Recycled Materials in Geotechnics*, vol. 127, pp. 61–74, Baltimore, MD, USA, October 2004.

[6] M. A. Warith and S. M. Rao, "Predicting the compressibility behavior of tire shred specimens for landfill applications," *Waste Management*, vol. 26, no. 3, pp. 268–276, 2006.

[7] H. M. Young, K. Sellasie, D. Zeroka, and G. Sabnis, "Physical and chemical properties of recycled tire shreds for use in construction," *Journal of Environmental Engineering*, vol. 129, no. 10, pp. 921–929, 2003.

[8] U. Balunaini, S. Yoon, M. Prezzi, and R. Salgado, "Pullout response of uniaxial geogrid in tire shred-sand mixtures," *Mechanical and Geological Engineering*, vol. 32, no. 2, pp. 505–523, 2014.

[9] D. P. Duffy, "Using tire chips as a leachate drainage layer," *Waste Age*, vol. 26, pp. 113–122, 1995.

[10] D. N. Humphrey, T. C. Sandford, and M. M. Cribbs, "Shear strength and Compressibility of tire chips for use as retaining wall backfill," *Transportation Research Record*, vol. 1422, pp. 29–35, 1999.

[11] S. Youwai and D. T. Bergado, "Numerical analysis of reinforced wall using rubber tire chips-sand mixtures as backfill material," *Computers and Geotechnics*, vol. 31, no. 2, pp. 104–114, 2004.

[12] N. Hataf and M. M. Rahimia, "Experimental investigation of bearing capacity of sand reinforced with randomly distributed tire shreds," *Construction and Building Materials*, vol. 20, no. 10, pp. 910–916, 2006.

[13] A. F. Cabalar, "Direct shear test on waste tires-sand mixtures," *Mechanical Geological Engineering*, vol. 29, no. 4, pp. 411–418, 2011.

[14] J. G. Zornberg, A. R. Cabral, and C. Viratjandr, "Behavior of tire shred-sand mixtures," *Canadian Geotechnical Journal*, vol. 41, no. 2, pp. 227–241, 2004.

[15] H. Cetin, M. Fener, and O. Gunaydin, "Geotechnical properties of tire-cohesive clayey soil mixtures as a fill material," *Engineering Geology*, vol. 88, no. 1-2, pp. 110–120, 2006.

[16] H. Sellaf, H. Trouzine, and M. Hamhami, "Mechanical properties of rubber tires and sediments mixtures," *Engineering, Technology and Applied Science Research*, vol. 4, pp. 618–624, 2014.

[17] M. Attom, T. Khedaywi, and S. A. Mousa, "The effect of shredded waste tire on the shear strength, swelling and compressibility properties of the clayey soil," *Journal of Solid Waste Technology and Management*, vol. 33, pp. 25–33, 2007.

[18] E. Kalkan, "Preparation of scrap tire rubber fiber-silica fume mixtures for modification of clayey soils," *Applied Clay Science*, vol. 80-81, pp. 117–125, 2013.

[19] The National Standard Compilation Group of People Republic of China, *Standard for Soil Test Method, GB/T50123 1999*, China Plans to Press, Beijing, China, 1999.

[20] ASTM, *Standard Test Method for Density, Relative Density (Specific Gravity), and Absorption of Coarse Aggregate. Annual Book of ASTM Standards*, Tech. Rep. ASTM C127–07 2007, ASTM International, Philadelphia, PA, USA, 2007.

[21] ASTM, *Standard Test Method for Particle-Size Analysis of Soils. Annual Book of ASTM Standards*, Tech. Rep. ASTM D422–63 1998, ASTM International, Philadelphia, PA, USA, 1998.

[22] ASTM, *Standard Test Method for Moisture-Density Relations of Soils and Soil Aggregate Mixtures using 5.5 lb (2.49 kg) Rammer and 12 in (305 mm) Drop. Annual Book of ASTM Standards*, Tech. Rep. ASTM D698–78, ASTM International, Philadelphia, PA, USA, 2007.

[23] S. Akbulut, S. Arasan, and E. Kalkan, "Modification of clayey soils using scrap tire rubber and synthetic fibers," *Applied Clay Science*, vol. 38, no. 1-2, pp. 23–32, 2007.

[24] A. Senol, "Effect of fly ash and polypropylene fibres content on the soft soils," *Bulletin of Engineering Geology and the Environment*, vol. 71, no. 2, pp. 379–387, 2012.

[25] ASTM, *Test Method for Direct Shear of Soils under Consolidated Drained Conditions. Annual Book of ASTM Standards*, Tech. Rep. ASTM D 3080 1998, ASTM International, Philadelphia, PA, USA, 1998.

[26] C. S. Tang, B. Shi, and L. Z. Zhao, "Interfacial shear strength of fiber reinforced soil," *Geotextiles and Geomenbranes*, vol. 28, no. 1, pp. 54–62, 2010.

2

Study on the Permeability Characteristics of Polyurethane Soil Stabilizer Reinforced Sand

Jin Liu,[1] Xiaohui Qi,[1] Da Zhang,[1] Qiao Feng,[1] Yong Wang,[1] and Debi Prasanna Kanungo[2]

[1] *School of Earth Sciences and Engineering, Hohai University, Nanjing 210098, China*
[2] *CSIR-Central Building Research Institute (CBRI), Roorkee 247667, India*

Correspondence should be addressed to Jin Liu; jinliu920@163.com

Academic Editor: Charles C. Sorrell

A polymer material of polyurethane soil stabilizer (PSS) is used to reinforce the sand. To understand the permeability characteristics of PSS reinforced sand, a series of reinforcement layer form test, single-hole permeability test, and porous permeability test of sand reinforced with PSS have been performed. Reinforcement mechanism is discussed with scanning electron microscope images. The results indicated that the permeability resistance of sand reinforced with polyurethane soil stabilizer is improved through the formation of reinforcement layer on the sand surface. The thickness and complete degree of the reinforcement layer increase with the increasing of curing time and PSS concentration. The water flow rate decreases with the increasing of curing time or PSS concentration. The permeability coefficient decreases with the increasing of curing time and PSS concentration and increases with the increasing of depth in specimen. PSS fills up the voids of sand and adsorbs on the surface of sand particle to reduce or block the flowing channels of water to improve the permeability resistance of sand. The results can be applied as the reference for chemical reinforcement sandy soil engineering, especially for surface protection of embankment, slope, and landfill.

1. Introduction

Natural sand material has certain deficiency in the permeability and stability of foundation, slope, and embankment for geotechnical engineering. The materials used in covering on the surface of sand or adding into sand body to reduce or prevent the water permeating into the sand body are mainly geomembrane, geotextile, concrete slab, plant, and so on [1–9]. Ghosh and Yasuhara [1] studied the flow characteristics of geosynthetic drain confined in soils undergoing consolidation. Raisinghani and Viswanadham [3] evaluated the permeability characteristics of a geosynthetic-reinforced soil. Palmeira and Gardoni [4] presented the drainage and filtration properties of nonwoven geotextiles under confinement using different experimental techniques. These materials can improve the permeability resistance and stability of the soil, but the properties of soil are not modified. While the material is moved or damaged, the high permeability or instability of soil will be reverted.

At present, the polymer materials of polyvinyl acetate (PVA), polyurethane, polyacrylamide, and have been considered as the new soil stabilizers to reinforce the soil [10–16]. Kukal et al. [10] studied the water-drop stability of PVA-treated natural soil aggregates from different land uses. Liu et al. [11] presented the polyurethane sand-fixing agent on sand fixation. Inyang and Bae [12] introduced the polyacrylamide sorption opportunity on interlayer and external pore surfaces of contaminant barrier clays. Yang et al. [14] studied the effects of aging tests on a novel chemical sand-fixing agent-polyaspartic acid. Telysheva and Shulga [15] introduced the silicon-containing polycomplexes for protection against wind erosion of sandy soil. Orts et al. [16] proposed the biopolymer additives to reduce erosion-induced soil losses during irrigation. These research results indicated that the polymer materials can improve the strength, water stability, antierosion, and permeability of soil through filling up the soil voids and enwrapping the soil particles.

In this paper, the polyurethane soil stabilizer (PSS) is sprayed on sand surface to form a reinforcement layer, and

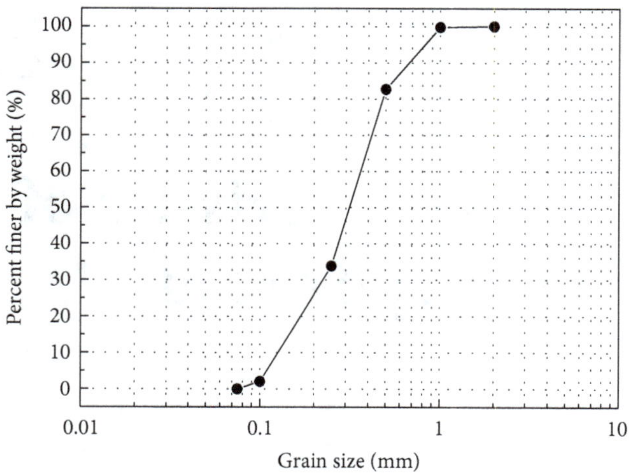

FIGURE 1: Grain size distribution of sand.

the reinforcement layer form test, single-hole permeability test, and porous permeability test are performed in laboratory to analyze the effects of PSS reinforced on sand. Its reinforcement mechanism is discussed with images of scanning electron microscope (SEM). Based on the analyses of interaction between PSS and sand, the optimal curing time and PSS concentrations are suggested. The results can be applied as the reference for PSS reinforced sand engineering, especially for surface protection of embankment, slope, and landfill.

2. Experimental Outlines

2.1. Materials. The sand selected from Nanjing City of Jiangsu Province, China, was used in this study. Its particle size distribution is shown in Figure 1. It has a specific gravity (Gs) of 2.64, maximum dry density (ρ_{max}) of 1.66, minimum dry density (ρ_{min}) of 1.34 g/cm^3, maximum void ratio (e_{max}) of 0.970, and minimum void ratio (e_{min}) of 0.590. Its particles have a mean grain size (D_{50}) of 0.30 mm, gradation coefficient (C_g) of 1.13, and uniformity coefficient (C_u) of 2.77.

A new type of polyurethane soil stabilizer (PSS) was used in this study (Figure 2). It has a main constituent of polyurethane resin and contains enormous amount of functional group -NCO. PSS is light yellow of oil liquid with a pH of 6-7, viscosity of 650–700 mPa·s, specific gravity of 1.18 g/cm^3, solid content of 85%, coagulation time of 30–1800 s, and hold water content of larger than 40 times. The coagulation time decreases with the increase in PSS concentration. The PSS concentrations of 1%, 3%, 5%, 7%, and 9% used in this study have the coagulation times about 30 min, 8 min, 6 min, 4.5 min, and 3.5 min, respectively. PSS is mainly used for the reinforcement of sandy soil. The sand particle contents of reinforced soil are more than 50%.

PSS has the primary advantages: (a) it reacts with water to form elastic and viscous reinforced layer on soil surface with excellent mechanical property; (b) it is an environment-friendly product with no additional pollution, and it is a kind of biodegradable water-soluble polymer; (c) it is easy to produce and has a low cost.

FIGURE 2: Photo of polyurethane soil stabilizer (PSS).

2.2. Experimental Methods. PSS as a soil reinforcement material is sprayed on sand surface to form a reinforcement layer. The initial permeability time and the permeability coefficient of sand are considered as important factors to evaluate the permeability characteristics of sand reinforced with PSS. In this study, the laboratory tests of reinforcement layer form test, single-hole permeability test, and porous permeability test were performed to evaluate permeability characteristics of PSS reinforced sand.

2.2.1. Reinforcement Layer Form Test. The formation of reinforcement layer on sand surface is important to slope surface protection. In reinforcement layer form test, the reinforcement layer thickness of the specimen was measured. Sand specimens taken from slope surface were first oven dried and then filled in a container. The container has a diameter of 10 mm and height of 6 mm. The dry weight of each specimen was 500 g. Six groups of different concentrations of 0%, 1%, 3%, 5%, 7%, and 9% of PSS solution and 24 specimens of each group were proposed. The amount 1.6 L/m^2 of each solution was sprayed on the specimen surface uniformly. After spraying, the reinforced specimens were kept in curing box with a temperature around 20°C, and the reinforcement layer thickness was tested with different curing time of 0 h, 0.1 h, 0.5 h, 1 h, 3 h, 6 h, 12 h, and 24 h. The block unit of quadrate area 3 × 3 cm was taken from the reinforcement layer on specimen surface and its thickness was measured by vernier caliper. Additionally, three specimens were measured for each test and their average values were used.

2.2.2. Single-Hole Permeability Test. In single-hole permeability test, the permeability characteristic of PSS reinforced surface sand was tested by TST-70 permeameter (Figure 3). This apparatus has a top inlet and a bottom outlet tube. The top inlet tube is connected with water tank which keeps a constant head.

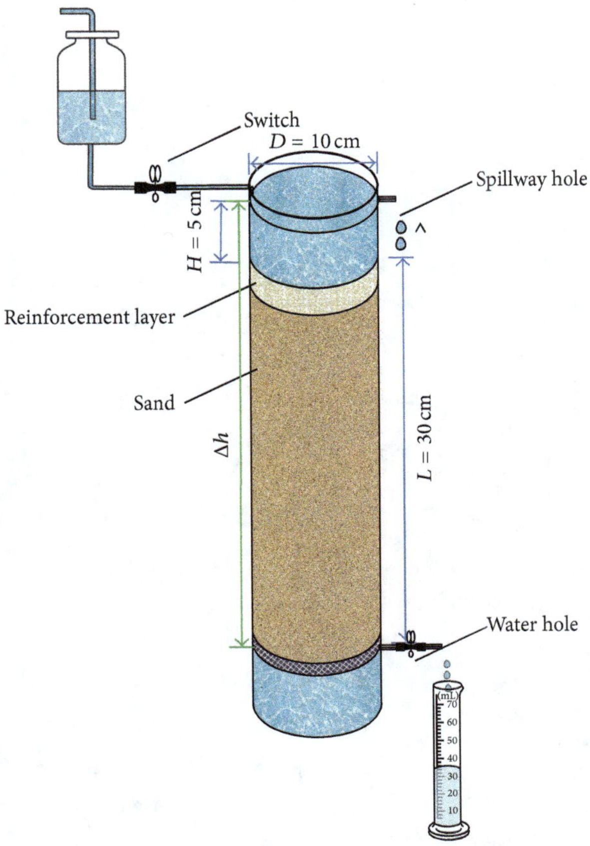

FIGURE 3: TST-70 permeameter used in single-hole permeability tests.

The specimen was directly formed in the testing apparatus. The dry sand was divided into 10 equal parts and each part was put into the box of permeameter and compacted. Specimens for this test with a diameter 10 cm and a height 30 cm were prepared in ten layers of equal height to achieve the proposed density 1.44 g/cm^3. After that, six concentrations of 0%, 1%, 3%, 5%, 7%, and 9% of PSS solution were sprayed on sand surface. The amount 1.6 L/m^2 of each solution was sprayed on the specimen surface uniformly. After spraying, the specimens were kept in curing box with a temperature around 20°C, and the permeability test was carried out with different curing time of 0 h, 3 h, 6 h, 12 h, and 24 h. Firstly, the top inlet and bottom outlet tube were kept open; the water level of specimen surface was maintained. The initial time of water running from specimen surface to the bottom tube was recorded. The running water volume in every 3 minutes was recorded until the water volume difference for successive two times is less than 3%. The average water flow rate for each 3 minutes was given by the following equation [17]:

$$F = \frac{Q}{\Delta t}. \quad (1)$$

In (1), F (ml/min) is defined as the average water flow rate during the designed record time; Q (ml) is water volume during the designed record time interval; Δt (min) is the designed record time interval. In this study, the average water flow rate (F) represents the water flow rate of the middle of record time. Therefore, F of record times of 0–3 min, 3–6 min, 6–9 min, 9–12 min, and 12–15 min are defined as the water flow rate of test time being 1.5 min, 4.5 min, 7.5 min, 10.5 min, and 13.5 min, respectively.

The permeability coefficient of specimens is obtained by

$$K = \frac{QL}{A\Delta h\Delta t}. \quad (2)$$

In (2), K (cm/s) is defined as the permeability coefficient of specimen, L (cm) is the height of the specimen, A is the cross sectional area of the specimen (cm^2), and Δh (cm) is water head difference.

The permeability resistance of reinforced specimens was evaluated using

$$R = \frac{K_u}{K_r}. \quad (3)$$

In (3), R is relative permeability resistance of the reinforced sand, K_u is permeability coefficient of the unreinforced sand (cm/s), and K_r is permeability coefficient of the PSS reinforced sand (cm/s). The R values of unreinforced specimens are 1 and the specimens with no water flow are designed as ∞. The higher the permeability resistance, the more effective the PSS.

2.2.3. Porous Permeability Test. The permeability resistance of sand with different depth is an effective parameter to evaluate the reinforcement layer. The permeability coefficient of PSS reinforced surface sand with different depth was studied by porous permeability test. In this test, a self-manufactured permeameter shown in Figure 4 was used. This apparatus with diameter of 20 cm and height of 100 cm has 10 outlet tubes in one side. The interval between outlet tubes is 10 cm. The outlet tube number 0 is a spillway hole with a distance of 5 cm from the top.

The specimen was directly formed in the apparatus. All the tubes were closed at the beginning. The dry sand was divided into 9 equal parts and each part was put into the box of permeameter and compacted. Specimens for this test with a diameter of 20 cm and a height of 90 cm were prepared in 9 layers of equal height to achieve the proposed density 1.44 g/cm^3. After that, six concentrations of 0%, 1%, 3%, and 5% of PSS solution were sprayed on sand surface. The amount 1.6 L/m^2 of each solution was sprayed on the specimen surface uniformly. After spraying, the specimens were kept in curing box with a temperature around 20°C for 24 h to form the reinforcement layer. After the specimens were formed, the inlet tube and the outlet tubes numbers 0 and 9 were kept open, and the initial time of water running from specimen surface to the outlet tube number 9 and its running water volume in every 3 minutes were recorded. After that, the operation of close number 9 and open number 8 was performed and its running water volume of every 3 minutes with a number of 5 times was recorded. Then, the outlet tubes numbers 7 to 1 were open in sequence to record their running water volume. It should be noted that only one number of outlet tubes was open to record its running water

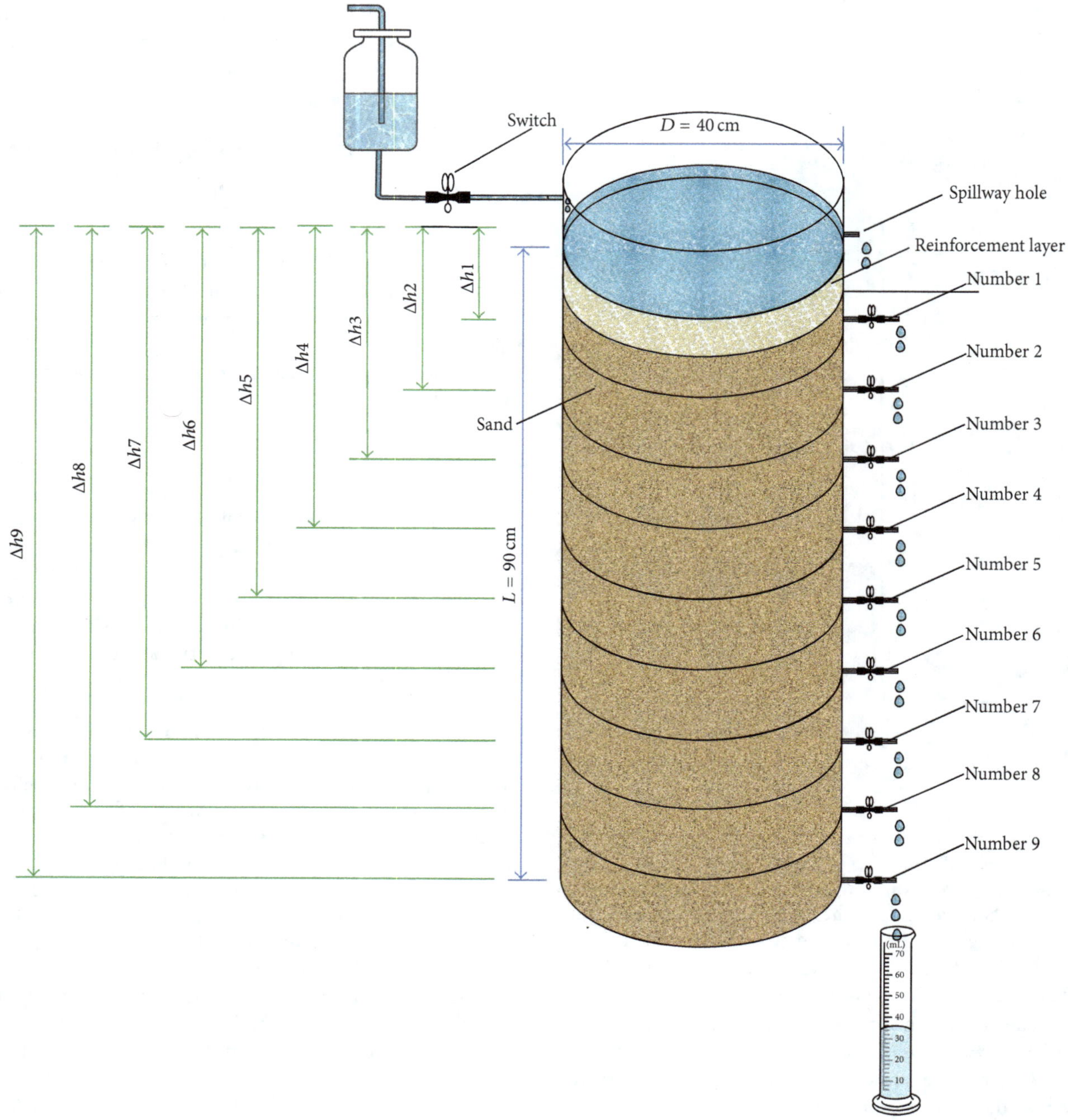

FIGURE 4: Porous permeameter used in porous permeability test.

volume every time. Finally, the initial time of water running from number 9 and the running water volume of number 9 to number 1 were taken to evaluate the permeability resistance of specimens. The data of each outlet tube was recorded until the water volume difference for successive two times is less than 3%. The average water flow rate (F) for each 3 minutes was given by (1). The average water flow rate (F) is also defined as the water flow rate of test time being 1.5 min, 4.5 min, 7.5 min, 10.5 min, and 13.5 min for each outlet tube, respectively.

3. Test Results

3.1. Reinforcement Layer Form Test. The thickness of reinforcement layer measured with vernier caliper was presented in Figure 5. The results of reinforcement layer form test were presented in Table 1. The reinforcement layer thicknesses of specimens with different curing times are shown in Figure 6.

As seen in Table 1 and Figure 6, with the unreinforced sand as a control specimen it is impossible to form an aggregation due to its little cohesive force. The sand reinforced with PSS 1% is only able to from a bigger aggregation but not the reinforcement layer. It is different in obtaining an intact reinforcement layer from the specimen surface. The reinforcement layer of specimens reinforced with PSS 3%, 5%, 7%, and 9% had not been formed without curing. After the specimens curing, the reinforcement layer thickness of treated specimens increased with the increasing of curing

Table 1: The reinforcement layer thickness of samples tested (cm).

Number	PSS concentrations (%)	Curing time 0 h	0.1 h	0.5 h	1 h	3 h	6 h	12 h	24 h
Group 1	0	/	/	/	/	/	/	/	/
Group 2	1	/	/	/	/	/	/	/	/
Group 3	3	/	0.28	0.44	0.95	1.87	2.09	2.13	2.12
Group 4	5	/	0.34	0.72	1.28	2.04	2.53	2.55	2.57
Group 5	7	/	0.45	1.15	1.56	2.54	3.00	3.02	3.05
Group 6	9	/	0.86	1.30	1.87	2.98	3.58	3.58	3.56

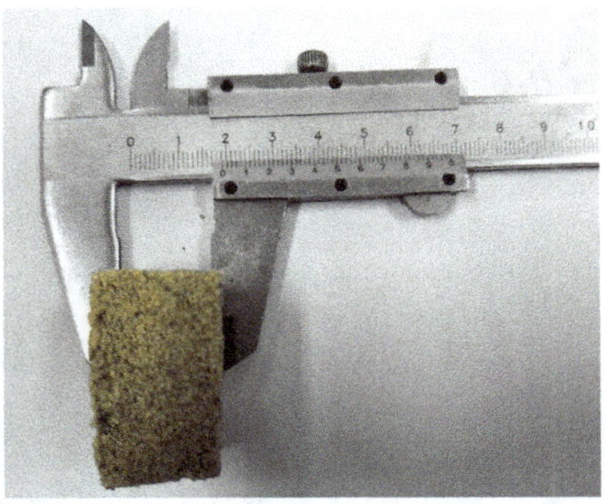

Figure 5: Photo of reinforcement layer measured with vernier caliper.

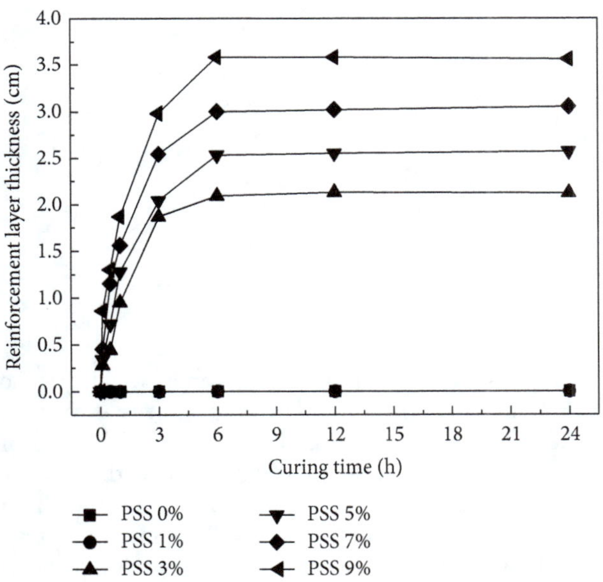

Figure 6: The reinforcement layer thickness of specimens with different curing times.

time at the beginning period from curing time 0 to 6 h. Their thickness was kept at a relatively stable value while the curing time was more than 6 h. The reinforcement layer thicknesses of specimens modified by concentration of 3%, 5%, 7%, and

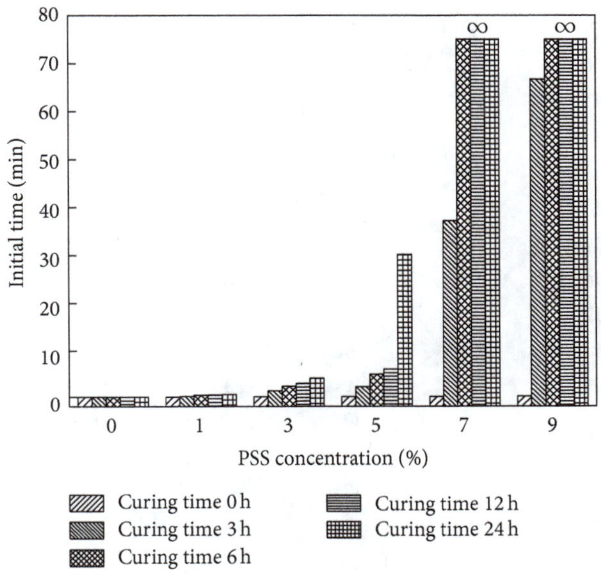

Figure 7: Initial time of running water from bottom outlet.

9% with curing time of 24 h were approximately 2.12, 2.57, 3.05, and 3.56 cm, respectively.

3.2. Single-Hole Permeability Test. In single-hole permeability test, the specimens reinforced with PSS 0%, 1%, 3%, 5%, 7%, and 9% at curing time of 0 h, 3 h, 6 h, 12 h, and 24 h were considered. The initial time of water running from the bottom tube and its running water volume of every 3 minutes with a number of 5 times were recorded. The test results are presented in Table 2.

The initial times of running water from bottom tube that is the time of water flowing from top to bottom of specimen are given in Figure 7. As seen in Figure 7, the presence of PSS increased the initial time of specimens. The specimens reinforced with higher PSS concentrations have larger initial times than those reinforced with smaller PSS concentrations. For the specimens unreinforced and reinforced with PSS 1% and 3% under different curing times, the water flowed from bottom tube immediately at the beginning of test. With the curing time increasing, the initial time increased, especially for the specimens with PPS 7% and 9%. While the curing time is larger than 6 hours, specimens with PPS 7% and 9%

TABLE 2: Results of single-hole permeability test.

Number	PSS concentrations (%)	Curing time (h)	Initial time (s)	Water volume of 0–3 min (ml)	Water volume of 3–6 min (ml)	Water volume of 6–9 min (ml)	Water volume of 9–12 min (ml)	Water volume of 12–15 min (ml)
S-1	0	0	2	215.35	283.34	274.67	282.57	278.85
S-2	1	0	2	216.84	274.57	272.41	275.11	277.33
S-3	3	0	4	115.18	120.31	110.93	147.83	147.62
S-4	5	0	5	103.92	105.97	98.54	105.41	108.54
S-5	7	0	6	59.38	78.64	77	80.11	83.93
S-6	9	0	8	46.63	67.8	53.01	51.51	64.59
S-7	0	3	2	213.46	280.24	270.78	274.70	281.36
S-8	1	3	8	145.06	161.43	168.5	160.77	160.61
S-9	3	3	82	14.35	43.69	45.17	43.96	43.57
S-10	5	3	135	2.71	3.85	4.24	4.21	4.31
S-11	7	3	2230	2.28	2.39	2.22	2.14	1.91
S-12	9	3	4000	0.91	1.04	1.15	1.08	1.22
S-13	0	6	3	212.16	276.84	274.45	282.42	278.58
S-14	1	6	25	107.12	139.05	136.68	137.36	138.87
S-15	3	6	143	11.94	38.03	41.6	40	37.48
S-16	5	6	302	2.96	3.8	3.94	3.8	3.96
S-17	7	6	∞	0	0	0	0	0
S-18	9	6	∞	0	0	0	0	0
S-19	0	12	3	215.46	281.25	275.28	281.30	278.37
S-20	1	12	27	104.32	122.08	134.74	135.66	137.07
S-21	3	12	180	9.55	30.22	30.7	28.95	27.24
S-22	5	12	366	2.16	3.15	3.94	3.83	3.9
S-23	7	12	∞	0	0	0	0	0
S-24	9	12	∞	0	0	0	0	0
S-25	0	24	3	213.27	284.56	280.35	283.35	279.47
S-26	1	24	5	70.51	129.58	125.17	126.6	123.52
S-27	3	24	248	7.64	13.91	13.58	14.37	13.51
S-28	5	24	1810	2.41	3	3.35	3.25	3.38
S-29	7	24	∞	0	0	0	0	0
S-30	9	24	∞	0	0	0	0	0

formed a water-resisting layer on surface to prevent the water infiltration.

Based on the data of water volumes in Table 2, the water flow rates (F) of specimens reinforced using different PSS concentrations are calculated with (1) and presented in Figure 8. It can be seen from Figure 8 that the F values of specimens reinforced with PSS are affected strongly by curing time and decrease with the curing time increasing, especially for the ones reinforced with higher PSS concentrations. While the curing time is 0 h (see Figure 8(a)), the F values of specimens increase with the flow time and then trend to relatively stable values, and the ones unreinforced and reinforced with PPS 1% have higher F values. As shown in Figures 8(b)–8(d), the F values of specimens with PSS 5%, 7%, and 9% are negligible. The F values of specimens reinforced with PSS 0%, 1%, and 3% increase with the water flow time at the start period of 0–6 min; after that, it approximately trends to be stable. But all the unreinforced specimens as reference ones kept the same change trend during the different curing times. It also can be seen from Figure 8 that the F values of specimens at the same curing time reduce with the increasing of PSS concentration. And all the specimens reach a stable flow rate at the stage of 12–15 min, and their F values are considered as parameters to evaluate the permeability resistance of specimens reinforced with PSS.

The variations of water flow rate (F) of specimens with PSS concentrations at flow time of 12–15 min are presented in Table 3 and Figure 9. As seen, the F values of specimens reduce with the increasing in PSS concentration. While the curing time is larger than 3 hours, the F values of specimens reinforced with lower PSS concentrations reduce rapidly. The water flow of specimens with higher PSS concentrations of

Table 3: The evaluation parameters of specimens with different curing times in single-hole permeability tests.

PSS (%)	(F) water flow rate at time 12–15 min (ml/min)					(K) permeability coefficient (cm/s)					(R) relative permeability resistance				
	0 h	3 h	6 h	12 h	24 h	0 h	3 h	6 h	12 h	24 h	0 h	3 h	6 h	12 h	24 h
0	92.95	93.79	92.86	92.79	93.16	1.69$E-2$	1.71$E-2$	1.69$E-2$	1.69$E-2$	1.70$E-2$	1	1	1	1	1
1	92.44	53.54	46.29	45.69	41.17	1.68$E-2$	9.74$E-3$	8.42$E-3$	8.31$E-3$	7.49$E-3$	1.01	1.75	2.01	2.03	2.26
3	49.21	14.52	12.49	9.08	4.50	8.96$E-3$	2.64$E-3$	2.27$E-3$	1.65$E-3$	8.19$E-4$	1.89	6.46	7.43	10.22	20.70
5	36.18	1.44	1.32	1.30	1.13	6.58$E-3$	2.62$E-4$	2.40$E-4$	2.37$E-4$	2.06$E-4$	2.57	65.13	70.35	71.38	82.44
7	27.98	0.64	0	0	0	5.09$E-3$	1.16$E-4$	0	0	0	3.32	146.55	∞	∞	∞
9	21.53	0.41	0	0	0	3.92$E-3$	7.46$E-5$	0	0	0	4.32	228.76	∞	∞	∞

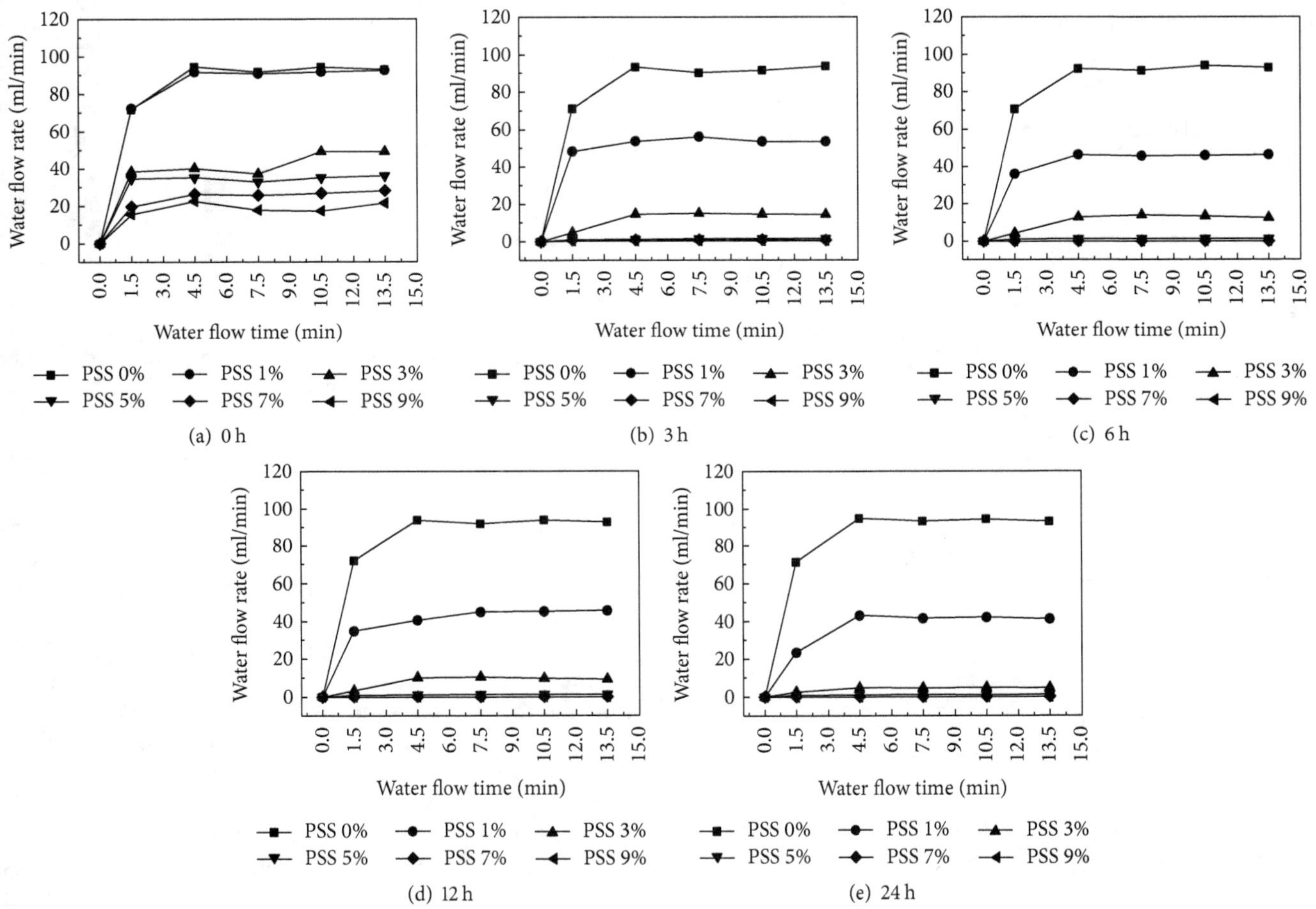

Figure 8: Water flow rate (F) of specimens reinforced using different PSS concentrations at (a) curing time of 0 h (b), curing time of 3 h, (c) curing time of 6 h, (d) curing time of 12 h, and (e) curing time of 24 h.

7% and 9% with curing time larger than 6 hours cannot be recorded and their F values are 0 ml/min due to the water-resisting layer on the surface of specimens. The permeability coefficients (K) of specimens calculated with (2) are shown in Table 3. They have the same change characteristics with the F values.

The relative permeability resistance (R) calculated with (3) is shown in Table 3. The R values of specimens with PSS concentrations of 1%, 3%, and 5% at different curing times are presented in Figure 10. As seen, the relative permeability resistances of all reinforced specimens with curing time 0 hours are not improved. The R values of the ones with curing times larger than 3 increase with the PSS concentrations, especially for the specimens with PSS 3% and 5%. It also can be seen from Table 3 and Figure 10 that the R value is mainly improved within the first 3 hours of curing time. The photos of reinforcement layers taken from the specimens with curing time 24 h after the tests are presented in Figure 11.

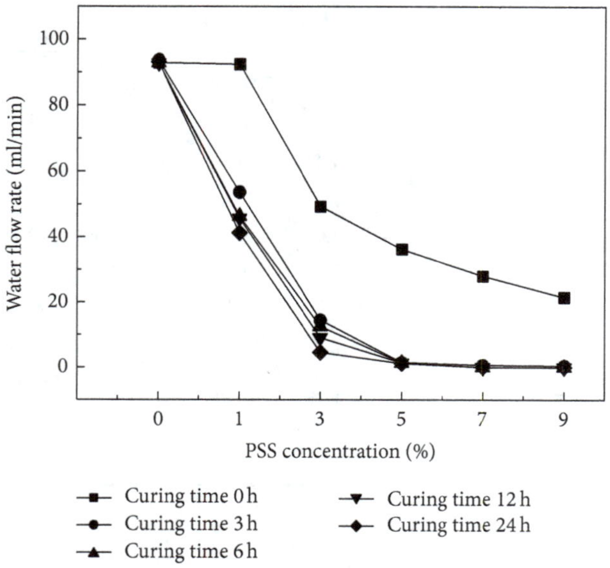

FIGURE 9: Variations of water flow rate (F) of specimens with PSS concentrations at flow time of 12–15 min.

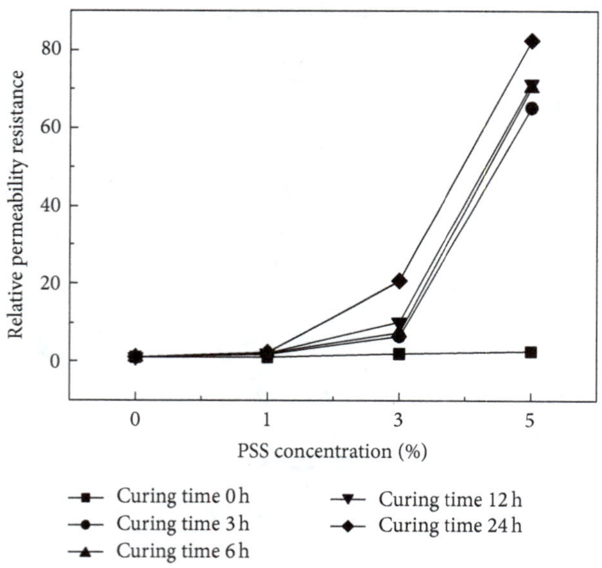

FIGURE 10: The relative permeability resistance (R) of specimens with PSS concentrations of 1%, 3%, and 5% at different curing times.

As seen, the thickness and complete degree of reinforcement layer taken from the surface of specimens with curing time 24 h increase with the increasing of PSS concentration. The reinforcement layers of specimens with PSS 1%, 3%, and 5% are states of being loose, less completed, and completed, respectively. The ones of specimens with PSS 7% and 9% have excellent complete degrees to form the water-resisting layer on sand surface.

3.3. Porous Permeability Test. In porous permeability test, the specimens reinforced with PSS 0%, 1%, 3%, and 5% at curing time of 24 hours were considered to study the influence depth of sand surface reinforcement. The initial times of water

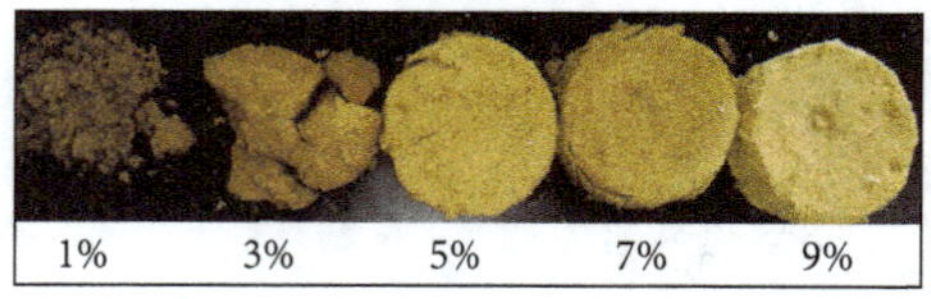

FIGURE 11: Photos of reinforcement layers taken from the specimens with curing time of 24 h after single-hole permeability tests.

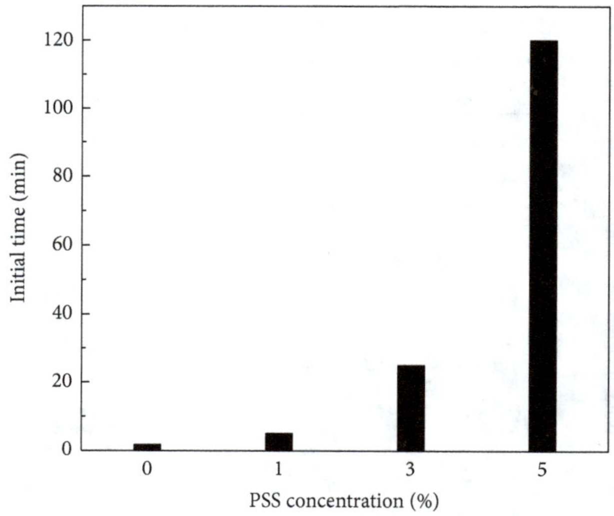

FIGURE 12: Initial time of running water from outlet number 9 of specimens with different PSS concentrations.

running from the outlet tube number 9 and the running water volume of outlet tubes numbers 1 to 9 of every 3 minutes with a number of 5 times were recorded. The test results are presented in Table 4. The initial times of running water from outlet tube number 9 of specimens with different PSS concentrations are given in Figure 12. As seen in Figure 12, the presence of PSS increased the initial time of water running from number 9 of specimens, especially for PSS 3% and 5%. The time of water flow from surface to 90 cm depth of specimens with PSS 3% and 5% reaches up to 25 min and 120 min, respectively.

Based on the data of water volumes in Table 4, the water flow rate (F) of specimens is calculated with (1). The F values of different depth in specimens reinforced with different PSS concentrations of 0%, 1%, 3%, and 5% are presented in Figure 13. It can be seen from Figures 13(a) and 13(b) that the F values of depth 90 cm in specimens unreinforced and reinforced with PSS 1% quickly reach the stable values, except that the ones of depth 90 cm reach the stable values after water flow time larger 3 min and 9 min, respectively. As seen in Figures 13(c) and 13(d), the F values of specimens reinforced with PSS 3% and 5% increase with the water flow time at the start period of 0–6 min; after that, it approximately trends to be stable. It also can be seen from Figure 13 that the F values of specimens are affected strongly by PSS concentrations and decrease with the increasing of PSS concentration, especially for higher PSS concentrations of 3% and 5%. And all the specimens reach a stable value at the stage of 12–15 min, and

TABLE 4: Results of porous permeability test.

Number	PSS concentrations (%)	Sand depth (cm)	Initial time (s)	Water volume of 0–3 min (ml)	Water volume of 3–6 min (ml)	Water volume of 6–9 min (ml)	Water volume of 9–12 min (ml)	Water volume of 12–15 min (ml)
P-1	0%	90	108	163.58	1152.47	1158.19	1164.67	1152.22
	0%	80	/	1113.69	1119.76	1110.97	1115.56	1122.39
	0%	70	/	1018.21	1081.89	1037.63	1047.71	1059.72
	0%	60	/	858.27	871.75	864.56	853.86	854.19
	0%	50	/	724.82	733.72	718.24	720.44	786.06
	0%	40	/	699.01	710.60	677.15	702.00	697.38
	0%	30	/	698.38	676.47	692.47	691.87	695.30
	0%	20	/	501.95	508.12	486.43	469.03	489.07
	0%	10	/	365.28	419.73	392.43	383.27	388.21
P-2	1%	90	305	147.07	277.42	989.39	1062.42	1041.25
	1%	80	/	885.82	865.74	866.86	880.41	881.39
	1%	70	/	753.78	762.15	755.60	777.70	773.99
	1%	60	/	749.71	754.97	747.34	758.36	758.73
	1%	50	/	743.83	734.2	726.61	737.77	707.91
	1%	40	/	570.61	569.03	562.11	553.26	554.50
	1%	30	/	568.32	574.71	584.79	544.77	549.69
	1%	20	/	413.01	395.92	398.83	392.69	393.34
	1%	10	/	282.60	293.46	307.17	286.12	289.92
P-3	3%	90	1505	72.13	329.33	360.59	360.39	363.18
	3%	80	/	245.77	334.64	350.96	357.55	359.86
	3%	70	/	188.36	292.52	324.62	348.77	350.26
	3%	60	/	217.59	303.63	328.66	346.53	337.17
	3%	50	/	219.46	296.48	326.05	338.52	323.58
	3%	40	/	269.97	301.23	319.64	313.52	321.66
	3%	30	/	234.20	267.63	278.76	292.08	293.12
	3%	20	/	197.11	213.06	216.26	213.55	214.65
	3%	10	/	128.55	129.40	148.07	144.43	141.15
P-4	5%	90	7200	17.19	35.50	33.24	33.46	34.26
	5%	80	/	16.75	30.07	31.74	32.34	33.75
	5%	70	/	15.72	30.72	32.75	32.76	32.96
	5%	60	/	14.36	29.61	30.48	30.19	32.41
	5%	50	/	14.61	29.69	29.13	28.58	28.33
	5%	40	/	12.88	28.64	28.25	27.58	27.78
	5%	30	/	10.89	24.42	26.34	26.49	25.97
	5%	20	/	10.36	21.02	23.04	22.27	22.83
	5%	10	/	8.62	19.16	18.53	18.43	18.93

the F values at 12–15 min are considered as parameters to evaluate the permeability resistance of different depths in specimens reinforced with PSS.

The variations of water flow rate (F) of different depths in specimens at flow time 12–15 min are presented in Table 5 and Figure 14. As seen, the F values of every depth in specimens decrease with the increasing of PSS concentration. The F values of specimens unreinforced and reinforced with PSS 1% gradually increase with the increasing of depth. The F values of one reinforced with PSS 3% increase with the depth during the depth range of 0–40 cm and then keep a stable value. The F values of one reinforced with higher PSS concentration 5% are very small and not affected by the depth of specimen. The permeability coefficients (K) of specimens calculated with (2) are shown in Table 6. They have the same change characteristics with the F values. The K values decrease with the increasing of PSS concentration and increase with the increasing of depth in specimen.

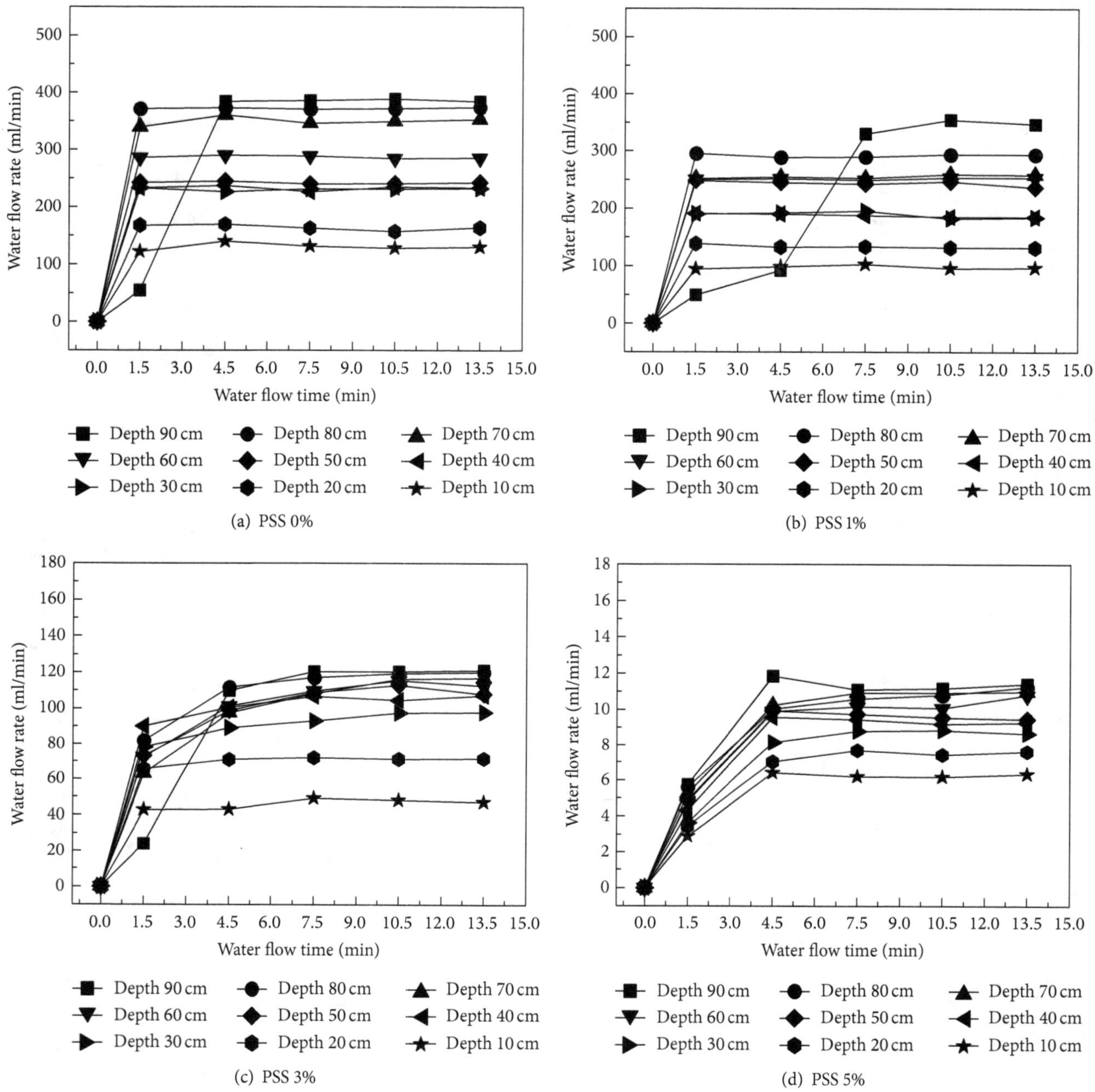

Figure 13: The water flow rate (F) of different depth in specimens reinforced with different PSS concentrations.

Table 5: The water flow rate at time of 12–15 min of specimens with different depth (ml/min).

Number	PSS concentrations (%)	Depth (cm)								
		90	80	70	60	50	40	30	20	10
G-1	0	384.07	374.13	353.24	284.73	262.02	232.46	231.77	163.02	129.4
G-2	1	347.08	293.8	258.00	252.91	245.97	184.83	183.23	131.11	96.64
G-3	3	121.06	119.95	116.75	112.39	107.86	107.22	97.71	71.55	47.05
G-4	5	11.42	11.25	11.00	10.80	9.44	9.26	8.66	7.61	6.31

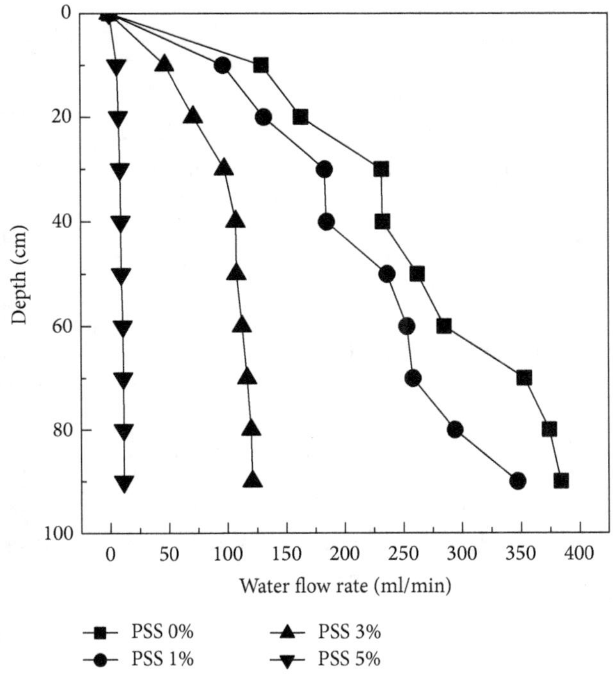

Figure 14: Variations of water flow rate (F) of different depths in specimens at flow time of 12–15 min.

The relative permeability resistance (R) calculated with (3) is shown in Table 7. As seen, the relative permeability resistance of each depth of reinforced specimens with curing time of 24 hours is improved. The R values of the specimens are mainly affected by the PSS concentration, especially for the specimen with PSS 5%. The R values of specimen with depth of 10 cm, 30 cm, 60 cm, and 90 cm are 20.51, 26.76, 26.36, and 33.63, respectively. The photos of reinforcement layers taken from the specimens with curing time of 24 h after the porous permeability tests are presented in Figure 15. As seen, the thickness and complete degree of reinforcement layer taken from the surface of specimens increase with the increasing of PSS concentration. The reinforcement layers of specimens with PSS 1%, 3% and 5% are states of being loose, less completed, and completed, respectively.

4. Discussion

The polyurethane soil stabilizer (PSS) contains a significant proportion of the long-chain macromolecule of polyurethane resin and enormous amount of isocyanate group (–NCO). The structural formula of PSS is represented by formula (4) and its reinforcement reaction process is represented by formulas (5) and (6). When PSS solution is diluted and sprayed on sand, a part fills up the voids of sand and others adsorb on the surface of sand particle. The active groups -NCO in formula (4) react with water in voids and on surface of sand through chemical formulas (5) and (6) to form the reinforcement layer. These reactions create physicochemical bonds between molecules and sand particle. With these bonds, the PSS enwrap the sand particles and interlink them to form a reinforcement layer on the sand surface. The SEM images of specimen reinforced with PSS 5% are presented in Figure 16. As seen in Figures 16(a)–16(d), the sand particles are enwrapped and connected by the PSS to form a stable structure. This structure may increase in bonding and interlocking forces between sand particles and decrease the void ratio of sand.

$$\mathrm{O{=}C{=}N}\left[\mathrm{R_1{-}\overset{H}{\overset{|}{N}}{-}\underset{O}{\underset{\|}{C}}{-}O{-}R_2{-}O{-}\overset{O}{\overset{\|}{C}}{-}\underset{H}{\underset{|}{N}}}\right]_n\mathrm{{-}R_1{-}N{=}C{=}O} \quad (4)$$

$$\mathrm{O{=}C{=}N{-}R{-}N{=}C{=}O + 2H_2O \longrightarrow HO{-}\overset{O}{\overset{\|}{C}}{-}\underset{H}{\underset{|}{N}}{-}R{-}\overset{H}{\overset{|}{N}}{-}\underset{O}{\underset{\|}{C}}{-}OH \longrightarrow H_2N{-}R{-}NH_2 + 2CO_2} \quad (5)$$

$$(n+1)\mathrm{H_2N{-}R{-}NH_2} + n\,\mathrm{O{=}C{=}N{-}R{-}N{=}C{=}O \longrightarrow H_2N}\left[\mathrm{R{-}\overset{H}{\overset{|}{N}}{-}\underset{O}{\underset{\|}{C}}{-}\overset{H}{\overset{|}{N}}}\right]_{2n}\mathrm{{-}R{-}NH_2}. \quad (6)$$

The specimen surface is sprayed by PSS dilution to form the reinforcement layer. The curing time and dilution concentration are two important factors to the formation of reinforcement layer on specimen surface. This forming process requires a certain amount of curing time. This is the reason why the thickness of reinforcement layer reaches a stable value when the curing time is larger than 3 hours (see Figure 6). The higher PSS concentration, the more content of long-chain macromolecule to fill the sand voids and enwrap the sand particles with a resultant increase of the thickness of reinforcement layer on specimen surface (see Figure 6). And the complete degree of reinforcement layer taken from the surface of specimens also increases with the increasing of PSS concentration (see Figures 11 and 15). The sand voids filled with PSS reduce or block the flowing channels of water, resulting in the fact that the time in which water flow passes through the reinforcement layer from the surface of specimen is increased. This permeability time of reinforcement layer increases with the PSS concentration (see Figures 7 and 12).

TABLE 6: The permeability coefficient of specimens with different depth (cm/s).

Number	PSS concentrations (%)	Depth (cm) 90	80	70	60	50	40	30	20	10
G-1	0	$1.93E-2$	$1.87E-2$	$1.75E-2$	$1.40E-2$	$1.26E-2$	$1.10E-2$	$1.05E-2$	$6.92E-3$	$4.58E-3$
G-2	1	$1.75E-2$	$1.47E-2$	$1.28E-2$	$1.24E-2$	$1.19E-2$	$8.72E-3$	$8.33E-3$	$5.57E-3$	$3.42E-3$
G-3	3	$6.09E-3$	$5{,}99E-3$	$5.78E-3$	$5.51E-3$	$5.20E-3$	$5.05E-3$	$4.45E-3$	$3.04E-3$	$1.66E-3$
G-4	5	$5.74E-4$	$5.62E-4$	$5.45E-4$	$5.29E-4$	$4.55E-4$	$4.37E-4$	$3.94E-4$	$3.23E-4$	$2.23E-4$

TABLE 7: The relative permeability resistance of specimens with different depth.

Number	PSS concentrations (%)	Depth (cm) 90	80	70	60	50	40	30	20	10
G-1	0	1	1	1	1	1	1	1	1	1
G-2	1	1.11	1.27	1.37	1.13	1.07	1.26	1.26	1.24	1.34
G-3	3	3.17	3.12	3.03	2.53	2.43	2.17	2.37	2.28	2.75
G-4	5	33.63	33.26	32.11	26.36	27.76	25.10	26.76	21.42	20.51

FIGURE 15: Photos of reinforcement layers taken from the specimens with curing time of 24 h after the porous permeability tests.

The flowing water rate of reinforcement layer on specimen surface reduces with the decreasing of its void radio. This decrement of void radio is improved by the curing time or PSS concentration. The flowing water rate of tested part of sand is mainly controlled by the sand layer with lowest void radio. So the flowing water rate decreases with the increase in curing time or PSS concentration (see Figures 8, 9, and 13). While there is lower PSS concentration or shorter curing time, there is not enough macromolecule or time to form the reinforcement layer. Therefore, the relative permeability resistances of specimens were mainly improved with the curing time larger than 3 hours and PSS concentrations higher than 3% (see Figure 10 and Table 7). The water-resisting layer on sand surface is formed with the curing time and PSS concentration larger than 6 hours and 7%, respectively. The upper limit of PSS concentration for improving the antipermeability of sand is 9%, due to its good water-resisting layer on sand surface.

As seen in Figure 16, the polymers enwrap the sand particle's surface and interlink them to form an elastic and viscous cover on the sand surface to improve the interlocking forces between sand particles. This void filling and physicochemical bonding may lead to a resulting improvement in the mechanical properties such as compressive strength, tensile strength, erosion resistance, and cohesion of sand. This paper is only limited to evaluate the effect of PSS on the permeability characteristics of sand with the particle size of 0.1–1 mm. The results indicated that the curing time and PSS concentration are important consideration factors in reinforcing process. While the PSS is used to improve different mechanical properties, sand content, and soil particle size, reinforcement method, curing time, and optimal concentration should be adjusted to meet the different engineering requirements in practical application.

PSS is a kind of waterborne polyurethane polymer and a form of oil liquid. When polymer as a reinforcing material is used in practical engineering, the environmental concerns must be taken seriously. The potential environmental concerns are mainly because of volatile organic compounds (VOC) and residual PSS solution, which will cause air pollution and water resource pollution. In order to reduce or avoid the environmental concerns, the VOC of PSS must be strictly controlled according to international regulations; the residual PSS solution must be recycled with effective methods. The sunny day should be selected for spraying PSS in field to ensure sufficient reaction of polymer and sand particle to reduce the polymer loss and water pollution.

The aging degradation of polymer is mainly affected by temperature, ultraviolet ray, water, and chemical mediator. Reactions (5) and (6) during the reinforcement process of PSS are not reversible. The reinforcement layer on sand will be damaged due to the aging degradation of polymer. The running water channel formed in reinforcement layer will lead to increase in the permeability of reinforced sand. The durability of PSS on permeability characteristics of sand with the condition temperature −10~50°C, ultraviolet intensity ≤ 250 mW/m^2, and no chemical mediator is about 10 years. However, while the PSS is applied in engineering area with

FIGURE 16: SEM images of specimen reinforced with PSS 5%: (a) 50 times magnification, (b) 100 times magnification, (c) 150 times magnification, and (d) 200 times magnification.

extreme weather, high ultraviolet intensity, or polluted soil, the long term effects of polymer on permeability characteristics will be shortened.

5. Conclusions

In order to evaluate the permeability characteristics of PSS reinforced sand, the laboratory tests of reinforcement layer form test, single-hole permeability test, and porous permeability test were considered. The test results and reinforcement mechanism were analyzed. Based on the results of the tests presented herein, the main conclusions can be summarized as follows:

(1) The permeability resistance of sand reinforced with polyurethane soil stabilizer (PSS) is improved through the formation of reinforcement layer on the sand surface. The thickness and complete degree of the reinforcement layer increase with the increasing of curing time and PSS concentration. The reinforcement layer thicknesses of specimens modified by concentrations of 3%, 5%, 7%, and 9% with curing time of 24 h were approximately 2.12, 2.57, 3.05, and 3.56 cm, respectively.

(2) Laboratory tests of single-hole permeability test and porous permeability test indicated that the permeability time of reinforcement layer increases with the PSS concentration. The flowing water rate decreases with the increasing of curing time or PSS concentration. The permeability coefficient decreases with the increasing of curing time and PSS concentration and increases with the increasing of depth in specimen. The relative permeability resistances of specimens were mainly improved with the curing time larger than 3 hours and PSS concentrations higher than 3%. The water-resisting layer on sand surface is formed with the curing time and PSS concentration larger than 6 hours and 7%, respectively.

(3) The PSS contains a significant proportion of the long-chain macromolecule of polyurethane resin and enormous amount of isocyanate group (–NCO). Microscale SEM observations show that the PSS enwrap the sand particle and interlink them to form a reinforcement layer on the sand surface. PSS fills up the voids of sand and adsorbs on the surface of sand particle to reduce or block the flowing channels of water to improve the permeability resistance of sand. The results can be applied as the reference for PSS reinforced sand engineering, especially for surface protection of embankment, slope, and landfill. The reinforcement method, curing time, and optimal concentration should be adjusted to meet the different engineering requirements in practical applications.

Conflicts of Interest

The authors declare that there are no conflicts of interest regarding the publication of this paper.

Acknowledgments

This research was financially supported by the National Natural Science Foundation of China (Grants nos. 41472241 and 41202208) and Natural Science Foundation of Jiangsu Province, China (Grant no. BK20141415).

References

[1] C. Ghosh and K. Yasuhara, "Clogging and flow characteristics of a geosynthetic drain confined in soils undergoing consolidation," *Geosynthetics International*, vol. 11, no. 1, pp. 19–34, 2004.

[2] C. Yoo and H.-Y. Jung, "Case history of geosynthetic reinforced segmental retaining wall failure," *Journal of Geotechnical and Geoenvironmental Engineering*, vol. 132, no. 12, pp. 1538–1548, 2006.

[3] D. V. Raisinghani and B. V. S. Viswanadham, "Evaluation of permeability characteristics of a geosynthetic-reinforced soil through laboratory tests," *Geotextiles and Geomembranes*, vol. 28, no. 6, pp. 579–588, 2010.

[4] E. M. Palmeira and M. G. Gardoni, "Drainage and filtration properties of non-woven geotextiles under confinement using different experimental techniques," *Geotextiles and Geomembranes*, vol. 20, no. 2, pp. 97–115, 2002.

[5] K. R. Lekha, "Field instrumentation and monitoring of soil erosion in coir geotextile stabilised slopes—a case study," *Geotextiles and Geomembranes*, vol. 22, no. 5, pp. 399–413, 2004.

[6] A. Tolooiyan, I. Abustan, M. R. Selamat, and S. Ghaffari, "A comprehensive method for analyzing the effect of geotextile layers on embankment stability," *Geotextiles and Geomembranes*, vol. 27, no. 5, pp. 399–405, 2009.

[7] U. Smoltczyk and K. Malcharek, "Slope protection by membrane structures," *Geotextiles and Geomembranes*, vol. 2, no. 4, pp. 323–336, 1985.

[8] N. Pollen, "Temporal and spatial variability in root reinforcement of streambanks: accounting for soil shear strength and moisture," *Catena*, vol. 69, no. 3, pp. 197–205, 2007.

[9] D. Liu and Y. Li, "Mechanism of plant roots improving resistance of soil to concentrated flow erosion," *Journal of Soil and Water Conservation*, vol. 17, no. 3, pp. 34–37, 2003.

[10] S. S. Kukal, M. Kaur, S. S. Bawa, and N. Gupta, "Water-drop stability of PVA-treated natural soil aggregates from different land uses," *Catena*, vol. 70, no. 3, pp. 475–479, 2007.

[11] J. Liu, B. Shi, Y. Lu et al., "Effectiveness of a new organic polymer sand-fixing agent on sand fixation," *Environmental Earth Sciences*, vol. 65, no. 3, pp. 589–595, 2012.

[12] H. I. Inyang and S. Bae, "Polyacrylamide sorption opportunity on interlayer and external pore surfaces of contaminant barrier clays," *Chemosphere*, vol. 58, no. 1, pp. 19–31, 2005.

[13] J. Yang, L. Fang, and T. Tan, "Synthesis and characterization of superabsorbent hydrogels composites based on polysuccinimide," *Journal of Applied Polymer Science*, vol. 102, no. 1, pp. 550–557, 2006.

[14] J. Yang, F. Wang, L. Fang, and T. Tan, "The effects of aging tests on a novel chemical sand-fixing agent—polyaspartic acid," *Composites Science and Technology*, vol. 67, no. 10, pp. 2160–2164, 2007.

[15] G. Telysheva and G. Shulga, "Silicon-containing polycomplexes for protection against wind erosion of sandy soil," *Journal of Agricultural Engineering Research*, vol. 62, no. 4, pp. 221–227, 1995.

[16] W. J. Orts, R. E. Sojka, and G. M. Glenn, "Biopolymer additives to reduce erosion-induced soil losses during irrigation," *Industrial Crops and Products*, vol. 11, no. 1, pp. 19–29, 2000.

[17] Y. Qin, H. Yang, Z. Deng, and J. He, "Water permeability of pervious concrete is dependent on the applied pressure and testing methods," *Advances in Materials Science and Engineering*, vol. 2015, Article ID 404136, 6 pages, 2015.

3

The Compressibility and Swell of Mixtures for Sand-Clay Liners

Muawia A. Dafalla

Bugshan Research Chair in Expansive Soils, Department of Civil Engineering, King Saud University, Riyadh 11421, Saudi Arabia

Correspondence should be addressed to Muawia A. Dafalla; mdafalla@ksu.edu.sa

Academic Editor: Aboelkasim Diab

Sand-clay liners utilize expansive clay to act as a filler to occupy the voids in the sand and thus reduce the hydraulic conductivity of the mixture. The hydraulic conductivity and transfer of water and other substances through sand-clay mixtures are of prime concern in the design of liners and hydraulic barriers. Many successful research studies have been undertaken to achieve appropriate mixtures that satisfy hydraulic conductivity requirements. This study investigates compressibility and swelling properties of mixtures to ensure that they were acceptable for light structures, roads, and slabs on grade. A range of sand-expansive clay mixtures were investigated for swell and compression properties. The swelling and compressibility indices were found to increase with increasing clay content. The use of highly expansive material can result in large volume changes due to swell and shrinkage. The inclusion of less expansive soil material as partial replacement of bentonite by one-third to two-thirds is found to reduce the compressibility by 60% to 70% for 10% and 15% clay content, respectively. The swelling pressure and swell percent were also found significantly reduced. Adding less expansive natural clay to bentonite can produce liners that are still sufficiently impervious and at the same time less problematic.

1. Introduction

The demand for liners in environmental and containment projects is growing [1]. Sand-clay liners utilize highly plastic bentonite clay to act as a barrier and fill the voids in the sand and thus reduce the hydraulic conductivity of the mixture. The use of bentonite can result in large volume changes as a result of swell and shrinkage. In order to improve compressibility and swell it is suggested to add a portion of clay with less plasticity than bentonite. This is expected to produce sufficiently impervious liner with less compressibility and less potential for expansion leading to utilizing the natural local clays of less plasticity in liners and thus reducing the demand for bentonite and improving the liner performance. The purpose of this paper is to investigate the influence of adding natural clay soils on the compressibility and swell. Dafalla and Al-Mahbashi [2] investigated the effect of adding natural clay to bentonite on the water retention curve of sand-bentonite mixtures. This study is focused on the use of expansive clay as part of sand-clay liners in geoenvironmental projects. The work carried out in this research is part of a project financed by NPST (National Plan for Science and Technology, Saudi Arabia) to investigate using local clays in liners in eastern parts of Saudi Arabia.

2. Background

2.1. Expansive Soils and Their Use. Extensive research work has been carried out on expansive soils as they are a serious hazard to buildings and light structures due to their volume changes when subject to moisture content alterations causing movement of these structures. The behaviour of such clays when subjected to moisture changes is controlled by some intrinsic and external factors. The intrinsic factors include the type of clay (mineralogy and chemical equilibrium), density and packing state, initial moisture content, pore water pressure, and pore air pressure, while the external factors include vertical and horizontal stresses, the source of moisture, hydraulic gradient, and rate at which water is introduced in addition to the chemistry of the water. Many of these factors are interrelated [3].

Expansive soils, sometimes known as black cotton soil [4], are good for agriculture due to the level of nutrients and minerals they contain. This clay is also used in medical products and used extensively in cosmetics. Clay nanocomposite studies for producing materials of enhanced performance are currently attracting the attention of many researchers [4].

2.2. Sand-Clay Mixtures. Naturally occurring sand-clay soils are not usually efficient for the intended purpose of, or usage as, impermeable liners. Geotechnical and geoenvironmental engineers have found it necessary to devise an engineering approach to achieve the appropriate hydraulic conductivity and other required properties. The sand-clay liners can also control the movement of other materials and toxic pollutants due to their low permeability and ionic absorption properties. The optimum ratio of clay to sand depends mainly on the porosity of the granular material. Compacted sand-clay mixtures were introduced to act as barriers in waste disposal following increased awareness and rising environmental concerns during the second half of the twentieth century [1]. The control of waste leachate can be contained using liners of low hydraulic conductivity [5].

The sand-clay mixtures are commonly used as liners in waste disposal and in the protection of strategic power projects. Designs using bentonite have been found successful in many applications. The cost of processed clay and bentonite prompted many researchers to investigate the use of local materials. Rawas et al. (2005) investigated the use of Oman shale in liners. Obrike et al. [6] investigated the use of Auchi Shale and Imo shale in waste disposal landfills in Nigeria. Langdon et al. [7] studied the permeability of clay liners of the same geological formation and different depositional basins in Turkey. These are only examples and other research is currently going on.

Geosynthetic material can also be used in combination with sand-clay liners. The ASTM standards introduced a new test to determine the hydraulic conductivity values [8] which describes the laboratory measurement of both flux and hydraulic conductivity of GCL-specimens utilizing a flexible wall permeameter.

2.3. Literature Review on the Compressibility of Sand-Clay Mixtures. The compressibility and swell nature of sand-clay mixtures is of great importance. Highly expansive material can expand to a stage that is likely to cause excessive distortion to surfaces or provide an uneven support for foundations and light structures. The response to loads and stresses on sand-clay mixtures has been studied by many researchers. Wasti and Alyanak [9] studied sand-clay mixtures and observed that when the initial, unswelled, clay content is sufficient to fill the voids within the sand at its maximum porosity, the overall behaviour tends to be similar to clay. The influence of the clay material on the overall behaviour of soil mixtures has been raised by many researchers (e.g., [10–12]). Tsotsos et al. [13] introduced a new experimental and numerical concept to deal with mixed soils. They concluded that deformational behaviour of mixed soils depends strongly on the percentage of clay in the mixture and the mechanical properties of each component. The modelling approach was based on the behaviour of well-defined types of structures. This is unlikely to work for all mixtures or for variable ranges and types of sand-clay mixtures.

The factors, including placement conditions, affecting swelling in bentonite and other clays are also the same factors affecting sand-clay mixtures. Dafalla [14] discussed the role of initial moisture content and dry density on the behaviour of clays. Dixon [15] worked on the swelling behaviour of bentonite based backfill used for nuclear structures in Canada. He showed that swelling pressure increases with the increase of effective clay dry density.

Research on sand-clay mixtures in semiarid areas, however, has been limited. Dafalla [3] presented a model for predicting the behaviour of artificial sand-clay mixtures using the fall cone testing method. Alawaji [16] studied the swell and compressibility characteristics of sand-bentonite mixtures wetted with liquids with two types of commercial bentonite and investigated the compressibility behaviour when the mixture is subjected to liquids with variable concentrations of $Ca(NO_3)_2$ and $NaNO_3$. Alawaji [16] used these chemicals to investigate their influence on the process of swelling and compressibility. His results indicated that swell potential (SP), swell time, swell pressure, and volume compressibility decrease with the increase in chemical concentrations. Mollins et al. [17] found that the compaction method does not affect the final clay void ratio of the tested samples. Benson and Boutwell [18] investigated compaction conditions and scale dependent hydraulic conductivity of compacted clay liners.

Phanikumar et al. [19] conducted compressibility and swell tests for clay-sand mixtures and concluded that as the sand content in the blends increased from 0% to 30%, swell potential decreased by 71% and 50% and swelling pressure decreased by 67% and 57%, respectively, for soil fractions passing the 425 μm and 75 μm sieve. They also stated that the coefficient of volume compressibility reduced by 30% and compression index decreased by 50% as the sand content increased from 0% to 30% for soil fractions passing the 425 μm sieve.

The main objective of this study is to examine the compression and swell of sand-expansive clay liners in order to help designers to optimize and select the appropriate mix. Use of clay mixtures consisting of natural clay and commercial bentonite instead of only bentonite in sand-clay mixtures is investigated as a new approach to produce a less problematic and more economical mixes. This is considered to be a significant problem in semiarid areas where significant drying of these mixes may occur and lead to cracking.

3. Materials and Test Methods

Natural clay materials of high plasticity are abundant in many semiarid regions. The properties of most of these clays do not satisfy the requirements for use in sand-clay liners. This can be attributed to low plasticity or failure to provide the required hydraulic conductivity. It was decided to investigate utilizing these materials in reducing the amount of processed bentonite and enhancing the shrink-swell properties of sand-clay liners. Al-Qatif clay in Saudi Arabia is believed to have good potential as an additive to enhance the performance

TABLE 1: Physical properties of Al-Qatif clay.

Property	Range
Material passing sieve number 200	>90%
Liquid limit	130–150
Plastic limit	60–70
Plasticity index	70–80
Maximum dry density	1.150–1.200 gm/cm^3
Optimum moisture Content	32–40%
Swell percent (ASTM D4546)	16–18%
Swelling pressure (ASTM D4546)	500–800 kN/m^3 (γ = 12 kN/m^3)

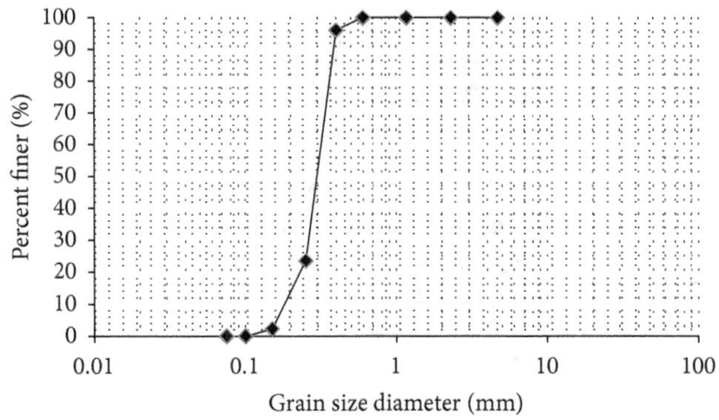

FIGURE 1: Grain Size distribution of the uniform sand used in this research.

of sand-clay liners. Commercial bentonite and Al-Qatif clay were chosen to study the swell and compressibility properties of selected sand-clay liners.

3.1. Al-Qatif Clay. The unprocessed, natural expansive clay used in this research was obtained from the city of Al-Qatif located on the Arabian Gulf shoreline 400 km from Riyadh, the capital of Saudi Arabia. Several researchers have investigated the swelling characteristics of Al-Qatif expansive clay [20–22]. Based on these investigations, Al-Qatif clay is typically characterized as highly expansive soil due to its high montmorillonite mineral content. Soil samples were obtained from open pits excavated to a depth of 1.5–3.0 m below ground surface. Samples were transferred to the laboratory and complete geotechnical characterization and chemical composition were performed. Shamrani et al. [23] indicated that natural Al-Qatif clay is characterized among problematic clays due to its high plasticity. A summary of geotechnical characterization results is presented in Table 1. Chemical composition of Al-Qatif clay is presented in Table 2.

3.2. Sand. Sand used in this study was commercially available uniform sand that is locally used in concrete mixes in Riyadh. It is typically known as "concrete sand" and is abundant in Saudi Arabia. The grain size ranges from 0.6 to 0.1 mm. According to the Unified Soil Classification System (USCS, [24]), this sand is classified as poorly graded sand (SP).

3.3. Bentonite. The bentonite used in this study was HY OCMA obtained from a local supplier. The index properties of HY OCMA bentonite used in the present study are given in Table 3. The chemical composition of bentonite is given in Table 4.

3.4. Sand-Clay Mixture Preparation and Compaction Tests

3.4.1. Sample Preparation. Samples of Al-Qatif expansive clay obtained from the field were air dried, pulverized, and sieved using sieve size 425 μm (number 40). The oven dried sand and Al-Qatif clay were mixed thoroughly and then the required water content was added and mixed in and the sample stored in plastic bags for 24 hours to mature. Sand-bentonite clay mixes were prepared in a similar way.

3.4.2. Compaction Tests. Compaction tests were carried out to assess the optimum water contents and maximum dry unit weights of sand-Al-Qatif expansive clay mixtures and sand-bentonite mixtures. Al-Qatif expansive clay contents used were 0%, 5%, 10%, 15%, 20%, and 25% by dry weight of sand and the bentonite contents used were 0%, 5%, 10, and 20% by dry weight of sand. The mix proportions quoted are for clays in the air dried state. For each mixture, the optimum water content and the maximum dry unit weight were determined using standard Proctor compaction method (ASTM D698 [25], method A). Samples of mixtures were prepared with water contents ranging from 3 to 22%. Distilled water was added to the mixtures to obtain the desired water contents.

A 2.5 kg (5.5 lb) hammer was used to compact the mixtures into the 101.6 mm (4 in) mold (inside diameter) with 113.9 mm (4.5 in) in height in order to ensure the uniform compaction for each layer. Three layers were compacted in each mold. After compaction and leveling, the weight of the compacted mixtures was determined along with their water contents. The maximum dry unit weight and optimum water content of the compacted sand-expansive clay and compacted sand-bentonite mixtures were determined from the compaction curve. The sand gradation and porosity are prime factors in designing sand-clay mixtures. Figure 1 presents grain size distribution of the uniform sand used in this research. Figure 2 presents the maximum dry density at optimum moisture content for Al-Qatif clay-sand mixtures. Addition of 5%, 10%, and 12% bentonite to the sand was found to shift the maximum dry density to 17.7, 18.2, and 18.5 kN/m^3 and the optimum moisture content to 11, 11, and 12 percent. Figure 3 presents the maximum dry density and optimum moisture content for the selected clay-sand mixtures used in this study.

3.5. Compressibility and Swell Testing. One-dimensional consolidation and swelling tests were carried out using conventional oedometer methods. The fixed ring apparatus is an oedometer in which the ring holding the sample is not allowed to move during testing. A chamber surrounding the ring is used for submergence of the sample. Two porous disks with Whatman filter papers next to the sample were placed at the bottom and top of the sample. The oedometer cell was

TABLE 2: Chemical composition of Al-Qatif clay.

K^+ (%)	K_2O (%)	Al (%)	Al_2O_3 (%)	Si (%)	SiO_2 (%)	Ca^{2+} (%)	CaO (%)
1.8	2.2	3.3	6.3	8.1	17.3	0.7	0.9

TABLE 3: Properties of HY OCMA bentonite.

Property	Value
Specific gravity, GS	2.76
Liquid limit, LL (%)	480
Plastic limit, PL (%)	49.6
Plasticity index, PI (%)	430

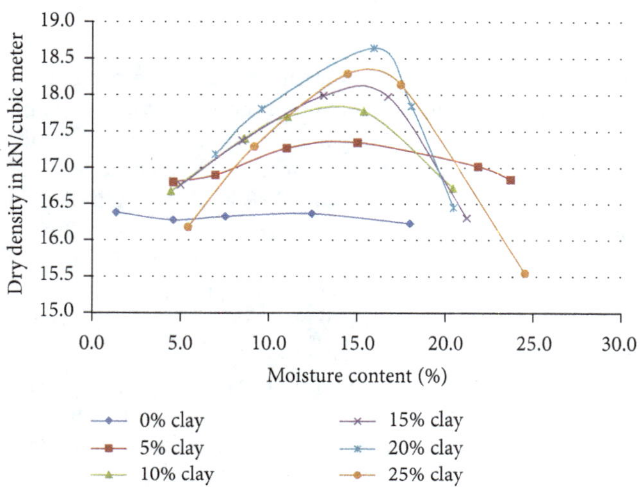

FIGURE 2: Dry density versus moisture content for Al-Qatif clay-sand mixtures.

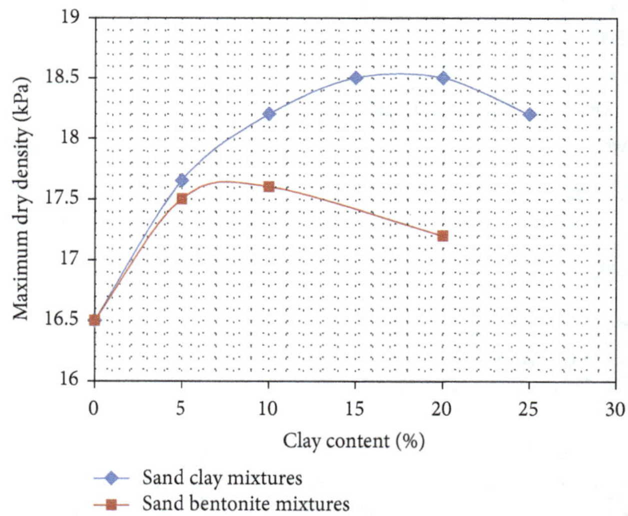

FIGURE 3: Maximum dry density-clay content relationships of sand-clay mixtures and sand-bentonite mixtures.

placed within a loading frame, which transmits vertical loads through a specially designed lever arm. A dial gauge of an accuracy of 0.01 mm was used to monitor the change of the sample height. Digital dial gauges connected to a data logger were used in some of the experiments. Front loading frames were used for most of the tests. The diameter of the ring used was 50 mm.

The one-dimensional swell testing method used for determining swell potential was carried out in accordance with ASTM D4546-96 [26] (method A). The variation from this method was the use of a seating pressure of 7 kN/m^2. In this method the sample was wetted and allowed to swell vertically at the seating pressure until primary swell was complete.

The oedometer tests performed on samples of expansive clay, bentonite powder, or mixtures were carried out using a similar approach. The amount of soil needed to fill up the consolidation ring is worked out when the dry density is known. The calculated amount is placed in three lifts in the ring using hand tamping. The samples prepared were placed in the oedometer ring and leveled off prior to placement in the oedometer cell. The weights of the ring and the sample were determined. The initial height of the sample was recorded. The initial moisture content and specific gravity were obtained. These measures enabled the height of solids, Hs, to be computed.

4. Test Program

The test program included testing three clay mixtures described as A, B, and C. Mixture referred to as (Form A) clay is commercial bentonite without any addition of Al-Qatif clay. This gives a typical sand-bentonite mixture as used in practice. For this mixture, clay contents were 5, 10, and 15% by weight of sand. The tests for Form A were carried out at a dry density of 17.5 kN/m^3. Swell and compressibility properties of these mixtures were used as reference for the other mixtures. Mixture referred to as (Form B) is a clay mixture of commercial bentonite and Al-Qatif clay in which one-third is bentonite and two-thirds are Al-Qatif clay. Mixture referred to as (Form C) is a mixture in which two-thirds are bentonite and one-third is Al-Qatif clay. These ratios are based on air dried weights. The initial dry densities for Forms B and C were the maximum dry density obtained from the compaction tests. Moisture contents at the optimum moisture content were selected. These selections were made because the liner material on site is normally compacted to approximately the maximum dry density and optimum moisture content by the contractors. Variations of +/− 2% moisture are normally acceptable and 95% level of compaction is also accepted.

For Form B and Form C, total clay contents of 5, 10, 15, 20, 25, and 30% were tested for swell and compressibility. A total of 12 samples were tested for these two forms. Each sample is represented by two specimens. Three samples were tested

TABLE 4: Typical chemical composition of OCMA grade bentonite.

FeO_3 (%)	K_2O (%)	Na_2O (%)	Al_2O_3 (%)	MgO (%)	SiO_2 (%)	TiO_2 (%)	CaO (%)
2.9	0.1	1.9	17.0	4.6	55.2	<0.1	0.9

Source Ore-Arabian Gulf Region-OCMA Grade.

TABLE 5: Data of swell percent and swelling pressure for the bentonite sand mixtures.

Clay content	Mixture Form	Cc	Cs	e_o	Swell %	Swell pressure kN/m^2	Dry density kN/m^3	m.c
5% (S R1)	Form A	0.0362	—	0.5167	1.55	95	1.75	0
10% (S R2)	Form A	0.0986	—	0.5182	4.57	175	1.75	0
15% (S R3)	Form A	0.1883	—	0.5198	8.23	200	1.75	0

for the bentonite mixtures (Form A). The main parameters measured in these tests include the swell percent, the swelling pressure, the compressibility index, and the swelling index. The initial void ratio was reported for each test.

Swell and compressibility data and plots for Forms A, B, and C clays were constructed and compared. Trends and behaviour as shown by test results should provide information about the general behaviour and a guide to select the most reliable mixture for the intended liner.

5. Results and Discussion

5.1. Moisture-Dry Density Relationship. The main reason for adding clay is to reduce the mixture permeability. Adding too little clay will result in an unacceptable high permeability and may result in the fines washing away when subjected to water flow under significant hydraulic gradients. As part of this study, moisture-density relationships were investigated for the sand-clay mixtures. The maximum dry density is found to increase with the increase of clay content up to a certain limit and then decreases. This is due to more fines replacing sand grains when the voids are fully occupied. At the optimum moisture content the degree of saturation is less than 100% due to air voids present in the system. The compacted soil is a three-phase system consisting of air, water, and solids. A clay paste formed within the pores can expand and fill up all air spaces when 100% impervious mixture is to be attained. Figure 3 presents moisture-density relationships of sand-clay mixtures and sand-bentonite mixtures. From Figure 2 it can be seen that 15% clay is just enough to fill up the voids at the optimum moisture content. Smaller amount of clay can still fill up the voids provided that enough expansion can be generated due to moisture increase beyond the optimum moisture content. In semiarid areas it is advisable to consider less expansive material as dry seasons may result in shrinkage and cracking. Clays of high plasticity can result in high linear shrinkage and expected to produce severe cracking in the field. For bentonite mixes it can be seen that less than 15% can fill up the gaps of the sand. In fact 5% or little more can be sufficient to fill the gap when the clay is fully saturated.

5.2. Compressibility and Swell of Sand-Bentonite Mixtures. Three different bentonite contents were considered to investigate the compressibility and swell of sand-bentonite mixtures. The tests were carried out on initially dry mixtures

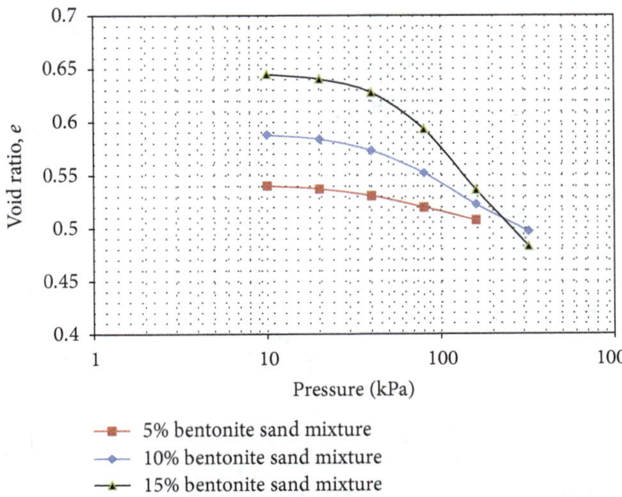

FIGURE 4: Void ratio versus log pressure curves for sand-bentonite mixtures.

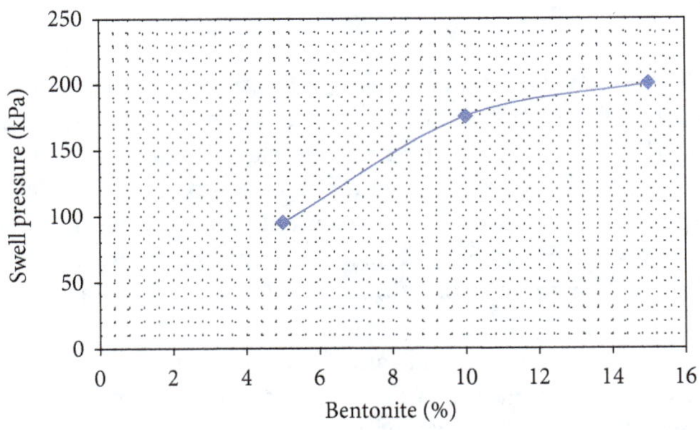

FIGURE 5: Swelling pressure for different bentonite contents in the mixture.

compacted to densities close to the maximum dry density and void ratios. Table 5 and Figures 4, 5, and 6 present the compressibility, swell pressure, and swell percent profiles for the sand-bentonite mixtures tested. The swelling pressure and swell percent indicated a clear increase with increase. The compressibility index is high for higher bentonite content.

Table 6: Summary of swell and compression data for Form B clay.

Clay content	Mixture form	Compression index Cc	Swell index Cs	Void ratio e_o	Swell %	Swell pressure (kPa)	Dry density kN/m^3	m.c. %
5% (S1 -a)	Form B	0.0179	—	0.5167	0.20	17	17.5	0
5% (S1 -b)	Form B	0.0374	0.0126	0.4911	0.00	0	17.8	11
10% (S3 - a)	Form B	0.0392	—	0.5198	1.67	70	17.5	0
10% (S3 -b)	Form B	0.0492	0.0159	0.4620	0.51	11	18.2	11
15% (S5- a)	Form B	0.0734	—	0.5198	3.83	100	17.5	0
15% (S5 -b)	Form B	0.0377	0.0154	0.4706	2.17	50	18.1	13
20% (S7 -a)	Form B	0.0754	0.0277	0.4404	6.16	85	18.5	13
20% (S7 -b)	Form B	0.0447	0.0183	0.4423	7.23	100	18.5	13
25% (S9 -a)	Form B	0.0691	0.0223	0.4739	6.43	90	18.1	17
25% (S9 -b)	Form B	0.0497	0.0199	0.4730	5.15	60	18.1	17
30% (S11 -a)	Form B	0.0860	0.0297	0.5096	11.30	170	17.7	16
30% (S11 -b)	Form B	0.0726	0.0202	0.5102	11.44	150	17.7	16

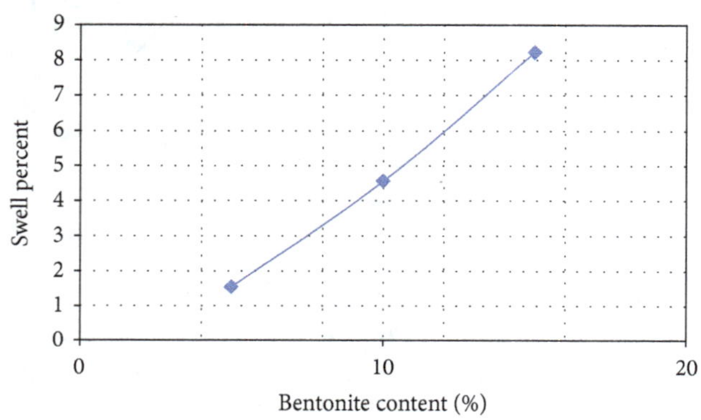

Figure 6: Swelling percent for different bentonite contents in the mixture.

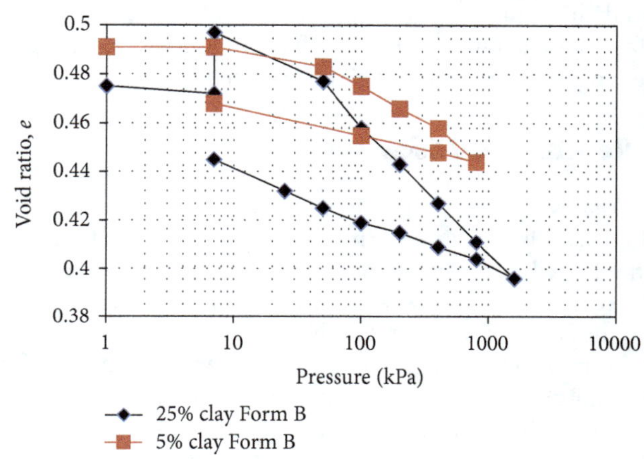

Figure 7: Swell-compressibility profiles for two selected clay ratios of Form B clay.

5.3. Compressibility and Swell of Mixtures of Bentonite and Natural Clay (1 : 2 Ratio). This section discusses material described as Form B in which the clay added to the sand is composed of one-third bentonite and two-thirds Al-Qatif expansive clay. The overall clay content by weight of sand was in the range 5% to 30%. Table 6 presents summary of data obtained for the combination of bentonite and natural clay (1 : 2 ratio).

Figure 7 presents void ratio versus the semilogarithm of pressure for 5% and 25% Form B added clay. This data was selected as typical representation of the general behaviour. The swell percent is negligible for 5% clay and equals 5.15% for the 25% clay. The compressibility is much higher for 25% clay and this is reflected by the steeper slope shown for different loading stages. The swelling pressure and the swell percent are plotted for all Form B clay ratios in Figures 8 and 9. The swelling pressure for 5%, 10%, and 15% was obtained for samples of different initial moisture content and this is reflected in the wider variation of the results.

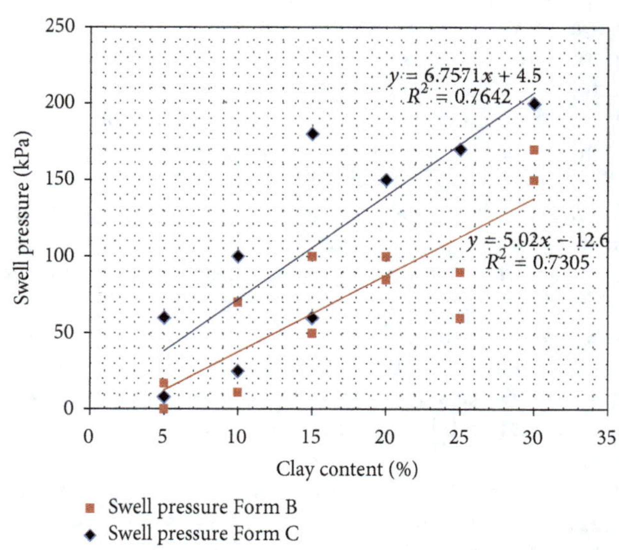

Figure 8: Swelling pressure of sand-clay mixtures.

5.4. Compressibility and Swell of Mixtures with Bentonite and Natural Clay (2 : 1 Ratio). This section discusses material

TABLE 7: Summary of swell and compression data for Form C clay.

Clay content	Mixture form	Compression index Cc	Swell index Cs	Void ratio e_o	Swell %	Swell pressure (kPa)	Dry density kN/m^3	m.c. %
5% (S2 -a)	Form C	0.0275	—	0.5167	0.92	60	17.5	0
5% (S2 -b)	Form C	0.0347	0.0141	0.4925	0.24	8	17.8	11
10% (S4 - a)	Form C	0.0438	—	0.5198	1.91	100	17.5	0
10% (S4 - b)	Form C	0.0392	0.0114	0.4620	2.26	25	18.2	11
15% (S6 - a)	Form C	0.0721	—	0.5198	4.48	180	17.5	0
15% (S6 - b)	Form C	0.0352	0.0219	0.4706	4.09	60	18.1	13
20% (S8 -a)	Form C	0.0718	0.0274	0.4417	10.80	150	18.5	13
20% (S8 -b)	Form C	0.0573	0.0209	0.4490	18.15	150	18.5	13
25% (S10 -a)	Form C	0.0666	0.0127	0.4739	13.19	170	18.1	17
25% (S10 -b)	Form C	0.0605	0.0206	0.4773	16.56	170	18.1	17
30% (S12 -a)	Form C	0.1370	0.0302	0.5569	34.74	200	17.7	16
30% (S12 -b)	Form C	0.0870	0.0249	0.5523	28.11	200	17.7	16

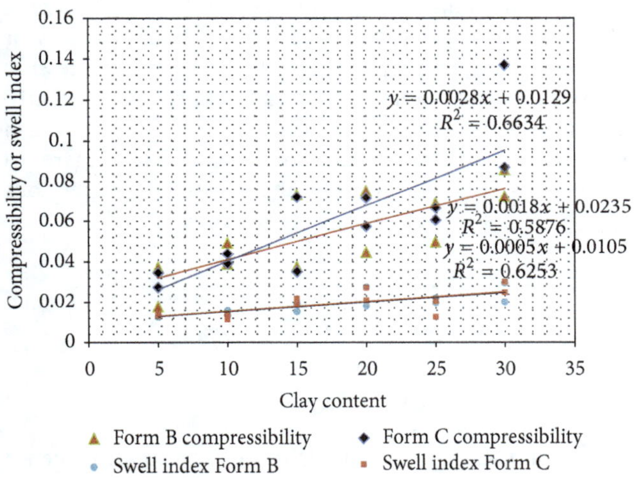

FIGURE 9: Swelling percent of sand-clay mixtures.

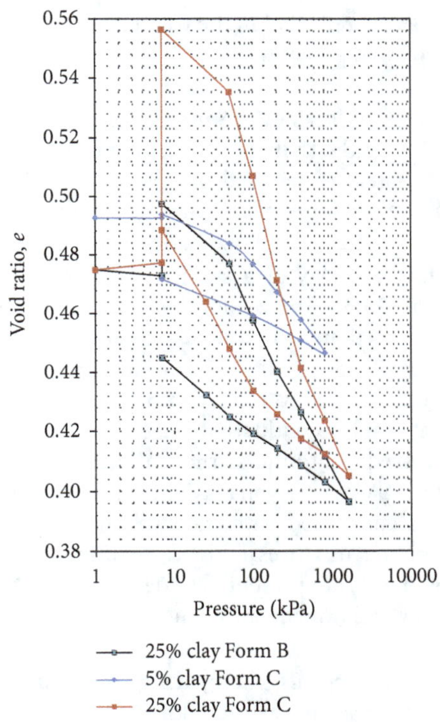

FIGURE 10: Swell-compressibility profiles for selected ratios (25% of Form C and 25% of Form B clay compared to 5% clay Form C).

described as Form C in which the clay added to the sand is composed of two-thirds bentonite and one-third Al-Qatif expansive clay. Overall clay content by weight of sand was in the range of 5% to 30%. The samples were subjected to swell and compression tests using oedometers. Table 7 presents summary of data obtained for the combination of bentonite and natural clay (2 : 1 ratio).

The selected, typical ratios of clay mixtures shown in Figure 10 demonstrate that the compressibility of bentonite rich material is high when compared to the other less expansive mixtures. The compressibility index at an intermediate point (15%) can be taken as 0.045. Figure 11 provides an overall view of compressibility index and swell index for all mixtures tested regardless of the type of clay used.

The compressibility index can be predicted based on the general clay content for the two proposed forms using the equation

$$\begin{aligned} Cc &= 0.0018\ CL + 0.024\ \ (\text{Form B}) \\ Cc &= 0.0028\ CL + 0.013\ \ (\text{Form C}), \end{aligned} \tag{1}$$

where Cc is the compressibility index and CL is the clay content expressed as percentage.

The swelling index, Cs, can similarly be predicted using the equation

$$Cs = 0.0005\ CL + 0.011\ \ (\text{Form B and Form C}). \tag{2}$$

This suggests that variation in swell index is negligible for the two tested forms.

Swelling pressure and swell percent for Form C values are plotted in Figures 8 and 9. The higher percentage of bentonite

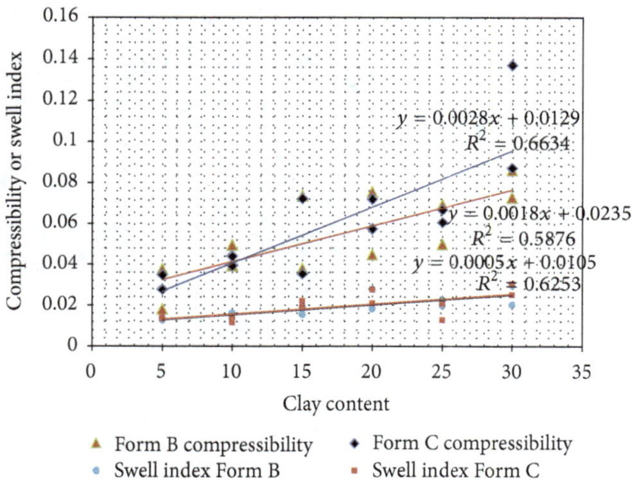

FIGURE 11: Compressibility and swell index for all tested clay ratios.

material is clearly reflected in the trend lines of both swelling pressure and swell percent.

5.5. Hydraulic Conductivity of Sand-Expansive Clay Mixtures. The hydraulic conductivity is a governing factor for the minimum clay content to be used. Al-Qatif clay-sand mixtures have been reported to give hydraulic conductivity values in the range of 7.20×10^{-8} cm/s for 5% clay and 3.54×10^{-8} cm/s for 25% clay at a surcharge pressure of 100 kPa (Dafalla et al. 2013). The bentonite-sand mixture with 10% bentonite resulted in a hydraulic conductivity of 2.719×10^{-6} cm/s (Dafalla et al. 2013). It can be noted that Al-Qatif clay can perform better than commercial bentonite with regard to hydraulic conductivity. However, the advantage of the processed bentonite is the uniformity of properties unlike natural clay where you may encounter wide variations. This paper is concerned with the swell and compressibility behaviour of the mixtures and the hydraulic conductivity values quoted here are presented to demonstrate that the use of natural clay will not affect the liner basic requirements.

5.6. General Comments. Comparing the compressibility indices it can be seen that the addition of Al-Qatif clay to bentonite will result in reducing the compression index from 0.0986 to 0.0400 for 10% clay content and from 0.1883 to 0.05 for 15% clay content. This is equivalent to 60 and 70 percent for 10% and 15% clay, respectively.

The swelling pressure and swell percent are reduced significantly by adding Al-Qatif clay. The swelling pressure of bentonite-sand mixture of 15% clay is reduced from 200 kPa to 100 kPa and 185 kPa for initially dry mixtures. The swell percent is reduced by nearly 50%.

It can be seen that using bentonite alone may not be advisable due to the expected high swelling pressure and swell percent. The minimum clay ratio that will just satisfy the permeability required should be used. This is expected to be achieved using mixtures of bentonite and local natural clay of less expansion. From this study it seems that the use of 15% clay consisting of one-third bentonite is an appropriate choice for a clay liner with a hydraulic conductivity of 1×10^{-7} cm/s. Going for higher clay ratios will increase the compressibility and swelling of the liner material and this may result in problems for light structures or the general profile of the ground level.

6. Conclusions

The swell and compressibility behaviour of sand-expansive clay mixtures should be considered along with the other design requirements for liners. The appropriate clay ratio that satisfies the hydraulic conductivity and other factors should be selected in view of the swell and compression properties of the mixture. Local clay material of lower expansion properties can be considered as partial replacement for bentonite within the sand-clay mixtures. This study investigated adding Al-Qatif clay to commercial bentonite to compose a clay material with 1 : 2 and 2 : 1 Al-Qatif clay to bentonite ratios and then use it in sand-clay mixtures for liners.

It was found that addition of less expansive soil material to bentonite can reduce the compressibility by 60% to 70% for 10% and 15% clay content, respectively. The swelling pressure and swell percent were also found reduced significantly. Adding local less expansive natural clay to bentonite can produce liners that are less problematic with regard to expansion and settlement and should still be impervious enough to satisfy normal design requirements.

Conflicts of Interest

The author declares that they have no conflicts of interest.

Acknowledgments

The author would like to thank Tawasol Research Excellence Program (TRE Program), King Saud University, Riyadh, Saudi Arabia, for support. The author would like to thank Dr. Terry Cousens, University of Leeds, United Kingdom, for his valuable consultation and guidance for this study. Thanks also extend to Eng. Abdulsattar Al Qatt and Eng. Galal Fathelrahman for conducting most of the laboratory tests.

References

[1] D. E. Daniel, *Geotechnical Practice for Waste Disposal*, Chapman & Hall, 1993.

[2] M. A. Dafalla and A. M. Al-Mahbashi, "Effect of adding natural clay on the water retention curve of sand—bentonite mixtures," in *Unsaturated Soils: Research & Applications*, pp. 1017–1022, CRC Press, 2014.

[3] M. A. Dafalla, *The use of the fall cone penetrometer in the characterization and assessment of swelling clays [Ph.D. thesis]*, University of Leeds, Leeds, UK, 2004.

[4] K. S. Katti, D. Sikdar, D. R. Katti, P. Ghosh, and D. Verma, "Molecular interactions in intercalated organically modified clay and clay-polycaprolactam nanocomposites: experiments and modeling," *Polymer*, vol. 47, no. 1, pp. 403–414, 2006.

[5] J. K. Mitchell, J. D. Bray, and R. A. Mitchell, "Material interactions in solid waste landfills," in *Proceedings of the Specialty*

Conference on Geotechnical Practice in Waste Disposal, pp. 568–590, New Orleans, La, USA, February 1995.

[6] S. E. Obrike, C. C. Osadebe, and S. S. Omoniyi, "Geotechnical analysis of two Nigerian soils for use as clay liners," *Bulletin of Engineering Geology and the Environment*, vol. 68, no. 3, pp. 417–419, 2009.

[7] N. J. Langdon, M. J. Al Hussaini, P. J. Walden, and C. M. Sangha, "An assessment of permeability of clay liners: two case histories," *Geological Society*, vol. 11, pp. 291–297, 2008.

[8] "Standard test method for evaluation of hydraulic properties of geosynthetic clay liners permeated with potentially incompatible liquids. Annual book of ASTM Standards," Tech. Rep. ASTM D 6766–2012, ASTM, Philadelphia, Penn, USA, 2012.

[9] Y. Wasti and I. Alyanak, *Kil Muhtevasının Zeminin Davranışına Tesiri. İnşaat Mühendisleri Odası*, Türkiye İnşaat Mühendisliği 4. Teknik Kongresi, Ankara, Turkey, 1968.

[10] R. E. Olson and G. Mesri, "Mechanisms controlling the compressibility of clay," *Journal of the Soil Mechanics and Foundations Division, A.S.C.E*, vol. 96, pp. 863–1878, 1970.

[11] A. K. Mundegar, *An investigation into the effects of platy mica particles on the behavior of sand [M.S. thesis]*, Imperial College, London, UK, 1997.

[12] P. V. Lade, C. D. Liggio Jr., and J. A. Yamamuro, "Effects of non-plastic fines on minimum and maximum void ratios of sand," *Geotechnical Testing Journal*, vol. 21, no. 4, pp. 336–347, 1998.

[13] S. Tsotsos, F. E. Karaoulanis, and T. Chatzigogos, "A new concept on the compressibility of mixed soils: experimental and numerical approach," *Geotechnical and Geological Engineering*, vol. 28, no. 2, pp. 199–207, 2010.

[14] M. A. Dafalla, "The influence of placement conditions on the swelling of variable clays," *Geotechnical and Geological Engineering*, vol. 30, no. 6, pp. 1311–1321, 2012.

[15] D. A. Dixon, "Porewater salinity and the development of swelling pressure in bentonite based buffer and backfill materials," Atomic Energy of Canada Limited. POSIVA Report 2000-04, Posiva Oy, Helsinki, Finland, 2000.

[16] H. A. Alawaji, "Swell and compressibility characteristics of sand-bentonite mixtures inundated with liquids," *Applied Clay Science*, vol. 15, no. 3-4, pp. 411–430, 1999.

[17] L. H. Mollins, D. I. Stewart, and T. W. Cousens, "Predicting the properties of bentonite-sand mixtures," *Clay Minerals*, vol. 31, no. 2, pp. 243–252, 1996.

[18] C. Benson and G. Boutwell, "Compaction conditions and scale-dependent hydraulic conductivity of compacted clay liners," in *Constructing and Controlling Compaction of Earth Fills*, D. Shanklin, K. Rademacher, and J. Talbot, Eds., ASTM STP 1384, pp. 254–273, ASTM, 2000.

[19] B. R. Phanikumar, C. Amshumalini, and R. Karthika, "Swell-consolidation characteristics of artificial sand clay mixes," in *Proceedings of the Indian Geotechnical Conference*, Geotrendz, IGS Mumbai Chapter & IIT Bombay, 2010.

[20] S. N. Abduljauwad, "Characteristics and chemical treatment of expansive clay in Al-Qatif, Saudi Arabia," *International Journal of Rock Mechanics and Mining Sciences & Geomechanics Abstracts*, vol. 29, no. 4, pp. 226–236, 1992.

[21] S. Azam, S. N. Abduljauwad, N. A. Al-Shayea, and O. S. B. Al-Amoudi, "Expansive characteristics of gypsiferous/anhydritic soil formations," *Engineering Geology*, vol. 51, no. 2, pp. 89–107, 1998.

[22] S. Azam, "Influence of mineralogy on swelling and consolidation of soils in eastern Saudi Arabia," *Canadian Geotechnical Journal*, vol. 40, no. 5, pp. 964–975, 2003.

[23] M. A. Shamrani, E. Mutaz, A. J. Puppala, and M. A. Dafalla, "Characterization of problematic expansive soils from mineralogical and swell ell characterization studies," in *Proceedings of the GeoFlorida 2010: Advances in Analysis, Modeling and Design*, pp. 793–903, West Palm Beach, Fla, USA, February 2010.

[24] ASTM D2487-11, *Standard Practice for Classification of Soils for Engineering Purposes (Unified Soil Classification System)*, vol. 15.09 of *Annual Book of ASTM Standards*, ASTM, West Conshohocken, Pa, USA, 2012.

[25] ASTM D698-02 2012, *Standard Test Methods for Laboratory Compaction Characteristics of Soil Using Standard Effort (12 400 ft lbf/ft3 (600 kN m/m^3))*, vol. 15.09, Annual book of ASTM Standards, West Conshohocken, Pa, USA, 2012.

[26] "Standard test methods for one dimensional swell or settlement potential of cohesive soils. Annual book of ASTM Standards," Tech. Rep. ASTM D4546-96 2012, ASTM, Philadelphia, Penn, USA, 2012.

4

In Situ Testing of Square Footing Resting on Geobelt-Reinforced Gravel Thin Cushion on Soft Silt

Wei Zhang[1], **David Du**[1,2] **and Xiaohong Bai**[1]

[1]*Department of Civil Engineering, Taiyuan University of Technology, Taiyuan, Shanxi 030024, China*
[2]*Subsea 7 (US) LLC, 17220 Katy Freeway, Suite 100, Houston, TX 77094, USA*

Correspondence should be addressed to Xiaohong Bai; bxhong@tyut.edu.cn

Academic Editor: Michael Aizenshtein

A series of in situ static loading tests of square footing were carried out on the geobelt-reinforced gravel cushion on soft silt. The reinforced gravel cushion was thin with the depth-to-width ratio of 0.2. A parameters study was conducted by considering the number of geobelt layers, the depth of the first geobelt layer beneath the footing, the vertical spacing between two geobelt layers, the linear density of reinforcement, and the material type of geobelt. The pressure distribution on bottom of the cushion was measured. The test results showed that the bearing capacity of reinforced gravel cushion was significantly larger than that of unreinforced gravel cushion, and the stress diffusion effect of reinforced gravel cushion was also more pronounced than that of the unreinforced cushion. The pressure distribution on bottom of reinforced gravel cushion was in a saddle shape. According to calculation and analysis, the stress diffusion angles of reinforced cushions were all larger than 20°.

1. Introduction

For using it as foundation, soft silt soil has to be treated due to its low bearing capacity and large settlement. An effective treatment is to replace a shallow depth of upper soft silt with geosynthetic-reinforced gravel cushion, which is cost-effective compared to other conventional methods [1, 2]. It has been demonstrated that geosynthetic reinforcement can improve the bearing capacity and reduce uneven settlement of the foundation [3–5]. By conducting 65 groups of model tests, Binquet and Lee [6, 7] first reported that, by reinforcing sand layer underneath the strip footing with strips of aluminum foil, the settlement and ultimate bearing capacity of the foundation was greatly improved. In literatures[8–14], similar results were reported by researchers. It should be noted that it is complicated to determine the ultimate bearing capacity of square footing on reinforced soil. In addition, there has been limited research investigating the stress diffusion effect of reinforced cushion.

For determining the bearing capacity and stress diffusion capacity, most studies have used small-scale models test, which would have size effect and limitations to reflect the actual behaviors of the deformation and bearing capacity of foundation [15–17]. Therefore, it is imperative to conduct full-scale in situ tests to investigate the effects of reinforcing geosynthetic materials. In the literature, geotextiles including geogrid [17–19], geonet [20], geocell [21, 22], and fiber [23, 24] are commonly used as reinforcing materials, while geobelt, a relatively new material, has been rarely utilized. Moreover, most of the previous studies have focused on the settlement and bearing capacity of foundation [1, 6, 9, 12, 15, 25], while little attention has been paid to the stress diffusion angle of reinforced cushion. In practical, due to presence of geobelt, the overall rigidity and stress diffusion capacity of the reinforced cushion can be greatly improved even for thin cushions. Therefore, based on the stress diffusion theory of double-layered foundation, this study aims to investigate the bearing capacity and earth pressure distribution of square footing resting on the thin gravel cushion through a series of in situ static loading tests. The stress diffusion effect of geobelt on the reinforced gravel cushion was also studied.

Table 1: Physical and mechanical parameters of silt soil.

γ(kN/m^3)	γ_D(kN/m^3)	G_S	e	ω(%)	ω_L(%)	ω_p(%)	E_S(MPa)	f(MPa)
18.9	14.5	2.69	0.881	31.8	32.9	23.7	3.93	70

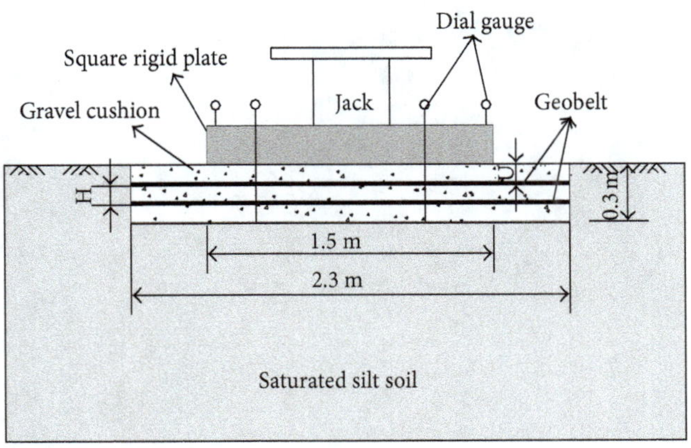

Figure 1: Setup of the loading test.

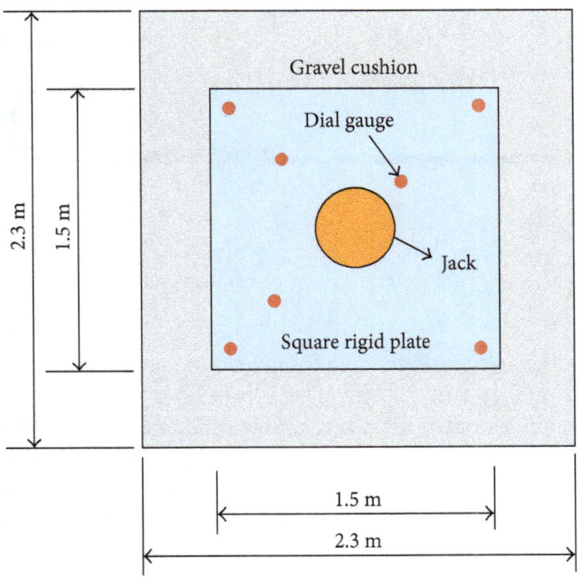

Figure 2: Top view of the testing setup.

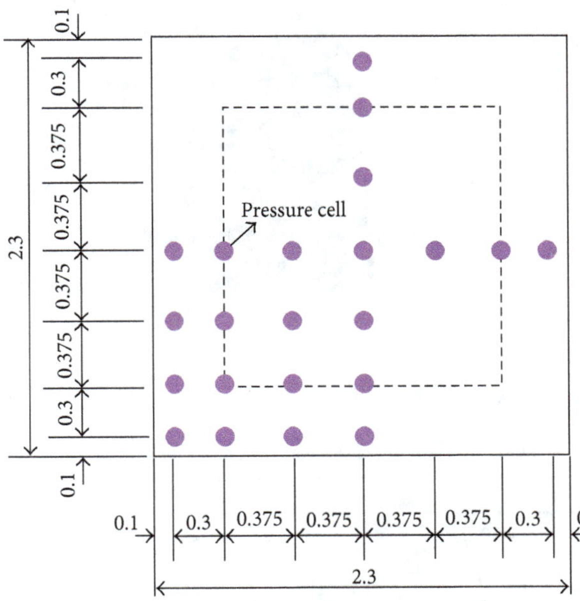

Figure 3: Distribution of pressure cells (unit: m).

Table 2: Testing program.

Test number	N^*—	U^* (mm)	H^* (mm)	LDR* (%)	Material type
A0	Unreinforced	—	—	—	—
A1-1	1	50	—	33.3	TG
A1-2	1	100	—	50.0	TG
A1-3	1	100	—	33.3	TG
A1-4	1	100	—	25.0	TG
A1-5	1	200	—	33.3	TG
A2-1	2	50	100	33.3	TG
A2-2	2	50	150	33.3	TG
B2-1	2	50	100	33.3	CPE
B2-2	2	50	150	33.3	CPE

$^*N$ represents the layers of geobelt set in the gravel cushion, U is the depth of the first geobelt layer beneath the footing, H refers to the vertical spacing between two layers of geobelt, and LDR is the linear density of reinforcement which means the ratio of the width of geobelt to the center spacing of two geobelts.

2. Experiment

2.1. Test Design and Setup. The testing site, with a dimension of 30 m × 17 m, is located in southwest part of Taiyuan city, China. In the testing site, the natural soil beneath the gravel cushion was saturated soft silt soil. The properties of the silt soil are listed in Table 1.

Each static loading test was conducted on the field test pit with a length of 2.3 m and width of 2.3 m within the testing site. Figure 1 shows the layout of the test setup. A steel square rigid plate with a length of 1.5 m, width of 1.5 m, and depth of 0.3 m was used as a footing to apply the loading. A gravel cushion with a dimension of 2.3 m × 2.3 m × 0.3 m with or without geobelt reinforcement was prepared before each loading test. The loading was applied by a mechanical jack-beam loading system on the gravel cushion through the square footing.

Seven dial gauges were deployed in each test, as shown in Figure 2. Four gauges were attached on the four corners of the square rigid plate to measure the overall settlement of the footing. Other three gauges were attached on the bottom of the gravel cushion to monitor the settlement of silt soil. The deformation of the gravel cushion can then be calculated according to the measurements of seven gauges. As shown in Figure 3, twenty-two pressure cells, with a measuring capacity of 0.6 MPa, were deployed on the bottom of the gravel cushion to measure the pressure of the cushion resting on silt soil during loading.

In total, 10 in situ tests were conducted on unreinforced and geobelt-reinforced gravel cushions over saturated silt soil. The testing programs are summarized in Table 2.

(a) (b)

FIGURE 4: Photo of geobelts: (a) TG geobelt; (b) CPE geobelt.

TABLE 3: Engineering properties of the geobelts.

	TG geobelt	CPE geobelt
Geometry size (mm × mm)	25 × 2.5	25 × 2.0
Ultimate tensile strength (MPa)	95.4	139.4
Modulus at strain = 2% (MPa)	10,373	15,064
Length per kilogram (m/kg)	>16	>12
Fracture strain (%)	0.85	1.89
Breaking load (kN)	5.96	6.97
Wrapping material	Polyolefin	Chlorinate polyethylene
Inner material	Glass fiber	Galvanized steel wire

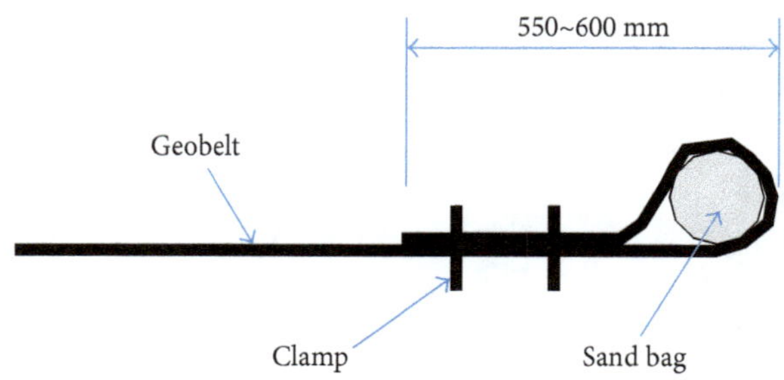

FIGURE 5: Fixing of geobelt ends.

2.2. Materials

2.2.1. Geobelts. Two types of geosynthetic materials used in this study are TG geobelt and CPE geobelt. TG geobelt, as shown in Figure 4(a), is mainly made of fiberglass coated with polyolefin. As shown in Figure 4(b), CPE geobelt is high-strength galvanized iron wires coated with chlorinated polyethylene. These two geobelts have high tensile strength, low expansion, and small creep and are resistant to bending fatigue, tension stress crack, bearing, and punching. Moreover, the geobelt has properties such as antiaging, acid/alkali resistance, and suitability for burying in soil. From construction perspective, the geobelts have the advantages of light-weight, simple construction, and shorter construction period. There are rough details and ribs on the surface of two types of geobelt materials for improving the bonding between geobelts and gravel. Their engineering properties are listed in Table 3.

In the test, in order to ensure strong bonding between geobelts and the gravel, the geobelts are bent at both ends by using a sand bag which is around 550~600mm and then fixed by two clamps, as shown in Figure 5. Therefore, the effective length of the geobelt used in the cushion is around 3.5m, including the width of the cushion and the fixing length at both ends.

2.2.2. Gravel Cushion. The thin cushion consisted of gravels with diameter in the range of 10mm– 30mm. Particle size distribution of the gravels is shown in Figure 6. The physical parameters obtained in laboratory tests are listed in Table 4, in which the maximum dry density and optimum moisture content were determined by Standard Proctor Test. According to USCS and AASHTO classification, the gravel was classified as poorly graded gravel (GP).

2.3. Test Procedure. To ensure that each test was conducted under same conditions, the following preparations have to be done. First, the testing pits were dug, cleaned, and leveled to make sure their sizes meet the requirements. Secondly, the sorted fine sand was paved, compacted, and leveled on the bottom of the pits with a thickness of 10 mm~15 mm for reducing stress concentration in the pressure cell. Then the pressure cells were put on the preset positions. The pressure cells should be waterproof and calibrated before each test. Six layers of gravel were paved on the pressure cell with each layer about 50 mm thick. Each layer of gravel in the loose state had the same weight and was then compacted to same compactness by using a wood hammer. Geobelts were laid on the specified position in the gravel cushion with the required linear density of reinforcement (LDR) in two dimensions. The geobelts should be tightened and straightened. A layer of fine sand with a thickness of 10 mm~20 mm was paved above and beneath the geobelts for protecting

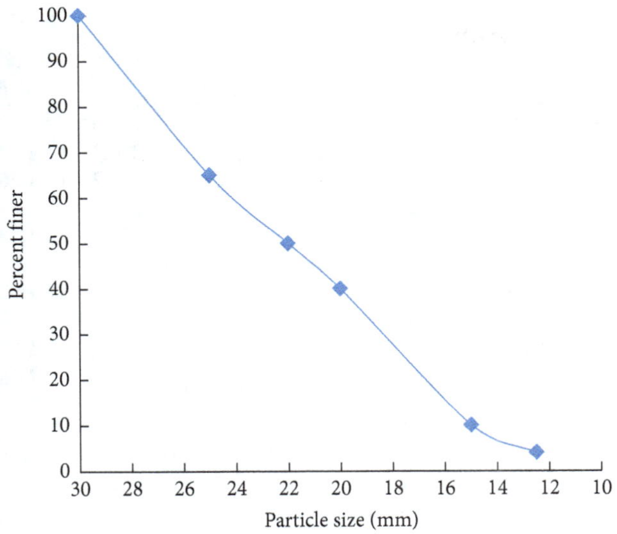

Figure 6: Particle size distribution of the gravel.

Table 4: Physical parameters of gravel.

ρ_{dmax}(kg/m^3)	c_c	c_u	D_{10}(mm)	D_{50}(mm)	ω_{op}(%)
1800	1.92	1.02	15	22	6.8

them from not being punctured by gravel. Geobelts and gravel were laid alternately according to the requirements until the cushion height reached 300 mm. Then the loading plate was put on the top of the reinforced gravel cushion, leveled, and centered properly, to ensure that the load can be evenly distributed on the gravel cushion.

The static loading was applied by using a hydraulic jack. The loading method and stability criteria followed the *Code for Design of Building Foundation* (GB 50007-2011) [26]. The loading was applied with an increment of 20 kPa. If it was less than 0.1 mm/h in two hours, the settlement was recognized to meet the criteria and the next increment of loading can be applied. The test was terminated when the total settlement reached 0.06 *B*, that is, 90 mm in this program.

After one test was completed, a new test pit was dug in adjacent locations within the testing site. Identical testing procedures were used to perform each test.

3. Test Results and Analysis

3.1. Reinforcing Effect on Bearing Capacity of Foundation. Figure 7 shows pressure-settlement curves of foundations with and without geobelts. It can be seen that the settlement increased with the increase of pressure for both reinforced and unreinforced cushions. However, cushion reinforced with two geobelt layers showed the slowest increase rate of settlement, while unreinforced cushion showed the quickest increase rate. The results indicate that the bearing capacity can be effectively improved by geobelt reinforcement. The reason is that the lateral restraint of geobelt on gravel cushion can decrease the settlement of the gravel cushion, and thus improve the bearing capacity of the foundation compared to the unreinforced cushion. At early stage, the

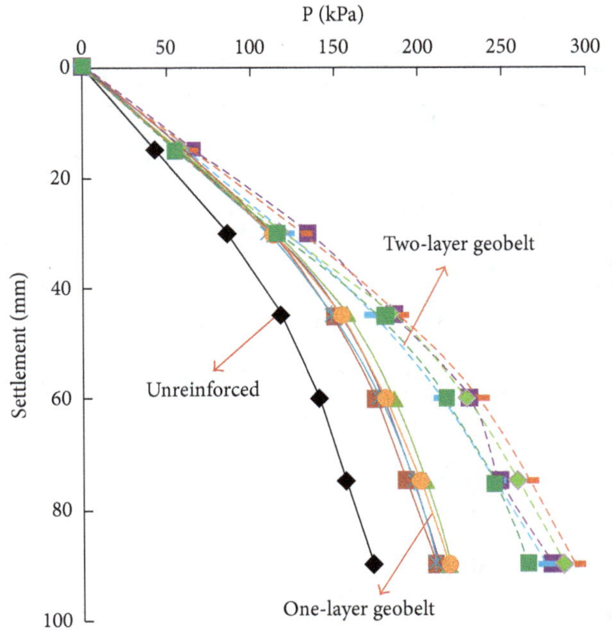

Figure 7: Pressure-settlement curves of cushions with and without geobelts.

settlement increased linearly with pressure. As pressure increased, geobelt comes into play and decreases settlement effectively.

Since no abrupt increase of settlement was observed in all tests, the ultimate bearing capacity of foundation was determined when the settlement of foundation reached 0.06 *B* (*B* is the length of the square footing).

To account for the size effect of footing, the bearing capacity ratio (BCR) recommended by Binquet and Lee was calculated as following:

$$\text{BCR} = \frac{q_R}{q_0}, \tag{1}$$

in which q_R and q_0 are the bearing capacity for reinforced and unreinforced soil, respectively. For convenience, the test results were analyzed according to the BCRs calculated at different settlement ratios (s/B). The settlement ratio is calculated by dividing settlement of footing (s) with the width of footing (B). The values of BCR at settlement ratios (s/B) of 0.01, 0.02, 0.03, 0.04, 0.05, and 0.06 are listed in Table 5. The BCR for one-layer geobelt reinforcement was between 1.22 and 1.37, and gradually decreased with the increase of the settlement ratio; the BCR of two-layer geobelt reinforcement was between 1.34 and 1.70 and gradually increased with the increase of the settlement ratio. In current study, the values of BCR are lower than that proposed by Adams and Collin [27], and Chen et al.[28]. This may be due to the relatively small thickness of the cushion used in this study.

Figures 8(a) and 8(b) show the relationship between the number of reinforced layers (N) and BCR. As can be seen, two-layer geobelt reinforcement was much better than one-layer geobelt reinforcement, especially at the late stage of loading. At s/B of 0.06, BCRs of one-layer reinforcement were from 1.22 to 1.26, while for two-layer reinforcement, they were from 1.62 to 1.70, which indicates that the bearing

Table 5: The summary of test results.

Test number	$s/B=0.01$		$s/B=0.02$		$s/B=0.03$		$s/B=0.04$		$s/B=0.05$		$s/B=0.06$	
	q (kPa)	BCR	q (kPa)	BCR	q (kPa)	BCR	q (kPa)	BCR	q (kPa)	BCR	q (kPa)	BCR
A0	43.4	1.00	86.2	1.00	117.7	1.00	140.6	1.00	156.7	1.00	172.6	1.00
A1-1	59.5	1.37	115.4	1.34	150.5	1.28	174.2	1.24	192.5	1.23	210.3	1.22
A1-2	58.2	1.34	114.3	1.33	157.4	1.34	185.2	1.32	204.2	1.30	218.2	1.26
A1-3	56.6	1.30	112.7	1.31	153.0	1.30	180.0	1.28	198.4	1.27	213.4	1.24
A1-4	56.1	1.29	110.2	1.28	148.5	1.26	177.1	1.26	198.1	1.26	210.6	1.22
A1-5	55.2	1.27	112.7	1.31	154.2	1.31	180.0	1.28	201.1	1.28	218.1	1.26
A2-1	67.3	1.55	134.5	1.56	183.9	1.56	230.9	1.56	249.0	1.59	279.8	1.62
A2-2	63.8	1.47	131.4	1.52	188.5	1.60	236.1	1.68	265.5	1.69	293.6	1.70
B2-1	60.3	1.39	120.8	1.40	173.3	1.47	214.4	1.53	245.8	1.57	276.3	1.60
B2-2	58.1	1.34	118.4	1.37	184.3	1.57	228.5	1.63	258.3	1.65	286.7	1.66

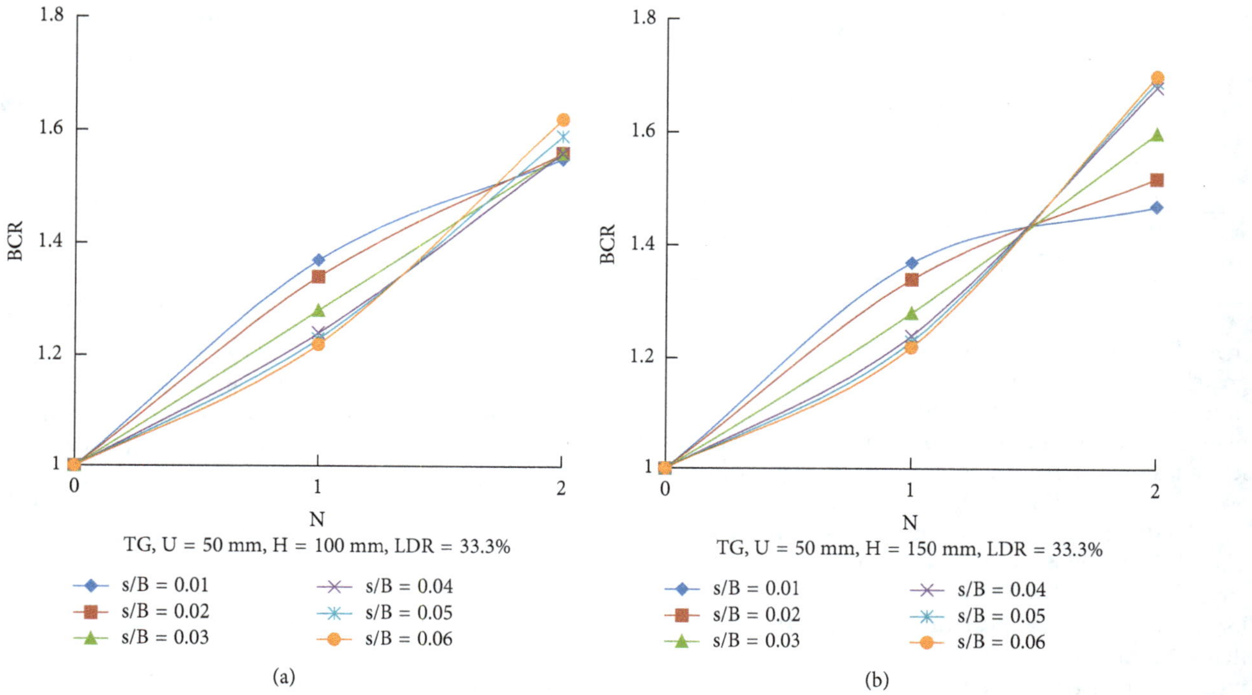

Figure 8: The relationship between BCR and N: (a) TG, $U=50$ mm, $H=100$ mm, and LDR = 33.3%; (b) TG, $U=50$ mm, $H=150$ mm, and LDR = 33.3%.

capacity of soil reinforced with two-layer geobelt can be increased more at the limit state.

Figure 9 shows the relationship between BCR and the depth of the first geobelt layer beneath the footing (U). At the initial stage of loading ($s/B\le 0.2$), BCR reached the maximum at 50 mm (U), and gradually decreased as U increased. At the later stage of loading ($s/B\ge 0.3$), BCR slightly increased and the bearing capacity of cushion was improved. This can be explained that the geobelt comes into play after certain amount of settlement happened. In addition, the geobelt could take the effect earlier if it is closer to the footing.

Figure 10 shows the relationship between BCR and the linear density of the reinforcement (LDR). As can be seen, BCR increased slightly with the increase of LDR. When LDR increased from 25% to 50%, the value of BCR increased from (1.22~1.29) to (1.26~1.34). This can be ascribed to the increase of shear strength since the interactions between the geobelt and gravel is improved when LDR is increased which provides more lateral restraint to the gravel cushion.

Figures 11(a) and 11(b) demonstrate the relationship between BCR and vertical spacing (H) for two-layer geobelt-reinforced cushion. At the initial stage of loading ($s/B\le 0.2$), BCR decreased with the increase of H, while at the later stage ($s/B\ge 0.3$), BCR increased with the increase of H. The results also indicate that the geobelt takes effect when certain amount of settlement happened. At the later stage of loading, reinforcing effect of the geobelt became more pronounced as there was a continuous increase of deformation of the cushion.

Figures 12(a) and 12(b) show the bearing capacity ratio against the settlement ratio for the cushion reinforced with the TG geobelt and CPE geobelt to compare their reinforcing effects. At the initial stage of loading, in terms of bearing

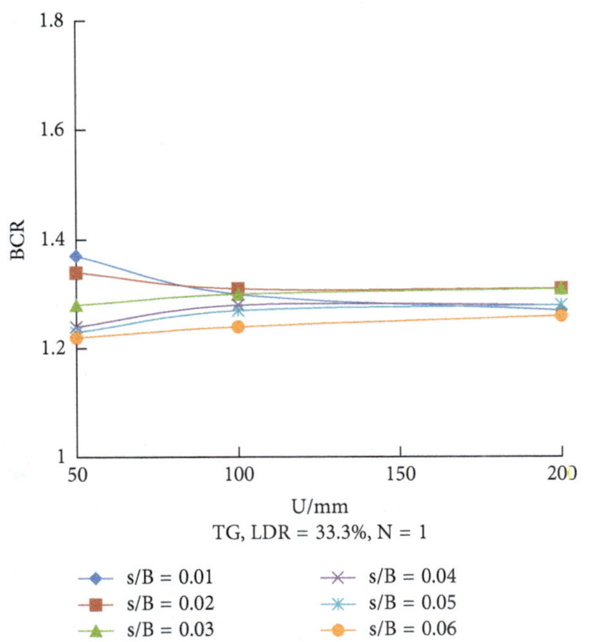

FIGURE 9: The relationship between BCR and U (TG; LDR = 33.3%; N = 1).

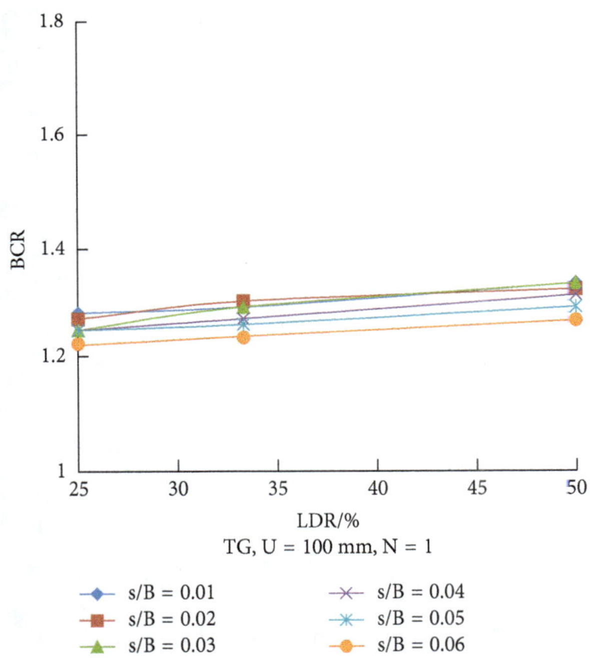

FIGURE 10: The relationship between BCR and LDR (TG; U = 100 mm; N = 1).

capacity, the TG-type geobelt performed better than the CPE-type geobelt. However, the difference between them became unnoticeable with the increase of loading.

3.2. Stress Distribution at the Bottom of Reinforced Cushions. Figure 13(a) shows the stress distribution on bottom of the unreinforced gravel cushion. Figures 13(b) and 13(c) show stress distribution on bottom of the reinforced gravel cushion under different pressures. As can be seen in Figure 13(a), for the unreinforced gravel cushion, the stress distribution on bottom of the cushion was parabolic with maximum net earth pressure reached at the center of the cushion. The stress distribution curves on bottom of the reinforced cushion were both in saddle shape and the maximum earth pressures were found to be about 370 mm (one-layer geobelt) and 750 mm (two-layer geobelt) away from the center with the minimum pressure at the center. For the unreinforced cushion, surrounding soft silt soil could not provide strong constrain. As loading increases, the gravels on bottom edges of the footing could be easily pushed out laterally since the pressure on bottom could not be adjusted by the unreinforced cushion itself. While for reinforced cushions, the geobelt restricted the lateral displacement of the cushion owing to frictions between the geobelt and gravels. Therefore, it was not easy to push out the soils located on bottom edges of the footing. The results indicate that, due to using the geobelt, the stress distribution was improved. The central reaction force was decreased, and the edge reaction force was increased. For one-layer and two-layer reinforced cushions (Figures 13(b) and 13(c)), it can be found that, compared to the one-layer cushion, the pressure on the edges of the two-layer reinforced cushion dramatically increased and the pressure distribution became smoother. It is clear that in terms of the reinforcing effect, the two-layer reinforcement is much better than the one-layer reinforcement.

3.3. Stress Diffusion Angle of Cushion. As shown in Figure 13, the earth pressure measured on bottom of the cushions was not evenly distributed. So in this section, the average pressure on the bottom of cushion was calculated according to the measured settlement results which are more accurate and uniform. In same stress path, there exists a one-to-one relationship between settlement and pressure of soil layers. Therefore, the average pressure (P_z) under the cushion can be obtained by using the pressure-settlement curves of natural ground without cushion and with upper cushion, as shown in Figure 14. When the settlement of natural ground under the cushion was equivalent to that of natural ground without the cushion, the corresponding relationship between the pressure P_0 under the cushion and the load P_y on the top of natural ground was obtained; that is, P_y is equal to the mean earth pressure of cushions P_z. Based on the principle of stress diffusion (Figure 15), the stress diffusion angle of the reinforced cushion can be calculated by the following formula:

$$P_0 B^2 = P_Z (B + 2Z \tan\theta)^2,$$
$$\tan\theta = \frac{B}{2Z}\left(\sqrt{\frac{P_0}{P_z}} - 1\right). \qquad (2)$$

To investigate the effect of geobelt on stress diffusion, a ratio SDR was defined, in this study, by dividing the stress diffusion angle of the geobelt reinforced cushion with that of the unreinforced cushion. Table 6 lists stress diffusion angles

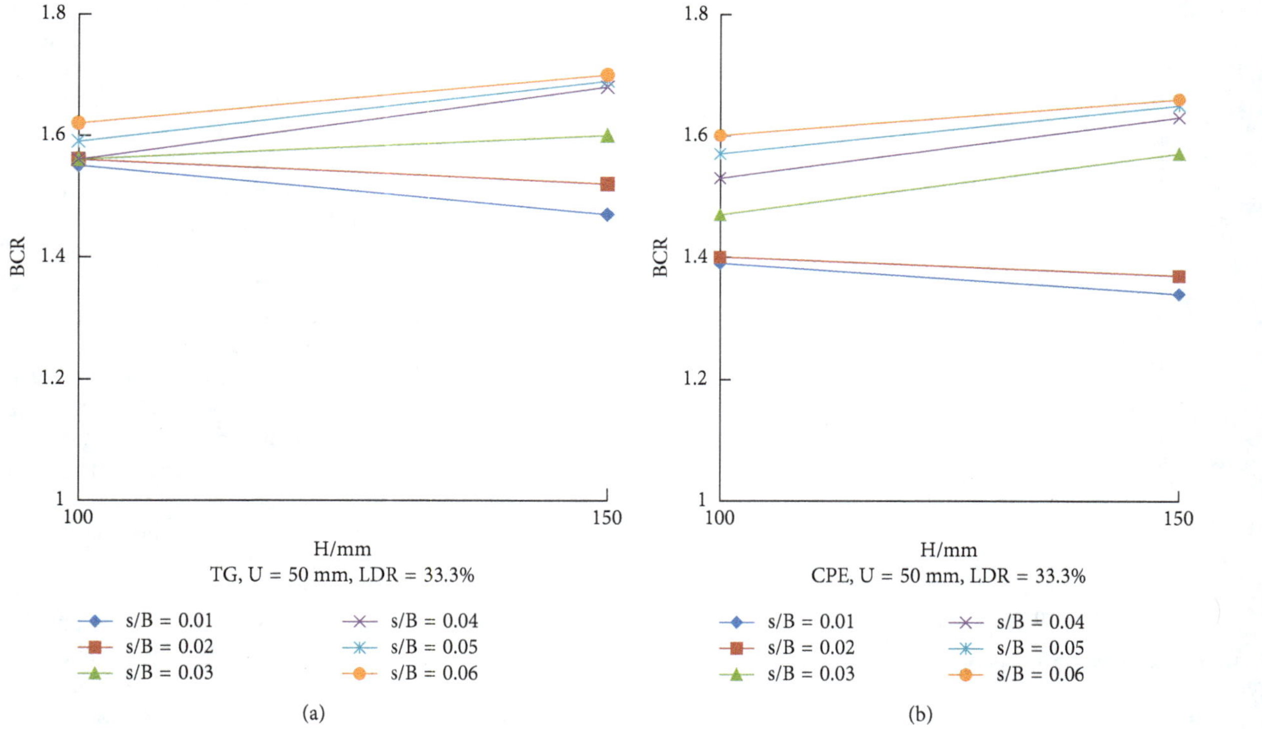

FIGURE 11: The relationship between BCR and H: (a) TG geobelt ($U = 50$ mm; LDR = 33.3%); (b) CPE geobelt ($U = 50$ mm; LDR = 33.3%).

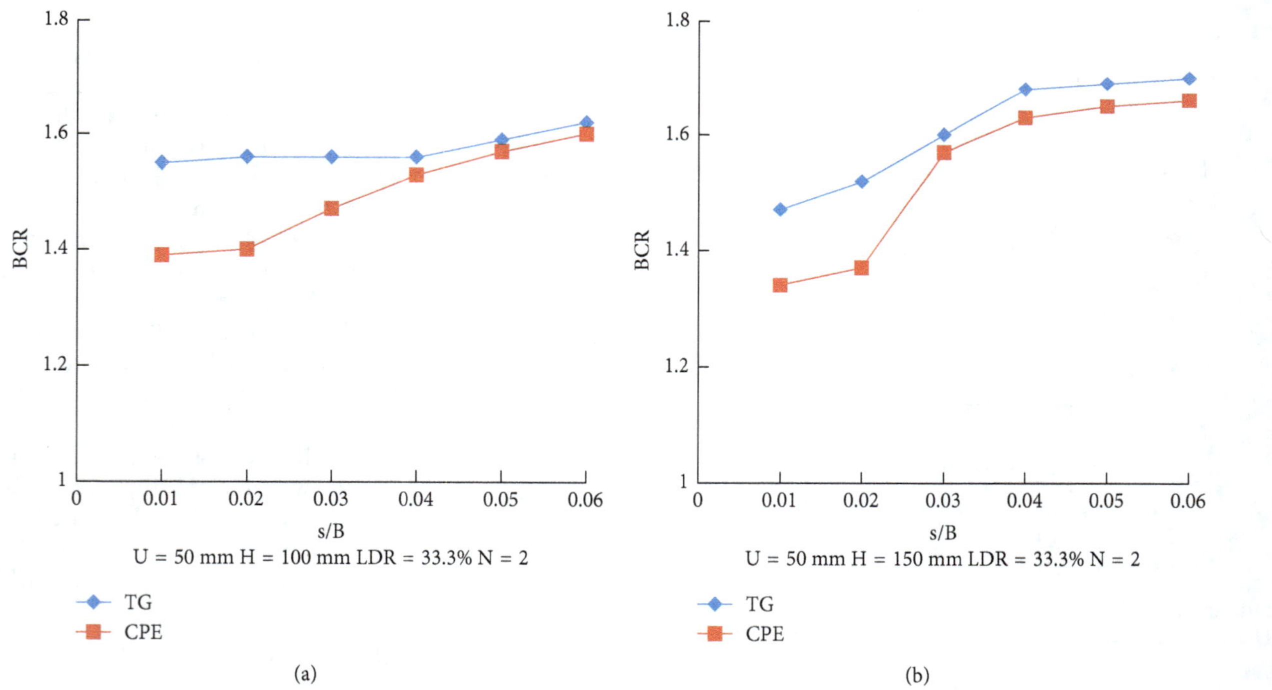

FIGURE 12: The relationship between BCR and geobelt type: (a) $U = 50$ mm, $H = 100$ mm, LDR = 33.3%, and $N = 2$; (b) $U = 50$ mm, $H = 150$ mm, LDR = 33.3%, and $N = 2$.

of the cushion and the corresponding SDR values at the proportional limit state and stable state. As can be seen in the table, the stress diffusion angle of reinforced cushions was obviously larger than that of unreinforced cushions. At the proportional limit state, SDR values spanned from 1.46 to 1.60, which means the stress diffusion angle was improved by 1.46– 1.60 times with geobelt reinforcement. At the limit state, SDR values spanned from 1.76 to 2.63, which means the stress diffusion angle was improved by 1.76–2.63 times. Current results are consistent with the result of Gabr et al. [29, 30] that the stress diffusion angle at the stable state was smaller than that at the proportional limit state.

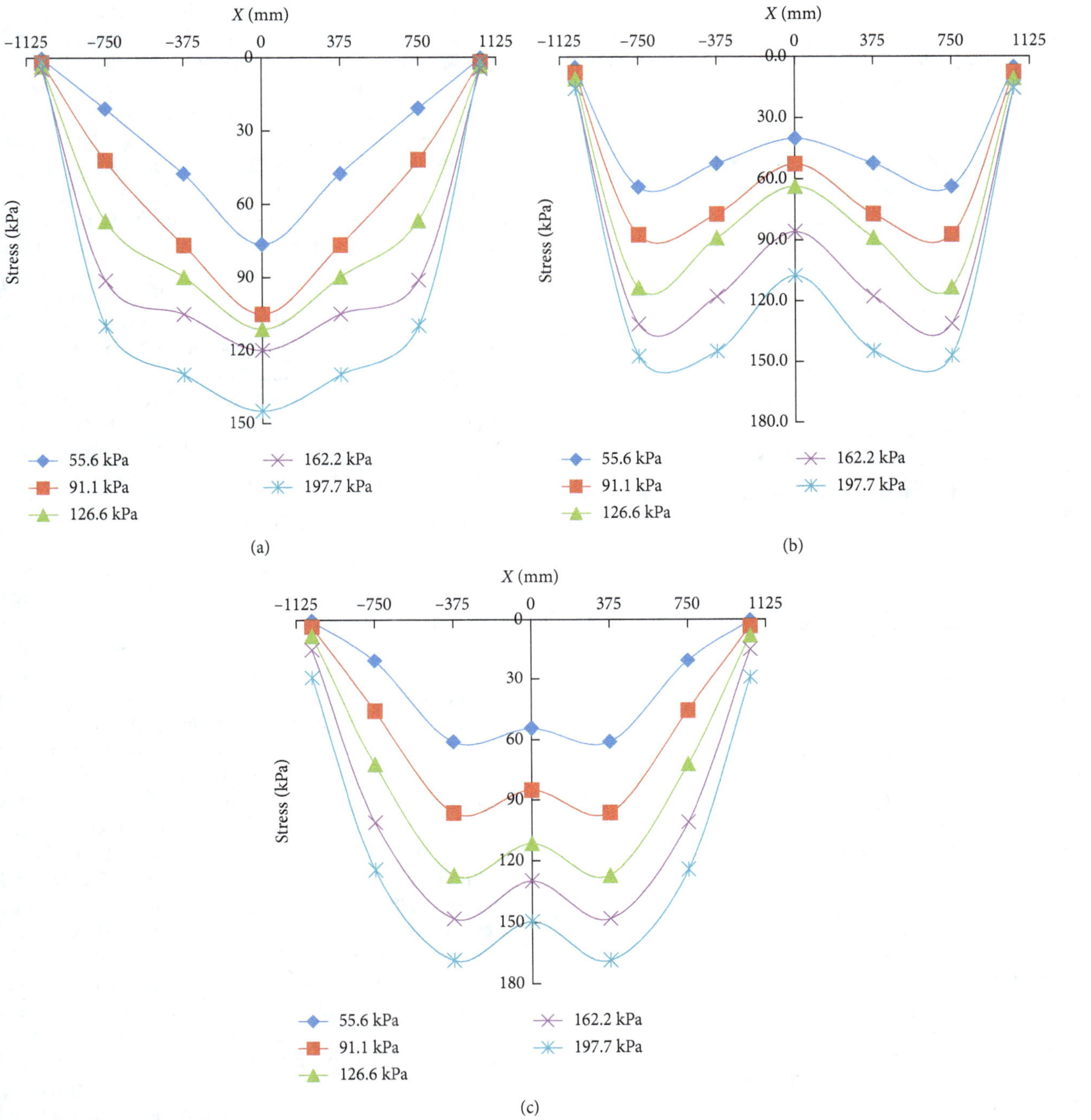

FIGURE 13: Stress distribution curves under the gravel cushion: (a) A0; (b) A1-1; (c) A2-2.

Figures 16–19 show the relationships between different parameters and the stress diffusion angle ratio. As can be seen in Figures 16(a) and 16(b), under same reinforcement conditions, the two-layer geobelt-reinforced cushion showed larger stress diffusion angle compared to the one-layer geobelt reinforced one. At the proportional limit state, SDR of the two-layer reinforced cushion was similar to that of the one-layer reinforced cushion, while at the stable state, there was a big difference between them. The reason is that, for two-layer geobelt reinforcement, first geobelt closer to the footing takes effect when the loading is small. As the loading increases, the second geobelt would then take its effect. Therefore, at the later stage of loading, the reinforcing effect of two-layer reinforcement became more pronounced. As can be seen in Table 6, at the stable state, SDRs spanned from 1.76 to 1.84 for the one-layer geobelt reinforced cushion, while for the two-layer geobelt reinforced cushion, the ratios were ranging from 2.46 to 2.63.

Figure 17 shows the relationship between the depth of the first geobelt layer beneath the footing (U) and SDR. Figure 18 shows the relationship between the linear density of reinforcement (LDR) and SDR. As can be seen, U and LDR had little impact on stress diffusion of the cushion. The possible reason is that the depth of the cushion adopted in

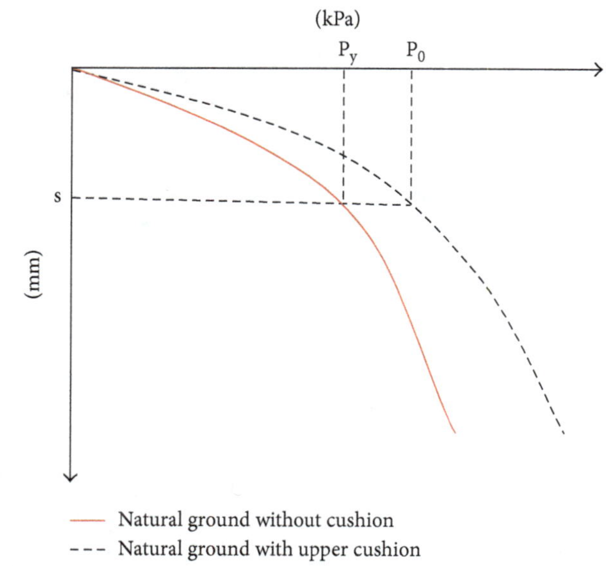

FIGURE 14: Calculation sketch of average earth pressure under the cushion.

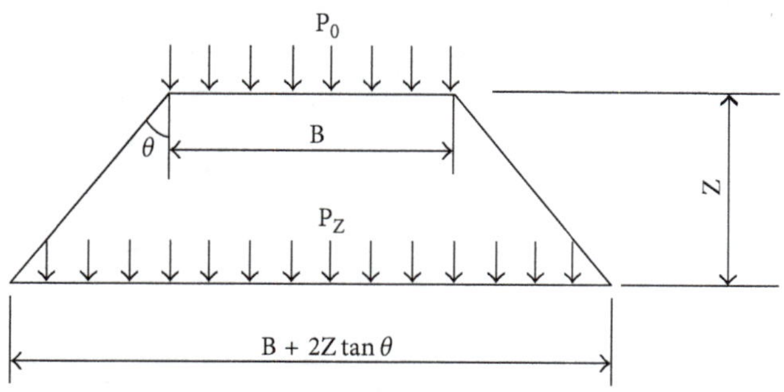

FIGURE 15: Stress diffusion of the cushion.

TABLE 6: The summary of stress diffusion angles.

Test number	Proportional limit state		Stable state	
	Angle (°)	SDR	Angle (°)	SDR
A0	38.8	1	18.9	1
A1-1	58.2	1.5	33.3	1.76
A1-2	57.4	1.48	34.8	1.84
A1-3	57.0	1.47	33.6	1.78
A1-4	57.0	1.47	33.5	1.77
A1-5	56.6	1.46	34.8	1.84
A2-1	62.1	1.60	48.0	2.54
A2-2	60.9	1.57	49.7	2.63
B2-1	58.6	1.52	46.7	2.46
B2-2	58.2	1.51	49.0	2.60

this study was small. As can be seen in Table 6, at the proportional limit state, the SDR decreased from 1.50 to 1.46 as U increased from 50 mm to 200 mm, while at the stable state, the SDR increased from 1.76 to 1.84 as U increased from 50 mm to 200 mm. It should be noted that the relationship between SDR and U is consistent with that between BCR and U. This also indicates that the geobelt only takes its effect efficiently after certain amount of deformation of cushion happens. The SDR was found to increase from 1.47 to 1.48 at the proportional limit state and increase from 1.77 to 1.84 at the stable state as the LDR increased from 25% to 50%. Therefore, the LDR showed unnoticeable effect on the stress diffusion of the geobelt reinforced cushion.

Figures 19(a) and 19(b) show the effect of vertical spacing between two geobelt layers (H) on the stress diffusion angle. For different types of geobelt, the values of SDR showed a similar trend, which is the stress diffusion angle increased slightly with the increase of H. At the proportional limit state, the SDRs of the TG-reinforced cushion were 1.57 and 1.60, while for the CPE-reinforced cushion, the values were 1.51 and 1.52, which is slightly smaller than that of the TG-reinforced cushion. At the stable state, the SDRs of the TG-reinforced cushion were 2.54 and 2.63, while for the CPE-reinforced cushion, the values were 2.46 and 2.60, which was also slightly smaller than that of the TG-reinforced cushion. It can be concluded that the reinforcing effect of TG geobelt was slightly better than that of CPE geobelt. The results are also consistent with BCR results discussed in Section 3.1.

4. Engineering Applications

In this section, a practical project included in *Technical code for ground treatment of buildings* [31] was presented to demonstrate the effect of geobelt.

The site of the project is located on a saturated silt soil with a bearing capacity of 80 kPa. The footing has a length and width of 60.8 m and 14.9 m, respectively. The thickness of the footing is 2.73 m. An average pressure beneath the footing is 130 kPa. Two possible treatment solutions are (1) unreinforced gravel cushion with a density of 19.5 kN/m^3, and the thickness of the cushion is taken as 3.73 m by calculation and (2) two-layer TG geobelt-reinforced thin gravel cushion. The density of the cushion was 19.5 kN/m^3. The width (B) was 17 m and thickness (Z) was 1.5 m with a Z/B ratio of 0.088 (<0.2). The project adopted the second treatment solution by using the geobelt-reinforced cushion. The first layer geobelt beneath the cushion was 0.6 m, and the vertical spacing between two layers was 0.4 m. The linear density reinforcement is 17%. According to the test results, the stress diffusion angle takes 35° considering a certain safety factor. After a comparative analysis, the reinforced cushion used in the second solution reduced the cushion thickness by 60% compared to the unreinforced cushion. The deformation after engineering settlement was 3.9 mm which meets the requirements.

According to this project, the stress diffusion angle of the reinforced gravel cushion was larger than that of the unreinforced one, which satisfies the bearing capacity and deformation requirements. In addition, the thickness of the cushion and construction cost can be reduced by using geobelt reinforcement.

5. Conclusions

By conducting the in situ square footing loading test on the geobelt-reinforced thin gravel cushion, this study investigated the reinforcing effect of geobelt on bearing capacity of the foundation and stress diffusion of cushions. The following conclusions can be made:

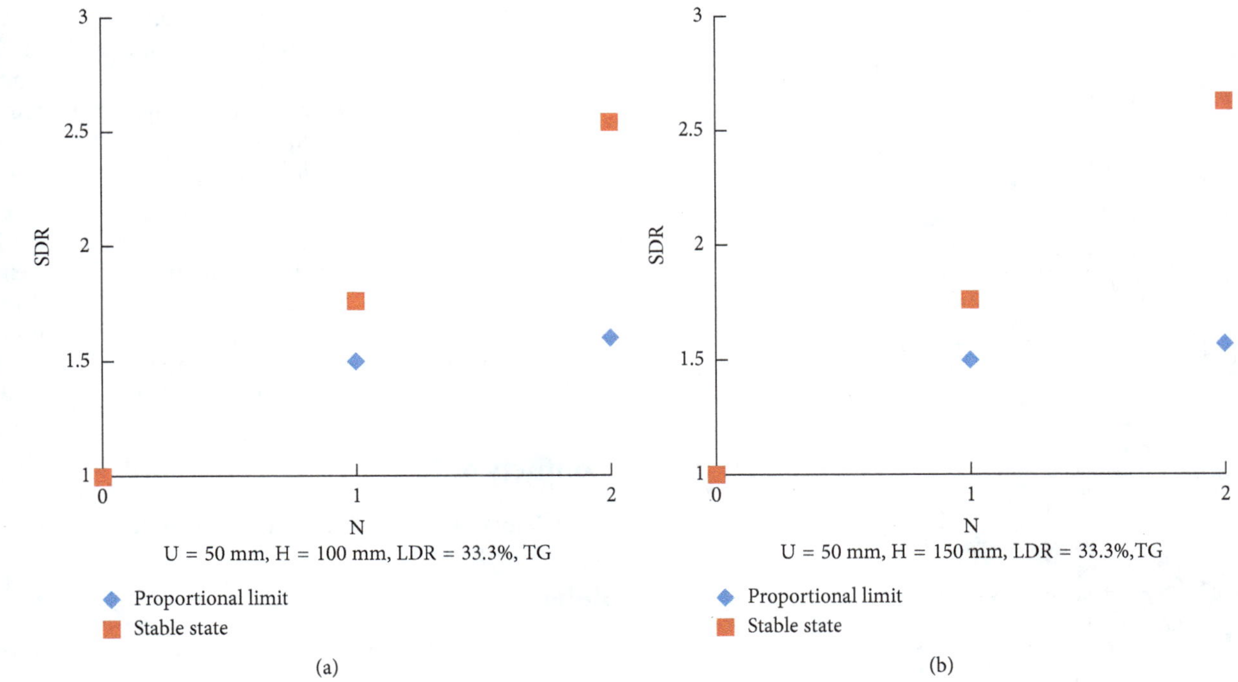

FIGURE 16: The relationships between SDR and N: (a) TG, $U = 50$ mm, $H = 100$ mm, and LDR = 33.3%; (b) TG, $U = 50$ mm, $H = 150$ mm, and LDR = 33.3%.

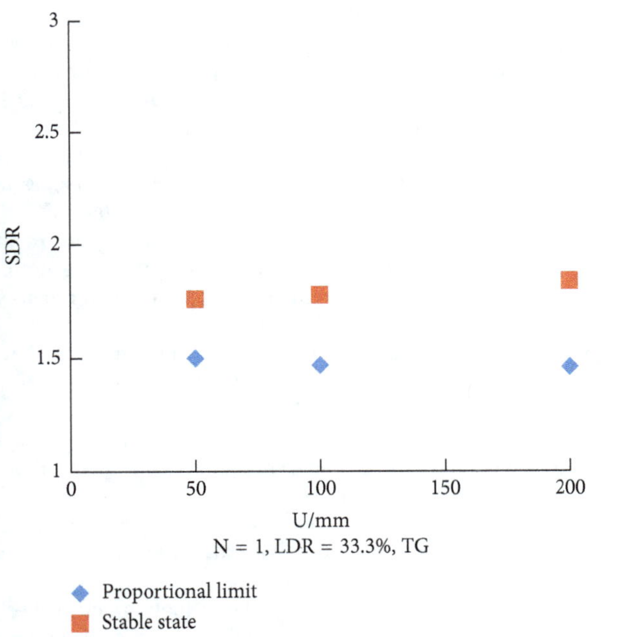

FIGURE 17: The relationship between SDR and U (TG; $N = 1$; LDR = 33.3%).

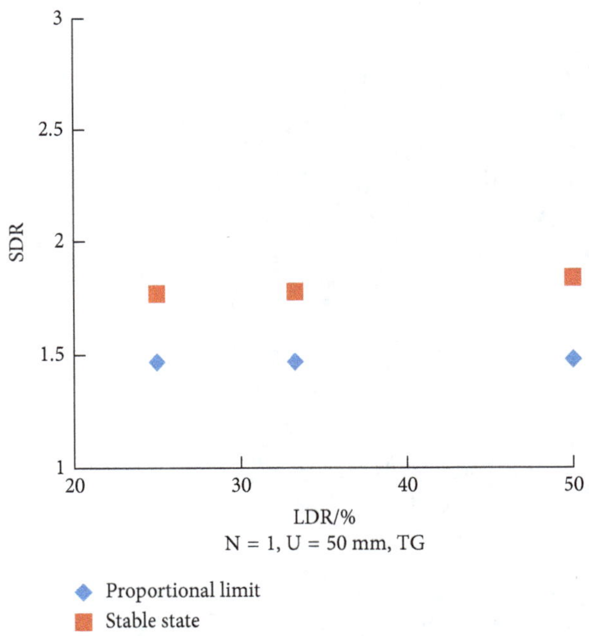

FIGURE 18: The relationship between SDR and LDR (TG; $N = 1$; $U = 50$ mm).

(1) Compared to unreinforced gravel cushions, geobelt reinforced cushions can improve the bearing capacity of foundation. The bearing capacity of one-layer geobelt-reinforced foundation was increased by 1.22–1.37 times while that of two-layer geobelt reinforced ones can be improved by 1.34–1.7 times.

(2) The earth pressure distribution on bottom of reinforced cushions was in a saddle shape, which was smoother than that of unreinforced cushions. When loaded, geobelt can restrict the movement of soils due to interfacial frictions between it and crushed gravels; therefore, the stress distribution of reinforced cushions becomes less varied compared to that of unreinforced cushions.

(3) The deformation and stress distribution of underlying soil layers under reinforced cushions were

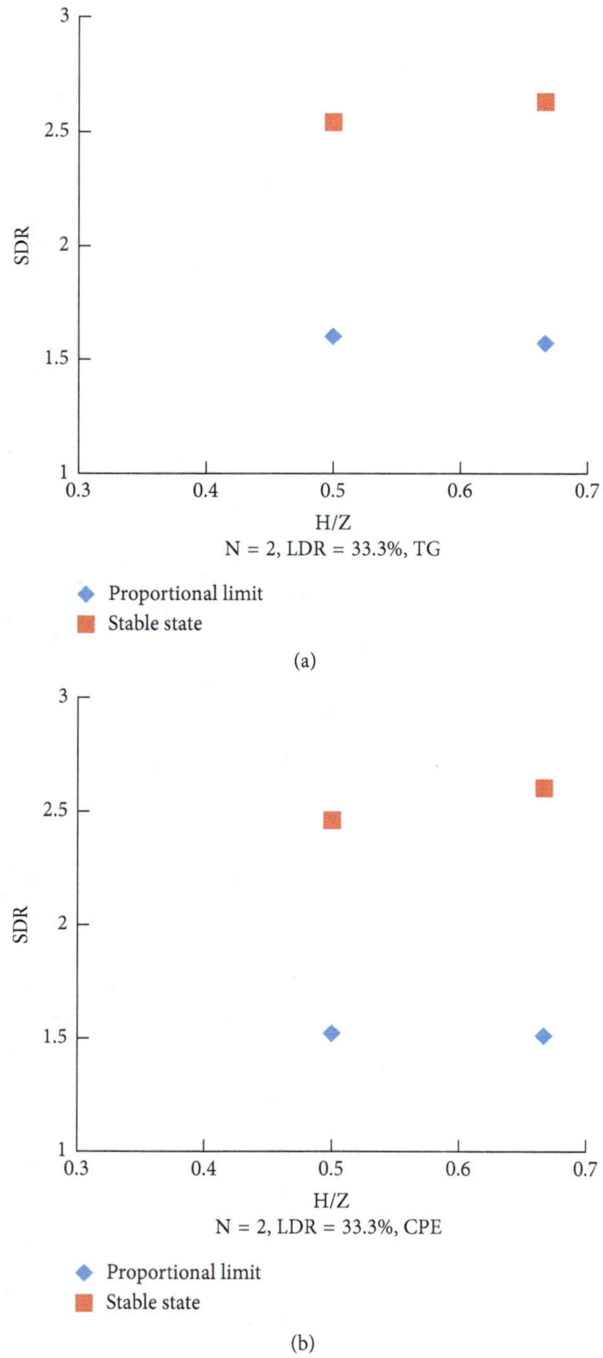

FIGURE 19: The relationship between SDR and H: (a) TG geobelt ($N=2$; LDR = 33.3%); (b) CPE geobelt ($N=2$; LDR = 33.3%).

much different from that under unreinforced cushions. Especially, the effect of geobelt on stress diffusion was pronounced. The stress diffusion angles of reinforced cushions were all larger than 20°.

(4) The reinforcing effect of two-layer geobelt reinforcement is obviously better than that of one-layer geobelt reinforcement. At the stable state, the stress diffusion angles of one-layer reinforced cushions were increased by 1.76–1.84 times while two-layer reinforcement was improved by 2.46–2.63 times, compared to unreinforced cushions.

(5) At the stable state, for the one-layer geobelt reinforced cushion, the stress diffusion and bearing capacity can be improved by increase of distance between the first geobelt and footing. A higher linear density of reinforcement can get a better bearing capacity and stress diffusion.

(6) At the stable state, for the two-layer geobelt reinforced cushion, the greater the vertical spacing between two layers, the better the stress diffusion effect and the larger the bearing capacity of the reinforced cushion. And the performance of the reinforced cushion with TG geobelt was better than that of the reinforced cushion with CPE geobelt.

Conflicts of Interest

The authors declare that they have no conflicts of interest.

References

[1] A. Kumar, M. L. Ohri, and R. K. Bansal, "Bearing capacity tests of strip footings on reinforced layered soil," *Geotechnical and Geological Engineering*, vol. 25, no. 2, pp. 139–150, 2007.

[2] A. Demir, A. Yildiz, and M. Laman, "Experimental and numerical analyses of circular footing on geogrid-reinforced granular fill underlain by soft clay," *Acta Geotechnica*, vol. 9, no. 4, pp. 711–723, 2014.

[3] B. M. Das, E. C. Shin, and M. T. Omar, "The bearing capacity of surface strip foundations on geogrid-reinforced sand and clay-a comparative study," *Geotechnical and Geological Engineering*, vol. 12, no. 1, pp. 1–14, 1994.

[4] G. M. Latha and A. Somwanshi, "Bearing capacity of square footings on geosynthetic reinforced sand," *Geotextiles and Geomembranes*, vol. 27, no. 4, pp. 281–294, 2009.

[5] S. N. M. Tafreshi, P. Sharifi, and A. R. Dawson, "Performance of circular footings on sand by use of multiple-geocell or -planar geotextile reinforcing layers," *Soils and Foundations*, vol. 56, no. 6, pp. 984–997, 2016.

[6] J. Binquet and K. L. Lee, "Bearing capacity tests on reinforced earth slabs," *Journal of Geotechnical Engineering Division*, vol. 101, pp. 1241–1256, 1975.

[7] J. Binquet and K. L. Lee, "Bearing capacity analysis on reinforced earth slabs," *Journal of Geotechnical Engineering Division*, vol. 101, pp. 1257–1276, 1975.

[8] C. C. Huang and F. Tatsuoka, "Bearing capacity of reinforced horizontal sandy ground," *Geotextiles and Geomembranes*, vol. 9, no. 1, pp. 51–80, 1990.

[9] C. C. Huang and F. Y. Meng, "Deep-footing and wide-slab effects in reinforced sandy ground," *Journal of Geotechnical and Geoenvironmental Engineering*, vol. 123, no. 1, pp. 30–36, 1997.

[10] A. Ghosh and A. K. Bera, "Bearing capacity of square footing on pond ash reinforced with jute-geotextile," *Geotextiles and Geomembranes*, vol. 23, no. 2, pp. 144–173, 2005.

[11] A. B. Cerato and A. J. Lutenegger, "Bearing capacity of square and circular footings on a finite layer of granular soil underlain by a rigid base," *Journal of Geotechnical and Geoenvironmental Engineering*, vol. 132, no. 11, pp. 1495–1501, 2006.

[12] G. M. Latha and A. Somwanshi, "Effect of reinforcement form on the bearing capacity of square footings on sand," *Geotextiles and Geomembranes*, vol. 27, no. 6, pp. 409–422, 2009.

[13] A. Demir, M. Laman, A. Yildiz, and M. Ornek, "Large scale field tests on geogrid-reinforced granular fill underlain by clay soil," *Geotextiles and Geomembranes*, vol. 38, no. 4, pp. 1–15, 2013.
[14] L. D. Suits, T. C. Sheahan, M. Abu-Farsakh, Q. Chen, R. Sharma, and X. Zhang, "Large-scale model footing tests on geogrid-reinforced foundation and marginal embankment soils," *Geotechnical Testing Journal*, vol. 31, no. 5, pp. 413–423, 2008.
[15] T. G. Sitharam, S. Sireesh, and S. K. Dash, "Model studies of a circular footing supported on geocell-reinforced clay," *Canadian Geotechnical Journal*, vol. 42, no. 2, pp. 693–703, 2011.
[16] M. Kiran and N. Bacha, "An experimental study on behaviour of bearing capacity and settlement of circular and square footing resting on reinforced sand bed stratified with lateritic soil," *International Journal of Engineering and Technical Research*, vol. 4, no. 6, 2015.
[17] R. Sharma, Q. Chen, M. Abu-Farsakh, and S. Yoon, "Analytical modeling of geogrid reinforced soil foundation," *Geotextiles and Geomembranes*, vol. 27, no. 1, pp. 63–72, 2009.
[18] M. G. Hussein and A. A. Meguid, "A three-dimensional finite element approach for modeling biaxial geogrid with application to geogrid-reinforced soils," *Geotextiles and Geomembranes*, vol. 44, no. 3, pp. 295–307, 2016.
[19] A. Kumar and S. Saran, "Closely spaced footings on geogrid-reinforced sand," *Journal of Geotechnical and Geoenvironmental Engineering*, vol. 129, no. 7, pp. 660–664, 2003.
[20] M. O. A. Bazne, F. Vahedifard, and S. Shahrokhabadi, "The effect of geonet reinforcement on bearing capacity of low-compacted soft clay," *Transportation Infrastructure Geotechnology*, vol. 2, no. 1, pp. 47–63, 2015.
[21] K. S. Sherin, S. Chandrakaran, and N. Sankar, "Effect of geocell geometry and multi-layer system on the performance of geocell reinforced sand under a square footing," *International Journal of Geosynthetic and Ground Engineering*, vol. 3, no. 3, p. 30, 2017.
[22] S. Biswas and S. Mittal, "Square footing on geocell reinforced cohesionless soils," *Geomechanics and Engineering*, vol. 13, no. 4, pp. 641–651, 2017.
[23] A. Kumar and A. Kaur, "A model tests of square footing resting on fibre-reinforced sand bed," *Geosynthetics International*, vol. 19, no. 5, pp. 385–392, 2012.
[24] D. Lal, N. Sankar, and S. Chandrakaran, "Effect of reinforcement form on the behaviour of coir geotextile-reinforced sand beds," *Soils and Foundation*, vol. 57, no. 2, pp. 227–236, 2017.
[25] B. M. Das and M. T. Omar, "The effects of foundation width on model tests for the bearing capacity of sand with geogrid reinforcement," *Geotechnical and Geological Engineering*, vol. 12, no. 2, pp. 131–141, 1994.
[26] Chinese Ministry of Housing and Urban-Rural Development, *Code for Design of Building Foundation*, Chinese Ministry of Housing and Urban-Rural Development, Beijing, China, 2011.
[27] M.T. Adams and J. G. Collin, "Large model spread footing load tests on geosynthetic reinforced soil foundations," *Journal of Geotechnical and Geoenvironmental Engineering*, vol. 123, no. 1, pp. 66–72, 1997.
[28] Q. Chen, M. Abu-Farsakh, and R. Sharma, "Experimental and analytical studies of reinforced crushed limestone," *Geotextiles and Geomembranes*, vol. 27, no. 5, pp. 357–367, 2009.
[29] M. A. Gabr, R. Dodson, and J. G. Collin, "A study of stress distribution in geogrid-reinforced sand," in *Proceeding of Geosynthetic in Foundation Reinforcement and Erosion Control Systems, ASCE*, pp. 62–76, Boston, MA, USA, October 1998.
[30] R. Chaney, K. Demars, M. Gabr, and J. Hart, "Elastic modulus of geogrid-reinforced sand using plate load tests," *Geotechnical Testing Journal*, vol. 23, no. 2, pp. 245–250, 2000.
[31] Chinese Ministry of Housing and Urban-Rural Development, *Technical Code for Ground Treatment of Buildings*, Chinese Ministry of Housing and Urban-Rural Development, Beijing, China, 2013.

5

Estimation of Sand Production Rate Using Geomechanical and Hydromechanical Models

Son Tung Pham

Faculty of Geology & Petroleum Engineering, Department of Drilling & Production Engineering, Ho Chi Minh City University of Technology-Vietnam National University, Ho Chi Minh City, Vietnam

Correspondence should be addressed to Son Tung Pham; phamsontung@hcmut.edu.vn

Academic Editor: Antonio Riveiro

This paper aims to develop a numerical model that can be used in sand control during production phase of an oil and gas well. The model is able to predict not only the onset of sand production using critical bottom hole pressure inferred from geomechanical modelling, but also the mass of sand produced versus time as well as the change of porosity versus space and time using hydromechanical modelling. A detailed workflow of the modelling was presented with each step of calculations. The empirical parameters were calibrated using laboratory data. Then the modelling was applied in a case study of an oilfield in Cuu Long basin. In addition, a sensitivity study of the effect of drawdown pressure was presented in this paper. Moreover, a comparison between results of different hydromechanical models was also addressed. The outcome of this paper demonstrated the possibility of modelling the sand production mass in real cases, opening a new approach in sand control in petroleum industry.

1. Introduction

Sand production occurs in many oil fields across the world and it is especially common in the porous sediments. Sand production observed on the surface occurs as a series of three events that happen at downhole area: (1) formation failure, (2) sand erosion due to flow, and (3) sand transport.

(1) Formation Failure. In situ stresses and pore pressure act on formation sands and under certain conditions; the criteria for failure are met. The presence of the wellbore and perforations causes a concentration of stresses near these cavities and deformation and failure can occur under certain well-known conditions. This criterion is bottomhole pressure that makes the maximum effective tangential compressive stress equal or higher than the rock strength (failure criteria); significant sanding begins at some point (the onset).

(2) Sand Erosion due to Flow. Damaged regions that have failed (meeting the failure criteria) face additional stresses caused by pore pressure gradients. The process of sand erosion is essential for the sand to be removed from the failed region and then to be entrained with the fluid.

(3) Sand Transport. Sand erosion detaches sand grains into a perforating cavity or wellbore. Some of the grains are transported to the surface while others settle into the perforation tunnel or into the well hole.

During and after sand production, wells can sand-up and that has different effects on the productivity. At first, the productivity seems to increase due to the increase of permeability. However, after a while, the sand produced can obstruct the entrance of hydrocarbon into the wellbore, especially for cased perforated wells using sand screens or gravel pack. Disposal of produced sand is also a significant cost associated with sand production. Finally, sand can be transported to the surface which causes erosion of pipe lines, joints, chokes, and valves. So, if the prediction of sand production is identified, it will help operators to manage the situation properly and prepare suitable treatment methods for the well.

This study focuses on the first and second steps of sand production because they have the most impact and they are not well understood. The main reasons why in petroleum industry nowadays we still do not predict the mass of sand production are because of the complexity of numerical models (hence the lack of professional software in this domain) and the unavailable real data of sand production due to the difficulty in collecting this kind of data in the oil field. Therefore, most of the studies predicting the mass of produced sand still stay at the laboratory step. In real life, in petroleum companies' reports, only Geomechanical models are being used to predict the onset of sand production. This papers aims to bring the application of the Hydromechanical models into a real case in petroleum industry and to combine the use of Geomechanical model, which predicts the onset of sand production, and the Hydromechanical model, which predicts the mass of produced sand.

2. Literature Review

Several studies released models predicting the onset of sand production and the amount of sand produced. Parameters affecting sand production have been discussed for decades. However, there is no clear consensus. In this section, a brief review of major conclusions of these past studies is presented.

Willson et al. [1] developed a model to predict the sand production rate from the onset of sanding model. The onset of sanding is predicted using a stress-based model of shear failure around a perforation or an open hole. Sand production is assumed to occur once the maximum value of the effective tangential stress around the perforation exceeds the apparent rock strength (rock strength is $b_f \cdot 1.55 \cdot$ TWC strength with b_f is boost factor, for cased perforated wells $b_f = 2$; TWC is the Thick Wall Cylinder which is a measure of rock's strength and is used in sand production study instead of Unconfined Compressive Strength, because TWC reflects more closely the reality of in situ stress sustained by the borehole or the perforation channels than the UCS). No consideration was given to sand transport by drag forces. The model of Willson et al. [1] requires TWC test data for hole size the same as the perforation size. If the perforation is of different size, the application of the model must be careful. Hence the b_f boost factor exists to compensate for the difference between the real data and the test data.

The criterion for sanding is

$$\text{CBHP} \geq \frac{3\sigma_1 - \sigma_2 - \text{UCS}}{2 - A} - p_o \frac{A}{2 - A} \quad (1)$$

where CBHP is critical bottomhole pressure; σ_1 is maximum principal stress; σ_2 is minimum principal stress; p_o is pore pressure. A is poroelastic constant given by $A = (1 - 2\nu)\alpha/(1 - \nu)$ with ν: Poisson ratio and α: Biot's constant.

The model predicts the rate of sand production by utilizing the nondimensionalized concepts of Loading Factor, LF (near-wellbore formation stress normalized by strength), Reynolds number (Re), and water production factor. An empirical relationship between Loading Factor, Reynolds number, and the rate of sand production incorporating the effect of water production was proposed as follows:

$$\text{SPR} = f\,(\text{LF}, \text{Re}, \text{Water-cut})\,; \quad (2)$$

(see [1]).

In this formula given by Willson et al. [1], SPR is the sand production rate; water cut is the ratio of water produced compared to the volume of total liquids produced.

Although the Willson et al. model [1] takes into account the different phases of the fluid via water cut, the model did not give clear expression of the Sand Production Rate SPR in function of the Loading Factor LF, the Reynolds number, and the water cut.

In 1996, Vardoulakis et al. [2] proposed the following sand production model usingmixture theory, assuming that the sand in place is fully degraded from the beginning and the production is due only to the hydrodynamic forces. Equilibrium equation for solid phase is often ignored. The process is initiated with a very small solid concentration given as a boundary condition. The results are insensitive to this value as long as it is small:

$$\frac{d\phi}{dt} = \frac{\dot{m}}{\rho_s} = \lambda\,(1 - \phi)\,c\,\left\|q^i_{\text{fl}}\right\|, \quad (3)$$

where ϕ is the porosity, $\dot{m}$ is the rate of eroded solid mass per unit volume, ρ_s is the solid density, λ is the experimentally evaluated sand production coefficient, c is the transport concentration, q^i_{fl} is the specific discharge in the ith direction, and $\|\|$ is the notation representing the norm of a vector.

The Vardoulakis model is difficult to solve because of the complexity of the equations. Moreover, the model does not take into account the different phases of the fluid, so the fluid is considered as single phase, which does not reflect the reality of petroleum fluid. Furthermore, the coefficients were not calibrated due to lack of experimental data.

Papamichos et al. [3] developed the following model based on the assumption that failure is due to erosion and porosity increases until it reaches unity. Below is the relation between sand mass and the porosity:

$$\frac{\dot{m}}{\rho_s} = \frac{\partial\phi}{\partial t}. \quad (4)$$

The dimension of the above equation is that $\dot{m}$ is the rate of eroded solid mass per unit volume ($\text{g}\cdot\text{s}^{-1}\cdot\text{m}^{-3}$), ρ_s is the solid density ($\text{g}\cdot\text{m}^{-3}$), ϕ is the porosity, and t is time in second (s).

Variation of sand mass due to erosion is given as

$$\frac{\dot{m}}{\rho_s} = \lambda\,(1 - \phi)\left\|q^i_{\text{fl}}\right\|, \quad (5)$$

where λ is sand production coefficient $\lambda = \lambda(\varepsilon_p)$ and ε_p is plastic shear strain.

The main advantage of Papamichos et al. model [3] is that the authors provided experimental data in order to calibrate the experimental parameters. In this model, not all the plastic deformation areas will produce sand. The sand is produced only when the plastic deformation reaches a limit value ($\varepsilon_p > \varepsilon_{\text{peak}}$). However, it is practically difficult to determine this limit value in reality. Besides, the system of differential equations of this model is extremely complex with plenty of empirical parameters.

Chin and Ramos [4] developed a sand production model considering that erosionoccurs during sand production as follows, where v_s is the solid velocity:

$$\frac{d\phi}{dt} = (1 - \phi)\,\nabla \cdot v_s. \quad (6)$$

The main inconvenient of this model [4] is that it was developed only for weak formation. Besides, the model retains only primary physics of rock failure and coupled rock deformation and fluid flow. The model does not include the effects of well configuration and completion, wellbore storage, erosion, interaction of disaggregated solid and flowing fluid, and solid transport through the porous medium.

The analytical model developed by Fjær et al. [5] combined a theoretical model with an empirical relationship as shown below:

$$\frac{\partial m_{\text{sand}}}{\partial t} = \lambda_{\text{sand}}\mu\frac{1-\phi}{\phi^3}\left(q_{\text{fl}} - q_{\text{fl}}^{\text{cr}}\right), \tag{7}$$

where the porosity increases with time:

$$\phi = \phi_0\left[1 + 4\frac{\lambda_{\text{sand}}\mu}{\rho_s\phi_0^4}\left(q_{\text{fl}} - q_{\text{fl}}^{\text{cr}}\right)(t - t_0)\right]^{1/4}, \tag{8}$$

where q_{fl} is the flow rate, $q_{\text{fl}}^{\text{cr}}$ is the critical flow rate, ϕ_0 is the initial porosity, t_0 is the initial time, and λ_{sand} is proportionality constant; λ_{sand} has the dimension of s/m^3.

Gravanis et al. [6] developed a coupled stress-fluid flow erosion model:

$$\frac{\dot{m}}{\rho_s} = \lambda\lambda_p(r,t)(1-\phi)^\beta q_{\text{fl}}, \tag{9}$$

where λ, with dimension of inverse length, represents the strength of the erosive processes that lead to sand production. β is an exponent parameter of the model. $\lambda_p(r,t)$ is an erosion function defined as follows:

$$\lambda_p(r,t) = \exp\left[-\Lambda(t)^{-a}\left[\Gamma\left(1+\frac{1}{a}\right)\right]^a \times (r - r_{\text{in}})^a\right]. \tag{10}$$

$\Gamma(x)$ is the usual Euler Gamma function and Λ is the depth of the plastic region. The profile function $\lambda_p(r,t)$ approaches a step function as a increases. The exponent a is fixed at the beginning of the analysis and can be tuned further as part of the calibration procedure. In this work we will fix $a = 2$, which is the same value used in the work of Fjær et al. [5].

In 2010, Isehunwa and Olanrewaju [7] proposed a new model for sand production, considering the effects of flow rate, fluid viscosity and density, grain size, and cavity height. Sand is produced by drag and buoyancy forces which predominantly act on the sand particles. The radius of sand production cavity is

$$R_a = \frac{q_{\text{fl}}\mu}{(4/9)R_s^2\pi H\rho_f g}. \tag{11}$$

The volume of sand produced can be expressed as

$$V_{\text{sp}} = \pi R_a^2 H, \tag{12}$$

where q_{fl} is the fluid flow rate, R_s is the grain radius, H is the cavity height, and g is the gravitational acceleration.

Among these models, the ones of Fjær et al. [5] and Gravanis et al. [6] were chosen for this study because of their

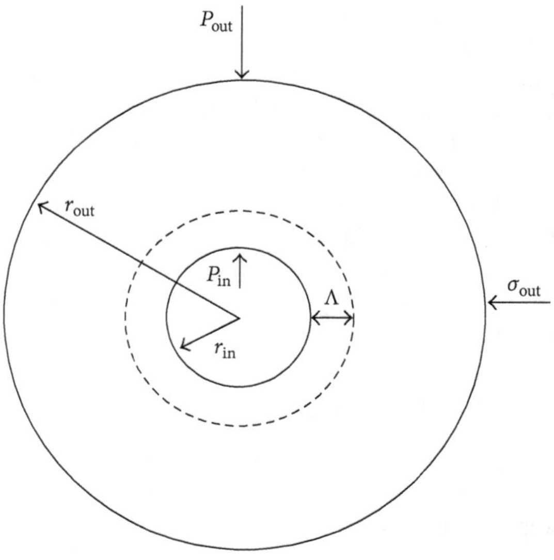

FIGURE 1: Schematic of the hollow cylinder [6].

clarity in the explanation and equations, which are necessary for us to be able to numerically solve the problems. In addition, these models have some advantages in comparison with others. They proposed the basic theory for hydrodynamic erosion of sandstone which is based on filtration theory. They adopted full strength hardening/softening behavior of reservoir stone. They also do take into account the grain size, the gradient elastoplasticity for thick wall cylinders and the cavity failure around boreholes.

The models were solved using the workflow developed in Section 3 of this paper; then the coding was made using MATLAB software.

3. Workflow for Calculation and Calibration

3.1. Hydromechanical Erosion Model of Gravanis et al. [6]. According to Gravanis et al. [6], the basic assumptions are as follows:

(1) Fluid flow can be described by Darcy's law.

(2) We define the mathematical time T by $dT = (\lambda/\Delta r)A(t)dt$ for simple calculations. t is the real time while T is the mathematical time which is introduced to facilitate the resolution of the problem. $\Delta r = r_{\text{out}} - r_{\text{in}}$ (Figure 1).

The function $A(t)$ is related to pressure drop $\Delta p = p_{\text{out}} - p_{\text{in}}$ as follows:

$$A(t) = \frac{\Delta p}{\mu\int_{r_{\text{in}}}^{r_{\text{out}}}(dr/k(\phi)r)}. \tag{13}$$

k is the permeability.

(3) The whole region is divided into a plastic region and an elastic region. In elastic region, we apply Hooke's law and in plastic region we consider the Mohr-Coulomb failure criterion.

(4) Under the condition that plasticity of the material is damaged and subject to decohesion, it can be eroded under weak hydrodynamic forces. It happens when drawdown pressure (DP) exceeds a critical drawdown pressure (CDP).

3.1.1. Calculation Workflow

Step 1 (at $T = 0$).

(i) Calculate function $A(T)$:

$$A(T) = \frac{\Delta p}{\mu \int_{r_{\text{in}}}^{r_{\text{out}}} (dr/k(\phi) r)}, \tag{14}$$

where $A(T)$ is related to the flow rate $q_{\text{fl}}(T)$ by the relation $q_{\text{fl}}(T) = 2\pi H A(T)$, $\Delta p = p_{\text{out}} - p_{\text{in}}$.

(ii) Calculate pressure $p(r,T)$:

$$p(r,T) = p_{\text{in}} + \frac{\mu}{k_0} A(T) \int_{r_{\text{in}}}^{r} \frac{(1-\phi)^2}{\phi^3} \frac{dr}{r}. \tag{15}$$

(iii) The depth of plastic region Λ: the elastic and plastic regions are presented in Figure 1 according to Gravanis et al. [6]. In elastic region, combining with the boundary condition $\sigma_{r\,\text{elastic}}(r_{\text{out}}) = \sigma_{\text{out}}$, we have

$$\begin{aligned}\sigma_r &= \sigma_{\text{out}} + C_2 2G\left[\frac{1}{r_{\text{out}}^2} - \frac{1}{r^2}\right] + \frac{\alpha 2G}{\lambda_{\text{Lame}} + 2G}\left[\frac{1}{r^2}\int_{r_{\text{out}}}^{r} r' p(r')\,dr'\right] \\ \sigma_\theta &= \sigma_{\text{out}} + C_2 2G\left[\frac{1}{r_{\text{out}}^2} + \frac{1}{r^2}\right] + \frac{\alpha 2G}{\lambda_{\text{Lame}} + 2G}\left[p(r) - \frac{1}{r^2}\int_{r_{\text{out}}}^{r} r' p(r')\,dr'\right].\end{aligned} \tag{16}$$

r varies from r_{in} to r_{out}.

In plastic region, we consider the Mohr-Coulomb failure criterion and combine with the boundary condition $\sigma(r_{\text{in}}) = \sigma_{\text{in}}$:

$$\begin{aligned}\sigma_r &= \sigma_{\text{in}}\frac{r^{K-1}}{r_{\text{in}}^{K-1}} - \frac{S_0}{K-1}\left[1 - \frac{r^{K-1}}{r_{\text{in}}^{K-1}}\right] - \alpha(K-1) \cdot \left[r^{K-1}\int_{r_{\text{in}}}^{r} (r')^{-K} p(r)'\,dr'\right] \\ \sigma_\theta &= \alpha p(1-K) + \text{UCS} + K\left[\sigma_{\text{in}}\frac{r^{K-1}}{r_{\text{in}}^{K-1}} - \frac{S_0}{K-1}\left(1 - \frac{r^{K-1}}{r_{\text{in}}^{K-1}}\right) - \alpha(K-1) r^{K-1}\int_{r_{\text{in}}}^{r} (r')^{-K} p(r)'\,dr'\right],\end{aligned} \tag{17}$$

where S_0 is material cohesion [MPa]. α is Biot's constant variant from 0 to 1; here we shall set this value to 1 meaning that we will neglect any compressibility effects. K is the principal stress ratio which is equal to σ_h/σ_v with σ_h being the horizontal stress and σ_v the vertical stress.

At this stage, there are two continuity conditions and two unknowns, the integration constant C_2 and the location of the plastic zone boundary location R. The conditions are shown below:

$$\begin{aligned}\sigma_{r\,\text{elastic}}(R) &= \sigma_{r\,\text{plastic}}(R) \\ \sigma_{\theta\,\text{elastic}}(R) &= \sigma_{\theta\,\text{plastic}}(R).\end{aligned} \tag{18}$$

C_2 is given explicitly in terms of R which is solved numerically for R [6]. Note that (18) are purely algebraic. Thus, the plastic region depth is determined as $\Lambda = R - r_{\text{in}}$.

Step 2 (at time $T = \delta T$).

(i) The porosity field is calculated from erosion model equation (9):

$$\phi(r,T) = 1 - \frac{1-\phi_0}{\left(1 + (\beta-1)(1-\phi_0)^{\beta-1}(\Delta r/r)\int_0^T \lambda_p(r,T)\,dT\right)^{1/(\beta-1)}} \quad \text{if } \beta \neq 1$$
$$\phi(r,T) = 1 - (1-\phi_0)\exp\left[-\frac{\Delta r}{r}\int_0^T \lambda_p(r,T)\,dT\right] \quad \text{if } \beta = 1. \tag{19}$$

The formula involves the integral $\int_0^T \lambda_p(r,T)dT$ which is approximated by $\lambda_p(r,0)\delta T$ and $\lambda_p(r,0)$. $\Lambda_0 = \Lambda(T=0)$ is known from the previous time step where β is an exponent parameter of the model and must be tuned by calibration.

(ii) Calculation of porosity and radius of sand production cavity:

$$\bar{\phi}(T) = \Lambda(T)^{-1}\int_{r_{\text{in}}}^{r_{\text{out}}} \lambda_p(r,T)\,\phi(r,T)\,dr. \tag{20}$$

Radius of sand production cavity r_a is calculated from wellbore to the distance where porosity of formation is constant and equal to initial porosity ϕ_0.

(iii) Calculate function $A(T)$, pressure $p(r,T)$, and the plastic region depth Λ_1 from (14) to (17).

(iv) The erosion strength is λ:

$$\lambda(\Lambda) = \begin{cases}\lambda_0 + (\lambda_2 - \lambda_0)\dfrac{\Lambda - \Lambda_0}{\Lambda_p - \Lambda_0}, & \Lambda \le \Lambda_p \\ \lambda_2, & \Lambda \ge \Lambda_p,\end{cases} \tag{21}$$

where λ_0, λ_2, and Λ_p are model parameters that must be tuned by calibration, λ_0 is initial erosion strength $[\mathrm{m}^{-1}]$, λ_2 is maximum erosion strength $[\mathrm{m}^{-1}]$, and Λ_p is threshold depth of the λ plateau.

(v) Calculate sand production mass rate:

$$\dot{M} = \int \dot{m} dV = \rho_s \cdot \overline{\frac{\partial \phi}{\partial t}} \cdot \pi \cdot \left(r_a^2 - r_{\mathrm{in}}^2\right) \cdot L. \quad (22)$$

Note that this equation must be used in the next time step along with Λ_p.

Step 3 (at the general time $T = i\delta T$).

(i) The porosity field is calculated from (20) and (21), where the integral $\int_0^T \lambda_p(r, T) dT$ is approximated by $\{\lambda_p(r, 0) + \cdots + \lambda_p(r, (i-1)\delta T)\}\delta T$.

This involves the values of the depth Λ at all the previous time steps, which are known. As explained above, the function A, the pressure, and the new value of the plastic region depth $\Lambda_i = \Lambda\ (T = i\delta \mathrm{T})$ can then be calculated.

3.2. Analytical Model of Fjær et al. [5]. The analytical model is based on these assumptions:

(1) The driving mechanism for continuous sand production is erosion from plastified material in the vicinity of the production cavity.

(2) The sand production rate depends on (1) how much the well pressure is reduced below the critical sand production pressure, (2) the fluid flow rate and the fluid viscosity, and (3) the cementation of the rock.

(3) The sand in place is fully degraded from the beginning and the production is due only to the hydrodynamic forces. These forces are proportional to the fluid pressure drop over the volume element, and the pressure drop is proportional to the fluid flow rate as specified by Darcy's law; moreover the permeability is given by the Kozeny-Carman equation; we have

$$\frac{\partial m_{\mathrm{sand}}}{\partial t} = \lambda_{\mathrm{sand}} \mu \frac{1-\phi}{\phi^3} \left(q_{\mathrm{fl}} - q_{\mathrm{fl}}^{\mathrm{cr}}\right), \quad (23)$$

where μ is the fluid viscosity. The proportionality constant λ_{sand} has the dimension of $\mathrm{s/m^3}$

(4) Firstly, it is required that DP > CDP, which expresses the fact that stress induced damage of the rock is a necessary condition for sand production. Secondly, it is required that $q_{\mathrm{fl}} > q_{\mathrm{fl}}^{\mathrm{cr}}$ (critical fluid flow rate), which expresses that the hydrodynamic forces must be strong enough to uphold the erosion process.

3.2.1. Calculation Workflow. Following Fjær et al. [5] we simply assume that the stiffness of the rock remains constant until the porosity has reached a critical value ϕ_{cr}. At that point, the entire part of the rock that has been producing sand collapses, and the remaining solid material in that part is produced in one burst. The stresses around the cavity are then redistributed, and the situation is similar to the initial state at $t = t_o$, except the cavity has become a little larger.

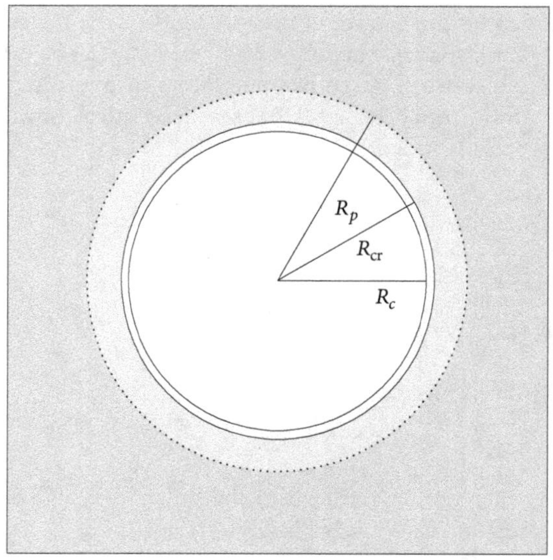

FIGURE 2: Illustration of the plastified zone and the sand producing zone around a cavity [8].

Step 1 (continuous sand production < ϕ_{cr}).

(i) Calculate porosity from (8):

$$\phi = \phi_0 \left[1 + 4\frac{\lambda_{\mathrm{sand}}\mu}{\rho_s \phi_0^4} \left(q_{\mathrm{fl}} - q_{\mathrm{fl}}^{\mathrm{cr}}\right)(t - t_0)\right]^{1/4}. \quad (24)$$

(ii) Calculate volumetric sand production V_{sp}: we assume that sand production is related to plastification of the rock, so that only parts of the rock that have suffered some plastic deformations are in a state where sand can be produced.

$$R_{\mathrm{cr}} = R_c e^{(\mathrm{DP}-\mathrm{CDP})/\mathrm{UCS}} \approx R_c \left(1 + \frac{\mathrm{DP} - \mathrm{CDP}}{\mathrm{UCS}}\right)$$
$$V_{\mathrm{sp}} = \pi \left(R_{\mathrm{cr}}^2 - R_c^2\right) L \approx 2\pi L R_c^2 \frac{\mathrm{DP} - \mathrm{CDP}}{\mathrm{UCS}}, \quad (25)$$

where R_c is radius of the cavity and R_{cr} is sand production zone (Figure 2).

(iii) Sand production mass:

$$M = \rho_s \left(\phi - \phi_0\right) V_{\mathrm{sp}} \quad (26)$$

Step 2 (instantaneous sand production $\phi = \phi_{\mathrm{cr}}$). Once the entire part of the rock that has been producing sand collapses, the remaining solid material in that part is produced in one burst.

Table 1: Experimental data of Papamichos et al. [3].

Variable	Value
Cylinder internal radius, r_{in} [m]	0.01
Cylinder external radius, r_{out} [m]	0.1
Cylinder height, H [m]	0.2
Internal pore pressure, P_{in} [Mpa]	0
External pore pressure, P_{out} [Mpa]	0.15
Inner radial stress, S_{in} [Mpa]	0
Outer radial stress, S_{out} [Mpa]	11
Flow rate, q_{fl} [l/min]	0.5
Ratio a_1	0.03
Young's modulus, E [MPa]	6750
Poisson ratio, ν [—]	0.19
Biot's ratio, α [—]	1
Cohesion, S_o [MPa]	3.7
Friction angle, Φ [°]	37.4
Initial porosity, ϕ_0 [—]	0.3
Permeability, k [mD]	500
Kozeny-Carman parameter, k_o [m2]	$8.96E-12$
Solids density, ρ_{solid} [kg/m^3]	2640
Dynamic viscosity, μ [MPa·s]	5E09

The collapse of the sand producing zone implies that the radius of the cavity increases, from R_c to $(1+a_1)\cdot R_c$, where a_1 can be found from (27):

$$a_1 = \frac{R_{cr}}{R_c} - 1 \approx \frac{\text{DP} - \text{CDP}}{\text{UCS}} \tag{27}$$

$$R_c \longrightarrow (1+a_1)\,R_c; \qquad R_{cr} \longrightarrow (1+a_1)\,R_{cr}. \tag{28}$$

(i) Cumulative sand production: sand mass is given as the initial amount of solid material in the sand producing zone; that is,

$$M = V_{sp}\rho_s\,(1-\phi_0)\,. \tag{29}$$

Now the stresses around the cavity are redistributed, and the situation is the same as it was at $t = t_o$. Thus, the erosion process starts a new cycle.

3.3. Calibration of Empirical Parameters of the Models of Fjær et al. [5] and Gravanis et al. [6].

We use experimental data profile of Papamichos et al. [3] to calibrate empirical parameters (Table 1).

3.3.1. Calibration of Empirical Parameters of the Gravanis et al. Model.

We have empirical parameters: β, λ_0, λ_2, and Λ_p and we slightly varied these parameters around the values obtained by trial and error from calibrating with test data of Papamichos et al., 2001 [3]. The calibration results are shown from Figures 3 to 6. For each calibration, the model curve is compared with the experimental curve.

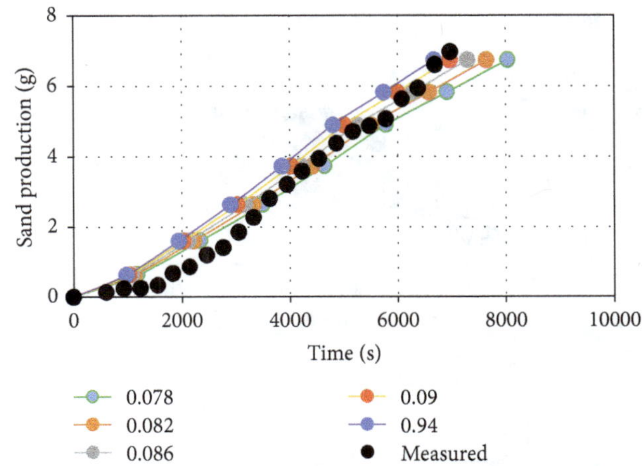

Figure 3: Sand mass produced over time while maximum erosion strength λ_2 varies from 0.078 to 0.94 m^{-1}. $\beta = 1$, $\lambda_0 = 0.42\lambda_2$, and $\Lambda_p = 1.1\Lambda_0$.

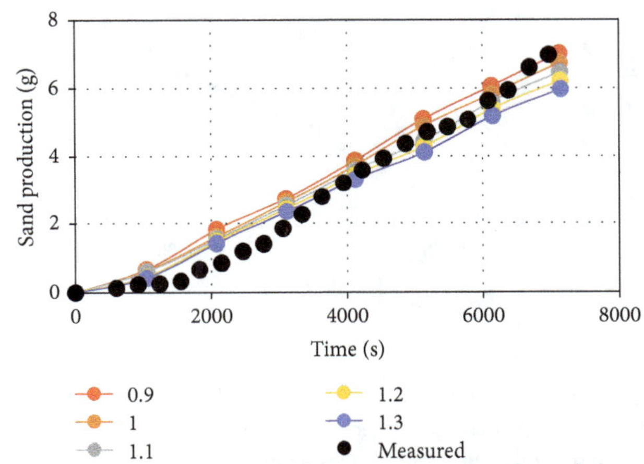

Figure 4: Sand mass produced over time while changing exponent coefficient β from 0.9 to 1.3. $\lambda_2 = 0.088$, $\lambda_0 = 0.42\lambda_2$, and $\Lambda_p = 1.1\Lambda_0$.

From the results presented from Figures 3–6, we choose suitable values of empirical parameters which are: the exponent coefficient ($\beta = 1$), maximum erosion strength ($\lambda_2 = 0.088\,\text{m}^{-1}$), initial erosion strength ($\lambda_0 = 0.42\lambda_2\,\text{m}^{-1}$) and threshold depth of the λ plateau ($\Lambda_p = 1.1\Lambda_o\,\text{m}$). The small discrepancy with the experimental data can be neglected.

3.3.2. Calibration of Empirical Parameters of Fjær et al. Model.

The model of Fjær et al. has three empirical parameters: q_{fl}^{cr}, ϕ_{cr}, and λ_{sand}. Similar to the model of Gravanis et al. [6], we use experimental data of Papamichos et al., 2001 [3] and then run the model and compare with the corresponding experimental curve to calibrate empirical parameters. Results are presented in Figures 7, 8, and 9.

Figures 7, 8, and 9 indicate that the erosional process gradually increases until critical porosity is reached. Each step is followed by a long period of continuous production at a low rate. When the porosity has reached a critical value ϕ_{cr}, at that point, the collapsed zone of the remained solids is produced,

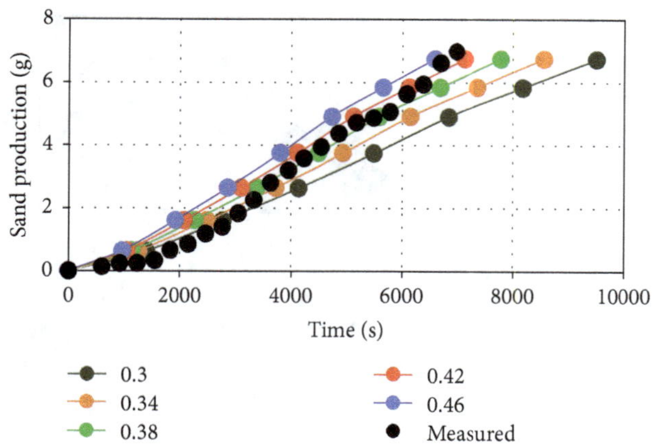

FIGURE 5: Sand mass produced over time while changing ratio of initial to maximum erosion strength λ_0/λ_2 from 0.3 to 0.46. λ_2 = 0.088, β = 1, and $\Lambda_p = 1.1\Lambda_0$.

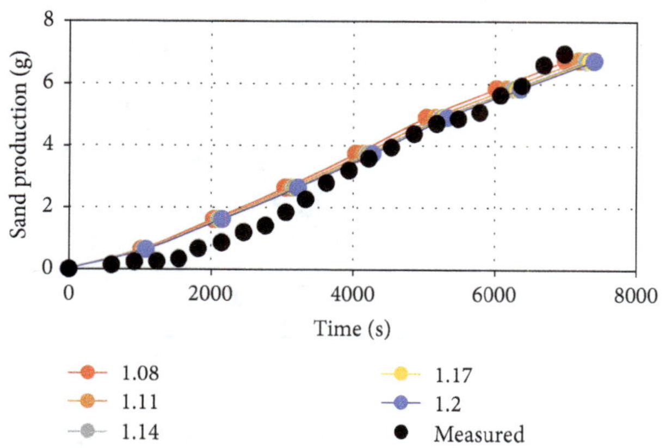

FIGURE 6: Sand mass produced over time while changing ratio of threshold depth and initial plastic region Λ_p/Λ_0 from 1.08 to 1.2. $\beta = 1$, $\lambda_2 = 0.088$, and $\lambda_0 = 0.42\lambda_2$.

and process of sand production is immediate and rapid. Then the erosion process starts on a new cycle.

In fact, the result of the model seems very different than the experimental result because Fjaer's model considers a step of "collapse" (Step 2: instantaneous sand production), but overall, the sand production mass over time is matched between the model and the experimental data and that is why we could calibrate the parameters.

These results also allow us to choose suitable values of empirical parameters which are critical porosity (ϕ_{cr} = 0.4), critical fluid flux (q_{fl}^{cr} = 0.001 m/s), and sand production coefficient ($\lambda_{sand} = 600\,m^3/s$).

4. Case Study

An application for the oilfield X in Cuu Long basin using Geomechanical model of Willson et al. [1] to predict the onset of sand production condition is now presented. In addition, it was combined with the Hydromechanical model of Fjær et al. [5] and Gravanis et al. [6] to predict sand production rate; then the results of the models will be compared.

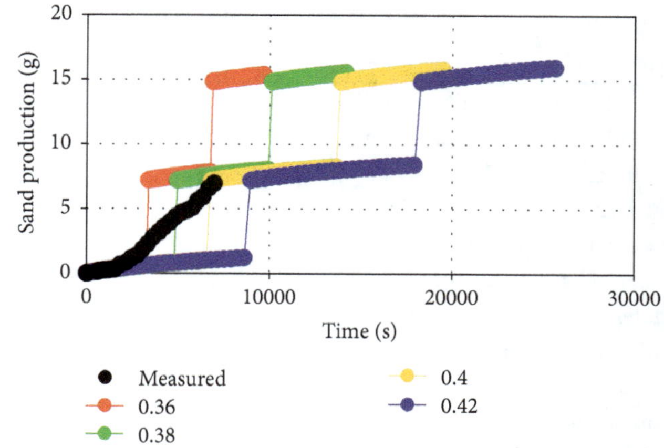

FIGURE 7: Sand mass produced over time while changing critical porosity ϕ_{cr} from 0.36 to 0.42.

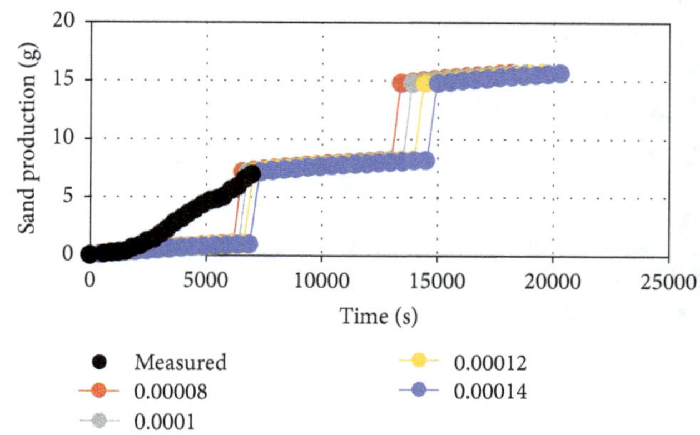

FIGURE 8: Sand mass produced over time while changing critical fluid flux q_{fl}^{cr} from 0.00008 to 0.00014 m/s.

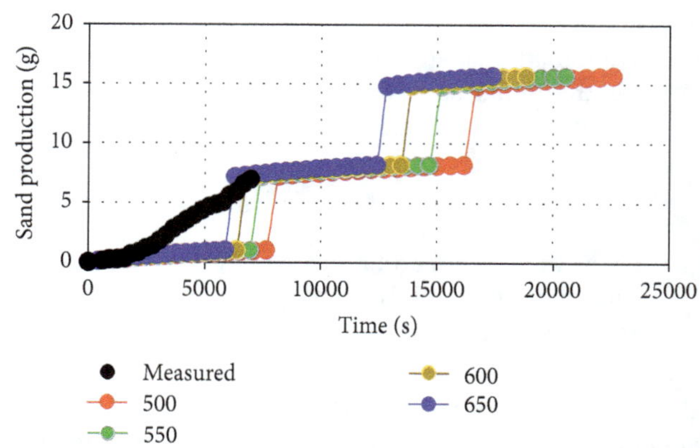

FIGURE 9: Sand mass produced over time coefficient sand production λ_{sand} from 500 to 650 m^3/s.

After collecting and processing data by analyzing log and other parameters, input data are shown in Table 2. The petroleum industry, for historical reasons, has been landed with a mixture of US, British, and SI units, which is often referred to by "oilfield units." In Vietnam they follow this

TABLE 2: Data of Well X1.

Input data	Value
TVD (ft)	9479.5
I (degree)	52
W_{az} (degree)	244
σ_v (psia)	9300
σ_h (psia)	8670
σ_H (psia)	9448
ν	0.3
UCS (psia)	2450
α	1
σ_{hz} (degree)	193
ϕ_{perf} (degree)	90
TWC (psia)	4936
k (mD)	60
ϕ_o	0.3
μ (cP)	3
$r_{perf} = r_{in}$ (ft)	0.0328
l_{perf} (ft)	1.6405
E (psia)	979004

tradition of using oilfield units, which is a requirement by the government; hence in Table 2 we do not use SI units.

4.1. Calculate Critical Bottomhole Pressure Using Geomechanical Model of Willson et al. [1]. The critical bottomhole pressure CBHP is important information during production phase because it indicates the lowest bottomhole pressure for sand production to not occur. For a specific well in oil field, the only production data that we control is the bottomhole pressure, which is adjusted using surface chokes. The pressure is controlled; hence the flowrate is controlled. For this reason, before predicting the sand production using Hydromechanical models of Gravanis et al. and Fjær et al., we firstly calculate the CBHP using Geomechanical model of Willson et al.

The testing showed that a relationship between the effective in situ strength of the formation U and the TWC (Thick wall cylinder) strength relative to a specimen with an OD/ID ratio (Outer/Inner diameters) would be equivalent to

$$\text{UCS} = b_f * 1.55 * \text{TWC}, \quad (30)$$

where b_f the boost factor is often taken as $b_f = 2$ for cased–hole perforated wells. Equation (1) considers effect of reservoir pressure decline [9] which can be accounted for by updating in situ stresses, while the vertical stresses are usually kept constant:

$$\begin{aligned}\sigma_h &= \sigma_h + A \cdot \Delta P \\ \sigma_H &= \sigma_H + A \cdot \Delta P,\end{aligned} \quad (31)$$

where $A = \alpha((1-2\nu)/(1-\nu))$ with α is the Biot's constant and ν is coefficient of Poisson, $\Delta P = P_c - P_i$ with P_c is current reservoir pressure, and P_i is initial reservoir pressure.

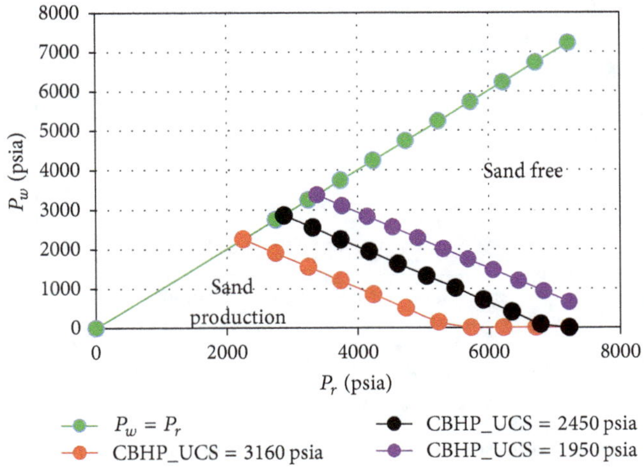

FIGURE 10: Prediction of sand production with variation of UCS (Unconfined Compressive Strength).

Figure 10 presents the bottomhole pressure P_w versus reservoir pressure P_r. In Figure 10 we also present the CBHP when the UCS changes. For example, for UCS = 1950 psia, if reservoir pressure P_r is 6000 psia, the CBHP is about 1500 psia; if the bottomhole pressure P_w is lower than this value 1500 psia, sand production will occur.

The results showed that if UCS gets smaller, the sand production window is greater. This can be explained by the reason that the formation rocks around the wellbore are weakened when UCS decreases. Thus, risk of sand production increases.

We consider a case study with UCS = 2450 psi, P_r = 4000 psi. From Figure 10, we see that if the current reservoir pressure is 4000 psi, then the critical bottomhole pressure is 2020 psia. For sand production to not occur, the bottomhole pressure must be greater than 2020 psi. Otherwise, to increase the flow rate, we must increase DP and if it is greater than CDP, it means that bottomhole pressure must be lower than critical bottomhole pressure (1000 psi); then the sand will be produced. So there is a compromise between increasing flowrate (production requirement) and sand production.

4.2. Calculate Sand Production Mass Using the Models of Gravanis et al. and Fjær et al. We use Hydromechanical Erosion models of Fjær et al. [5] and Gravanis et al. [6] to calculate sand production mass versus time. However, we do not have experimental data of Well X1 to calibrate empirical parameters, so we must use empirical parameters already calibrated in Section 3. In Figure 11 the results of sand mass produced over time calculated by the two models are presented. We also did a sensitivity study of the effect of drawdown pressure. Drawdown pressure, which is the difference between the reservoir pressure and the bottomhole pressure, is the most important data that we control during production phase. We control the drawdown; consequently, we control the production rate. The drawdown is typically controlled by surface chokes. The more the choke is open, the

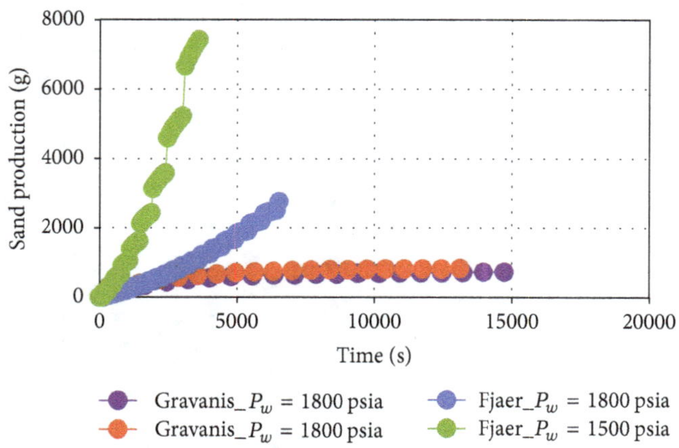

FIGURE 11: Results of two models in the case study of Well X1.

bigger the bottomhole pressure is and, hence, the smaller the drawdown is, and vice versa.

Figure 11 shows that sand production mass increases when drawdown pressure increases. However, results given by the model of Gravanis et al. are smaller than results given by Fjær et al. At each bottomhole pressure, sand mass given by the model of Gravanis et al. increases until it reaches constant value. On the other hand, the one of Fjær et al. increases continuously because the process of erosion repeats which causes the radius of cavity and production zone to increase, so sand production mass increases continuously. These two models show different results although they use the same physical mechanisms: the fluid flow can be described by Darcy's law and the driving mechanism for continuous sand production is erosion from plastified material in the vicinity of the production cavity under hydrodynamic forces. However, the model of Fjær et al. uses a step of instantaneous sand production which is not considered in the model of Gravanis et al. Moreover, while Gravanis et al. divide the whole region into a plastic region where Mohr-Coulomb failure criterion is considered, and an elastic region where Hooke's law is applied, Fjær et al. consider only the plastic region. Fjær et al. simply assume that the stiffness of the rock remains constant until the porosity has reached a critical value ϕ_{cr}. At that point, the entire part of the rock that has been producing sand collapses, and the remaining solid material in that part is produced in one burst. As a result, the equations used in these two models as well as the associated experimental parameters are quite different and eventually led to quite different results.

Finally, it is important to recall that, due to the lack of validation data in reality, it is impossible to choose which model to use in practical use. At this state, these results can only be used for illustration in study, and to demonstrate the possibility of using Hydromechanical models in real cases to predict eroded sand mass over time. In the future if real data is available, these models can be revised and recalibrated. Unfortunately, until now, measuring sand mass rate is technically impossible in the oilfields. An idea was proposed to solve this problem; the determination of real sand rate may be based on the erosion rate of the choke. However, this study has never been realized and may constitute a new objective in our future study.

5. Conclusion

This study combines Geomechanical model and the two Hydromechanical Erosion models of Fjær et al. [5] and Gravanis et al. [6] aiming to predict sand production. The proposed model can not only predict the critical pressure for onset of sand production but also estimate sand production mass and the variation of porosity in time and space. In addition, the study also gives detailed calculation steps for the two Hydromechanical models. Although each model has assumptions and empirical parameters, we can adjust them if we have experimental data. The results of this study will help to make an effective production planning that can avoid sand production when reservoir pressure declines. Hydromechanical model is still very new and not yet widely used, due to several reasons such as complicated calculation methods and parameters that need to be experimentally adjusted in practice or in the laboratory. Therefore, this initial computational workflow using MATLAB will significantly contribute to the development of future research in this domain in Petroleum Engineering.

The experimental data (sand production's mass in function of time) is rarely done not only in Vietnam but also in the world (only some laboratories and authors mentioned in the paper have done this kind of experience because of their research in this field). Hence, it is currently impossible to obtain empirical data for Cuu Long basin (sand mass in function of time), not only data in laboratory but also production data due to difficulties in collecting and measuring sand production rate. Therefore, application of the calibrated model for Cuu Long basin in this study is destined to demonstrate the possibility of using Hydromechanical models in real cases and the result can be used for illustration in research. In the next studies when we have available data from Cuu Long basin, we can do a revision/recalibration of these models. We already thought about the determination of real sand rate data based on the erosion rate of the choke. This idea will be the objective of our future study.

Notations

CBHP: Critical bottomhole pressure (psia)
DP: Drawdown pressure (psia)
LF: Loading Factor (near-wellbore formation stress normalized by strength)
SPR: Sand Production Rate (g/s)
TVD: True vertical depth (ft)
TWC: Thick wall cylinder strength (psia)
UCS: Unconfined compressive strength (psia)
$\|\|$: Notation representing the norm of a vector
α: Biot's constant
β: Exponent parameter
ε_p: Plastic shear strain

$\Gamma(x)$: Usual Euler Gamma function
λ: Experimentally evaluated sand production coefficient (1/m)
λ_{sand}: Proportionality constant (s/m^3)
λ_0: Initial erosion strength (m^{-1})
λ_2: Maximum erosion strength (m^{-1})
Λ: Depth of the plastic region (m)
Λ_0: Λ at $t = 0$ (m)
Λ_p: Threshold depth of the λ plateau
μ: Viscosity (MPa·s)
ϕ: Porosity
ϕ_o: Initial porosity
ϕ_{cr}: Critical porosity
ϕ_{perf}: Perforation orientation from the top of the wellbore in deviated wells (degree)
ρ_s: Solid density (g/m^3)
σ_1: Maximum principal stress (psia)
σ_2: Minimum principal stress (psia)
σ_v: Vertical stress (psia)
σ_h: Minimum principal horizontal stress (psia)
σ_H: Maximum principal horizontal stress (psia)
σ_{hz}: Minimum horizontal stress direction (degree)
ν: Poisson ratio
C_2: Integration constant
E: Young's modulus
I: Well inclination (degree)
k: Permeability (mD)
K: Principal stress ratio
l_{perf}: Perforation length (ft)
m_{sand}: Mass of produced sand
$\dot{m}$: Rate of eroded solid mass per unit volume ($g/s{\cdot}m^3$)
$\dot{M}$: Rate of eroded solid mass (g/s)
M: Initial amount of solid material in the sand producing zone (g)
p_o: Pore pressure (psia)
P_{in}: Internal pore pressure (Mpa)
P_{out}: External pore pressure (Mpa)
P_w: Bottomhole pressure
P_r: Reservoir Pressure
q_{fl}: Flow rate (m^3/s)
q_{fl}^i: Specific flow rate in the ith direction (g/s)
q_{fl}^{cr}: Critical flow rate (m^3/s)
r_{perf}: Perforation radius (ft)
r_{in}: Cylinder internal radius (m)
r_{out}: Cylinder external radius (m)
R : Location of the plastic zone boundary (m)
R_c: Radius of the cavity (m)
R_{cr}: Radius of the sand production zone (m)
R_p: Radius of the plastic zone (m)
Re: Reynolds number
S_o: Material cohesion (MPa)
t: Real time
T: Mathematical time
t_o: Initial time
v_s: Solid velocity (m/s)
V_{sp}: Volumetric sand production (m^3)
W_{az}: Well azimuth (degree).

Conflicts of Interest

The authors declare that there are no conflicts of interest regarding the publication of this paper.

References

[1] S. Willson, Z. A. Moschovidis, J. Cameron, and I. Palmer, "New model for predicting the rate of sand production," in *Proceedings of the SPE/ISRM Rock Mechanics Conference*, Irving, Tex, USA, October 2002.

[2] I. Vardoulakis, M. Stavropoulou, and P. Papanastasiou, "Hydro-mechanical aspects of the sand production problem," *Transport in Porous Media*, vol. 22, no. 2, pp. 225–244, 1996.

[3] E. Papamichos, I. Vardoulakis, J. Tronvoll, and A. Skjrstein, "Volumetric sand production model and experiment," *International Journal for Numerical and Analytical Methods in Geomechanics*, vol. 25, no. 8, pp. 789–808, 2001.

[4] L. Y. Chin and G. G. Ramos, "Predicting volumetric sand production in weak reservoir," in *Proceedings of the SPE/ISRM Rock Mechanics Conference*, Irving, Tex, USA, October 2002.

[5] E. Fjær, P. Cerasi, L. Li, and E. Papamichos, "Modeling the rate of sand production," in *Proceedings of the 6th North America Rock Mechanics Symposium (NARMS '04)*, Houston, Tex, USA, June 2004, https://www.onepetro.org/conference-paper/ARMA-04-588.

[6] E. Gravanis, E. Sarris, and P. Papanastasiou, "Hydro-mechanical erosion models for sand production," *International Journal for Numerical and Analytical Methods in Geomechanics*, vol. 39, no. 18, pp. 2017–2036, 2015.

[7] S. O. Isehunwa and O. M. Olanrewaju, "A simple analytical model for predicting sand production in a niger delta oil field," *International Journal of Engineering Science and Technology*, vol. 2, no. 9, pp. 4379–4387, 2010.

[8] E. Fjær, R. M. Holt, P. Horsrud, A. M. Raaen, and R. Risnes, *Petroleum Related Rock Mechanics*, Elsevier, 2008.

[9] K. Rahman, A. Khaksar, and T. J. Kayes, "An integrated geomechanical and passive sand-control approach to minimizing sanding risk from openhole and cased-and-perforated wells," *Society of Petroleum Engineers*, vol. 25, no. 2, pp. 155–167, 2010.

6

Experimental Investigation on the Deformation, Strength and Acoustic Emission Characteristics of Sandstone under True Triaxial Compression

Zhaolin Li, Lianguo Wang, Yinlong Lu, Wenshuai Li, and Kai Wang

State Key Laboratory for Geomechanics and Deep Underground Engineering, China University of Mining and Technology, Xuzhou, Jiangsu 221116, China

Correspondence should be addressed to Lianguo Wang; cumt_lgwang@163.com

Academic Editor: Carlo Santulli

The study of deformation, strength, and other mechanical characteristics of sandstone under true triaxial compression is significant for understanding failure mechanisms in rock and evaluating the stability of underground structures. Conventional and true triaxial compression tests for sandstone are conducted for different stress states in this study using the self-developed true triaxial electrohydraulic servo test system combined with acoustic emission (AE) testing. This study presents an in-depth and systematic investigation of deformation, strength, and AE characteristics. The results show significant differences in deformation, strength, and acoustic emission characteristics for the rock under conventional triaxial and true triaxial compression tests, respectively. The peak strength, axial strain, lateral strain, and incremental strain (in unstable crack growth stage) increase with increasing confining pressure under conventional triaxial compression, and the AE count gradually decreases while shear crack proportion gradually increases, indicating that increasing confining pressure gradually inhibits the shear slip effect along fractures, delays perforation of the rock shear fracture surface, and enhances the ability of the rock to withstand deformation and load. Under true triaxial compression, the peak strength increases and then decreases with increasing intermediate principal stress σ_2 and the axial strain ε_1 and lateral strain ε_2 gradually decrease; besides, the lateral strain (expansion) of the rock is mainly in the minimum principal stress σ_3 direction, and lateral expansion tends to decrease before increasing. AE events first weaken and then enhance with increasing σ_2, and the proportion of shear cracks increases first and then decreases, indicating that the confining pressure gradually changes from the shear slip effect that controls crack offset to the damage effect that promotes crack tension with increasing σ_2. In addition, the protective effect of confining pressure improves when σ_3 increases.

1. Introduction

Rock material is a kind of brittle geological material, which has a complex composition and contains numerous defects, such as joints and cracks. Rock mechanical properties change with variations in external load and environmental factors because of these inherent characteristics. Rock engineering is a fundamental part of construction and is an important support for social development and progress. These projects are all related to research on fundamental rock mechanics problems [1].

The mechanical properties of the rock in Earth's crust are closely related to the 3D stress state. The deformation, strength, and other mechanical properties can be obtained by studying the deformation and failure of various rocks. Understanding rock failure mechanisms is significant for geotechnical engineering. At present, the study of rock damage in a 3D stress state is primarily carried out using conventional triaxial tests. Domestic and foreign scholars have studied the mechanical properties of rock under conventional triaxial compression by using conventional testing machines. Karman obtained the earliest stress-strain curves for Carrara marble under different confining pressures on an ordinary testing machine [2]. Fredrich studied the mechanical properties of calcite marble under conventional triaxial compression and examined the effect of particle size on the brittle-ductile transition [3]. Yang et al. [4]. carried out conventional triaxial compression tests on sandstones by using a servo testing machine.

However, the influence of the intermediate principal stress is ignored due to the setup of the conventional testing machine. A cylindrical rock sample is subjected to different axial and lateral stresses such that the rock is in an axisymmetric stress state. So, only the strength and deformation for rocks in an axisymmetric stress state are measured, and the general stress states are not represented ($\sigma_1 \neq \sigma_2 \neq \sigma_3$). In a true triaxial compression test, each of the three principle stresses can be independently applied to the sample allowing for the failure mode to be accurately simulated and a more complex stress path to be measured. In recent years, Mogi [5, 6], Haimson et al. [7, 8], and Feng et al. [9] have carried out a large number of experimental studies on rocks using a true triaxial test machine. However, previous studies mainly focus on the effect of the intermediate principal stress on rock strength [10] and the establishment of true triaxial strength criterion to show the effect of the intermediate principal stress on rock strength [11, 12]. Furthermore, most true triaxial compression experiments examine strength and failure criteria for granite, marble, or other hard rocks, while failure and deformation characteristics for relatively soft rocks, such as sandstone and coal, remain largely unknown.

Underground projects such as roadways are mostly undertaken in sedimentary structures that consist of sandstone-filled channels, and the deformation problems in the roadways are more prominent. Due to the complex microstructure of and numerous defects, its physical and mechanical properties are more sensitive to external loads. Sandstone is one of the most widely distributed rocks in East, South, and Southwest China. With the rapid development of underground construction, understanding the response of sandstone to 3D stress states is of significance for the construction of underground foundations and roadways. Examining the deformation, strength, and other mechanical properties of sandstone under true triaxial compression can provide reference for the selection of rock mechanics parameters in future geotechnical engineering projects.

The rock damage induced by compression is essentially caused by the continuous development of microcracks, such that the rock fractures. Therefore, the combination of macroscopic mechanical properties such as strength, deformation of the rock, and the development of microcracks in the rock can reveal the cause of the deformation and failure. Acoustic emission (AE) is an effective method to observe the development of microcracks in rock. AE is caused by the generation and expansion of microfractures when rock materials are under an external stress, and AE is accompanied by elastic wave or stress wave release. AE information reflects the damage and failure state of the rock, which is closely related to the densification of the original internal fissures as well as the generation, expansion, and penetration of the new fissures. The influence of the intermediate principal stress on the development of internal microcracks in rocks can be revealed using AE information. Moradian et al. [13], Yang et al. [14], and Ranjith et al. [15] analyzed the evolution of internal cracks in rocks using AE. However, these studies focused on the evolution of rock fracturing under conventional triaxial compression. Therefore, it is necessary to explore the development of internal microcracks in rock based on AE to reveal the root cause of rock deformation and failure under true triaxial compression.

In this study, sandstone is subjected to conventional and true triaxial tests with varying stress conditions using the self-developed true triaxial electrohydraulic servo test system. The development of internal cracks within the rock is monitored using an AE monitoring system. The deformation, strength, and AE characteristics of sandstone under three triaxial stress states are discussed. The effects of different confining pressures under conventional triaxial compression and different stresses in true triaxial compression on the variation of rock deformation, strength parameters, and internal crack development are discussed.

2. Experimental Methodology

2.1. True Triaxial Electrohydraulic Servo Test System. The tests presented in this study are carried out using the self-developed true triaxial compression test system (Figure 1). The test system is primarily composed of a maximum principal stress, intermediate principal stress, minimum principal stress (σ_1, σ_2, σ_3) three-dimensional servo control load system, true triaxial pressure cell, automatic data acquisition system, and AE monitoring system. The σ_1, σ_2, σ_3 three-dimensional servo control load system forms an independent load frame. Piston rods are installed at one end of the three load frames, which is externally connected to the servo valve to produce independent servo loading. The three loading frames are perpendicular to each other. The positions of loading frames σ_1 and σ_3 are fixed, while the third loading frame (σ_2) is placed on a horizontal guide rail and has the able to slide along the guide rail to facilitate the operation and placement of the sample. The other end of the two loading frames (σ_2 and σ_3) is installed with reaction screws, which have an adjustable length to adapt to different specimen sizes.

Specimens used in this study are cuboids with a size of $50 \times 50 \times 100$ mm. The specimen is placed in a cubic pressure chamber composed of a fully rigid structure, which consists of a box-shaped body and a loading plate. In order to avoid deformation of the rock samples leading to mutual extrusion of the loading plate during the test, a gap is left between the vertical and horizontal plates in the traditional true triaxial pressure chamber. This gap causes the uneven stress and strain in the test, resulting in a decrease in peak strength and an abnormal failure pattern in the sample [16]. In order to overcome this defect, the true triaxial pressure cell adopts a rotary interlocking platen design, in which the loading plates on the 6 faces of the sample are arranged in a misplaced method. When the specimen is compressed, the loading plate moves synchronously with the specimen to avoid interference effects in each direction (Figure 1). The relative displacement between the loading plate and the specimen produces a friction force, which negatively influences the test result. To reduce the friction between the

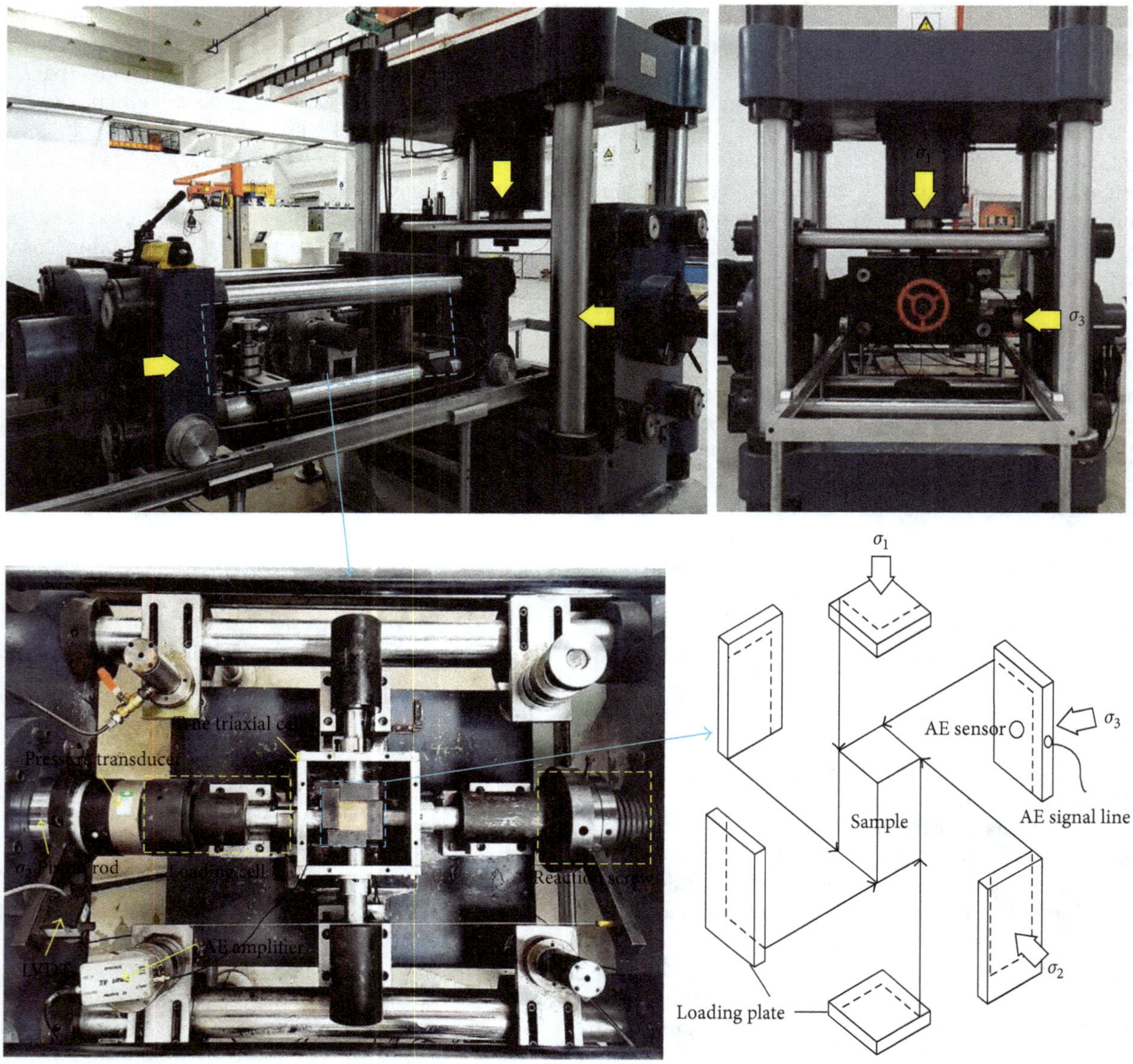

FIGURE 1: True triaxial compression test system.

loading plate and the sample, a 0.5 mm thick layer of lubricant is applied between the loading plate and the sample surface. The lubricant is composed of a 1 : 1 ratio of stearic acid and Vaseline [17].

A spoke-type pressure sensor with a precision of 0.01 kN is installed between the piston rod and loading block. A rope-type displacement sensor with an accuracy of 0.002 mm and a measurement range of 1,000 mm is fixed to the piston rod and reaction force screw. A DS2-8B full information acoustic emission signal analyzer is used to examine the AE activity during tests. The AE system used in this study mainly consists of the AE host, preamplifier, and AE sensor. The type of the AE sensor is RS-54A, and its monitoring frequency range is 100–900 kHz. The AE sensor is installed in the loading plate of the true triaxial pressure cell (an 8 mm diameter hole is set in the side of the plate). The AE signal line is drawn out from the hole in the side of the plate, and the probe is in direct contact with the surface of the sandstone to ensure the accuracy of the AE data. The model of AE preamplifier is 20/40/60 dB which can provide three different transmission gains; furthermore, it has the advantages of low noise, wide bandwidth, impact resistance, and small volume. During the test, the threshold is set to 40 dB to avoid the possibility of electronic/environmental noise. The AE signal is amplified by the preamplifier with the gain of 40 dB. The time parameters for AE waveforms included peak definition time (PDT), hit definition time (HDT), and hit lockout (HLT), which are set to 50, 100, and 100 μs, respectively.

2.2. Specimen Preparation. Sandstone is selected for examination in this study. Specimens are composed of feldspar and quartz, as well as a small amount of montmorillonite. The sandstone has a medium grain structure with grain size ranging from 0.1 to 0.35 mm and a relatively dense massive structure. Grains are relatively uniform with an identical macroscopic uniformity and an average density of 2380 kg/m^3. Samples are

TABLE 1: Tested sandstone specimens and conditions in this research.

Loading program	Group	Specimen number	Dimensions (mm)	Quality (g)	σ_2 (MPa)	σ_3 (MPa)
CTT	C1	C1-1	100.2 × 50.0 × 49.8	622.5	10	10
		C1-2	100.2 × 49.8 × 50.1	613.2		
		C1-3	100.3 × 50.1 × 49.9	612.9		
	C2	C2-1	100.0 × 49.5 × 50.0	613.0	20	20
		C2-2	100.5 × 49.5 × 50.0	613.5		
		C2-3	99.8 × 50.3 × 50.3	605.1		
	C3	C3-1	100.3 × 50.0 × 50.1	636.0	30	30
		C3-2	99.9 × 50.1 × 50.2	633.9		
		C3-3	99.6 × 50.3 × 49.6	602.3		
	C4	C4-1	99.9 × 50.1 × 50.2	625.3	40	40
		C4-2	99.8 × 50.2 × 50.1	629.5		
		C4-3	100.2 × 50.0 × 49.2	639.7		
TTT (σ_3 = 10 MPa)	T2	T2-1	100.1 × 50.3 × 49.8	628.1	20	10
		T2-2	100.4 × 50.5 × 50.1	636.0		
		T2-3	100.3 × 50.6 × 50.1	646.6		
	T3	T3-1	100.4 × 49.9 × 50.3	630.5	30	10
		T3-2	100.0 × 50.2 × 50.1	650.0		
		T3-3	99.8 × 50.4 × 50.5	624.7		
	T4	T4-1	100.4 × 49.6 × 49.6	618.8	40	10
		T4-2	100.3 × 49.7 × 49.9	626.7		
		T4-3	100.1 × 50.4 × 50.6	607.1		
	T5	T5-1	100.1 × 50.1 × 49.9	615.6	50	10
		T5-2	99.9 × 49.5 × 50.2	610.8		
		T5-3	100.2 × 50.1 × 49.9	614.5		
	T6	T6-1	99.7 × 49.7 × 50.3	610.4	60	10
		T6-2	99.8 × 49.9 × 50.2	606.3		
		T6-3	100.1 × 49.8 × 50.2	595.7		
TTT (σ_3 = 20 MPa)	T7	T7-1	100.3 × 49.9 × 50.2	630.5	30	20
		T7-2	99.8 × 50.4 × 50.5	624.2		
		T7-3	99.9 × 49.5 × 50.2	612.3		
	T8	T8-1	100.4 × 49.8 × 50.2	599.8	40	20
		T8-2	100.1 × 50.4 × 50.3	603.1		
		T8-3	99.7 × 50.1 × 50.3	615.4		
	T9	T9-1	100.3 × 49.7 × 49.9	621.7	50	20
		T9-2	100.1 × 50.1 × 49.9	615.6		
		T9-3	100.0 × 49.7 × 50.1	630.1		
	T10	T10-1	99.8 × 49.9 × 50.4	626.3	60	20
		T10-2	100.2 × 50.1 × 49.9	624.6		
		T10-3	100.3 × 49.8 × 49.9	621.7		

50 × 50 × 100 mm in size, and the end faces are carefully polished. Detailed sample information is given in Table 1.

2.3. Testing Procedure. In order to study the mechanical properties and AE characteristics of rock under true triaxial compression, the difference between the true triaxial test (TTT) and conventional triaxial test (CTT) is compared and the influence of different triaxial stresses on rock mechanical properties and AE characteristics is analyzed. The test setup is shown in Figure 2.

2.3.1. CTT. Pressure is applied to the specimen at a rate of 0.2 MPa/s up to a predetermined confining pressure (set values of σ_2 and σ_3), while the specimen is maintained in a hydrostatic pressure state (*oa*); then, with constant confining pressure, the axial stress is applied to the specimen at a loading rate of 0.002 mm/s, until the specimen failed (*ab*).

2.3.2. TTT. Pressure is applied to the specimen at a rate of 0.2 MPa/s up to a predetermined confining pressure (set value of σ_3), while the specimen is maintained in a hydrostatic pressure state (oa, oa'); then, with constant σ_3, the values of σ_1 and σ_2 are continuously increased to the set value of σ_2 (ab, $a'b'$) at the same rate. Afterwards with constant σ_2 and σ_3, the axial stress is applied to the specimen at a loading rate of 0.002 mm/s, until the specimen failed (bc, $b'c'$).

Since the test discreteness is considered in this study, each test is repeated 2 times. The specific test scheme is shown in Table 1.

3. Strength and Deformation Behavior

3.1. Strength and Criterion Verification. In order to analyze the strength characteristics of sandstone under conventional triaxial and true triaxial compression and to further verify the accuracy of the test results for the triaxial test system,

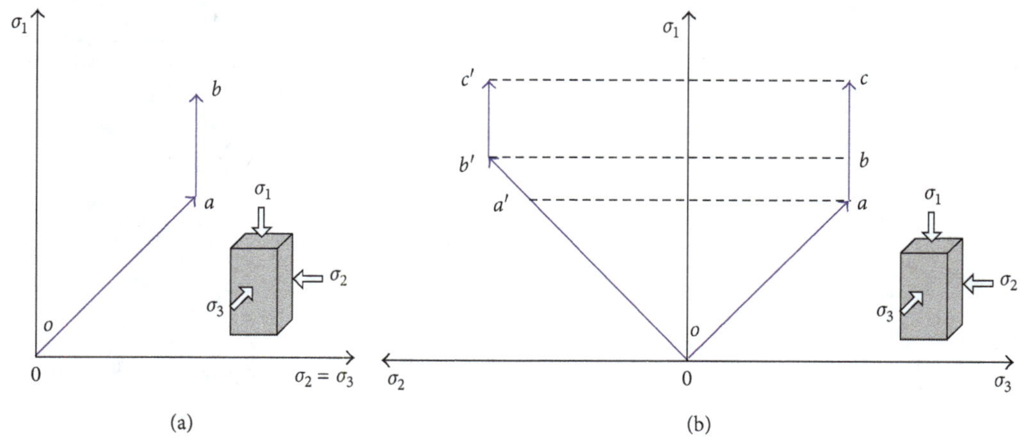

FIGURE 2: The two different loading programs: (a) CTT; (b) TTT.

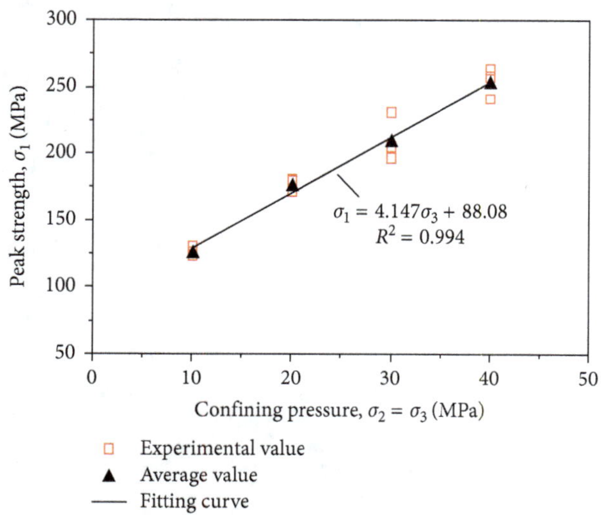

FIGURE 3: Peak strength σ_1 as a function of the minimum principal stress σ_3 under conventional triaxial stress and the best-fit linear function strength criterion (Mohr–Coulomb) for sandstone specimens.

rock strength and the strength-fitting curve under the two loading conditions are obtained (Figures 3 and 4). The peak strength increases linearly with increasing confining pressure under conventional triaxial compression (Figure 3). The Mohr–Coulomb criterion is used to fit the relationship between the confining pressure and rock strength under the conventional triaxial compression. The fitting correlation coefficient is $R^2 = 0.994$, which reflects the relationship between rock strength and varying confining pressure under conventional triaxial compression.

Because the influence of the intermediate principal stress on the rock strength is not considered in the Mohr–Coulomb criterion, the Mohr–Coulomb criterion is only applicable to the special case in which $\sigma_2 = \sigma_3$, and it cannot accurately reflect the strength characteristics of the rock. Domestic and foreign scholars have come to realize the significance of the intermediate principal stress on ultimate compressive strength. Mogi [5, 6], Al-Ajmi and Zimmerman

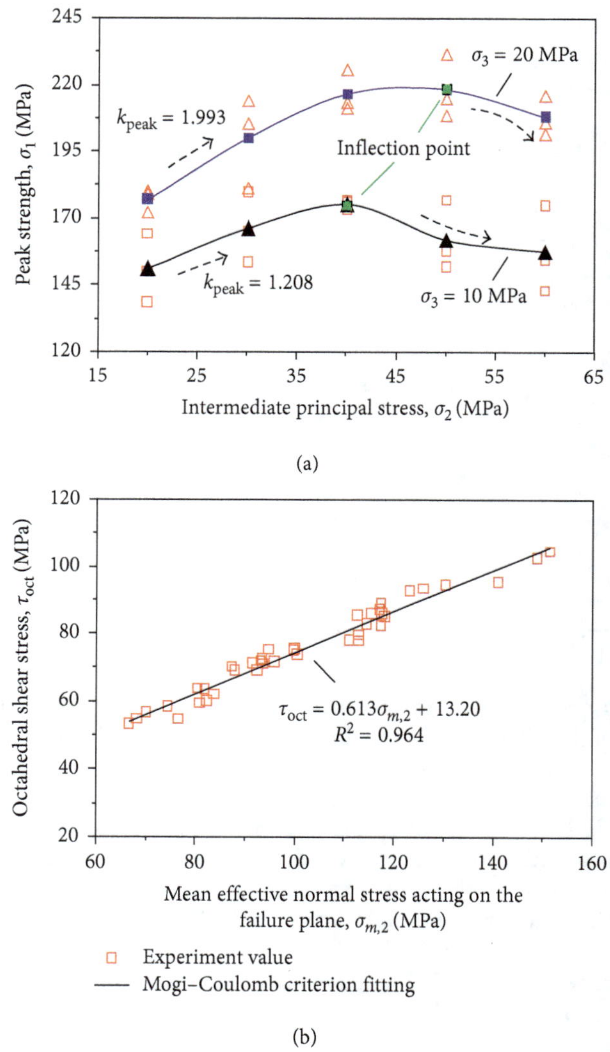

FIGURE 4: True triaxial strength for sandstone: (a) variation of peak compressive stress σ_1 as a function of σ_2 for the constant values of σ_3 = 10 MPa and σ_3 = 20 MPa; k_{peak} represents the slope of the peak strength with increasing σ_2; (b) the best-fit linear function true triaxial strength criterion (Mogi–Coulomb Criterion) based on all experimental results in terms of τ_{oct} versus $\sigma_{m,2}$.

[18], Oku et al. [19], Lee and Haimson [20], and Chang and Haimson [21] considered that a true triaxial failure criterion can be expressed as a linear function (the Mogi–Coulomb criterion) between octahedral shear stress τ_{oct} and the average effective normal stress acting on the shear plane $\sigma_{m,2}$:

$$\tau_{oct} = a + b\sigma_{m,2}, \tag{1}$$

$$\tau_{oct} = \frac{1}{3}\sqrt{(\sigma_1 - \sigma_2)^2 + (\sigma_2 - \sigma_3)^2 + (\sigma_1 - \sigma_3)^2}, \tag{2}$$

$$\sigma_{m,2} = \frac{\sigma_1 + \sigma_3}{2}, \tag{3}$$

where τ_{oct} and $\sigma_{m,2}$ are the octahedral shear stress and the average effective normal stress on the shear plane, respectively, and a is the intercept of Mogi's linear criterion fitting straight line and τ_{oct} axis, and b is the slope of straight line.

By comparing the Mohr–Coulomb criterion and Mogi–Coulomb criterion, the relationship between them is concluded:

$$a = \frac{2\sqrt{2}}{3} c \cos \varphi, \quad b = \frac{2\sqrt{2}}{3} \sin \varphi. \tag{4}$$

According to the suggestions from ISRM, the Mogi–Coulomb criterion is used to fit the strength of the rock (including conventional and true triaxial strength) (Figure 4(b)). According to the fitting results, the Mogi–Coulomb criterion ($R^2 = 0.964$) reflects the strength characteristics of the rock under true triaxial loading conditions relatively well and is consistent with the test results of Mogi, Haimson, and Chang. These results are further verified by the reliability and accuracy of the experimental system used in this study.

The values of c and φ of the sandstone under two different strength criteria are shown in Table 2 and are obtained by combining (1)–(4) according to the parameters in the fitting equations in Figures 3 and 4(b). The fitting coefficient of the two strength criteria is high. Therefore, the values of c and φ of the sandstone are relatively close to each other. The Mogi–Coulomb criterion applies to both the conventional and true triaxial test and is more universal. The Mohr–Coulomb criterion is a special case of the Mogi–Coulomb criterion.

The strength characteristics of rock under true triaxial compression are shown in Figure 4(a), and the variation is different from that of conventional triaxial condition. The peak rock strength increases and then decreases with increasing σ_2 under true triaxial loading. When σ_3 = 10 MPa and 20 MPa, the peak rock strength first increases to the maximum and then gradually decreases with increasing σ_2. Inflection points are located at 40 MPa and 50 MPa, respectively. The ascending segment tends to increase linearly with increasing σ_2, and the slopes of the peak strength are 1.208 and 1.993, indicating that improving σ_3 has a more significant protective effect on the rock.

TABLE 2: Fitting results and parameter values of two strength criteria of sandstone.

Group	Criterion	c (MPa)	φ (°)	R^2
CTT	Mohr–Coulomb	21.6	37.7	0.994
TTT	Mogi–Coulomb	18.4	40.5	0.964

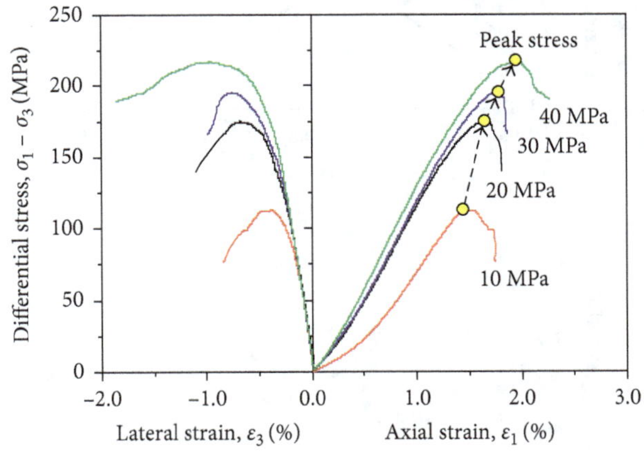

FIGURE 5: Stress-strain curves for sandstone from conventional triaxial compression experiments.

3.2. Stress-Strain Curve. In this study, the stress-strain curves under conventional and true triaxial compression with different confining pressure conditions are shown in Figures 5 and 6, where the yellow origin represents the peak strength of the rock. Rock stress-strain characteristics for four confining pressures under conventional triaxial compression are similar (Figure 5). The steepness of the curve in the elastic deformation stage is greatly affected by the confining pressure. The larger the confining pressure, the steeper the curve is, which indicates that the elastic modulus of the sandstone increases with increasing of confining pressure. The difference in steepness of the lateral curve is not obvious, which shows that the effect of confining pressure on the steepness of the lateral deformation of sandstone is small. The axial strain, lateral strain, and peak strength of the rock increase with increasing confining pressure, which shows that the increase in the conventional triaxial confining pressure improves the ability of the rock to withstand deformation and loading in both the axial and lateral directions.

Figures 6(a)–6(d), respectively, show the relationships between the differential stress ($\sigma_1 - \sigma_3$) and principal strain (ε_1, ε_2, and ε_3) as well as volumetric strain (ε_V) under true triaxial compression. Considering that different intermediate principal stresses under true triaxial compression are taken, the test process is divided into one section with a loading by force ($o \rightarrow a(a') \rightarrow b(b')$) and other section with a loading by displacement ($b(b') \rightarrow c(c')$). To facilitate the analysis, the rock's stress-strain characteristics are only analyzed for the loading by the displacement section.

The condition of the rock in the initial compaction stage is not obvious due to the effect of loading (Figure 6(a)). The rock enters the elastic stage quickly, and the axial curve is

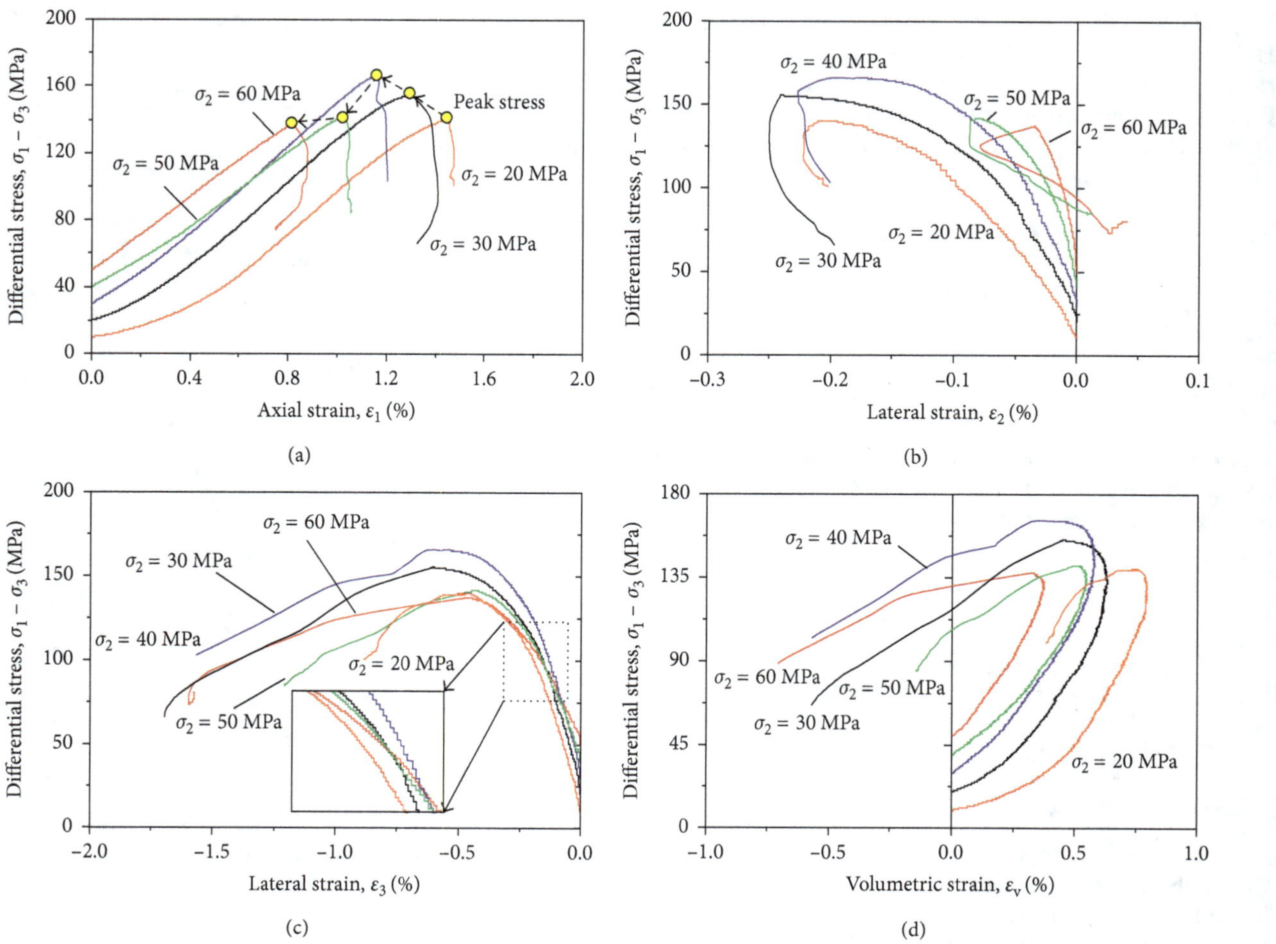

FIGURE 6: Differential stress ($\sigma_1 - \sigma_3$)-strain relations for sandstone under true triaxial compression conditions with a constant σ_3 of 10 MPa: (a) differential stress ($\sigma_1 - \sigma_3$)-axial strain (ε_1); (b) differential stress ($\sigma_1 - \sigma_3$)-lateral strain (ε_2); (c) differential stress ($\sigma_1 - \sigma_3$)-lateral strain (ε_2); (d) differential stress ($\sigma_1 - \sigma_3$)-volumetric strain (ε_2).

linear under different σ_2 conditions. The steepness of the axial curve is effected by σ_2. The steepness of the axial curve increases and gradually decreases with increasing σ_2. The variation law is different from that of conventional triaxial compression. The elastic modulus gradually increases and then decreases with increasing σ_2, and the variation law is consistent with peak strength. The strain during loading by displacement gradually reduces with increasing σ_2, and the lateral expansion of the rock gradually plays a more dominant role.

The strain curve in the direction of intermediate principal stress gradually steepens with increasing σ_2 (Figure 6(b)). The increase of σ_2 limits the deformation (expansion) in the σ_2 direction. After the peak stress is reached, the rock experiences compression in the intermediate principal stress direction especially when σ_2 equals 50 MPa or 60 MPa, indicating that brittle failure of the rock occurs under such intermediate principal stress conditions, and the rock extends rapidly along the minimum principal stress direction, which leads to the rapidly compressed in the intermediate principal stress direction after the peak.

The strain curve in the direction of minimum principal stress steepens and gradually shallows with increasing σ_2 (Figure 6(c)). The lateral expansion of the rock mainly in the direction of the minimum principal stress is due to the increase of σ_2. Because $\sigma_2 > \sigma_3$, the expansion of the rock in the direction of minimum principal stress is promoted by increasing σ_2. At the same time, the lateral expansion of rock could be limited by σ_2, which can be regarded as the confining pressure. Results show that when σ_2 is 20 MPa~40 MPa, the increase of σ_2 plays a protective role for the rock and limits deformation in the direction of σ_3 (lateral restraint effect). When σ_2 is 50 MPa or 60 MPa, the increase of σ_2 plays a damaging role for the rock and accelerates deformation in the direction of σ_3 (damage effect).

In Figure 6(d), the change in rock volumetric strain (ε_V) is shown. ε_V first increases and then decreases, showing that the rock undergoes gradual compression first, and then begins to expand with increasing differential stress. The inflection point of ε_V is the point where rock volume stops compression and begins to expansion. The inflection point of volumetric strain is more easily reached with the increasing σ_2.

The stress-strain curve of the rock under at σ_3 = 20 MPa is similar to that of σ_3 = 10 MPa (Figure 7). The rock axial strain curve is steeper, which indicates that the elastic

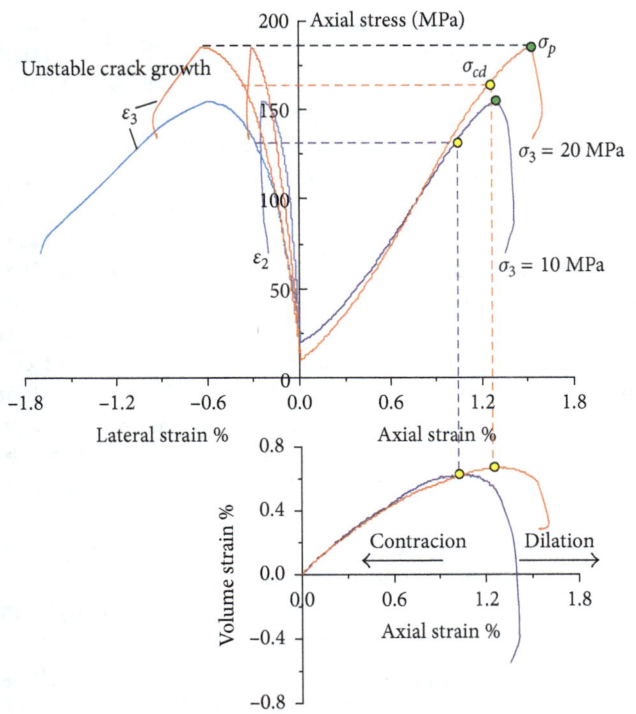

FIGURE 7: Stress-strain curve of rock at varying σ_3 (σ_2 = 30 MPa).

modulus of the rock increases. The peak strength, axial strain, and lateral strain of the rock increase with increasing σ_3, and the strain corresponding to the inflection point of ε_V greatly improved. The increase of σ_3 enhances the ability of the rock to bear stress and improves the carrying capacity of the rock.

The variation of elastic modulus under conventional triaxial and true triaxial compression is shown in Figure 8. The elastic modulus increases with increasing confining pressure under conventional triaxial compression (Figure 8(a)). The variation is similar to that of strength, which shows that there are microcracks in the rock and the confining pressure closes the cracks to some extent, causing some slippage between cracks. If the confining pressure increases, cracks would close more fully, and slippage between the cracks decreases. The increase of confining pressure makes slip along the cracks reduce due to the inhibition of the frictional force. Therefore, the axial deformation of the specimen decreases, and the elastic modulus of the rock increases.

The elastic modulus of the rock first increases and then decreases with increasing σ_2 under true triaxial compression, in which σ_3 is 10 and 20 MPa, and the values of the two peak points are both σ_2 = 40 MPa (Figure 8(b)). When σ_2 is 20~40 MPa, the number of slippage cracks in the rock is effectively reduced by improving σ_2, and slip along the microcracks is inhibited, which leads to an increased elastic modulus. The effect of confining pressure is similar to that of conventional triaxial compression. When σ_2 exceeds 40 MPa, the confining pressure promotes the expansion of microcracks in the rock, and the elastic modulus gradually decreases (Figure 6(c)). By comparing the variation laws of elastic modulus when σ_3 is 10 MPa and 20 MPa, it is apparent that the latter elastic modulus is larger, indicating that

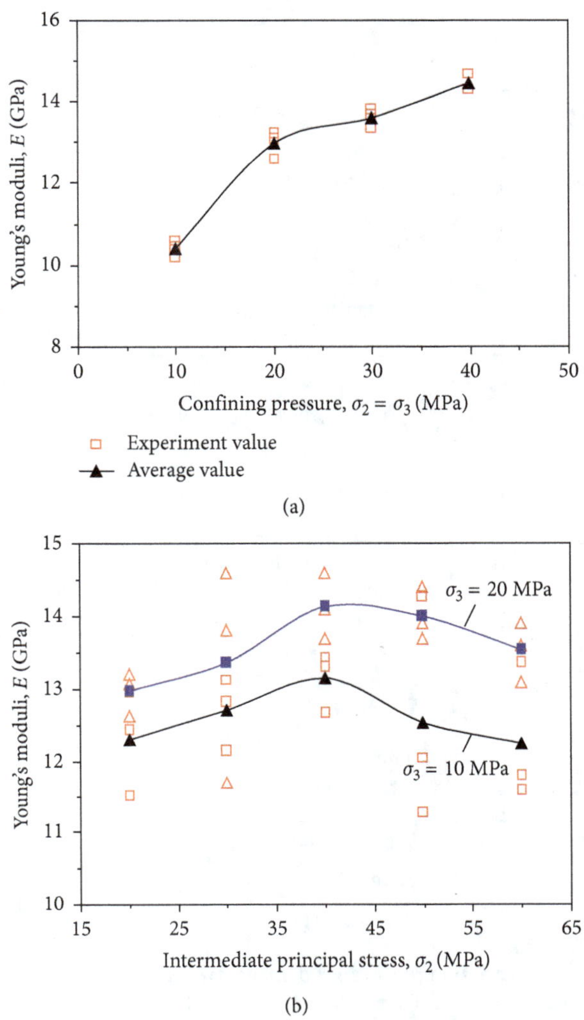

FIGURE 8: Influence of the confining pressure on Young's modulus of sandstone: (a) CTT; (b) TTT.

the increase of σ_3 makes the cracks close more fully and reduces slip. When σ_2 is 40~60 MPa, the decrease of the latter elastic modulus is slower than the former, which indicates that the increase of σ_3, to some extent, slows down the variation trend of "first increase and then decrease" under true triaxial compression and plays a protective role to the rock.

3.3. Strength and Deformation Characteristic in Unstable Crack Growth Stage. The deformation properties of rock are closely related to the growth and development of internal cracks. The rock failure process is associated with the closure, fracture initiation, expansion, and interactive perforation of cracks. The failure process is divided into the following stages [22]: (1) microcrack closure compaction stage; (2) elastic deformation stage; (3) stable crack growth stage; (4) unstable crack growth stage; (5) postpeaking stage. When the axial stress gradually increases to the rock damage stress σ_{cd}, the stress-strain curve enters the unstable crack growth stage, marking the beginning of shear and expansion of the rock sample. In this stage, the cracks are unstable.

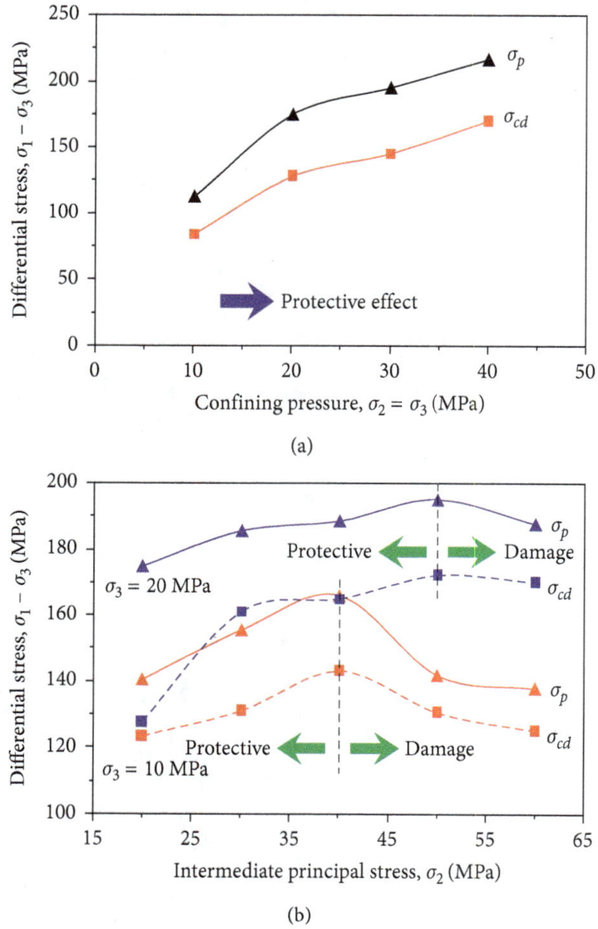

FIGURE 9: Variation rule of rock damage strength and peak strength under conventional triaxial and true triaxial conditions.

Even if the rock is no longer loaded, the internal cracks in the rock specimen cannot hold stable for a significant amount of time and continue to expand over time. Therefore, the damage stress is also called the long-term strength of the rock. Examination of rock deformation, strength, and crack development in this stage is of great significance to understanding its failure mechanism. This study focuses on the analysis of rock strength, deformation, and acoustic emission characteristics in this stage.

The variation rule of damage strength and peak strength under conventional triaxial and true triaxial compression is shown in Figure 9. Damage strength σ_{cd} and peak strength σ_p of the rock under conventional triaxial compression increase with increasing confining pressure (Figure 9(a)). Because there are microcracks in the rock, development of cracks is effectively prevented, and the expansion of internal cracks of the rock is restrained laterally by improving confining pressure.

Variation of damage strength σ_{cd} and peak strength σ_p for rocks under true triaxial compression (σ_3 = 10 MPa and 20 MPa) is different from rocks under conventional triaxial compression (Figure 9(b)). Damage strength σ_{cd} and peak strength σ_p both increase first and then decrease. When σ_3 is 10 MPa, the damage stress σ_{cd} and peak strength σ_p show an obvious trend of "first increasing and then decreasing," and the confining pressure constrains the expansion of internal cracks when σ_2 is small. While the lateral restraint gradually transforms into a damaging force on the rock with increasing σ_2, the expansion of internal cracks in the rock is promoted, leading to decreasing σ_{cd} and σ_p. By comparing variations in damage and peak strength under true triaxial compression, it is apparent that σ_3 equal to 10 MPa and 20 MPa corresponds to σ_2 equal to 40 MPa and 50 MPa, respectively. The latter values of σ_{cd} and σ_p are higher and vary more gradually than the former values. In particular, the values of σ_{cd} and σ_p decrease when the former σ_2 exceeds 40 MPa, while the values of σ_{cd} and σ_p reduce slightly when the latter σ_2 exceeds 50 MPa, indicating that the damage effect of the latter confining pressure is significantly weakened, and improving σ_3 limits the development of internal cracks.

The strain characteristics corresponding to damage and peak stress of the rock under conventional triaxial compression are shown in Figure 10. The axial strain and lateral strain corresponding to damage and peak stress of the rock under conventional triaxial compression vary linearly with confining pressure, all of which increase with increasing confining pressure. Volumetric strain fluctuates greatly. The lateral strain of the peak point ε_p^3 increases more quickly than the axial strain ε_p^1 with increasing confining pressure, which indicates that the lateral strain dominates during the unstable crack growth stage and the admission value of lateral expansion increases. At the same time, the volumetric strain increment of the rock $\Delta\varepsilon_v (= \varepsilon_p^v - \varepsilon_{cd}^v)$ in this stage increases with increasing confining pressure, which indicates that the ability of the rock to withstand deformation increases gradually due to increasing confining pressure under the conventional triaxial compression.

Incremental strain variation in the unstable crack growth stage under the conventional triaxial compression is shown in Figure 11. The axial strain increment $\Delta\varepsilon_1$ increases slightly with increasing confining pressure in this stage, while the lateral strain $\Delta\varepsilon_3$ increases quickly. The volumetric strain increment $\Delta\varepsilon_V$ increases quickly with increasing confining pressure due to $\Delta\varepsilon_v = \Delta\varepsilon_1 + 2\Delta\varepsilon_3$, which indicates that the lateral expansion of the rock in the unstable crack growth stage under the conventional triaxial compression is obvious, and the ability to withstand lateral deformation increases significantly due to increasing confining pressure.

The strain characteristics of rock damage point and peak point under true triaxial condition are shown in Figure 12. Because the deformation characteristics of the rock in the loading segment by displacement are only analyzed, the values of σ_3 and σ_2 are 20 MPa under the conventional triaxial compression in this study, and the strain law is not discussed because of its abnormality. Axial strains of the damage point and the peak point under true triaxial compression decrease with increasing σ_2. Variation is similar when σ_3 is 10 and 20 MPa, and the latter axial strains of the damage point and peak point ε_{cd}^1 and ε_p^1 are higher than the former ones. The lateral strains ε_{cd}^2 and ε_p^2 in the direction of σ_2 decrease gradually with increasing σ_2. Deformation

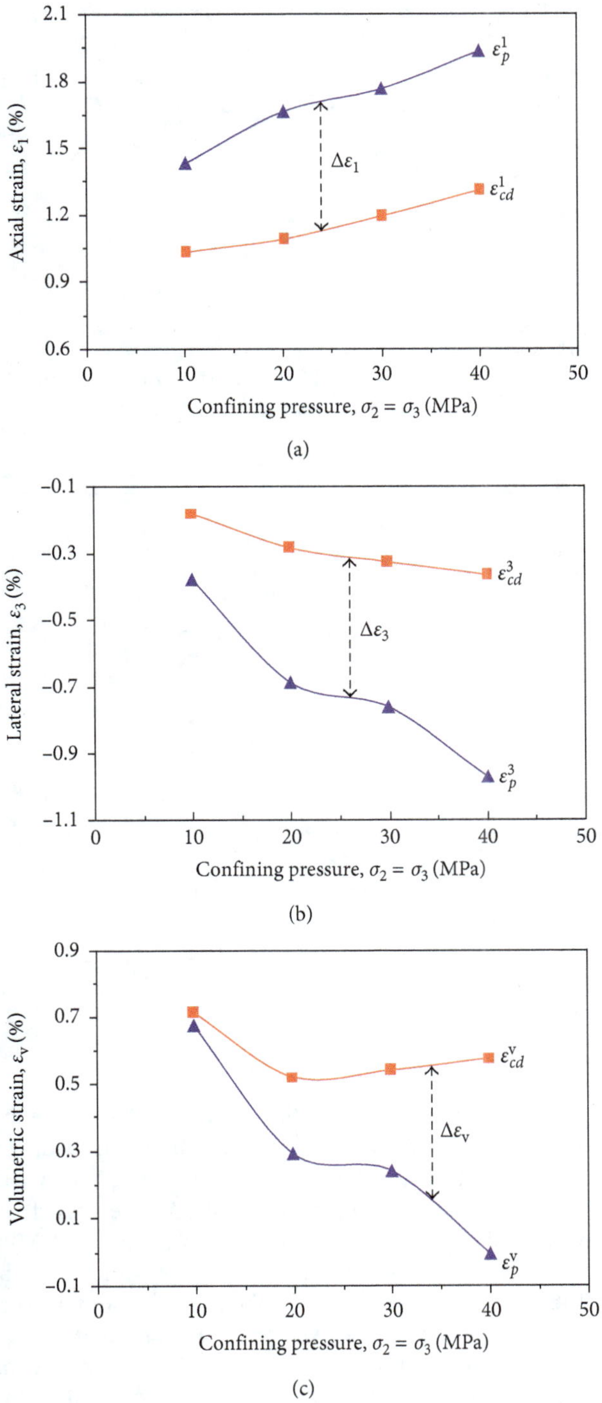

FIGURE 10: Influence of the confining pressure on strain corresponding to each characteristic stress under conventional triaxial compression experiments: (a) axial strain; (b) lateral strain; (c) volumetric strain.

in the rock reflects expansion in the σ_2 direction, and the expansion is limited by an increasing σ_2. The restriction is more pronounced when σ_3 equals 20 MPa (Figure 12(b)). The change laws of the two kinds of conditions where σ_3 is 10 or 20 MPa are different in the direction of σ_3 (Figure 12(c)). When σ_3 is 10 MPa, the lateral strain in the minimum principal stress direction ε_p^3 (expansion) of the rock peak

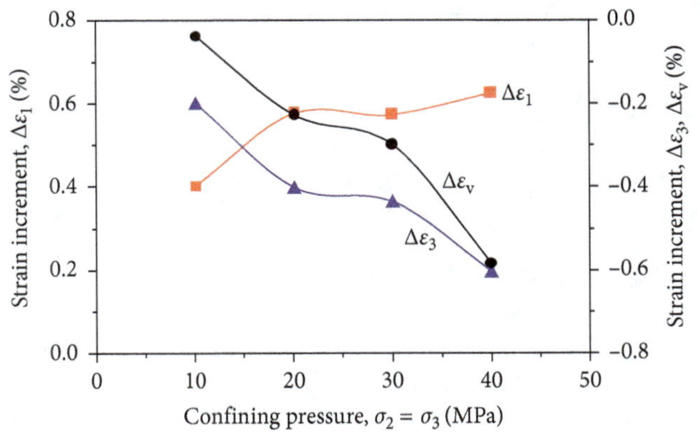

FIGURE 11: Variation in strain increment in the unstable crack growth stage under the conventional triaxial compression.

point first increases and then decreases with increasing σ_2. When σ_2 is 50 MPa or 60 MPa, the lateral strain exhibits a sharp decrease, indicating that the rock reaches the peak point at a fast speed and its ability to withstand deformation is significantly weakened. When σ_3 is 20 MPa, the lateral strain ε_p^3 (expansion) of the rock peak point first increases and then decreases with increasing σ_2. Here, while σ_2 is 60 MPa, the lateral strain decreases slightly, indicating that the lateral strain ε_p^3 of the peak point reduces the ability to withstand lateral strain. However, due to increasing σ_3, the rock suffers only a slight decrease in lateral deformation capacity. The law of rock volumetric strain is similar to that in the minimum principal stress direction (Figure 12(d)). When σ_3 is 10 MPa, the peak volumetric strain ε_p^v first increases and then decreases with increasing σ_2. When σ_3 is 20 MPa, the peak volumetric strain ε_p^v slightly decreases with increasing σ_2, whose value is higher than the former one (σ_3 = 10 MPa), which indicates that the rock's ability to support expansion and deformation could be improved by increasing σ_3.

The variation in strain increment in the unstable stage under true triaxial compression is shown in Figure 13. The axial strain increment $\Delta\varepsilon_1$ and lateral strain increment $\Delta\varepsilon_2$ in the intermediate principal stress direction in this stage are similar when σ_3 is 10 and 20 MPa and decreases with increasing σ_2. The difference in strain increment $\Delta\varepsilon_3$ in the direction of the minimum principal stress is large. Strain increments first increase and then decrease, but the former decreases significantly when σ_2 is 50 MPa and 60 MPa, which indicates that the ability to bear lateral deformation significantly decreases due to damage from confining pressure. The latter decreases slightly when σ_2 is 60 MPa, which indicates that the damage caused by confining pressure could be significantly weakened due to an increase in σ_3. The volumetric strain increment is similar to that in the direction of σ_3.

4. AE Behavior

AE is a phenomenon that microcracks generate during crack development, which is accompanied by the release of an elastic wave or stress wave. Deformation and failure of the

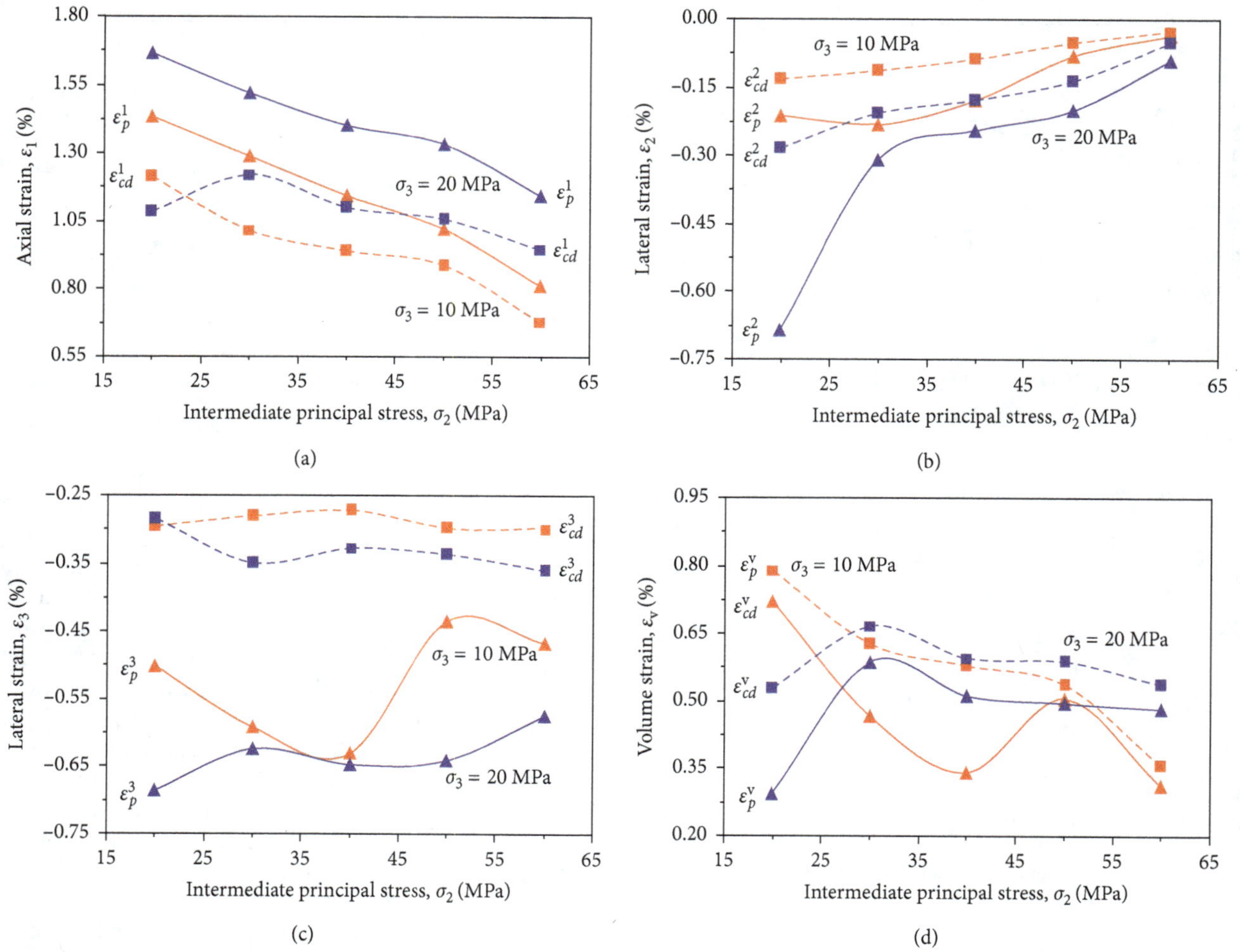

FIGURE 12: Strain characteristics of rock damage point and peak point under true triaxial compression: (a) axial strain; (b) lateral strain (the intermediate principal stress direction); (c) lateral strain (the minimum principal stress direction); (d) volumetric strain.

rock are accompanied by the generation of the AE. AE information contains precursor information for rock failure. The AE information during compression and failure of the rock is used to predict the rock failure process, which is of great importance to monitoring deformation in mines and other geotechnical engineering [23].

The typical curve of the cumulative AE events with the change of strain under different loading conditions is shown in Figure 14. The characteristics of the rock in unstable crack growth stage (the same below) are only selected in this study. AE events increase sharply in this stage, and interactions among cracks increase. AE events weaken gradually with increasing confining pressure under conventional triaxial compression, and the cumulative AE counts decrease gradually when the peak value is reached. With a confining pressure of 10 MPa, the cumulative AE counts are significantly higher than the one with higher confining pressure. Since the constraint on lateral deformation with low confining pressure under conventional triaxial compression is small, it is easy to produce new microcracks and generate slip along cracks, creating a more active acoustic emission signal. High confining pressure reduces the generation of internal fractures and shear slip, producing relatively low AE counts. The higher the confining pressure, the stronger the binding effect is, which enhances the rock's capacity to withstand deformation and loading.

AE count decreases first and then increases under true triaxial compression. When σ_3 is 10 MPa, AE events increase with increasing σ_2 (Figure 14(b)). When σ_2 is 20~40 MPa, the cumulative AE counts gradually reduce with increasing σ_2. There is no "sudden increase" phenomenon, and the rock produces many continuous AE counts. In combination with the AE characteristics under conventional triaxial compression, numerous shear slip cracks are produced in this stage (which would be analyzed in detail below), and microcracks converge and perforate within the rock, leading to the formation of macroshear and slip surfaces. When σ_2 is 50 and 60 MPa, the cumulative AE counts increase rapidly, and there is a "sudden increase" phenomenon indicating that the increase of σ_2 inhibits the deformation in the σ_2 direction but promotes the lateral expansion in the σ_3 direction. The rock produces many tensile cracks whose convergence and perforation cause fracture surfaces to form. The confining pressure gradually evolves from a constraining force to a damaging force, greatly weakening the carrying capacity of the rock. When σ_3 is 20 MPa, AE events change slightly with increasing σ_2. Here, when σ_2 is 30~50 MPa, the

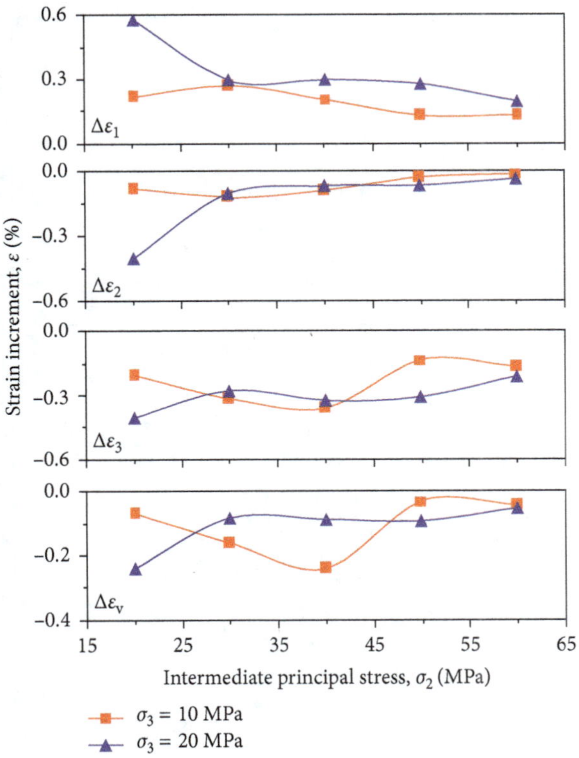

FIGURE 13: Variation in strain increment of rock in the unstable crack growth stage under true triaxial compression.

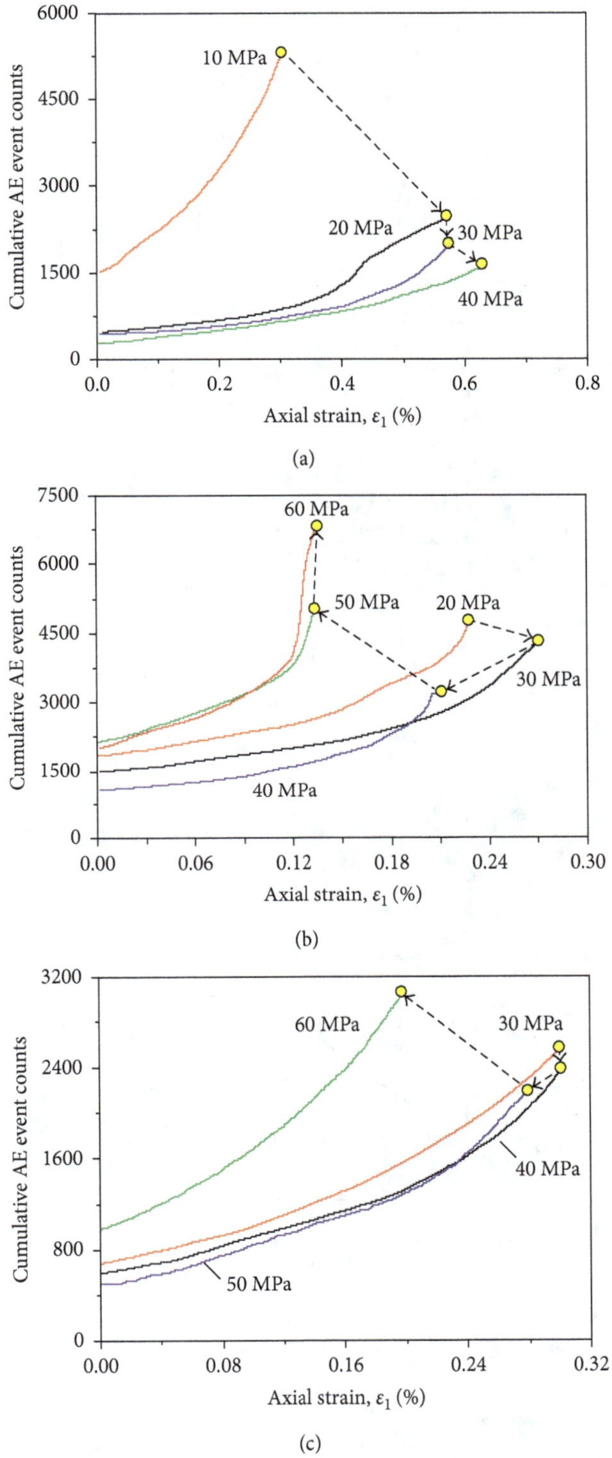

FIGURE 14: The typical curve of the cumulative number of AE events with the change in strain under different loading conditions: (a) CTT; (b) TTT (σ_3 = 10 MPa); (c) TTT (σ_3 = 20 MPa).

typical AE event curves (Figure 14(c)) are similar to each other, and more shear and slip cracks are formed. The constraining effect of confining pressure to internal cracks is enhanced, and the development of cracks is suppressed by increasing σ_2. When σ_2 is 60 MPa, the cumulative AE counts increase rapidly, but there is no "sudden increase" phenomenon, indicating that the confining pressure damages the rock and promotes the development of rock cracks. For σ_3 = 10 MPa, the damage effect is greatly weakened by improving σ_3 and the rock is protected.

In order to further analyze the destruction of sandstone under different loading conditions, the value of RA (the ratio of the rise time and the amplitude) and the average frequency value AF (the ratio of the ringing count and the duration) of the AE parameters are used to describe the type of microcracks within the rock. In general, AE signals with low AF and high RA or with high AF and low RA usually represent the generation or development of shear or tensile cracks, respectively [24–26]. Energy levels are combined to analyze the distribution law for RA under three kinds of AE levels which are 0.1~1, 1~10, and >10 mV^2·ms (ignore < 0.1) (Figure 15). Microcracks in the rock under the three loading conditions are stretched and expanded. The AE count which has an AE energy of over 10 mV^2·ms, 1~10 mV^2·ms, and 0.1~1 mV^2·ms is small, large, and dense. According to the energy level, the proportion rule of the rock shear cracks under different loading conditions is shown in Figure 16.

The proportion of the shear cracks under conventional triaxial compression with AE energy greater than 10 mV^2·ms varied little with increasing confining pressure, which is in 27%~32%. The proportion of shear cracks gradually increases (from 25% to 37%) with AE energy from 1 to 10 mV^2·ms. The proportion of the shear cracks is relatively small and irregular with AE energy varying from 0.1 to 1 mV^2·ms (only AE of more than 1 mV^2·ms would be analyzed in the following paper). Although a former study believed that the rock sample

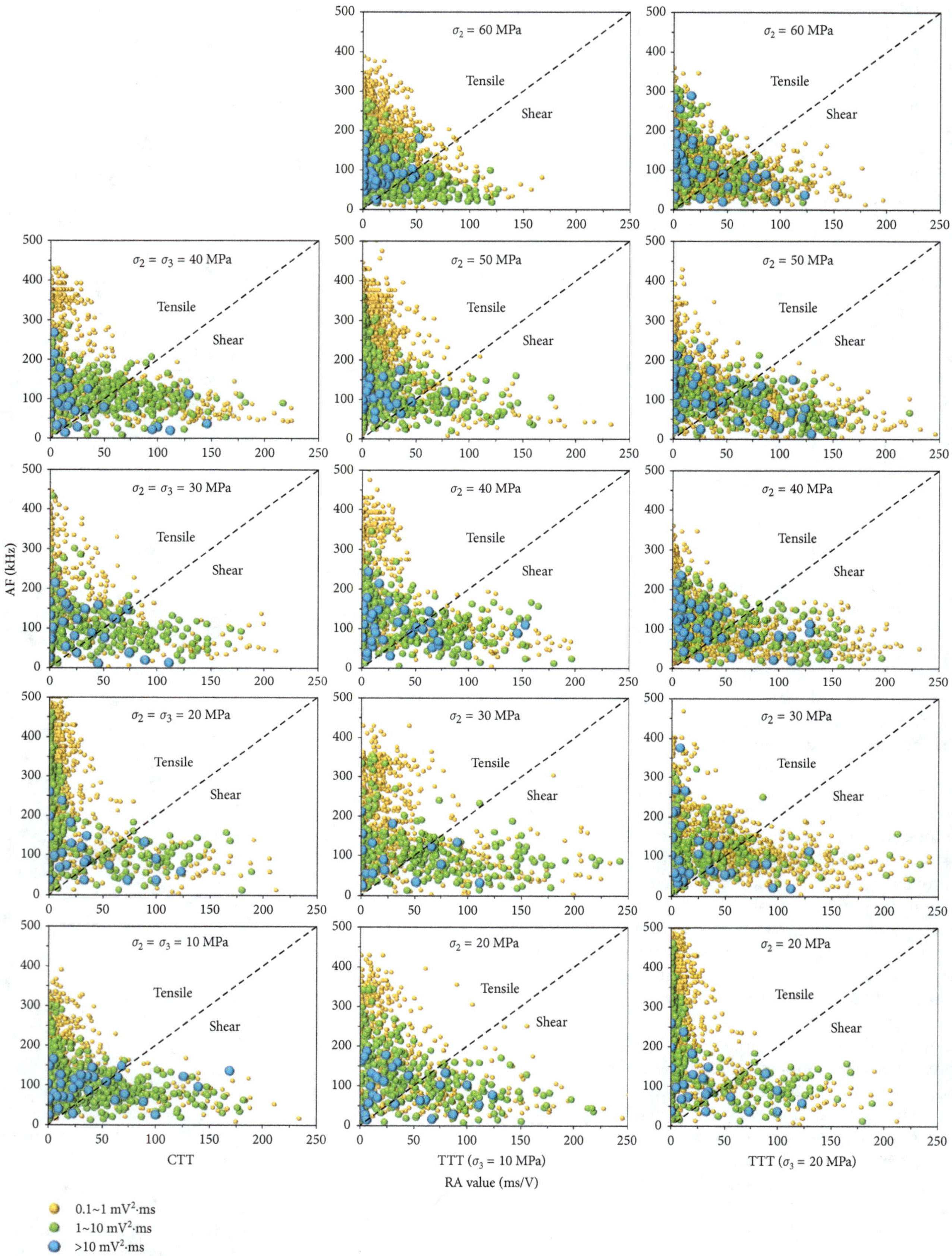

Figure 15: Variation of the RA and AF values under different loading conditions.

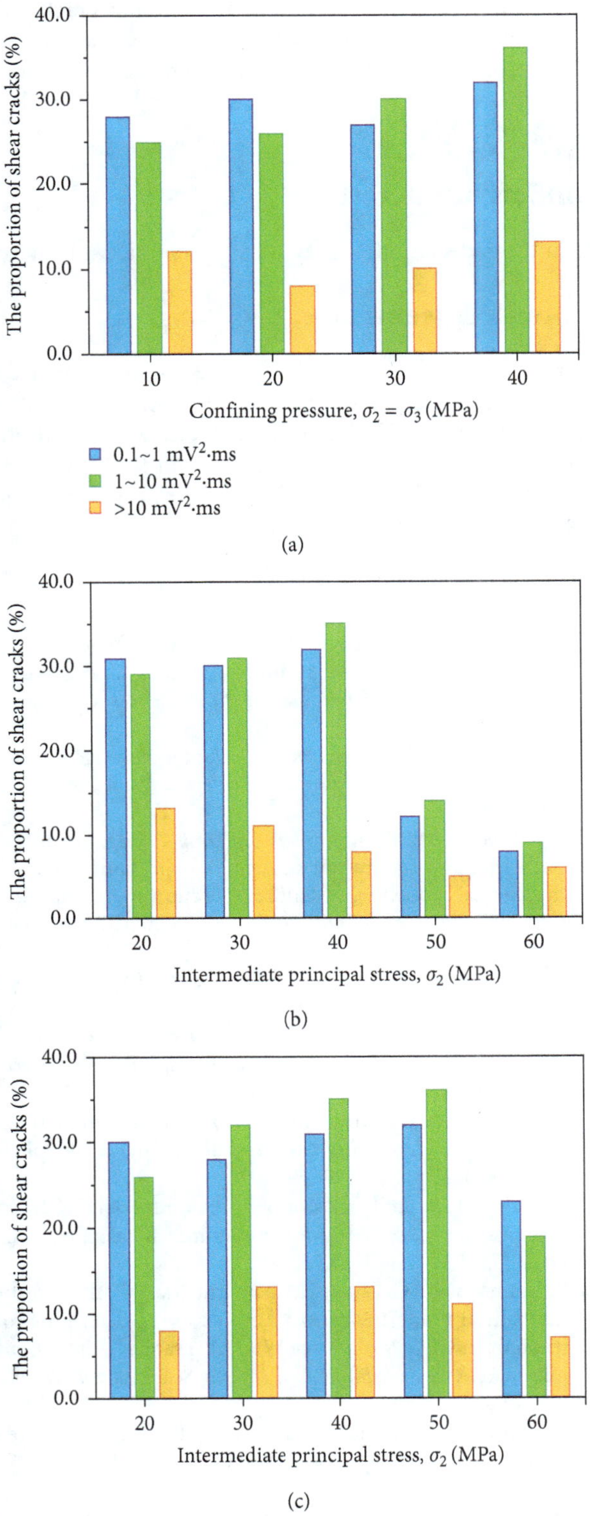

FIGURE 16: Proportion characteristics of shear cracks under different loading conditions: (a) CTT; (b) TTT (σ_3 = 10 MPa); (c) TTT (σ_3 = 20 MPa).

reflects the shear failure characteristics under conventional triaxial compression at the macrolevel, analyses in this study show that the macroshear failure of the rock samples at the microlevel is caused by tension, expansion, or a large number of connecting internal microcracks. The presence of shear cracks is often accompanied by a large number of tensile cracks, which is consistent with the existing literature [27, 28]. The cracks need to overcome the confining pressure. The law of the proportion of shear cracks in which AE is over 1 mV^2·ms increases with increasing confining pressure indicates that cracks gradually overcame the shear constraints of confining pressure. On the contrary, the shear slip is gradually inhibited due to increasing confining pressure and delaying penetration of the cracks and rupture of the rock surface, in affecting playing a protective role in rock.

There are significant differences in the change laws for RA under true triaxial compression in which σ_3 is 10 and 20 MPa. When σ_2 is 20~40 MPa, the former (σ_3 = 10 MPa) proportion of the shear cracks gradually increases (from 28% to 35%) with increasing $\Delta\varepsilon_3$ with AE energy greater than $1\Delta\varepsilon_v$. The role of confining pressure is similar to conventional triaxial compression. Slippage of rock cracks can effectively be inhibited by increasing σ_2, which delays the formation of macroshear and slip surfaces and plays a protective role in the rock. When σ_2 is 50 and 60 MPa, the proportion of shear cracks decreases dramatically (12% and 9% resp.) with its AE energy more than 1 mV^2·ms, indicating that the rock records mainly tension and expansion, accompanied by a small amount of shear cracks. Tension cracks within the rock are constantly fused, and the macroscopic tension and failure surface instantaneously forms, which explains why the cumulative AE counts appeared as a "sudden increase" (Figure 14(b)). The effect of confining pressure on the cracks within the rock gradually evolves from a constraining force to a damage force. The latter (σ_3 = 10 MPa) has a relatively smooth change in RA. When σ_2 is 20~50 MPa, the proportion of the shear cracks gradually increases (from 26% to 37%) with increasing σ_2 with AE energy of more than 1 mV^2·ms. By comparing with σ_3 = 10 MPa, the proportion of shear cracks increases, indicating that the slip effect may be inhibited by increasing σ_3 and the rock is restricted laterally. When σ_2 is 60 MPa, the proportion of shear cracks (20%) in which AE energy is more than 1 mV^2·ms decreases. It is significantly larger than the former proportion with conditions the same as σ_2, indicating that the damage effect of confining pressure is weakened because of increasing σ_3, although the rock is damaged by the confining pressure, to some extent, which led to the increase of the tensional cracks of the rock.

5. Conclusion

In this study, the conventional triaxial and true triaxial tests of sandstone under different stress states are carried out using the self-developed true triaxial electrohydraulic servo test system. Deformation, strength, AE characteristic parameters, and development of the internal cracks are studied under different confining pressures, intermediate principal stresses, and minimum principal stresses. The primary research results are summarized as follows:

(1) The strength of the sandstone increases with increasing confining pressure under conventional

triaxial compression, while the strength of sandstone increases first and then decreases with increasing intermediate principal stress under true triaxial compression. The strength under the true triaxial condition is consistent with the Mogi–Coulomb criterion. The rock is protected obviously, and the peak strength of the rock is greatly improved by increasing σ_3.

(2) The deformation characteristics of the rock under the two loading conditions are quite different. Under conventional triaxial compression, the axial strain and lateral strain as well as the elastic modulus increase with increasing confining pressure before failure, showing that the ability of the rock to withstand deformation capacity of the rock is gradually improved because of the increase in confining pressure under conventional triaxial compression. Under true triaxial compression (σ_3 equal 10, 20 MPa), the elastic modulus of the rock first increases and then decreases, and the axial strain ε_1 and lateral strain ε_2 decrease gradually with increasing σ_2. The lateral strain (swelling) of the rock is mainly along the σ_3 direction, and the tendency of the lateral expansion is to decrease first and then to increase. In the unstable crack growth stage, the lateral strain ε_p^3 and strain increment $\Delta\varepsilon_3$, $\Delta\varepsilon_V$ first increase and then decrease, but the former appears as "cliff" decrease when σ_2 is 50 and 60 MPa, while the latter decreases slightly when σ_2 is 60 MPa. The variation of strain increment $\Delta\varepsilon_3$, $\Delta\varepsilon_V$ in the stage is similar, indicating that when σ_2 is large, the former ability to bear lateral deformation is weakened significantly, while the latter can greatly reduce the role of confining pressure damage at this time and provide a protective effect to the rock due to increasing σ_3.

(3) In this study, the AE characteristics of rocks in unstable crack growth stage under conventional and true triaxial compression are analyzed. It is found that microcracks in the sandstone reflected mainly stretching and expansion under the two loading conditions (the characteristics of AE signals are high AF and low RA); however, the AE characteristics are significantly different. The trend of AE activity for rocks under conventional triaxial compression is "gradually reduced" with increasing confining pressure, and the shear and slip effect of the cracks is gradually inhibited. AE events first weaken and then enhance with increasing σ_2 under true triaxial compression in which σ_3 is 10 and 20 MPa. When the former σ_2 is 20~40 MPa, the cumulative AE counts gradually decrease with increasing σ_2, and the proportion of the shear cracks increases gradually. When σ_2 is 50 and 60 MPa, the cumulative AE counts increases rapidly, and there is a "sudden increase" phenomenon. The proportion of the shear cracks decreases dramatically, showing that the confining pressure gradually evolves from the shear and slip effect inhibiting shear cracks to damage the rock promoting the formation of tensional cracks with increasing σ_2, resulting in an instant of macrotension rupture formation. But, the damage of confining pressure is weakened by increasing σ_3 (the latter σ_3 = 20 MPa).

Conflicts of Interest

The authors declare that they have no conflicts of interest.

Acknowledgments

The authors gratefully acknowledge the support provided by the National Basic Research Program of China (Grant no. 2014CB046905), the National Natural Science Foundation of China (NSFC) (Grant no. 51404245), and the China Postdoctoral Science Foundation Funded Project (Grant nos. 2014M551699 and 2016T90527).

References

[1] H. Moomivand, "Effects of orientation, frequency, and number of sets of discontinuities on rock strength under triaxial stresses," *Arabian Journal of Geosciences*, vol. 7, no. 12, pp. 5345–5352, 2014.

[2] T. V. Karman, "Festigkeitsversuche unter allseitigem druck," *Mitteilungen Forschungsarbeit Gebiete Ingenieurs*, vol. 55, pp. 1749–1759, 1911.

[3] J. T. Fredrich, B. Evans, and T. F. Wong, "Effect of grain size on brittle and semibrittle strength: implications for micromechanical modelling of failure in compression," *Journal of Geophysical Research Solid Earth*, vol. 95, no. B7, pp. 10907–10920, 1990.

[4] S. Q. Yang, H. W. Jing, and S. Y. Wang, "Experimental investigation on the strength, deformability, failure behavior and acoustic emission locations of red sandstone under triaxial compression," *Rock Mechanics and Rock Engineering*, vol. 45, no. 4, pp. 583–606, 2012.

[5] K. Mogi, "Fracture and flow of rocks under high triaxial compression," *Journal of Geophysical Research Atmospheres*, vol. 76, no. 5, pp. 1255–1269, 1971.

[6] K. Mogi, "Flow and fracture of rocks under general triaxial compression," *Applied Mathematics and Mechanics*, vol. 6, no. 2, pp. 635–651, 1981.

[7] B. Haimson and C. Chang, "A new true triaxial cell for testing mechanical properties of rock, and its use to determine rock strength and deformability of Westerly granite," *International Journal of Rock Mechanics and Mining Sciences*, vol. 37, no. 1, pp. 285–296, 2000.

[8] B. Haimson, "True triaxial stresses and the brittle fracture of rock," *Pure and Applied Geophysics*, vol. 163, no. 5-6, pp. 1101–1130, 2006.

[9] X. T. Feng, X. Zhang, R. Kong, and G. Wang, "A Novel Mogi type true triaxial testing apparatus and its use to obtain complete stress–strain curves of hard rocks," *Rock Mechanics and Rock Engineering*, vol. 49, no. 5, pp. 1–14, 2015.

[10] T. Sriapai, C. Walsri, and K. Fuenkajorn, "True-triaxial compressive strength of Maha Sarakham salt," *International Journal of Rock Mechanics and Mining Sciences*, vol. 61, no. 10, pp. 256–265, 2013.

[11] M. You, "True-triaxial strength criteria for rock," *International Journal of Rock Mechanics and Mining Sciences*, vol. 46, no. 1, pp. 115–127, 2009.

[12] X. Ma, J. W. Rudnicki, and B. C. Haimson, "The application of a Matsuoka-Nakai-Lade-Duncan failure criterion to two porous sandstones," *International Journal of Rock Mechanics and Mining Sciences*, vol. 92, pp. 9–18, 2017.
[13] Z. A. Moradian, G. Ballivy, P. Rivard, C. Gravel, and B. Rousseau, "Evaluating damage during shear tests of rock joints using acoustic emissions," *International Journal of Rock Mechanics and Mining Sciences*, vol. 47, no. 4, pp. 590–598, 2010.
[14] S. Q. Yang, H. M. Ni, and S. Wen, "Spatial acoustic emission evolution of red sandstone during multi-stage triaxial deformation," *Journal of Central South University*, vol. 21, no. 8, pp. 3316–3326, 2014.
[15] V. Vishal, P. G. Ranjith, and T. N. Singh, "An experimental investigation on behaviour of coal under fluid saturation, using acoustic emission," *Journal of Natural Gas Science and Engineering*, vol. 22, pp. 428–436, 2015.
[16] J. Yin, C. Cheng, M. Kumruzzaman, and W. Zhou, "New mixed boundary, true triaxial loading device for testing three-dimensional stress-strain-strength behaviour of geomaterials," *Canadian Geotechnical Journal*, vol. 47, no. 1, pp. 1–15, 2010.
[17] J. F. Labuz and J. M. Bridell, "Reducing frictional constraint in compression testing through lubrication," *International Journal of Rock Mechanics and Mining Science*, vol. 30, no. 4, pp. 451–455, 1993.
[18] A. M. Al-Ajmi and R. W. Zimmerman, "Relation between the Mogi and the Coulomb failure criteria," *International Journal of Rock Mechanics and Mining Sciences*, vol. 42, no. 3, pp. 431–439, 2005.
[19] H. Oku, B. Haimson, and S. R. Song, "True triaxial strength and deformability of the siltstone overlying the Chelungpu fault (Chi-Chi earthquake), Taiwan," *Geophysical Research Letters*, vol. 34, no. 9, pp. 139–158, 2007.
[20] H. Lee and B. C. Haimson, "True triaxial strength, deformability, and brittle failure of granodiorite from the San Andreas Fault Observatory at depth," *International Journal of Rock Mechanics and Mining Sciences*, vol. 48, no. 7, pp. 1199–1207, 2011.
[21] C. Chang and B. Haimson, "A failure criterion for rocks based on true triaxial testing," *Rock Mechanics and Rock Engineering*, vol. 45, no. 6, pp. 1007–1010, 2012.
[22] C. D. Martin and N. A. Chandler, "The progressive fracture of Lac du Bonnet granite," *International Journal of Rock Mechanics and Mining Science*, vol. 31, no. 6, pp. 643–659, 1994.
[23] X. Kong, E. Wang, S. Hu, R. Shen, X. Li, and T. Zhan, "Fractal characteristics and acoustic emission of coal containing methane in triaxial compression failure," *Journal of Applied Geophysics*, vol. 124, pp. 139–147, 2016.
[24] D. G. Aggelis, E. Z. Kordatos, and T. E. Matikas, "Acoustic emission for fatigue damage characterization in metal plates," *Mechanics Research Communications*, vol. 38, no. 2, pp. 106–110, 2011.
[25] S. Shahidan, R. Pulin, N. M. Bunnori, and K. M. Holford, "Damage classification in reinforced concrete beam by acoustic emission signal analysis," *Construction and Building Materials*, vol. 45, no. 13, pp. 78–86, 2013.
[26] E. Tsangouri, D. Aggelis, T. Matikas, and A. Mpalaskas, "Acoustic emission activity for characterizing fracture of marble under bending," *Applied Sciences*, vol. 6, no. 1, p. 6, 2015.
[27] K. B. Broberg, "On crack paths," *Engineering Fracture Mechanics*, vol. 28, no. 5-6, pp. 663–679, 1987.
[28] Y. L. Lu, D. Elsworth, and L. G. Wang, "Microcrack-based coupled damage and flow modeling of fracturing evolution in permeable brittle rocks," *Computers and Geotechnics*, vol. 49, no. 4, pp. 226–244, 2013.

7

Electric Resistance Tests on Compacted Clay Material under Dynamic Load Coupled with Dry-Wet Cycling

Zheng Lu,[1] **Xiaowen Wu**,[1,2] **Zhi Hu**,[2] **Shaohua Xian**,[1] **and Ran Fang**[1]

[1]*State Key Laboratory of Geomechanics and Geotechnical Engineering, Institute of Rock and Soil Mechanics, Chinese Academy of Sciences, Wuhan 430071, China*
[2]*School of Civil Engineering, Architecture and Environment, Hubei University of Technology, Wuhan 430068, China*

Correspondence should be addressed to Zheng Lu; lzwhrsm@163.com

Academic Editor: João M. P. Q. Delgado

The study of compacted clay material under dynamic load coupled with dry-wet cycling is one of the most important areas in the field of transportation. In this paper, experiments in terms of compacted clay under dynamic load coupled with dry-wet cycling are performed, and synchronous resistivity tests are also conducted. According to the test results, the influences of cumulative plastic strain, dry-wet cycles, and amplitudes on the soil resistivity are analyzed. Then a new damage factor based on resistivity is proposed to evaluate the long-term performance of compacted clay material. The result of research shows that the evolution of the soil resistivity can be divided into two stages, which has a contrary tendency with that of cumulative plastic strain. The dry-wet cycles and amplitudes have a significant effect on the damage of the compacted soil, which indicates that the dry-wet cycling of compacted soil materials should not be ignored in road engineering, especially in rainy and humid areas.

1. Introduction

Roadways and railways commonly consist of upper pavement or ballast layers (track slab) over one or more silty clay layers which are together compacted over a suitable soil subgrade. As it is well known, compacted clay material provides the most important structural element in the road system because, in most cases, the cumulative deformation phenomenon takes place mainly in the subgrade layer causing progressive fatigue cracking of upper road layers [1–3]. So, it is essential to control the long-term performance of the subgrade so as to carry traffic safely, conveniently, and economically during its entire lifespan. However, the subgrade suffers from the humidity change and variation of temperature for a long term; besides, the cyclic traffic loads and other climate also bring adverse effects on it. Therefore, it is of great significance to study the long-term performance and damage laws of compacted clay material under the comprehensive effect of dynamic load and dry-wet cycling.

The damage of compacted soil is usually accompanied by the change of soil structure, and the electric resistance can reflect the feature of the composition of soil particles and pore structure well. Thus, resistivity can be used in the study of soil structure change and structural damage [4]. Archie [5] first studied the relationship between soil resistivity and its structure and proposed the soil structural factor-resistivity structure factor F (formation resistivity factor). Then, he established the exponential function relation between resistivity structure factor and sandy soil porosity. Miao et al. [6, 7] investigated the resistivity characteristic of cement-stabilized soil; the relationship between soil resistivity and unconfined compressive strength, cement mixing ratio, and age were obtained. Fukue and Liu et al. [8–10] developed a geotechnical resistivity test system and established the resistivity model of expansive soil and soft soil structure, and they put forward an average apparent structure factor concept for this type of soil. Subsequently, Lin et al. [11] conducted indoor experiments on the resistivity of loess, on the basis of which the influence of contact pressure, temperature, and measurement frequency on resistivity was studied and the variation rule of the resistivity of loess was analyzed. However, the literature on

TABLE 1: Material properties of the compacted clay.

Natural water content (%)	Natural density (g/cm^3)	Saturated water content (%)	Plastic limit (%)	Liquid limit (%)	Specific gravity	Maximum dry density (g/cm^3)	Optimum water content (%)	Permeability (cm/s)
23.7	1.68	23.8	21.1	34.8	2.72	1.90	14.1	6.5e-7

TABLE 2: Testing program of dynamic load coupled with dry-wet cycling.

Parameters	Values
Initial water content (%)	14.1
Number of drying-wetting cycles	0, 2, 4, 9
Amplitude of drying-wetting cycles (±%)	0, 2, 4, 6
Initial static deviatoric stress (kPa)	20
Confining pressure (kPa)	40
Amplitude of dynamic stress (kPa)	40
Loading frequency (Hz)	5
Number of loading cycles	10^5

electric resistance tests on compacted clay material under dynamic load coupled with dry-wet cycling is rarely published. Therefore, experiments of dynamic load coupled with dry-wet cycling are performed in this paper, and the synchronous resistivity tests on compacted clay material are conducted at the same time. According to the experimental results, the evolution of resistivity with cumulative plastic strain, dry-wet cycles, and amplitudes is analyzed, and then, a damage factor based on resistivity is proposed to evaluate the long-term performance of compacted clay material. The study can provide a reference for the damage assessment of the compacted subgrade soil under complex environment and traffic loads.

2. Experimental Procedures

The experimental clay material was taken from Sanmenxia along the Zhengxi Passenger Railway in China. The main material characteristics are listed in Table 1. All the test samples are reshaped in a cylinder-shaped sampler of 76 mm in height and 38 mm in diameter. The samples are compacted to an initial dry density of 1.81 g/cm^3 to meet the compaction requirement's degree of the specification [12].

In order to simulate the actual service condition of the road subgrade under traffic loads and climate changes, dynamic triaxial tests are conducted on the samples subjected to different dry-wet cycles and amplitudes. Aimed at simulating the field stress state of subgrade soil, the bias consolidation of the soil samples is carried out under a certain confining pressure for 15 hours before starting the triaxial tests. The experimental scheme for the dynamic load coupled with dry-wet cycling is designed and shown in Table 2. In fact, the testing program shown in Table 2 is designed according to the practical road subgrades subjected to traffic loads and climate changes. It is reported that the mechanical properties of compacted clay soil could finally tend to stabilize after 8~10 dry-wet cycles. Thus, the maximum number of drying-wetting cycles is fixed at 9 in this research. Moreover, in terms of clay soil material, the maximum water content of subgrades under climate changes is usually no more than OWC + 6% (here, OWC means the optimum water content of subgrades). The maximum amplitude of drying-wetting cycles is fixed at 6% in the experiment.

For the purpose of describing the damage state of compacted clay material under dynamic load coupled with dry-wet cycling, the soil sample resistivity is tested simultaneously with the dynamic triaxial test in this paper. According to the principle of measurement, the resistivity test method can be divided into the dipole method and the quadrupole method [13], and the two kind of measuring devices are shown in Figure 1. The quadrupole method provides a certain voltage through a type AB electrode, the resistance of soil sample can be obtained by the means of measuring the voltage of MN electrode and the current in the pathway, and then the soil sample resistivity will be obtained through the resistivity calculation formula. The dipole method is a special case that the electrode MN is shared with the electrode AB in the quadrupole method. When using the dipole method, the contact resistance will be greater because of the poor contact with the electrode and the soil sample; therefore, contact resistance size should be determined before the formal test [14]. The quadrupole method has the advantages of not considering the contact resistance compared to the dipole method, but the measuring electrode usually needs to be inserted into the soil sample. Thus, it is not suitable for the measurement of resistivity for which soil sample integrity requirements are high. For this purpose, the dynamic triaxial apparatus is upgraded and modified to satisfy the synchronous test requirements of dynamic characteristics and resistivity of the samples, as shown in Figure 2. In terms of the soil sample integrity, circuit installation, and operability of instrument, the dipole method is adopted to measure the soil sample resistivity in this paper. The resistivity is collected by WDJD-4 multifunctional digital DC exciter, as shown in Figure 3.

It is noted that temperature can affect the soil resistivity significantly; thus, the measured resistivity needs to be corrected. In this paper, the resistivity tests are firstly carried out at different temperatures to derive the temperature correction curve, as shown in Figure 4. In order to eliminate the effect of temperature, the resistivity of the soil sample in the standard state is adopted uniformly in the following analysis. According to the test results, the relationship between the resistivity of the standard state and the measured resistivity can be described by using the following formula:

$$\rho_{25} = \rho_T [1 + \alpha (T - 25)], \quad (1)$$

where ρ_{25} and ρ_T are the soil sample resistivity of the standard state (25°C) and the measured temperature, T (°C), respectively. T is the experimental temperature, α is the

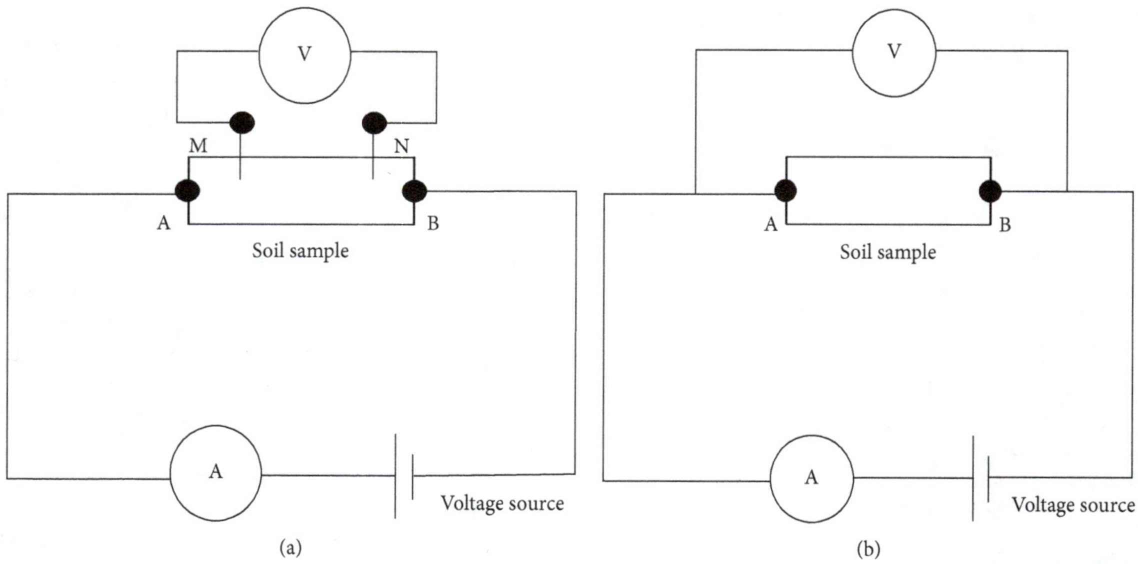

Figure 1: Resistivity test method for the soil samples: (a) the quadrupole method and (b) the dipole method.

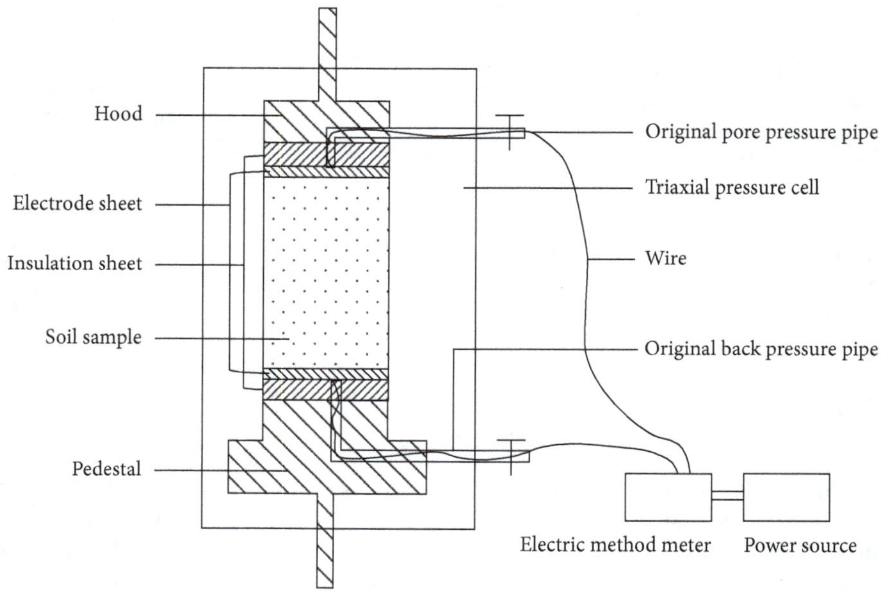

Figure 2: Schematic diagram of the soil sample resistivity test on the dynamic triaxial apparatus.

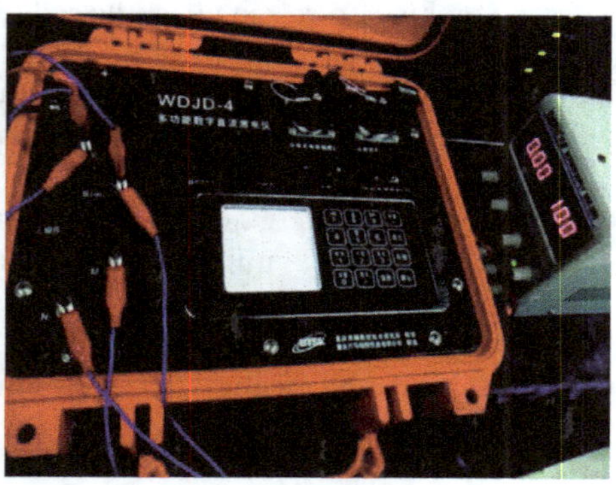

Figure 3: Soil resistivity acquisition instrument.

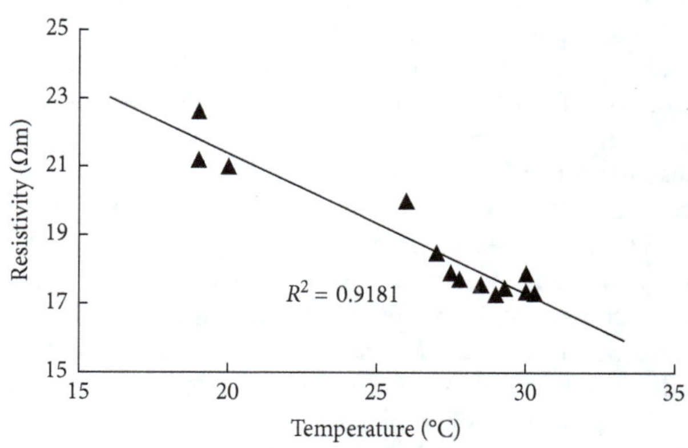

Figure 4: Temperature correction curve for soil sample resistivity.

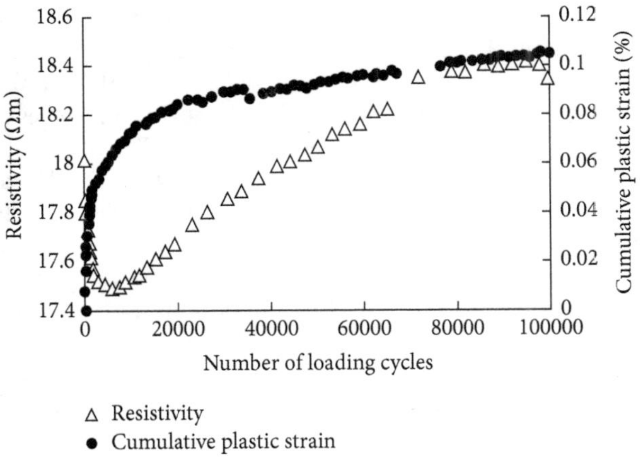

FIGURE 5: Variations of the soil sample resistivity and cumulative plastic strain with the number of loading cycles.

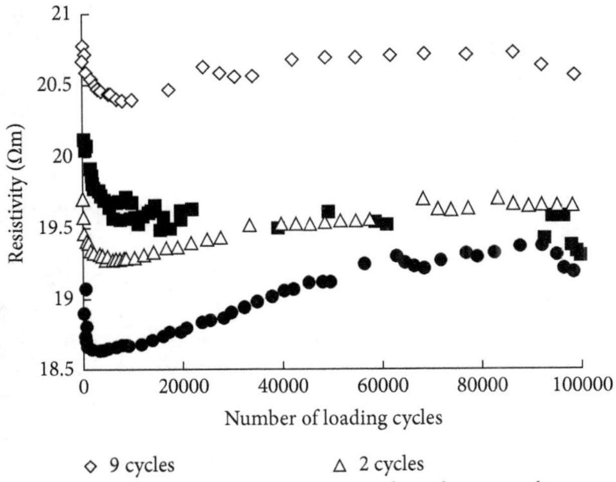

FIGURE 6: Effect of dry-wet cycles on the soil sample resistivity.

temperature correction coefficient, and $\alpha = 0.0207^{\circ}C^{-1}$ in this paper based on the test results.

The measured resistivity is calculated by the following equation:

$$\rho_T = \frac{\left(\Delta U - \Delta U_{sp}\right)S}{IL}, \tag{2}$$

where ΔU is the voltage between the two electrodes, ΔU_{sp} is the initial potential difference, I is the current value of the soil sample, and S and L are the surface area and height of the soil sample, respectively.

3. Results and Discussion

Figure 5 shows the variations of the soil sample resistivity and cumulative plastic strain with the number of loading cycles. It can be seen from Figure 5 that the evolution of the resistivity can be divided into two stages. The first stage is the early stage of cyclic loading: the cumulative plastic strain of soil sample increases rapidly and the resistivity decreases greatly. With the growth of the accumulative plastic strain,

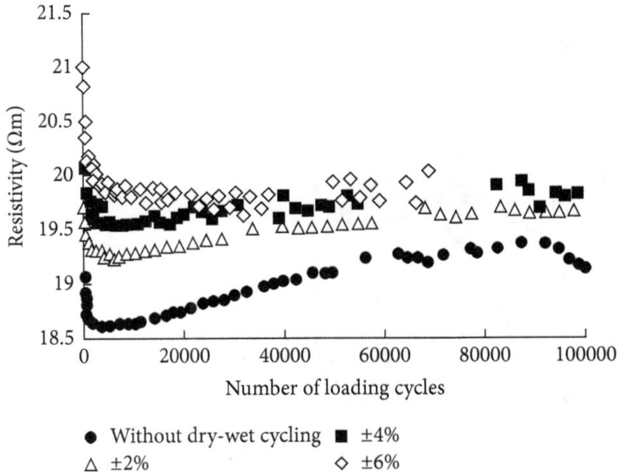

FIGURE 7: Effect of dry-wet cycling amplitudes on the soil sample resistivity.

the soil sample resistivity further reduces and then reaches the minimum value at a certain number of loading cycles. The second stage is the late cyclic loading: the growth rate of the accumulative plastic strain continues to decrease, and the values of accumulative plastic strain gradually stabilize. However, the soil sample resistivity increases inversely with the increasing loading cycles. At the first stage, the native micropores and microfractures of the soil sample gradually close because of the initial loadings, which causes larger soil compactness and better interconnectedness of pore water, thus leading to the increase of conductive passage in the soil sample [6]. When the resistivity reaches the minimum, the repair of the internal defect of the soil sample reaches its limit. At the second stage, the resistivity of the soil sample increases inversely, which indicates that the cyclic load causes new micropores or microcracks. The above results show that the resistivity can reflect the damage state and the process of damage evolution of the soil sample.

Figure 6 presents the variations of the soil sample resistivity with the number of loading cycles for different dry-wet cycles. It can be seen that the soil sample resistivity decreases initially, increases with the increasing number of loading cycles, and finally tends to stabilize, which are consistent with the rule of Figure 5. It also can be observed from Figure 6 that the soil sample resistivity increases gradually with the increasing dry-wet cycles, which indicates that the repeated dry-wet cycling can cause internal damage of the soil sample, thus leading to the density decrease. The effects of the dry-wet cycling amplitude on the soil sample resistivity are investigated in Figure 7. It can be seen from Figure 7 that the soil sample resistivity increases with the increase of the dry-wet cycling amplitude. This indicates that the different dry-wet cycling amplitudes can also cause internal damage and reduce compactness of the clay soil and thus results in a strength decreasing, which is consistent with those results obtained by Liu et al. [15, 16]. Therefore, the resistivity can be used to determine the strength of compacted soil materials indirectly, and then the internal damage of the subgrade under complex environment and

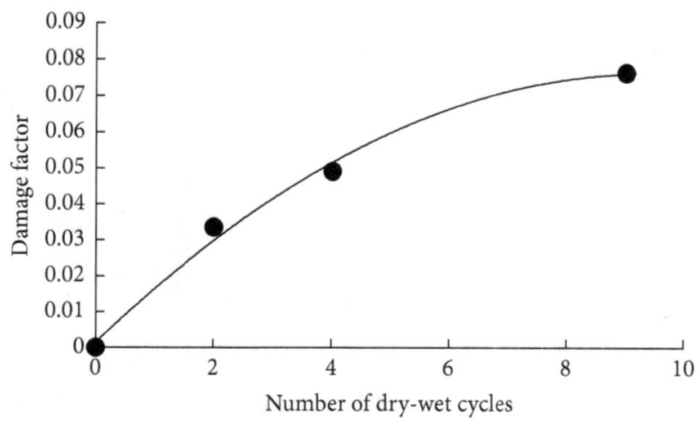

FIGURE 8: Variation of damage factor with the dry-wet cycles.

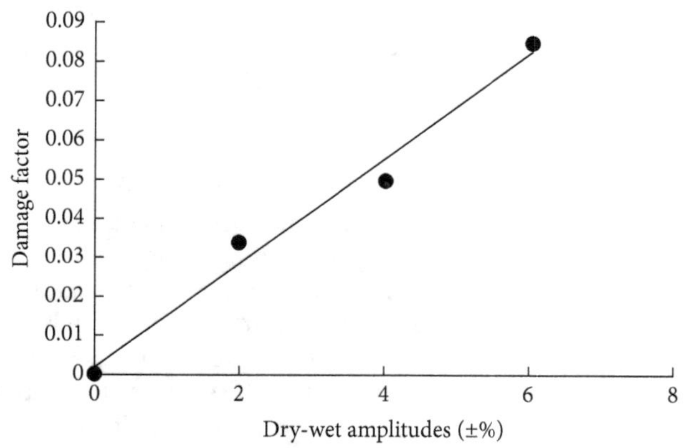

FIGURE 9: Variation of damage factor with the dry-wet amplitudes.

load can also be evaluated by adopting the resistivity measuring method.

In order to evaluate the damage property of compacted clay soil materials quantitatively using the resistivity measuring method, a damage factor based on resistivity is defined by the following equation:

$$D_\rho = 1 - \frac{\rho_0}{\rho}, \tag{3}$$

where D_ρ is the soil damage factor based on resistivity. Here, D_ρ is defined as a parameter to indicate the soil degradation. The bigger the parameter, the more serious the damage is. ρ_0 is the resistivity of the nondestructive soil sample. Here, the samples not experiencing the dynamic load coupled with dry-wet cycling are considered as nondestructive soil. ρ is the resistivity of the damage soil.

According to the above definition, the resistivity of the nondestructive soil sample can be measured, $\rho_0 = 19.07\ \Omega\text{m}$ in this paper. By employing (3), the damage characteristics of compacted soil materials under the dynamic load coupled with dry-wet cycling can be evaluated and analyzed. The measured sample resistivity increases with the increasing number and amplitude of drying-wetting cycles, but no more than 20.95 Ωm in this paper. Therefore, the values of the damage factor defined in (3) vary from 0 to 0.1, as far as Sanmenxia clay material is concerned. The variations in the damage factor of compacted soil with the dry-wet cycles and amplitudes are presented in Figures 8 and 9, respectively. As can be seen from Figure 8 that when there is a small number of dry-wet cycles, the damage factor increases dramatically with increasing dry-wet cycles. However, when the number of dry-wet cycles exceeds 4, the effect of the dry-wet cycles on the damage factor becomes relatively slight. This indicates that the long-term performance of compacted soil will tend to be stable after a certain number of dry-wet cycles. It can be observed from Figure 9 that the damage factor increases almost linearly with the increase of the dry-wet amplitudes. This result reveals that the dry-wet amplitudes have a significant effect on the damage of the compacted soil. The dry-wet cycling of compacted soil materials should not be ignored in road engineering, especially in rainy and humid areas.

4. Conclusions

In this paper, a series of experiments of dynamic load coupled with dry-wet cycling are performed, and the synchronous resistivity tests on compacted clay material are conducted. The influences of cumulative plastic strain, dry-wet cycles, and amplitudes on the soil resistivity are analyzed. Furthermore, a new damage factor based on resistivity is proposed to evaluate the long-term performance of compacted clay material. The main conclusions of this study can be summarized as follows:

(1) The evolution of the soil resistivity can be divided into two stages. The first stage is the early stage of cyclic loading: the cumulative plastic strain of soil sample increases rapidly and the resistivity decreases greatly. The second stage is the late cyclic loading: the soil sample resistivity increases inversely with the increasing loading cycles. It shows that the resistivity can reflect the damage state and the process of damage evolution of compacted clay material.

(2) The repeated dry-wet cycling can cause internal damage of the soil sample, thus leading to the density decrease. Therefore, the resistivity can be used to determine the strength of compacted soil materials indirectly, and then the internal damage of the subgrade under complex environment and load can also be evaluated by adopting the resistivity measuring method.

(3) The dry-wet cycles and amplitudes have a significant effect on the damage of the compacted soil, which indicates that the dry-wet cycling of compacted soil materials should not be ignored in road engineering, especially in rainy and humid areas.

Conflicts of Interest

The authors declare that they have no conflicts of interest.

Acknowledgments

This work was supported by the Youth Innovation Promotion Association, CAS, the outstanding youth fund of Hubei Province (2017CFA056), the Technology Service

Network Initiative (no. KFJ-STS-ZDTP-037), and the National Natural Science Foundation of China (nos. 41472286, 41472290, and 41672312).

References

[1] N. D. Beskou and D. D. Theodorakopoulos, "Dynamic effects of moving loads on road pavements: a review," *Soil Dynamics and Earthquake Engineering*, vol. 31, no. 4, pp. 547–567, 2011.

[2] A. Gomes Correia and J. Cunha, "Analysis of nonlinear soil modelling in the subgrade and rail track responses under HST," *Transportation Geotechnics*, vol. 1, no. 4, pp. 147–156, 2014.

[3] Z. Lu, H. Yao, J. Liu, and Z. Hu, "Experimental evaluation and theoretical analysis of multi-layered road cumulative deformation under dynamic loads," *Road Materials and Pavements Design*, vol. 15, no. 1, pp. 35–54, 2014.

[4] H. E. Stewart, "Permanent strains from cyclic variable-amplitude loadings," *Journal of Geotechnical Engineering*, vol. 112, no. 6, pp. 646–660, 1986.

[5] G. E. Archie, "The electrical resistivity log as an aid in determining some reservoir characteristics," *Transactions of the AIME*, vol. 146, no. 1, pp. 54–62, 1942.

[6] L. Miao, S. Liu, M. Yan et al., "Research on resistivity characteristics of water soil," *Engineering Survey*, vol. 5, pp. 32–34, 2000.

[7] S. Jie, L. Shucai, L. Bin, X. Xinji, W. Chuanwu, and N. Lichao, "Quantitative evaluation method of unsaturated soil compaction based on resistivity characteristics," *Journal of Chang'an University: Natural Science Edition*, vol. 35, no. 6, pp. 33–41, 2015.

[8] M. Fukue, T. Minato, H. Horibe, and N. Taya, "The microstructures of clay given by resistivity measurements," *Engineering Geology*, vol. 54, no. 1-2, pp. 43–53, 1999.

[9] S. Liu, S. Zhan, and X. J. Yu, "Research on the resistivity of soil test," *Journal of Engineering Geology*, vol. 14, no. 2, pp. 216–222, 2006.

[10] S. Chunni, F. Xiang, W. Hewen, S. Shuguo, and G. Jianfeng, "The influence of suction, moisture content and dry density on the shear strength of unsaturated soil," *Rock and Soil Mechanics*, vol. 30, no. 5, pp. 1347–1351, 2009.

[11] W. Lin hao, B. Xiaohong, and F. Junqin, "Discussion on the influencing factors of the shear strength index of compacted loess soil," *Journal of Geotechnical Engineering*, vol. 32, no. S2, pp. 132–135, 2010.

[12] Ministry of Communications of the China, *JTG D30-2015, Specifications for Design of Highway Subgrades*, China Communication Press, Beijing, China, 2015.

[13] L. Miao, S. Liu, M. Yan et al., "Resistivity characteristics of water clay and its engineering application," *Journal of Rock Mechanics and Engineering*, vol. 20, no. 1, pp. 126–130, 2001.

[14] X. Yu, *Research on the Application of Soil Mechanics of Resistivity Structure Model*, Southeast University, Nanjing, China, 2004.

[15] S. Liu, L. Han, and Y. Du, "The resistivity characteristics and application of water soil," *Journal of Geotechnical Engineering*, vol. 28, no. 11, pp. 1921–1926, 2006.

[16] Z. Liu, Y. Zhang, W. Fang et al., "Research on the relationship between the resistivity of loess and its compaction characteristics," *Journal of Xi'an University of Science and Technology*, vol. 33, no. 1, pp. 84–90, 2013.

8

Intelligent Machine Vision Based Modeling and Positioning System in Sand Casting Process

Shahid Ikramullah Butt,[1] Umer Asgher,[1] Umar Mushtaq,[1] Riaz Ahmed,[2] Faping Zhang,[3] Yasar Ayaz,[1] Mohsin Jamil,[1] and Muhammad Kamal Amjad[1]

[1]*School of Mechanical & Manufacturing Engineering, National University of Sciences and Technology (NUST), Islamabad, Pakistan*
[2]*Directorate of Quality Assurance, National University of Sciences and Technology (NUST), Islamabad, Pakistan*
[3]*School of Mechanical Engineering, Beijing Institution of Technology, Beijing, China*

Correspondence should be addressed to Shahid Ikramullah Butt; drshahid@smme.nust.edu.pk

Academic Editor: Charles C. Sorrell

Advanced vision solutions enable manufacturers in the technology sector to reconcile both competitive and regulatory concerns and address the need for immaculate fault detection and quality assurance. The modern manufacturing has completely shifted from the manual inspections to the machine assisted vision inspection methodology. Furthermore, the research outcomes in industrial automation have revolutionized the whole product development strategy. The purpose of this research paper is to introduce a new scheme of automation in the sand casting process by means of machine vision based technology for mold positioning. Automation has been achieved by developing a novel system in which casting molds of different sizes, having different pouring cup location and radius, position themselves in front of the induction furnace such that the center of pouring cup comes directly beneath the pouring point of furnace. The coordinates of the center of pouring cup are found by using computer vision algorithms. The output is then transferred to a microcontroller which controls the alignment mechanism on which the mold is placed at the optimum location.

1. Introduction

The casting process is used for forming metals into the desired shapes and sizes. It involves pouring boiling hot metal from a ladle into a mold by the workers. Handling molten metals at such a high temperature is an extremely delicate task and requires a lot of skill. Accidents occurring during the casting process can result in severe injuries or even death of the skilled workers. Therefore, a lot of research has been done to introduce automation in the casting process to improve the safety at the working environment and ergonomics. Some of the efforts in this area can be seen in the work by different scholars [1, 2]. In this paper, a novel scheme is presented for introducing automation in the sand casting process. There may be two categories of automation in sand casting process as follows:

(i) The automation of the metal pouring, where a lot of research and development has already been done in this area.

(ii) To develop such a system in which casting molds of different sizes, having different pouring cup location and radius, position themselves in front of the induction furnace such that the center of pouring cup comes directly beneath the pouring point of furnace (very latest research is being carried out in this area).

The idea of sand casting automation is a new concept and the same has been successfully implemented in this research project. The process includes a vision based system, in which a single camera takes the image of the casting mold, Harris Corner Detector technique is used for detecting a reference point on the pallet (on which the mold is placed), then the pouring cup center is found using Hough Transform, explained in [3–5], and the data is transferred to the controller which controls the alignment mechanism of the mold. To verify the concept a small scale model of the system has also been developed; it consists of three main parts which include design and fabrication of the mechanical model, design and fabrication of electronics hardware, and vision based algorithm and microcontroller algorithm. All the parts of the small scale model have been successfully completed and integrated to produce an automated mold positioning system.

The rest of the paper has been arranged as follows. Section 2 deals with the implemented vision system methodology, Section 3 addresses the design of the test platform, and Section 4 describes experimental results and analysis. Discussion and conclusions are drawn in Section 5.

Many studies have been conducted to introduce automation in the process of casting. Researchers [6, 7] have worked on a teaching/playback method to ensure that molten metal is quickly and precisely poured into a mold and does not spill over the sprue cup. The filling weight control has been implemented by using the state estimation of the pouring process and the predictive sequence control. Furthermore, to realize highly accurate flow rate control, a feed forward control employing the inverse dynamics of the flow rate model is also proposed [8]. Similarly, the presented Melpore system used teaching/playback method to automate the pouring system of the tilting-ladle type pouring system applying fuzzy control to the flow rate control system in tilting-ladle type automatic pouring system. Noda et al. [9] have optimized the sequence control of automatic pouring system in press casting process; recently modern manufacturing and automation [10, 11] have achieved the high precision positioning control of the pouring liquid while avoiding contact with the mold. However, all these techniques focus on the improvement and automation of the metal pouring. The problem of high precision positioning of molds of different sizes and having different pouring cup locations has not been addressed so far. In computer and machine vision's image capturing techniques, image is often corrupted while acquisition and some techniques like Normalized Convolution (NC) and High Resolution Normalized Convolution (HR-NC) are applied on the corrupted image in order to regain the missing and corrupted minor details which are lost in data acquisition [12].

2. Methodology and Experimental Setup

2.1. Machine Vision System, Its Alignment, Feedback, and Robot Guidance. Machine vision systems are being deployed as a mandatory requirement worldwide to ensure quality products. In our setup, need of four cameras which are generally used in the system to capture the complete surface of the product and to generate a composite surface is eliminated. Instead, each image is individually processed and if any individual part fails the inspection, it is tracked and rejected at the end of the process. To ensure that the top of the product maintains critical dimensions, the vertically positioned camera obtains an image of the top of the product. Using Mil measurement (The American System of Measures used in manufacturing dimensions like distance and length, etc.) for thickness functions, the inner and outer circumferences of the top can be located and used to plot the circumference. After this, features such as inner diameter, outer diameter, and ovality can be calculated and compared with known good measurements [10]. 2D representation of the system is shown in Figure 1.

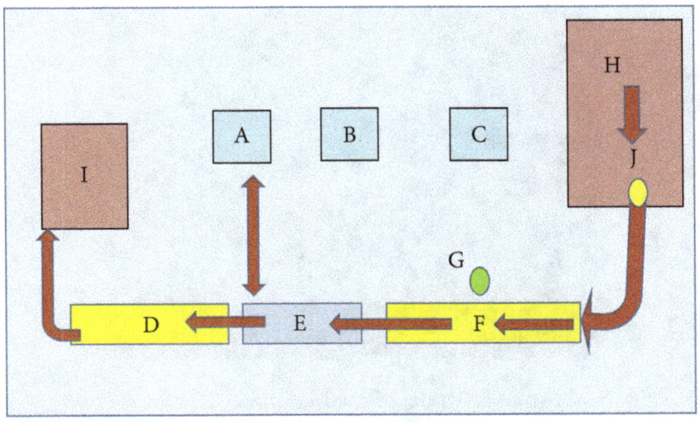

FIGURE 1: 2D representation of system, where A is induction furnace, B is induction heating generator, C is transformer, D is conveyor 1, E is conveyor 2, F is conveyor 3, G is camera, H is mold preparation area, I is cooling area, and J is mold and pallet.

The process starts from H which is the mold preparation area; inputs to this section are the raw material for mold preparation (sand, binder), mold parts (cope, drag), and pattern. Using these, molds are prepared and stacked. J represents a final prepared mold ready to enter the conveyor system. F is a transport conveyor which takes the mold from the preparation area to the module E. Within conveyor 3 a camera is mounted, the mold is detected with the help of proximity sensors, and when it is under the camera the conveyor stops for a moment while the camera takes an image of the mold, after which the conveyor restarts. The picture taken by the camera is sent to the computer where it undergoes an image processing algorithm which tells exactly the location in X and Y coordinates of the center of the pouring cup. This information will be used later to align the mold beneath the pouring point of the furnace. After passing the camera, mold reaches the end of conveyor 3, where it enters conveyor 2. The distance it travels on conveyor 2 is given by the controller from the output of the image processing algorithm; therefore in Y-direction the pouring cup center is aligned to the pouring point of furnace; now conveyor 2 moves forward by the screw mechanism on its base towards the furnace and the distance it moves is again given by the controller from the output of the image processing algorithm; therefore now the mold gets aligned in X-direction also, so when conveyor 2 stops at the furnace, the pouring cup center will be exactly beneath the pouring point of furnace. When the operator gets the signal that the mold is aligned, he manually pours the metal in the mold by simply rotating the handle on the induction furnace which tilts the crucible forward; after pouring is done the operator sends a signal to the system and conveyor 2 moves back in X-direction. After it reaches the end position, the mold moves from conveyor 2 to conveyor 1, and when conveyor 1 gets fully stacked, the operator shifts the molds to the cooling area (I). After molds are cooled, they are broken and the parts are retrieved while the sand, cope, and drag are sent back to the mold preparation area. Thus the same cycle repeats again and again.

When assembling components into a larger product, it is often necessary to align one part with another prior to

Figure 2: Model of casting mold on pallet.

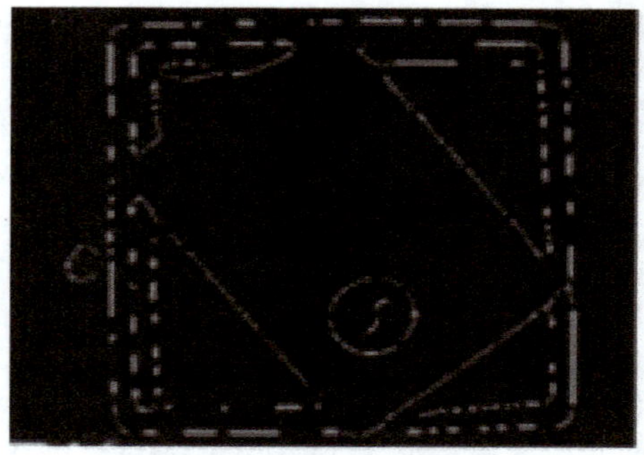

Figure 3: Output of canny edge detector.

insertion or attachment in some way. Machine vision systems can easily provide the alignment systems with the positional feedback they require to make sure the components fit together in the proper manner as shown in Figure 1. The algorithm for vision system has been developed in MATLAB, which is interfaced to the external microcontroller based hardware; thus the output of the vision system acts as an input to the microcontroller system. The aim of the vision system is to find the location/coordinates of the center of pouring cup of the casting mold (Figures 2 and 6). The mold can be of different sizes, with different positions and orientations on the pallet, and can have different position and radius pouring cup. There are three main sections of the image processing algorithm as presented in the succeeding sections.

2.2. Circular Hough Transform: Part Presence and Orientation. The pouring cup of a sand casting mold is actually a circular hole; thus the Circular Hough Transform [11] technique is implemented to detect the coordinates of the center of this hole. The steps performed in the developed algorithm are given below.

Convert RGB image to grey scale:

(a) Define and apply averaging filter.

(b) Apply canny edge detector, giving an output in the form of binary image, shown in Figure 2.

(c) Define a 3-dimensional accumulator array, with 2 dimensions representing the rows and columns of original image and the third dimension representing the range of radius of circle to look for.

(d) Set the 3D accumulator array to zero.

(e) Scan entire image (from step (c)):

   (i) If edge is detected (i.e., a value of 1 is found), proceed to next step; else keep scanning.

   (ii) Define limits to draw a circle in the accumulator array.

   (iii) Draw a circle in the accumulator array centered at the index of the edge point detected.

   (iv) Draw circles in the accumulator array shown in Figure 3 for all the values of radius defined earlier in the accumulator array, centered at the index of the edge detected; this is also called voting (drawing a circle means to add 1 to the corresponding array elements). The modern research on machine vision for tolerance detection [13] suggests novel methods to detect the dimensions with accuracy and we tried to implement our algorithm on similar lines. This system (as shown in Figures 4(a) and 4(b)) is designed such that the pouring cup of any radius should be accommodated. To detect the radius of circle, we draw circles on the accumulator array of that known diameter; if the diameter of circle is unknown, then for that purpose some modifications need to be done in the existing algorithm. Firstly, the accumulator array is of three dimensions, previously it was 2D (rows and columns), and now the third dimension in the array is of the radius. As the dimension has to be finite, so for the algorithm to complete its work a range of radius needs to be given to look for in the image. This range can be as big as we desire, but the bigger it gets the more computational power and time it would take to achieve the output; thus it is better to give an optimized range of radius of which the pouring cup can exist. Now, instead of drawing circles in the accumulator array we draw cones, with the starting radius of the cone as the staring value given in the third dimension of the array and the ending radius as the maximum value given in the 3D array. To simplify, it can also be looked as a stack of 2D arrays with the radius incrementing with each 2D array; all the arrays stacked on top of each other will give a cone. Now, the point where all the cones intersect will have the maximum magnitude and thus similarly scanning the 3D array for the index of maximum value will give us the location of the center point of the pouring cup. In this case one more additional piece of information will be retrieved; when the scanning of the 3D array will give output of the index of the maximum value pixel, it will contain three values as it is a 3D array, first two will be defining the position (columns and rows), and the third would be radius; thus in this we can also tell the radius of the detected circle along with the coordinated of the center.

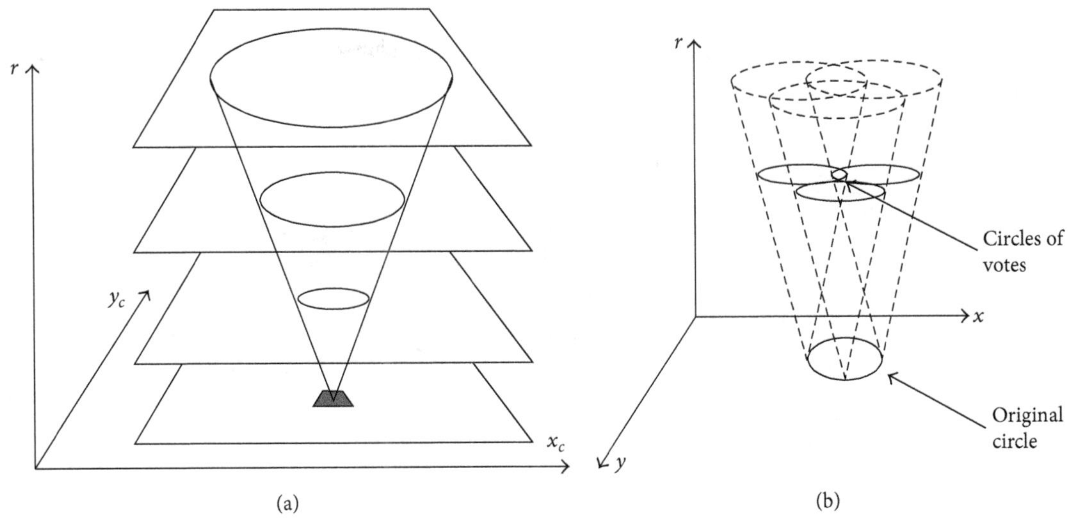

FIGURE 4: (a) 3D accumulator array in planner form. (b) 3D accumulator array in multiconical form.

(f) After complete image has been scanned and the accumulator array has been filled, scan the array for the maximum value.

(g) The index of the max value tells the coordinates of the center of actual circle detected in rows and columns, and as the index has 3 values, it also tells the radius of detected circle.

(h) The coordinates found are in pixels right now and will later be used to find the location of pouring cup center in millimeters.

2.3. Harris Corner Detectors. Finding just the coordinates of the pouring cup center is not enough for mold positioning. A reference point is needed from here; center measured is selected from the top right corner of the pallet. Harris Corner Detector technique [11, 14] is used to detect and locate the corner. The following is the final equation on which the algorithm is developed:

$$E(u,v) = (u,v)\,M\begin{pmatrix} u \\ v \end{pmatrix}$$

$$M = \sum_{x,y} w(x,y)\begin{pmatrix} Ix \cdot Ix & Ix \cdot Iy \\ Ix \cdot Iy & Iy \cdot Iy \end{pmatrix}, \quad (1)$$

where

$E(u,v)$ is change in intensity by moving a filter window over the image,

u and v are shifts in the window.

An extended form of General Hough Transform, Circular Hough Transform (CHT), is used to detect circles. In this the "Image" is the input image, "Output" is an edge detected binary image, "thresh" is a two-element vector in which the first element is the low threshold, and the second element is the high threshold, and sigma is the standard deviation of the Gaussian filter. Similar image processing techniques are also used in food processing industry for detection of dimension and quality [15]. In CHT, we first initiate an accumulator array which is of the same size as the edge detected image. Then this image is scanned and at each point where an edge is detected it votes for a possibility of a circle in the accumulator array. Let us first take example where the radius of circle to be detected is known and is constant. Figure 5(a) is an ideal edge detected image from the canny detector and Figure 5(b) is the accumulator array of the same size. Initially the accumulator array is initialized to zero. The binary image is scanned and wherever an edge is detected some operation is done on the accumulator array. An edge is simply a point where the value of pixel is 1 and 0 (zero) where there is no edge. During scanning point 1 is detected as an edge; now the index of point 1 (the row and column position where it lies) is noted and on the same index on the accumulator array a circle is drawn with the center point of the circle on that index. Drawing a circle basically means adding 1 to the pixels of the circle boundary/circumference. Now, similarly point 2 and point 3 will also be detected as an edge and same operation will be done on the accumulator array as done due to point 1. Over here as an example only 3 points are shown, while during scanning each point of the circle edge will be detected and the same operation as of point 1 explained will be done in the accumulator array. As each circle drawn in the accumulator array adds 1 to the circle boundary, then there is a point in the array which will have a maximum magnitude; this is the point where all the drawn circles intersect. As each drawn circle passes through this intersection point and each circle votes/adds 1 to this point, in the end this intersection point has the maximum magnitude. If you notice, now this intersection point of all drawn circles is basically on the same index as that of the center point of the actual circle in the original image. Therefore, after the accumulator has been completely filled we just scan it to find the index of the pixel with maximum magnitude and that index is basically the center of the circle we were looking for.

The Eigen values of matrix M can help us locate the possible corners in an image. The steps occurring in the

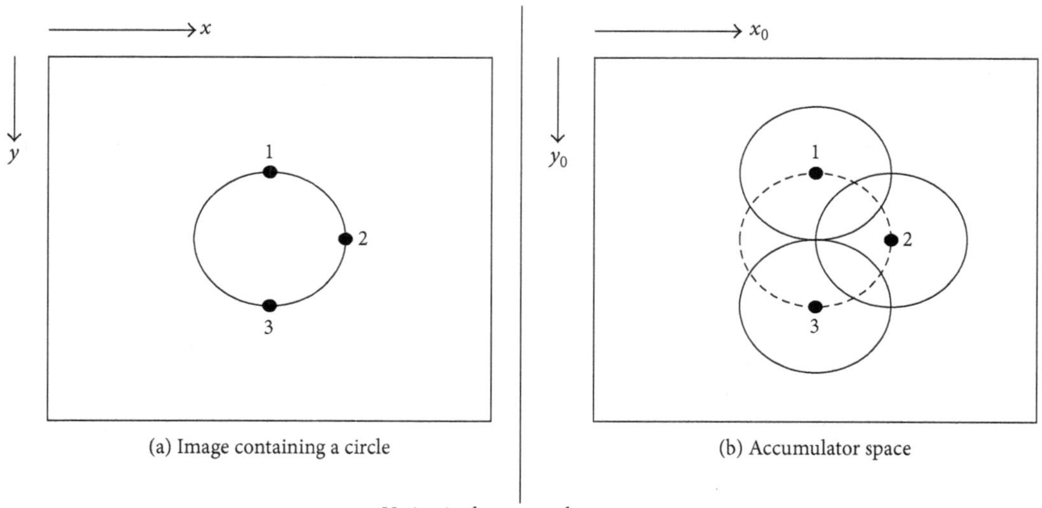

FIGURE 5: (a) Image containing circle. (b) Accumulator space.

developed algorithm are as follows:

(a) Define and apply derivative mask

$$Hx = [-1\ 0\ 1; -2\ 0\ 2; -1\ 0\ 1]$$
$$\text{on image in } x\text{-direction to get } Ix. \tag{2}$$

(b) Define and apply derivative mask Hy (Hy = transpose of Hx) on image in y-direction to get Iy.

(c) From Ix and Iy calculate $Ix \cdot Ix$, $Iy \cdot Iy$, and $Ix \cdot Iy$.

(d) Convert original color image to gray scale.

(e) Apply Gaussian filter to image obtained from step (d).

(f) Initiate corner detection array (R) to zeros with size of R equal to the size of original image.

(g) Scan entire image from step (e):

(i) For each pixel define a 2×2 matrix

$$M = [Ix \cdot Ix(m,n)\, Ixy(m,n);\ Ixy(m,n)\, Iy \cdot Iy(m,n)], \tag{3}$$

where m and n are indices of the specific pixel.

(ii) Find the two Eigen values of matrix M, which are $E1$ and $E2$.

(iii) Use $E1$ and $E2$ in the formula $r = (E1 * E2)/(E1 + E2)$.

(iv) Add the r value obtained to the matrix R on the same index as of the pixel for which matrix M was formed in the original image.

(v) Repeat steps (i)–(iv) for each pixel in the image, to get a completely filled matrix R (finally matrix R is filled with r values).

(h) Apply nonmaxima suppression to the matrix R; this eliminates the possibility of detecting a specific corner more than once. (In this, the neighborhood of a corner is scanned and the pixel with the largest magnitude which represents the strongest corner position is left while others in its neighborhood with lesser magnitudes are erased.)

(i) Scan matrix R from step (h), using a threshold which dictates how strong corner is required.

(j) Scan locally matrix R to find the coordinate of the required reference point. The output of this algorithm is coordinates (x_2, y_2) in pixels of the top right corner of the pallet. Now this corner coordinates and the pouring cup center coordinates (x_1, y_1) found earlier will be used to find the distance (shown in Figure 6) in millimeters of the pouring cup center from the reference point.

2.4. Scaling. Scaling is used to convert image distance into spatial distance, that is, in millimeter; there can be two approaches to it. First is to use a formula shown below:

$$D_m = \frac{D_i \cdot D_c}{f}, \tag{4}$$

where

D_m is distance required in millimeter (mm),

D_i is distance in the image (in this case it is the distance between the two points shown in Figure 4),

D_c is distance between the camera and mold surface, and f is the focal length of the camera.

If the focal length of camera is unknown, a practical way can be adopted to formulate the same relationship by finding the constant value (D_c/f). One side of the pallet is measured physically in millimeter; then the same side is measured in

FIGURE 6: Distance to find in scaling.

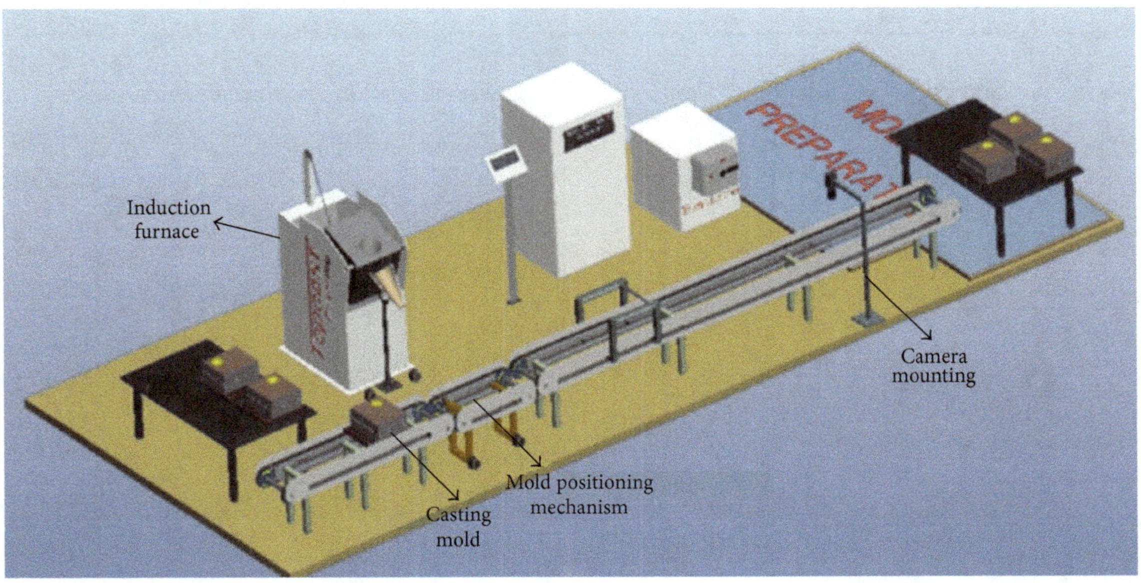

FIGURE 7: System model for automated sand casting.

pixels from the image and a relationship is made between millimeter and pixel:

$$\begin{aligned} n \text{ pixel} \quad & r \text{ mm} \\ 1 \text{ pixel} \quad & \frac{r}{n} \text{ mm.} \end{aligned} \tag{5}$$

Taking average of several iterations of the above practical approach gives an accurate result. After the distance in x- and y-direction between (x_1, y_1) and (x_2, y_2) has been found in s, the two values are sent to the external microcontroller based hardware through serial communication between the PC and the microcontroller. The microcontroller then uses these values to control the mechanical model in order to position the mold as desired.

3. Designing of Test Platform and Process Automation

3.1. System Designing. A complete system design mentioned in Figure 7, for the sand casting automation, was made in Pro-Engineer software. The model is to the scale and is designed according to an induction furnace setup available in a manufacturing facility.

After successful model design an animation of the same is also made to visualize the concept of automation by mold positioning. The process starts by a mold preparation area, where casting molds are prepared; these are then transferred via a conveyor. During this first conveyor a camera takes the snapshot of the mold and processes it to find the location of pouring cup center. An alternate to the vision system is also shown in the model, in which a sensor scans the mold to find the pouring cup center, though this is not implemented here but is given just to show an alternate technique for future work. After conveyor 1 the second conveyor comes which can also move towards and back to the induction furnace. This conveyor is responsible for the actual mold positioning; a small scale model of it is also made to verify the concept and test the algorithm. After mold positioning and metal pouring this second conveyor moves the mold to the last conveyor which takes the mold to the cooling area.

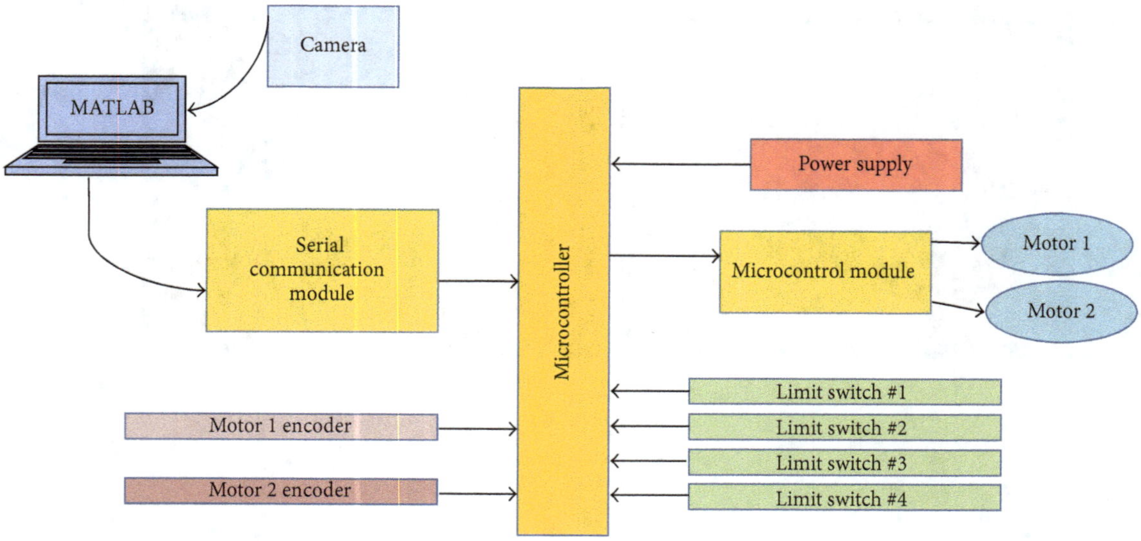

FIGURE 8: System electronic control with microcontroller nucleus linked with architecture.

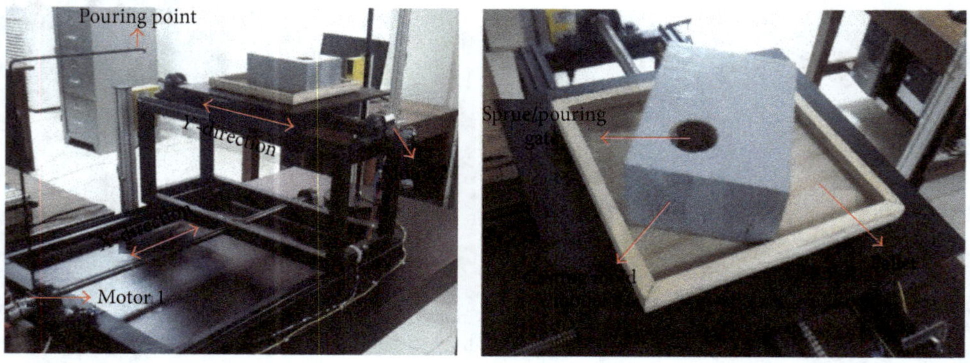

FIGURE 9: Mechanical model and positioning system.

3.2. Scale Prototype Modeling. The conveyor in the middle portion is the main part of the system which is responsible for carrying the mold towards the induction furnace and positioning it accordingly. Other two conveyors are just meant for transport; thus the small scale model (Figures 8, 9, 10, and 11) is made for them. The fabricated model consists of two lead screw mechanisms which give the mold place on top a two-degree-of-freedom positioning. The lead screws are run by DC-gear motors with encoder. Encoder provides pulses proportional to the rotation of lead screw and hence the linear distance traveled; hence they are used as a feedback for the microcontroller on the distance traveled by the mold as shown in Figure 9. Figure 8 gives a complete view of the system and it shows the linkage of different components.

The system consists of two portions; one is the vision system which is run on a computer, while the other is the microcontroller based hardware. To ensure successful operation of the system there is a serial communication link between these two portions. The output of the vision system is one of the inputs to the microcontroller hardware. The vision based system detects the position of the center of the mold pouring cup; this is basically an algorithm developed with a camera connected to it. Picture is taken in real time and the algorithm processes it to find the center point. The second system is a microcontroller based embedded hardware; it takes the output of the vision system and transforms it into mechanical movement of the model. The microcontroller based embedded hardware is the communication module and a protocol between these two systems so that the data can be sent quickly and reliably.

3.3. Mechanical Model and Positioning System. The first step in the project was to develop to the scale model of the system. Designing has been done by keeping in mind the actual equipment already present in industry and the dimensions of the actual area available for the setup of the new system. Initially there is an induction furnace, an induction heating generator, and a transformer present on the shop floor. Before starting modeling, the dimensions of the equipment were taken so as to produce a close to real life model. The software chosen for the modeling is Pro-Engineer v5. After modeling of the entire system to the scale, an animation of the same model has also been developed which shows the real time working of the model as an automated casting

FIGURE 10: Camera positioning mount above pallet.

FIGURE 11: Mold and pallet on mechanical model.

process. The animation has been developed in the animation module of PTC Pro-Engineer v5. The main components of the animation code are as follows:

(i) Define initial condition of all mechanisms and moving parts.

(ii) Define servo motors on all mechanisms and moving parts.

(iii) Add servo motors on the timeline with defining start time, end time, speed, and initial position.

(iv) Define body locks where required, and add them on time line with definition of start time, end time, lead body, and follower body.

(v) Define viewing angles for each motion and add them over the entire timeline accordingly.

After successful model design, animation of the entire system was done. The next step was the designing and fabrication of a small scale prototype model. The purpose of this prototype is to

(i) verify the feasibility of the concept given in the model,

(ii) check the working of the system in real time,

(iii) get to know about the constraints and variables involved in the system,

(iv) develop and test the algorithm which will run in the full scale system.

The main modules in the Prototype model are

(1) mechanical model,

(2) electronics hardware,

(3) image processing algorithm.

As this is a small scale model, it is not possible to add an induction furnace to it; hence a fixed point is made with help of a rod which represent the pouring point of the induction furnace, and the molds thus have to position themselves such that the pouring cup center comes directly beneath this pouring point. Two side views of the system show the fabricated mechanical model (Figure 9).

After successful fabrication of the mechanical model, the electronics hardware needs to run with the main requirements as follows:

(a) Microcontroller module.

(b) Two motor control modules.

(c) Module to allow communication between vision system and microcontroller.

For ease of understanding the practical system model is shown in Figures 10 and 11.

As shown in Figures 10 and 11, the mold which is placed on the top can be moved in X- and Y-direction for which two motors couples to the lead screws are present. The motors are mounted on a bracket which is fixed with the frame while the shaft on the motor fits directly in a hole made on the end of lead screw; the motor shaft is tightened with the help of a screw on the lead screw end. When the system would run the mold would be aligned automatically such that the center of the pouring cup would come beneath the pouring point as shown in Figure 11. A complete simulation of the system is also shown in Figure 12.

3.4. Circuit Layout Designing. Using all of the designing shown in Figure 13 the final layout of the printed circuit board was made. Lab Center Electronics Proteus has been used for the simulation, testing of the schematic, and the final designing of the PCB Layout. Firstly, the layout of each module has been designed individually and then they are merged together to form the final layout. Different color traces are used for ease of troubleshooting and understanding of Figure 13 is explained in Table 1.

The white connector on the left side is the DB9 connector which connects to the USB to serial converter, which is connected to the USB port of the PC. The wiring at the top

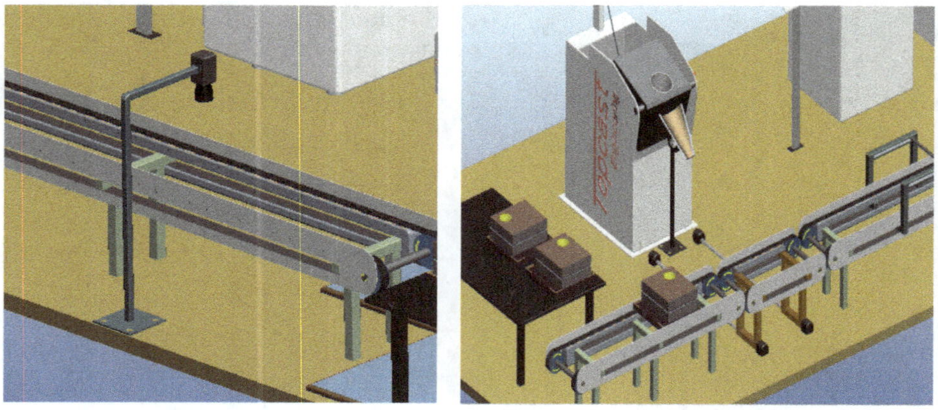

Figure 12: Simulated camera positioning on mold-pallet mechanical system.

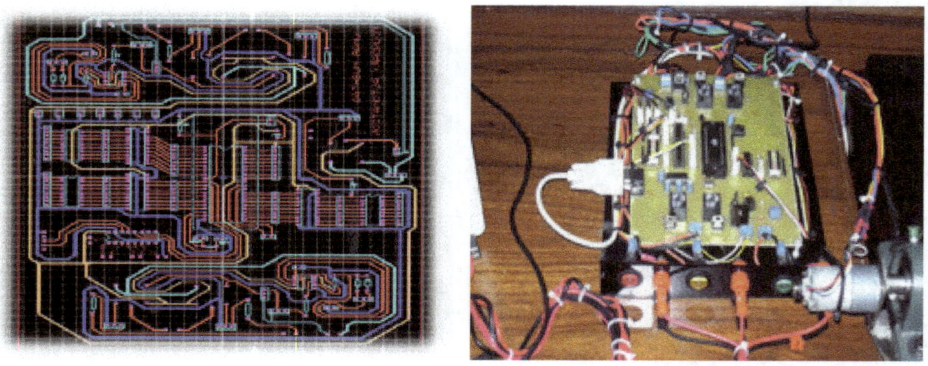

Figure 13: Proteus PCB layout fabricated electronics hardware.

Table 1: Color codes of layout.

Color	Function
Red	Data
Orange	Hardware power (5 V)
Blue	Ground of 5 V
Light green	Motor power (12–24 V)
Dark green	Ground for motor power

is going to the mechanical model; as shown in Figure 13 the entire wiring passes through the two sides of the housing; this makes the system neater and easier to troubleshoot if any problem occurs. The entire wiring of the circuit to the mechanical model has been done in a professional way, there is no soldering or hand joints in the entire wiring; wherever a joint is required between two or more wires, thimbles have been used and on top of those heat sleeves are inserted for insulation. Such wiring though takes time and patience but the outcome not only is a clean and tidy wiring job but also ensures the reliability of the system, as most of the problems occurring in such automation projects are due to poor wiring. Moreover as thimbles are used on every wiring joint, the system can be very easily dismantled for transportation purposes and the same wiring harness can be used again to assemble it.

The final fabricated circuit having all the required modules and being wired up to the mechanical model is shown in Figure 13. After a lot of rechecking and verification of the layout with the original schematics has been done the layout was fabricated on a fiber PCB.

4. Algorithm and Analysis

All the components of the prototype, mechanical model, electronics hardware, and vision system, were interfaced together as one system. The system was then fine-tuned by adjusting several parameters of the vision system and the microcontroller algorithm, to produce an automated mold positioning system. The flow of information in the entire system is worth mentioning; Figure 14 not only shows the flow of information but also shows all the modules and functions happening in the system in the correct sequence. The boxes in yellow show the functions executed in the vision system by MATLAB and in the green boxes are those executed by the microcontroller. After the start command is given in MATLAB till the actual physical mold positioning the process is fully automated without any human intervention.

After the distance values are in mm (millimeter) and sent to the microcontroller, it further modifies them into

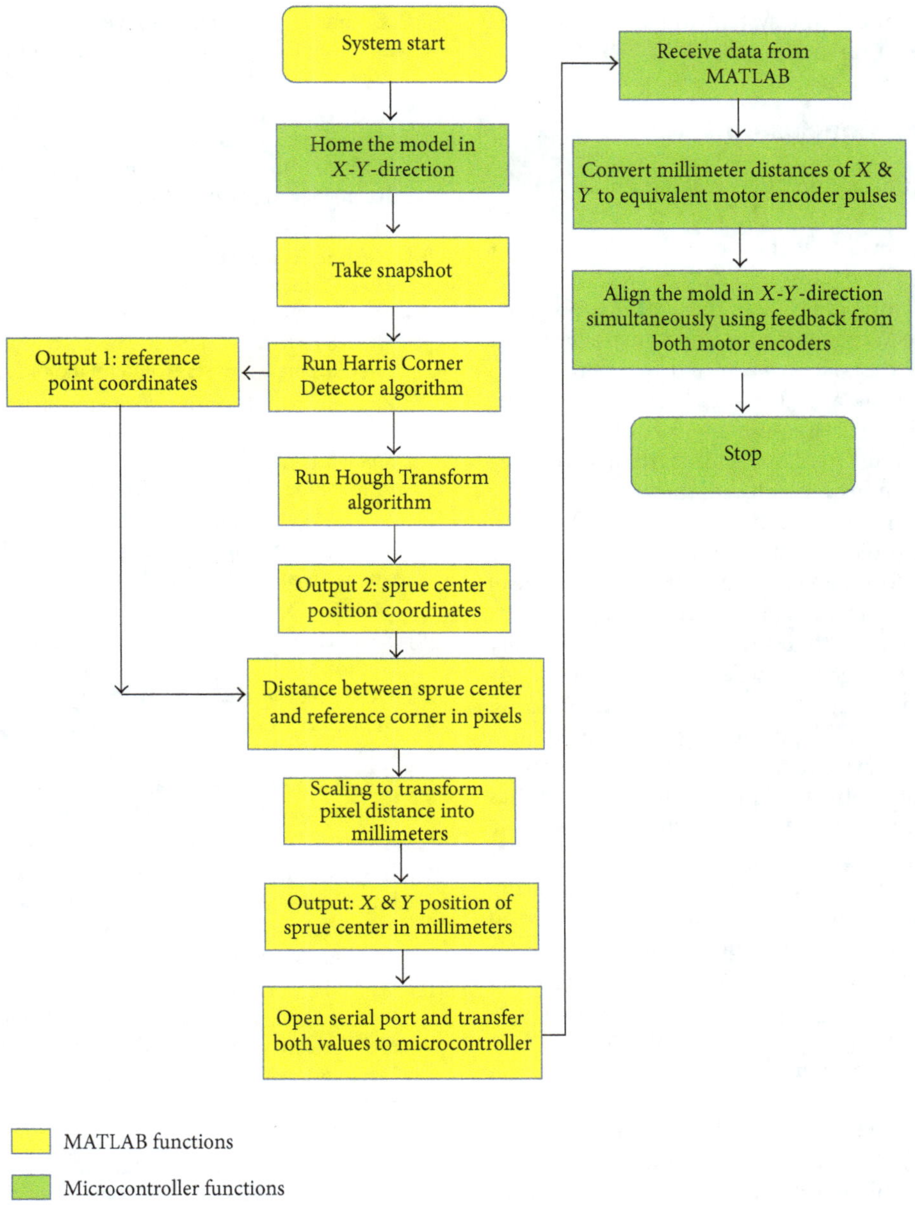

FIGURE 14: System flow of information.

coding that it can understand. The feedback from the two motors is in the form of pulses from the encoder, where the number of pulses is dependent on the number of rotations of the motor shaft. The pitch of the lead screw tells the linear distance traveled in one rotation of the lead screw; thus a relationship is formed between the encoder pulses and the linear distance to travel. The values thus sent by MATLAB are converted to equivalent motor encoder pulses, which the microcontroller monitors to achieve the distance desired. The system developed here is unique and better than other sand casting automations setups as it can accommodate the following variations in the molds without any human interventions or hardware changes to mechanism shown in Figure 6 and summarized in Figure 12 in form of flow graph:

(a) Different mold position on the pallet

(b) Different mold orientation on the pallet

(c) Molds of different sizes

(d) Molds with different pouring cup positions

(e) Molds with different pouring cup radius.

In a sand casting setup where wide variety of parts are made on the same assembly line, it requires the molds to be of different sizes, pouring cup location, or radius to vary; thus this system comes into play in such situations, and it autonomously handles the variations in the molds. Moreover, the worker who is preparing the molds and putting them on the conveyor belt does not have to think or put any effort on the placement of the molds; the system will accept molds in any orientation or position on the pallet

5. Discussion and Conclusion

In previous sand casting automation research studies, implementation of precise pouring of the molten metal requires molds on the conveyor line to be same size, having same pouring cup location and radius. But if molds have to be casted which are of different sizes, the entire production line has to be stopped, and after physical adjustment to the pouring mechanism the system is started to cater for this limitation. Our present machine vision based sand casting automation system does not require any sort of production stopping or manual adjustment once it is started; it will keep the casting process running regardless of the variations in the geometry and position of the casting mold.

In this research project, first a complete model of the automation system is formulated by which this new scheme of automation can be implemented in the sand casting industry. Then as a proof of concept, to test the mechanical feasibility and test and verify the algorithms developed, a small scale prototype model is made on which successful demonstration is given on automatic mold positioning to be used in a complete sand casting system model with machine vision methodology. The automatic mold positioning prototype has proved to be an achievement due to its sharp and reliable mold positioning capability. The system presented here can be coupled with already presented automatic metal pouring systems in the industry, thus giving a fully automatic sand casting system. The introduction of mold positioning system presented here will increase the flexibility, safety, and mold handling capability of existing automated sand casting setups, thus allowing different parts to be casted on the same conveyor line without any human intervention or process stopping.

The system presented here is not limited just to a sand casting process; other industries requiring an autonomous positioning system for their parts/items on a conveyor system can also benefit from this research.

Competing Interests

The authors declare that they have no competing interests.

References

[1] H. Kosler, U. Pavlovčič, M. Jezeršek, and J. Možina, "Adaptive robotic deburring of die-cast parts with position and orientation measurements using a 3D laser-triangulation sensor," *SV-Journal of Mechanical Engineering*, vol. 4, pp. 207–212, 2016.

[2] E. Neuman and D. Trauzeddel, "Pouring systems for ferrous applications," *Foundry Trade Journal*, pp. 23–24, 2002.

[3] A. Ito, Y. Noda, and K. Terashima, "High-precision liquid pouring control while keeping lower ladle position and avoiding clash with mold," in *Proceedings of the IEEE International Conference on Control Applications (CCA '12)*, pp. 246–251, Dubrovnik, Croatia, October 2012.

[4] X. Zhu, R. Chen, and Y. Zhang, "Automatic defect detection in spring clamp production via machine vision," *Abstract and Applied Analysis*, vol. 2014, Article ID 164726, 9 pages, 2014.

[5] L.-H. Zou, C. Jie, Z. Juan, and L.-H. Dou, "The comparison of two typical corner detection algorithms," in *Proceedings of the 2nd International Symposium on Intelligent Information Technology Application (IITA '08)*, vol. 2, pp. 211–215, Shanghai, China, December 2008.

[6] Y. Choe, H.-C. Lee, Y.-J. Kim, D.-H. Hong, S.-S. Park, and M.-T. Lim, "Vision-based estimation of bolt-hole location using circular hough transform," in *Proceedings of the ICROS-SICE International Joint Conference*, pp. 4821–4826, Fukuoka International Congress Center, Fukuoka, Japan, August 2009.

[7] Z. Zhifeng, "Measuring diameter of non-threaded hex bolts based on hough transform," in *Proceedings of the 3rd International Conference on Measuring Technology and Mechatronics Automation (ICMTMA '11)*, pp. 526–528, Shanghai, China, 2011.

[8] U. Asgher, R. Ahmad, and S. I. Butt, "Mathematical modeling of manufacturing process plan, optimization analysis with stochastic and DSM modeling techniques," *Advanced Materials Research*, vol. 816-817, pp. 1174–1180, 2013.

[9] Y. Noda, K. Terashima, M. Suzuki, and H. Makino, "Weight control of pouring liquid by automatic pouring robot," in *Proceedings of the IFAC Workshop on Automation in Mining, Mineral and Metal Processing*, vol. 70, pp. 1–6, Vina del Mar, Chile, 2009.

[10] Y. Noda and K. Terashima, "Modeling and feedforward flow rate control of automatic pouring system with real ladle," *Journal of Robotics and Mechatronics*, vol. 19, no. 2, pp. 205–211, 2007.

[11] A. Wilson, Ed., *System Checks Plastic Molding Consistency*, 2007, http://www.vision-systems.com/articles/print/volume-12/issue-2/features/profile-in-industry-solutions/system-checks-plastic-molding-consistency.html.

[12] U. Asgher, H. Muhammad, M. M. Hamza, R. Ahmad, S. I. Butt, and M. Jamil, "Robust hybrid normalized convolution and forward error correction in image reconstruction," in *Proceedings of the 10th International Conference on Innovations in Information Technology (IIT '14)*, pp. 54–59, Al-Ain, UAE, November 2014.

[13] J. Chi, L. Liu, J. Liu, Z. Jiang, and G. Zhang, "Machine vision based automatic detection method of indicating values of a pointer gauge," *Mathematical Problems in Engineering*, vol. 2015, Article ID 283629, 19 pages, 2015.

[14] H. M. Yip, J. C. Li, K. Xie et al., "Automated long-term monitoring of parallel microfluidic operations applying a machine vision-assisted positioning method," *The Scientific World Journal*, vol. 2014, Article ID 608184, 14 pages, 2014.

[15] S. Srivastava, S. Boyat, and S. Sadistap, "A robust machine vision algorithm development for quality parameters extraction of circular biscuits and cookies digital images," *Journal of Food Processing*, vol. 2014, Article ID 376360, 13 pages, 2014.

9

Investigation into the Heat of Hydration and Alkali Silica Reactivity of Sustainable Ultrahigh Strength Concrete with Foundry Sand

Federico Aguayo, Anthony Torres, Tate Talamini, and Kevin Whaley

Department of Engineering Technology, Texas State University, San Marcos, TX, USA

Correspondence should be addressed to Federico Aguayo; fred.aguayo@txstate.edu

Academic Editor: Yao Luan

This study presents the hydration reactivity and alkali silica reaction (ASR) of ultrahigh strength concrete (UHSC) that has been made more sustainable by using spent foundry sand. Spent foundry sand not only is sustainable but has supplementary cementitious material (SCM) characteristics. Two series of UHSC mixtures were prepared using a nonreactive and reactive sand (in terms of ASR) to investigate both the impact of a more reactive aggregate and the use of spent foundry sand. Conduction calorimetry was used to monitor the heat of hydration maintained under isothermal conditions, while ASR was investigated using the accelerated mortar bar test (AMBT). Additionally, the compressive strengths were measured for both series of mixtures at 7, 14, and 28 days to confirm high strength requirements. The compressive strengths ranged from 85 MPa (12,345 psi) to 181.78 MPa (26,365 psi). This result demonstrates that a UHSC mixture was produced. The calorimetry results revealed a slight acceleration in the heat of hydration flow curve compared to the control from both aggregates indicating increased hydration reactivity from the addition of foundry waste. The combination of foundry sand and reactive sand was found to increase ASR reactivity with increasing additions of foundry sand up to 30% replacement.

1. Introduction

The challenges facing the concrete industry in the 21st century are greater than ever due to familiar problems such as various durability issues plaguing our infrastructure and their associated life-cycle impacts. These issues, coupled with increasing stringent environmental regulations, have resulted in the need of a sustainable ultrahigh strength concrete (UHSC) that can achieve the goal of safer, durable, and more environmentally-friendly infrastructure. For instance, more effective utilization of locally available industrial wastes/byproducts in concrete such as foundry sand, fly ash, and slag can impart enhanced fresh and hardened properties to concrete, while significantly reducing the disposal volume of these materials. Given the fact that concrete is the most widely used man-made building material in the world, there is an opportunity for significant economical savings and environmental benefits from increased incorporation of waste material in concrete.

The American Concrete Institute (ACI) has defined UHSC as a concrete meeting high strength that cannot always be achieved routinely, using conventional constituents and normal mixing, placing, and curing practice [1]. In the 1950s, concrete with compressive strength of 34 MPa (5000 psi) was considered high strength [1]. Today, high strength concrete is defined as concrete with specified compressive strength of 55 MPa (8000 psi) or higher [1]. In many markets today, concrete having specified compressive strength in excess of 69 MPa (10,000 psi) is routinely produced on a daily basis [1]. UHSC itself is considered a sustainable construction building material due to the high requirement of silica fume and fly ash, both of which are waste products from other industries, which provides beneficial mechanical and durability properties for concrete [1]. Not only does UHSC utilize a high percentage of fly ash and/or silica fume, the high strength requirement results in higher specific strength of the material. The specific strength of a material is the ratio of the strength to its density [1]. The density of conventional concrete and UHSC is relatively the same, but the strength significantly increases with UHSC; therefore the specific strength is much higher. This fact impacts sustainable construction through decreased transportation cost and emissions as thinner structural elements can be produced out of UHSC, which will require less fuel and produce less carbon emissions during transportation.

Spent foundry sand (FS) is a by-product of ferrous and nonferrous metal casting industries primarily consisting of high quality silica sand with uniform physical characteristics. Approximately 6–10 million tons of this waste is produced annually by the foundry industry in the U.S. alone. Spent foundry sand, when used as partial replacement of fine aggregate in concrete, has been shown to improve the mechanical properties of concrete [2–5]. Existing studies, however, limit their focus to conventional concrete and few report on the inherent durability and hydration properties of these sustainable mixtures.

Alkali silica reaction (ASR) has been one of the most destructive distress mechanisms dominating our infrastructure for several decades. The reaction occurs due to the high pH of the concrete pore solution as a result of the alkalis from the cementitious material. This aggressive solution attacks certain siliceous minerals found in coarse and fine aggregates that may result in a hydrophilic gel that absorbs water causing abnormal expansion. This expansion can lead to excessive cracking, often resulting in a reduction in the service life of affected structures. Since FS is highly siliceous, ASR expansion can be accelerated from the addition of FS in the mixture.

In the present research, multiple UHSC mixtures with locally available foundry sand were produced and analyzed for their hydration and alkali silica reactivity. Two control UHSC mixtures were developed based on a literature survey and bench line strength. One control used manufactured sand as the sole aggregate, which is a commonly used fine aggregate and typically not reactive (in terms of ASR). In order to assess the impact of foundry waste on the reactivity sequestration of UHSC mixtures, a second control mixture was developed that consisted of reactive sand as the sole aggregate. Foundry sand was then used to partially replace the fine aggregate in the control mixtures in 10, 20, and 30% by volume increments and their strength, hydration, and ASR performance were compared to the control values.

2. Literature Review

2.1. Background Information. UHSC, like conventional concrete, can have a broad range of designs. Most UHSC are being designed without coarse aggregate, or coarse aggregate that is ≤9.5-mm (0.375-in.) [1]. In the case that coarse aggregate is omitted, fine aggregate is the sole aggregate in the mixtures, which makes for a more homogenous mixture. Most UHSC mixtures also contain supplemental cementitious materials (SCMs) such as silica fume, fly ash, granulated ground blast furnace slag (GGBFS), and quartz powder, fibers, and high range water reducing admixture (HRWRA). When used in optimum dosages, the HRWRA reduces the water-to-cement (or water-to-cementitious) ratio while improving the workability (viscosity) of the concrete. The addition of the SCMs enhances the mechanical properties of the cement paste by producing secondary hydrates, filling voids, and enhancing rheology [6, 7]. Due to the burgeoning large structure industry, there are more and more requirements for higher strength concrete. UHSC has been in development since the mid 1950s and many researchers are still investigating the optimum and most efficient manner to produce this material [2]. However, modern construction practices are moving towards sustainable construction through lower cost and sustainable materials; therefore, UHSC improvements should also be focused on becoming more sustainable and more affordable.

In recent years, there has been extensive research conducted to study the effect of FS in concrete production. Fiore and Zanetti [8] categorized the various ways and means of reusing foundry waste. The research conducted by the authors describes the potential for using spent foundry sand from molds in concrete. The focus of the study by Fiore and Zanetti [8] is to describe the potential that foundry sand possesses to ultimately benefit concrete construction materials. The authors begin by describing the particle size of spent foundry sand and the specific size range recommended for concrete. The authors comment that the foundry sand fraction between 0.1 mm and 0.0025 mm (0.0039-in and $9.84E-9$-in.) can be used as an SCM and any particle size larger than that can be used as fine aggregate in the production of concrete. The authors then focus on the chemical nature of foundry sand and its suitability in concrete. Concrete manufacturers generally require that the chemical percentage of cement be approximately 65% CaO, 25% SiO_2, 4–10% Al_2O_3, and 1–3% Fe_2O_3 [8]. Therefore, Fiore and Zanetti [8] conducted a chemical test on foundry sand and the results showed there was over 95% SiO_2 in all samples and due to the high silica content and its inert and fine particulate nature, foundry sand is suitable in concrete as either an SCM or a fine aggregate replacement.

Ganesh Prabhu et al. [9] experimented on the use of FS in lieu of fine particulate aggregates in the production of concrete. Density, slump cone, split tensile strength, flexural strength, and compressive strength tests were used in ascertaining the effect of FS in concrete. Five different concrete mixtures with and without foundry sand (control sample) were prepared. Those five sample mixtures were at 10%, 20%, 30%, 40%, and 50% FS replacement levels. The authors also used blue-gray sandstone as coarse aggregate and fine sand sourced from the bank of rivers as fine aggregates. The foundry sand had a specific gravity value of 2.24 and density of 1576 kg/m^3. Test results also showed that FS had water absorption value (1.13%), which is considered to be higher than normal. This can be attributed to the ash and wood contents of the foundry sand. Grain size distribution analysis conducted via sieve analysis showed that 8% of the FS is less than 75 μm. According to stipulated standards (IS 2720), this is a satisfactory equivalency value for fine aggregates size. The compressive strength test was carried out at curing ages of 7, 28, 90, and 180 days. The authors also evaluated splitting tensile and flexural strength at 28, 90, and 180 days. A slump cone test was used in analyzing the workability of the concrete mixtures. This was conducted at initial time of 0 minutes and duration/interval of 30 and 60 minutes. The results show that an increase in the replacement level of FS negatively impacts the workability of concrete. This is because water absorption increases with an incremental change in FS proportion. Increase in water absorption also has a direct correlation with the fineness of foundry sand. Alternatively, the authors also postulated that the high absorption properties could possibly be attributed to the ash and clay content of FS. However, the results also show negative impact on the workability of the concrete mixture is insignificant at 10% replacement level. The authors postulated that a modification of water quantity could potentially counteract the slump loss attributed from the fineness of FS and its ash and clayey content. The authors were able to experimentally show that, in all ages of curing, there was a consistent decrease in compressive strength. However, the compressive test result also showed that the sample mix with 20% FS has compressive strength value that is similar to that of the control mix. The authors postulated that the reduction of the specific gravity of the concrete due to its ash and clay content might explain the resultant reduction in compressive strength. The flexural strength and splitting tensile strengths results were also similar to the compressive strength results. This shows a marginally incremental change in flexural strength, with progression in the age of curing. However, there was no remarkable enhancement in flexural strength for all tested samples. Once again the only exception is with the concrete mixture of 20% FS replacement level, which showed a distinct similarity in flexural strength as the control mix. The splitting tensile strength result shows that high replacement proportion (40%, 50%) will enormously decrease the tensile strength of concrete. The authors concluded that the fineness and impurity of FS are underlying factors that impact the compressive, flexural, and tensile strength of a concrete mix with FS as replacement for fine aggregates. However, appropriately determining the right proportion (20%) will inevitably counteract any negative impact on the mechanical properties of a concrete mix.

Siddique et al. [2] have shown the potential of using FS in the production of concrete. Their experimental investigation showed how different replacement levels of FS as fine aggregate affect the mechanical properties of concrete. The mechanical properties investigated are compressive strength, splitting tensile strength, flexural strength, and modulus of elasticity. The values of the investigated properties were determined at different ages of concrete samples (28 days, 56 days, 91 days, and 365 days). The authors categorized their experiment at 3 replacement level percentages (10%, 20%, and 30%). Based on binder system content, FS can be classified into clay bonded FS, otherwise known as green sand, and chemically bonded FS. This classification is centered on physical and environmental factors. Previous researches have also shown that the naturally occurring components such as high quality silica sand (85–95%), bentonite clay (4–10%), and carbonaceous additive (4–10%) make clay bonded FS ideal for use as replacement material. This is largely because of high silica content and also the adhesive properties and fine particulate nature of its clayey content. The locally sourced FS was also chemically tested as per standards specified by IS 383-1970. Physical analysis shows that FS used has lower unit weight and fineness modulus, in comparison to Portland cement. Siddique et al. [2] also used locally sourced coarse aggregates of 10 mm and 20 mm gradation as per IS: 383-1970 23 standards. Melamine-based superplasticizers were used in the mixing and preparation of all tested samples. The authors prepared four different concrete mixtures. Three of the mixtures were of 10%, 20%, and 30% FS percentage weight. All concrete samples were stored at 23°C temperature to minimize loss of moisture, demolded after 24 hours, and then put in a water-curing tank. The authors investigated properties such as slump, unit weight, temperature, and air content as stipulated by Indian Standard Specifications IS: 1199-1959. The results showed that all specimens attained a marginally significant increase in compressive strength at 28 days of curing. The control mixture showed compressive strength value of 28.5 MPa. The other test samples (20% and 30%) had compressive strength values of 30.0 MPa and 31.3 MPa, respectively. Quantitatively, test samples of 20% replacement and 30% replacement showed compressive strength increases of 5.2% and 9.8%, respectively. All tested concrete mixtures showed increases in compressive strength that is directly proportional to curing age. The consistency in the results of all tested concrete mixtures showed that FS can positively impact the compressive strength of concrete. The authors postulated that this could be attributed to the fineness of FS. Splitting tensile strength, flexural strength, and elasticity modulus test result showed a lot of similarity to the compressive strength result. At 28 days of curing, concrete mixtures of 0%, 10%, 20%, and 30% replacement showed splitting tensile strength values of 2.75 MPa, 2.85 MPa, 2.9 MPa, and 3.0 MPa, respectively. Concrete mixtures of 0%, 10%, 20%, and 30% replacement achieved flexural strength of 3.41 MPa, 4.0 MPa, 4.1 MPa, and 4.18 MPa, respectively. Lastly, the concrete mixtures of 0%, 10%, 20%, and 30% replacement showed modulus of elasticity values of 25.1 GPa, 26.75 GPa, 27.60 GPa, and 28.4 GPa. This marginal increase in tensile strength and flexural strength was directly proportional to increase in curing age. The authors were able to experimentally show that an increase in FS as replacement for fine aggregate has a positive impact on the mechanical properties of concrete.

Mohammed Viquaruddin and Abdullah Mohammed [10] were able to show the impact of foundry sand on the mechanical and durability properties of concrete. The authors studied the intrinsic physical effects of foundry sand when used as partial replacement material. The mechanical properties studied were compressive strength, split tensile strength, and flexural strength. The tested samples were categorized according to percentage weight (0%, 10%, 20%, 30%, 40%, 50%, 60%, 70%, 80%, 90%, and 100%) of foundry sand used as replacement material. The authors carried out an all-encompassing analysis that ranged from partial to total percentage replacement. Previous researches enunciated by the authors show a definitive pattern of the impact of FS

TABLE 1: Chemical and mineral composition of cement and silica fume.

Compound	Cement	Silica fume
SiO_2	20.23%	95.8%
Al_2O_3	5.79%	0.18%
Fe_2O_3	1.97%	0.19%
CaO	62.02%	0.30%
K_2O	NA	0.29%
Na_2O	0.77% (equiv.)	0.20%
MgO	1.4%	0.20%
SO_3	3.3%	0.11%

TABLE 2: Mineral composition of foundry sand.

Compound	Foundry sand
MgO	0
Al_2O_3	1.7
SiO_2	94.1
K_2O	0
P_2O_5	0
CaO	0.2
TiO_2	0
MnO	0
Fe_2O_3	5.8

TABLE 3: Grain size distribution for the sand.

Sieve number	Sieve size, mm (in.)	Percentage passing
	Manufactured sand	
30	0.60 (0.0236)	98.19
50	0.30 (0.0118)	34.45
100	0.15 (.00591)	3.14
	Reactive sand	
30	0.60 (0.0236)	98.39
50	0.30 (0.0118)	20.4
100	0.15 (.00591)	3.5
	Foundry sand	
30	0.60 (0.0236)	98.59
50	0.30 (0.0118)	34.81
100	0.15 (.00591)	4.24

on concrete. This established postulation is centered on experiments buffering the notion that FS positively impacts the mechanical properties of concrete. However, some results have also shown that, above or below a particular percentage replacement value (20%), there is a reduction in strength performance and density of concrete mixtures tested. The authors postulated that this could be because of the increased water absorption ratio of the concrete mixture with percentage weight of FS. Mohammed Viquaruddin and Abdullah Mohammed [10] also investigated the absorptivity, initial surface absorption, and water absorption properties of all concrete samples tested. The results showed that increase in FS negatively impacted the workability of all tested concrete mixes. The results also showed a consistent increase in the compressive strength of concrete mixture samples up to 90% FS replacement level. However, concrete samples with 100% FS replacement showed a decrease in compressive strength. The concrete samples also showed increase in splitting tensile strength up to 70% FS replacement level. However, other samples showed reduced split tensile strength. Optimum split tensile strength was achieved at 40% FS replacement. The results also showed that increased age of curing positively impacts split tensile strength of all tested samples. All tested samples exhibited an increase in flexural strength with the exception of concrete mixture samples with FS replacement value more than 50%. The duration of curing also enhanced the flexural strength of all tested concrete samples.

3. Experimental Program

3.1. Cementing Materials. A single ASTM C150 [11] Type I/II ordinary Portland cement (OPC) clinker produced locally was used as a part of this study. The cement has been used extensively within this lab as a control binder for various studies; therefore, the long-term performance of this material is well documented and understood. The total alkali equivalent (Na_2O_{eq}) for this cement is 0.77%. Additionally, a densified silica fume was used at a 25% replacement by mass of cement in order to produce the UHSC requirements (≥70 MPa [10 ksi]) [1]. The silica fume used in this study conformed to ASTM C1240 [12] standards. The chemical and physical properties of the cementing materials are given in Table 1.

3.2. Aggregates. A reactive (siliceous) and a nonreactive (manufactured limestone) fine aggregate were used as a part of this study to produce two separate control UHSC mixtures. The siliceous sand contained quartz (64.0%), chert (17.1%), and feldspar (11.5%) and was procured from El Paso, TX. Previous research completed by Folliard et al. [13] showed this aggregate to be highly ASR reactive based on ASTM C1260 [14] and ASTM C1293 [15] test methods. The limestone aggregate was procured from within the central Texas region. The spent foundry sand (FS) was obtained from a local metal casting company. The sand is spent sand from molds used in their facility and is classified as a green sand (clay bonded system) [2]. Since the FS is unique and has prior processing, the chemical composition of the obtained FS was determined through X-ray fluorescence (XRF) and the results are shown in Table 2.

Based on a literature survey, it is recommended to only use fine aggregate in UHSC mixtures in order to increase the particle packing density and therefore increase strength [6]. It is specifically recommended to use fine aggregate passing ASTM number 30 sieve. Therefore, the two fine aggregates and FS were sieved and collected past the number 30 sieve. The grain size distribution for each aggregate can be seen in Table 3 while the physical properties are presented in Table 4. The specific gravity and densities were very close in value ranging within 2.43–2.62 and 1,411–1,510 kg/m^3, respectively. However, the water absorption for the FS was very low compared to the control measuring only 0.08%.

TABLE 4: Physical properties of aggregates and foundry sand.

Property	Standard	Unit	Manufactured sand	Siliceous sand	Foundry sand
Unit weight	ASTM C29	kg/m^3 (lb/ft^3)	1,432 (89.4)	1,510 (94.3)	1,411 (88.0)
Water absorption	ASTM C127	%	1.94	0.99	0.08
Bulk specific gravity$_{ssd}$ [a]	ASTM C127	—	2.52	2.56	2.62
Bulk specific gravity$_{od}$ [b]	ASTM C127	—	2.43	2.52	2.62

[a] ssd, saturated surface dry condition; [b] od, oven dried condition.

TABLE 5: Concrete mixture proportions.

Mixture name	Cement	Silica fume	Manufactured sand	Reactive sand	Foundry sand	Steel fibers	HRWRA	Water
	kg/m^3 (lb/yd^3)	kg/m^3 (lb/yd^3)	kg/m^3 (lb/yd^3)	kg/m^3 (lb/yd^3)	kg/m^3 (lb/yd^3)	kg/m^3 (lb/yd^3)	l/m^3 (gal/yd^3)	kg/m^3 (lb/yd^3)
NR-Cont	890 (1500)	222 (375)	821 (1384)	—	—	119 (200)	29.7 (6)	222 (375)
NR-10FS	890 (1500)	222 (375)	739 (1246)	—	90 (152)	119 (200)	29.7 (6)	222 (375)
NR-20FS	890 (1500)	222 (375)	657 (1108)	—	180 (303)	119 (200)	29.7 (6)	222 (375)
NR-30FS	890 (1500)	222 (375)	575 (455)	—	270 (455)	119 (200)	29.7 (6)	222 (375)
RA-Cont	890 (1500)	222 (375)	—	842 (1419)	—	119 (200)	29.7 (6)	222 (375)
RA-10FS	890 (1500)	222 (375)	—	757 (1276)	90 (152)	119 (200)	29.7 (6)	222 (375)
RA-20FS	890 (1500)	222 (375)	—	673 (1134)	180 (303)	119 (200)	29.7 (6)	222 (375)
RA-30FS	890 (1500)	222 (375)	—	589 (992)	270 (455)	119 (200)	29.7 (6)	222 (375)

NR: nonreactive (manufactured sand); RA: reactive aggregate (siliceous sand); FS: foundry sand.

3.3. UHSC Mixtures. The concrete mixtures were developed based on the literature [1, 6, 7]. In these studies, UHSC mixtures incorporating high volumes of cement and silica fume, as well as finely graded natural sand, were produced. In this study, the final UHSC mix designs consisted of fine aggregate sieved and collected past the #30 sieve (<0.595-mm [0.0234-in]), OPC, silica fume, and steel fibers. The steel fibers are a common addition for providing additional ductility and fracture toughness in UHSC mixtures. Two series of UHSC mixtures utilizing both fine aggregates (designated as NR and RA for the nonreactive and reactive aggregates, resp.) were prepared at a water-cementitious ratio (w/cm) of 0.20 to monitor the impact of spent foundry sand on the hydration and alkali silica reactivity. Spent foundry sand, with a size of <0.595-mm [0.0234-in] to match the size of the virgin fine aggregates, was utilized at 10, 20, and 30% replacement by volume of fine aggregate, where the 100% virgin aggregate represented the control. A polycarboxylate-based HRWRA was used to achieve desired workability. The mixture proportions of the two mixtures including their 10, 20, and 30% foundry waste replacements are shown in Table 5.

3.4. Specimen Preparation. The aggregates used in this study were sieved to obtain the desired size needed as described previously. The aggregates were then thoroughly washed over a number 200 sieve to remove any fine dust or debris. After washing, the aggregates were oven-dried at 44°C (110°F) for a minimum of 24 hours to achieve a 0% moisture content.

The constituents of each mixture were then mixed for approximately 20 minutes using a laboratory pan mixer. The dry constituents (aggregate, cement, and silica fume) were mixed for the first 2 minutes and then 75% of the water was added. After thorough mixing, the HRWRA was added with the remaining 25% of the water. This preparation method was used based on the literature and experience [1, 6, 7].

3.5. Curing Regimens. In order to minimize as many variables as possible, one curing regimen was tested for the compression testing samples, which was selected based on the literature review developed by Shaheen and Shrive [16]. This curing regimen consists first of curing the samples at room temperature 23°C (73°F) for the first 24 hours. After demolding, the specimens were then heat cured in a water bath at 50°C (122°F) until 2 days prior to testing. At two days prior to testing, the specimens were removed from the water bath and dry cured at 200°C (392°F). The dry curing was added based on recommendation from the literature review [16], such that a high dry curing 2 days prior to testing accelerates strength gain.

3.6. Compression Testing. Compressive strength specimens were molded using 50-mm (2-in.) cube molds. Cubes specimens were used to avoid problems with end preparation of cylindrical specimens [2]. After the specimens were properly cured, they were individually tested according to BS 12390-3-2009 [17]. The British Standard was used as it provides greater detail to testing hardened concrete cubes in compression than ASTM C 39-15a [17, 18]. An average of three samples was tested per data point reported in the Results.

3.7. Calorimetry. Mortar samples from each mixture were used to monitor the heat development using a Calmetrix isothermal calorimeter suitable for paste, mortar, and concrete samples. A standard 23°C isothermal temperature was used in this experimental program. Immediately after mixing, roughly 50 grams was weighed and placed in plastic cup molds designed specifically for use in the calorimeter. The average hydration curves of two samples from each batch were used and are presented in the Results.

3.8. Alkali Silica Reaction. The accelerated mortar bar method (AMBT) ASTM C1567 [19] was used for evaluating the potential reactivity of the aggregate in different foundry sand blends. However, modifications to the standard were used in this experimental program. The standard specifies a standard gradation for the fine aggregate and w/cm ratio to be used during testing. In this program, mixtures were proportioned according to the details mentioned in Section 3.2 to achieve the UHSC requirement. The modifications to ASTM C1567 included using a lower water-cement (w/c) ratio and fine aggregate sieved and collected past the #30 sieve (<0.595-mm [0.0234-in]).

A minimum of four prisms 25 × 25 × 285-mm [1 × 1 × 11.25-in] were used to determine the change in length as outline in ASTM C1567 [19]. Immediately after casting, the molds were covered and placed in a moist curing room at 23°C for 24 hours. Thereafter, the bars were preconditioned for 24 hours in water maintained at 80 ± 2°C. The length of the mortar bars was measured, and then the bars were immediately transferred to storage filled with 1 N NaOH solution maintained at 80°C for testing. The length of the prisms was periodically measured over a 28-day period instead of the normal 14-day period recommended in ASTM C1569 procedure as outlined by Thomas et al. [20]. This decision was made due to the unique mixture proportion nature of UHSC and the use of fibers, which Thomas et al. [20] do not have recommendations for; therefore measurements were taken over a longer period.

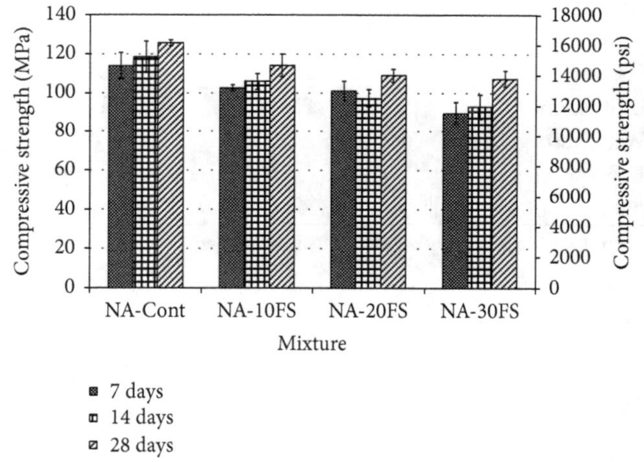

FIGURE 1: Compressive strength of manufactured (nonreactive) sand: foundry sand mixtures at 7, 14, and 28 days.

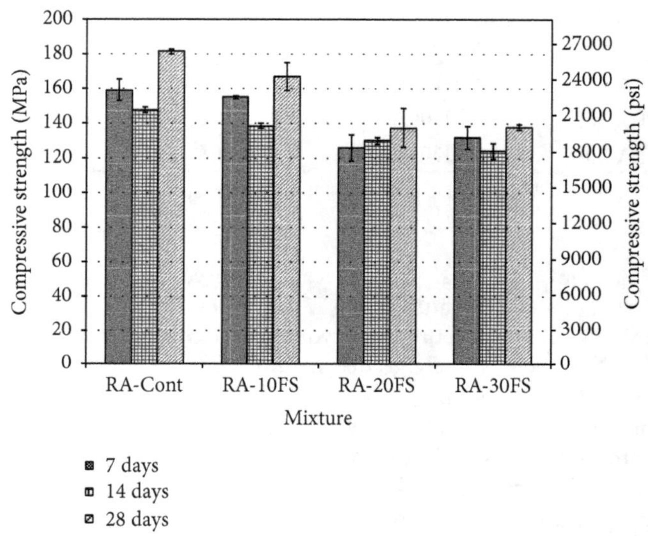

FIGURE 2: Compressive strength of reactive sand: foundry sand mixtures at 7, 14, and 28 days.

4. Results and Discussion

4.1. Strength Development. The compressive strengths for the UHSC mixtures with different percentages of spent foundry sand at 7, 14, and 28-day curing are shown in Figures 1 and 2 for the nonreactive and reactive aggregates, respectively.

The average control compressive strength for all ages was approximately 119 MPa (17,300 psi), which is well above the minimum compressive strength threshold set for UHSC mixtures. This result demonstrates that, at minimum, a UHSC mixture was produced and the impact of FS replacement is investigated. The mixtures incorporating FS ultimately demonstrate a loss in compressive strength with an increase in FS percentage. At 10% inclusion of FS, the average compression strength for all ages dropped to approximately 107 MPa (15,586 psi), which is approximately a 10% reduction in strength from the control. At a 20% fine aggregate replacement with FS, the average compressive strength for all ages dropped to approximately 102 MPa (14,849 psi), which is a 15% drop in compressive strength when compared to the control. Lastly, at a 30% FS replacement level, the compressive strength dropped to approximately 96 MPa (14,000 psi), which is 21% loss in compressive strength when compared to the control samples. To confirm if these results are statistically significant, Student's t-test was completed. This statistical evaluation was performed at 95% confidence interval and a statistical significance (p value) of 0.05 level of confidence. This analysis was completed comparing each FS replacement level to the control strength to determine if the noticed trend is significant. The statistical analysis shows that the 10%, 20%, and 30% FS replacement mixtures are statistically significant with a p value of 0.041, 0.015, and 0.005, respectively. This result confirms that an increase in FS replacement, up to the 30% replacement level tested, yields a loss in compression strength. However, it should be noted that at the highest replacement level tested (30%) the average compressive strength for all ages was still quite high (96 MPa (14,000 psi)) and still considered UHSC; therefore there is still application for this mixture, which is more sustainable than the control. Lastly, it should be noted that the nonreactive sand showed an increase in strength with age aside from the 20% FS replacement, which showed a drop in strength of approximately 4% from the 7-day strength to the 14-day strength; however, the 28-day strength remained the highest obtained strength. Although the average value of the NR-20FS was lowest at 14 days, this minor deviation from the overall trend is still within the standard deviation of the increasing strength with age trend of all other samples. Student's t-test was also performed on this value to further demonstrate that the 14-day strength (NR-20FS) value was not significant when compared to the 7-day strength (NR-20FS) with a p value of 0.18. Overall, the compression strength performance of the nonreactive (manufactured sand) demonstrated that all mixtures produced were of UHSC quality with varied degrees of strength and sustainability.

On the other hand, mixtures incorporating reactive siliceous sand showed substantially higher strength performance. The control mixture showed the highest compressive strength overall, exceeding 150 MPa (21,755 psi) after only 7 days of curing. Interestingly, a sharp reduction in compressive strength was noted after 14 days of curing for all mixtures with the exception of 20% foundry sand. This may be attributed to the self-desiccation of the low w/cm concrete mixture limiting the continued hydration for the additional formation of hydration product [21]. However, a notable change in the strength development after 28 days of heat curing was observed showing the highest compressive strength overall for all UHSC mixtures. Similarly to the mixtures incorporating the nonreactive manufactured limestone, a drop in compression strength was observed with increasing replacement of foundry sand. To determine if these results are statistically significant, Student's t-test was completed similarly as before. The statistical analysis shows that the 10%, 20%, and 30% FS replacement mixtures are statistically significant with a p value of 0.045, 0.027, and 0.019, respectively. The t-test confirms that there is a reduction in compressive strength relative to the control with increasing replacement of FS at up to 30% replacement.

The control mixture reached a compressive strength of 181.78 MPa (26,365 psi), whereas mixtures RA-10FS, RA-20FS, and RA-30FS observed a compressive strength of 167.21 (24,252 psi), 137.70 (19,972 psi), and 137.90 MPa (18,550.33), respectively. This is contrary to what previous studies have shown that spent foundry sand used as a partial replacement of fine aggregate can improve the mechanical properties of concrete [2, 3]. However, these past studies limited their focus to normal-strength concrete. Thus, the results may indicate that the addition of foundry sand may contribute to additional imperfections between the paste-aggregate interphase and limiting the concrete's ability to achieve higher compressive strengths. Nonetheless, the compressive strength results between all the mixtures demonstrate that up to 30% of foundry waste can be used to replace the virgin fine aggregate while still exceeding the UHSC requirements (≥70 MPa (10,000 psi)) [1].

4.2. Alkali Silica Reactivity. The potentials for alkali silica reactivity for both the nonreactive and reactive aggregates containing foundry sand are presented in Figures 3 and 4.

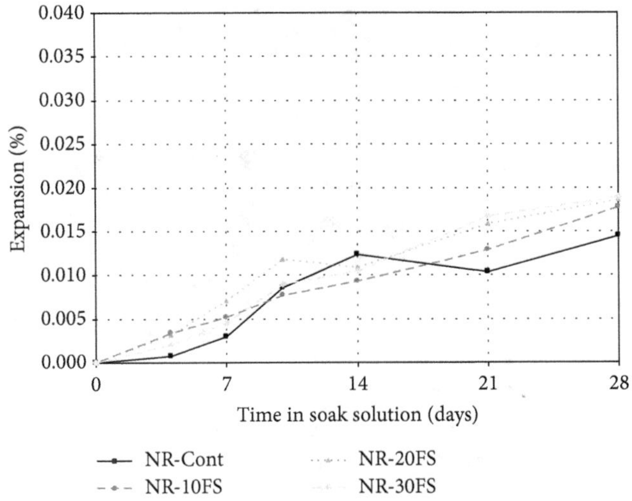

FIGURE 3: Expansion curves of manufactured (nonreactive) sand: foundry sand mixtures.

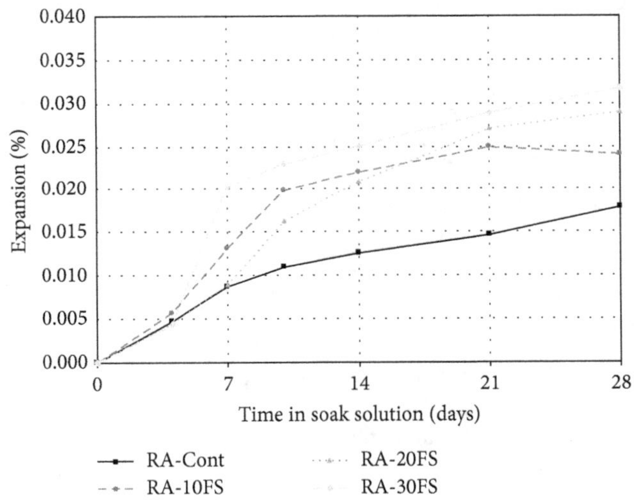

FIGURE 4: Expansion curves of reactive sand: foundry sand mixtures.

After 28-day storage in NaOH solution, mixtures containing foundry sand showed a slight increase in expansion compared to their control, although the final expansion value between the two aggregates is not very remarkable. The results indicate that as higher replacements of foundry sand are incorporated into the mixture, expansion of the prisms is also higher. The effects are pronounced for those mixtures utilizing the reactive siliceous aggregate. This may be attributed to the additional silica content coming from the foundry sand waste [22].

Note that the prisms were cast using high capacity steel fibers as a part of the matrix. Thus, only minor changes in length were observed for both series of mixtures and well under the expansion limit of 0.10% for deleterious alkali silica reactivity as suggested by [23, 24]. As discussed previously, the steel fibers provide extra restraint against tensile stresses and thus contributed to reduced expansion observed in the prisms. However, cracking was still apparent surrounding

FIGURE 5: Cracking of reactive bars observed after 28 days in NaOH solution: (a) RA-30FS; (b) RA-10FS.

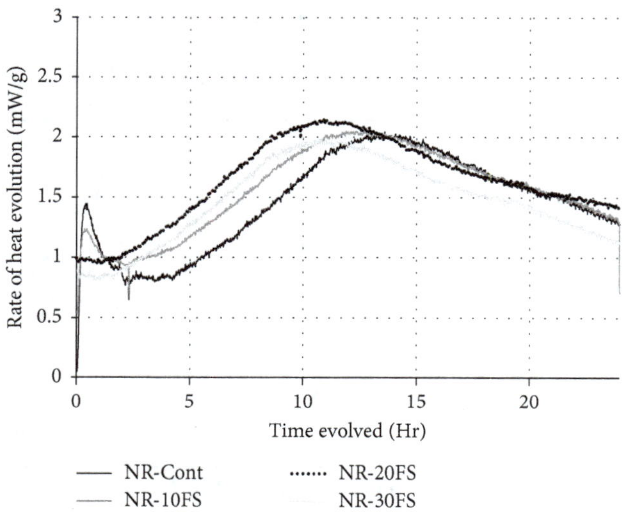

FIGURE 6: Heat flow curves for UHSC-FS mixtures using nonreactive aggregate.

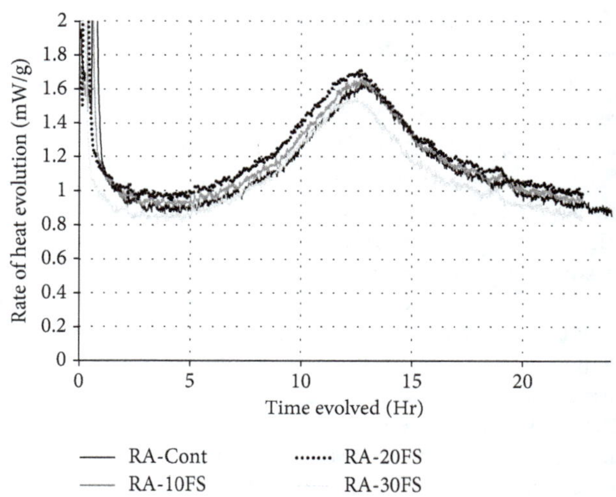

FIGURE 7: Heat flow curves for UHSC-FS mixtures using reactive aggregate.

the edges of the prisms for the mixtures incorporating the reactive aggregate (see Figure 5) indicating there still exists a large potential for ASR reactivity.

4.3. Heat of Hydration. Isothermal calorimetry was chosen as the method for investigating the hydration of UHSC mixtures with various replacement of foundry waste. Isothermal calorimetry measures the heat flow from cement hydration reactions by differential heat flow sensors that allow for comparison of different mixtures and provide a time-lapsed understanding of the hydration mechanisms. The timing and shape of heat flow curves can provide an understanding of the performance of different cementitious systems. Additionally, heat flow curves combined with compressive strength data can provide insight on the mechanisms through which the addition of mineral admixtures such as foundry waste influences strength development.

The data from isothermal calorimetry test were used to plot curves showing the rate of heat produced during hydration and are presented in Figures 6 and 7 for the nonreactive and reactive sand, respectively. A summary of the peak heat flow for each mixture compared to that of the control is also listed in Table 6. Each heat flow curve has been normalized to the amount of cement in the mortar (i.e., 50-gram mortar sample). The shape and position of the heat flow curves for the foundry waste UHSC mixtures relative to the curve of the control provide valuable information regarding the effect of the foundry waste on cement hydration. A leftward shift of the heat flow curve relative to the control curve indicates an advancement in the onset of cement hydration which in turn suggests an acceleration of cement hydration kinetics. From Figure 6, it can be observed that the rate of heat curve for the foundry waste UHSC mixtures using nonreactive aggregate shifted significantly to the left of the control curve. On the other hand, the rate of heat curves for similar mixtures using reactive aggregate did not show a significant deviation to the left of the control curve. In fact, mixture RA-10FS showed slight delay in activity with a slight shift to the right from the control (see Table 6). The increase activity from the addition foundry waste is most likely attributed to additional nucleation sites and increased reactivity of hydration products, or the "size filler effect"

Table 6: Heat of hydration summary for concrete mixtures.

Mix-ID	Avg. peak heat flow (mW/g cement)	Deviation from control sample (mW/g cement)	Avg. time until peak heat flow (Hr)	Deviation from control sample (Hr)
NR-Cont	2.04	—	13.58	—
NR-10FS	2.06	0.02	12.30	1.28
NR-20FS	2.15	0.11	10.92	2.67
NR-30FS	1.98	−0.06	10.77	2.82
RA-Cont	1.64	—	12.65	—
RA-10FS	1.66	0.03	12.90	−0.25
RA-20FS	1.71	0.07	12.63	0.02
RA-30FS	1.55	−0.09	12.40	0.25

[25–27]. The finer particle distribution of the foundry waste compared to that of the reactive sand (see Table 3) provided a better particle packing of materials. This may have resulted in an increase in the number of foundry sand particles available per grain of cement for nucleation and increased hydration reactivity. For the foundry waste UHSC mixtures using nonreactive sand, the average deviation in time to reach the peak heat flow was 1.28, 2.67, and 2.82 hours relative to the control, which is approximately a 9.4%, 19.6%, and 20.7% reduction for 10%, 20%, and 30% foundry waste replacement, respectively. For the equivalent mixtures using reactive siliceous sand, the deviation in time to reach the peak heat flow was −0.25, 0.02, and 0.25 hours relative to their control, which is approximately a 2.0% increase and 0.0016% and 2% reduction for 10%, 20%, and 30% foundry waste replacement, respectively. Note that a negative sign indicates a delay in hydration reactivity (rightward shift) relative to the control and vice versa for a positive value.

From Figures 6 and 7 and Table 6, it can be observed that the average measured peak heat flow between each foundry waste replacement relative to their control did not show a significant deviation. However, there was some notable differences in the peak heat flow between the two aggregate types (see Table 6). On average, mixtures using the nonreactive (manufactured) sand exhibited a 24.4%, 24.1%, 25.7%, and 27.7% increase in average peak heat flow for the control, 10%, 20%, and 30% foundry waste replacement compared to the reactive sand mixtures, respectively. This suggests that the foundry waste UHSC mixtures containing nonreactive aggregate, which has a similar particle size distribution to the foundry waste, exhibited enhanced nucleation and growth of hydration products, whereas the size filler effects are minimal in the mixtures containing reactive aggregate. Interestingly, the measured compressive strength between the two aggregate types also observed apparent differences. Note that all mixtures used equivalent cementitious contents and only the influence of foundry waste was investigated in this study.

5. Conclusions

A sustainable, high strength concrete mixture using waste foundry sand was investigated in this study. Mechanical properties, hydration reactivity, and alkali silica reactivity were monitored on two series of UHSC mixtures blended with spent foundry sand up to 30% replacement of fine aggregate by volume. Based on the investigation, the following conclusions can be drawn:

(i) For both series of UHSC mixtures using nonreactive and reactive fine aggregate, a decrease in compressive strength was observed with the replacement of natural fine aggregate with that of foundry waste up to the 30% replacement investigated.

(ii) UHSC mixtures using the reactive siliceous sand yielded the highest compressive strengths compared to those using the nonreactive manufactured sand. However, the minimum compressive strength observed was 85 MPa (12,345 psi) at the highest replacement level tested (30%) for the series of mixtures using nonreactive sand and is still considered UHSC.

(iii) The alkali silica reactivity for mixtures containing foundry sand showed an increase in expansion with increasing replacement of foundry waste up to 30%. The effects are pronounced for those mixtures blended with foundry waste and reactive siliceous aggregate.

(iv) The calorimetry data showed that the addition of foundry waste enhanced the growth of hydration products by providing extra nucleation and growth sites with their finer particle size distribution. This was very pronounced for the UHSC mixtures using nonreactive fine aggregate.

Acronyms

ACI: American Concrete Institute
ASTM: American Society for Testing and Materials
ASR: Alkali silica reaction
AMBT: Accelerated mortar bar test
CIM: Concrete Industry Management
FS: Foundry sand
GGBFS: Granulated ground blast furnace slag
HRWRA: High range water reducing admixture
NR: Nonreactive
OD: Oven dry

OPC: Ordinary Portland cement
RA: Reactive aggregate
REP: Research Enhancement Program
SCM: Supplementary cementitious material
SSD: Saturated surface dry
UHSC: Ultrahigh strength concrete
XRF: X-ray fluorescence.

Conflicts of Interest

The authors declare that there are no conflicts of interest regarding the publication of this paper.

Acknowledgments

The authors of this study would like to thank Research Enhancement Program (REP) at Texas State University for the support of this research. The authors would also like to thank Mr. Brady Gillar from Southwest Steel Casting Company for donating the foundry waste to their study. Lastly, thanks are due to the several undergraduate researchers in the Concrete Industry Management (CIM) program for their assistance in this work.

References

[1] ACI Committee 363, "Report on High-Strength Concrete," Tech. Rep., American Concrete Institute Committee 363, Farmington Hills, Mich, USA, 2010.

[2] R. Siddique, G. D. Schutter, and A. Noumowe, "Effect of used-foundry sand on the mechanical properties of concrete," *Construction and Building Materials*, vol. 23, no. 2, pp. 976–980, 2009.

[3] R. Siddique and A. Noumowe, "Utilization of spent foundry sand in controlled low-strength materials and concrete," *Resources, Conservation and Recycling*, vol. 53, no. 1-2, pp. 27–35, 2008.

[4] T. Abichou, C. Benson, and T. Edil, "Database on beneficial reuse of foundry by-products," in *Geotechnical Special Publication No. 79. ASCE*, C. Vipulanandan and D. Elton, Eds., Recycled Materials in Geotechnical Applications, pp. 210–223, American Society of Civil Engineers, 1998.

[5] Ministry Of Electricity & Renewable Energy, "Spent foundry sand: alternative uses study," report prepared by john emery geotechnical engineering limited for ontario ministry of the environment and energy and the Canadian foundry association, Queen's Printer for Ontario, 1993.

[6] S. Collepardi, L. Coppola, R. Toli, and M. Colepardi, "Mechanical properties of modified reactive powder cocnrete," *Mechanical Properties of Modified Reactive Powder Cocnrete*, 2007.

[7] A. S. Dili and M. Santhanam, "Investigations on reactive powder concrete: A developing ultra high-strength technology," *Indian Concrete Journal*, vol. 78, no. 4, pp. 33–38, 2004.

[8] S. Fiore and M. C. Zanetti, "Foundry wastes reuse and recycling in concrete production," *American Journal of Environmental Sciences*, vol. 3, no. 3, pp. 135–142, 2007.

[9] G. Ganesh Prabhu, J. W. Bang, B. J. Lee, J. H. Hyun, and Y. Y. Kim, "Mechanical and durability properties of concrete made with used foundry sand as fine aggregate," *Advances in Materials Science and Engineering*, vol. 2015, Article ID 161753, 2015.

[10] Mohammed Viquaruddin and Abdullah Mohammed, "An Experimental Study on Mechanical Properties of Concrete Incorporating Foundry Sand," *International Journal of Engineering Research and*, vol. V6, no. 3, pp. 83–88, 2017.

[11] "Standard Specification for Portland Cement ASTM International," ASTM Standard C150/C150M-15 (2015), West Conshohocken, Pa, USA, 2015, https://www.astm.org.

[12] "Standard Specification for Silica Fume Used in Cementitious Mixtures ASTM International," ASTM Standard C1240/C618M-15 (2015), West Conshohocken, Pa, USA, 2015, https://www.astm.org.

[13] K. Folliard, R. Barborak, T. Drimalas et al., "Preventing ASR/DEF in new concrete: final report," Report No. FHWA/TX-06-0-4085-5, 2006.

[14] "Standard Specification for Potential Alkali Reactivity of Aggregates (Mortar-Bar Method," in *Proceedings of the ASTM Standard C1260/C1260M-14 (2014) Standard Specification for Potential Alkali Reactivity of Aggregates (Mortar-Bar Method) ASTM International*, vol. C1260/C1260M-14, pp. 10–1520, West Conshohocken, Pa, USA, 2014, https://www.astm.org.

[15] "Standard Specification for Determination of Length Change of Concrete Due to Alkali-Silica Reaction ASTM International," ASTM Standard C1293/C1293M-15 (2008), West Conshohocken, Pa, USA, 2015, https://www.astm.org.

[16] E. Shaheen and N. G. Shrive, "Optimization of mechanical properties and durability of reactive powder concrete," *ACI Materials Journal*, vol. 103, no. 6, pp. 444–451, 2006.

[17] "Testing hardened concrete compressive strength of test specimens B/517/1," British Standard Institution BS EN 12390-3-2009, pg 22.

[18] "Standard Test Method for Compressive Strength of Cylindrical Concrete Specimens ASTM International," ASTM Standard C 39-15a (2015), West Conshohocken, Pa, USA, 2015, https://www.astm.org.

[19] "Standard Test Method for Determining the Potential Alkali-Silica Reactivity of Combination of Cementitious Materials and Aggregte (Accelerated Mortar-Bar Method) ASTM International," ASTM Standard C1567/C1567M-13 (2013), West Conshohocken, Pa, USA, 2015, https://www.astm.org.

[20] M. D. A. Thomas, B. Fournier, K. J. Folliard, M. H. Shehata, J. H. Ideker, and C. Rogers, "Performance limits for evaluating supplementary cementing materials usina accelerated mortar bar test," *ACI Materials Journal*, vol. 104, no. 2, pp. 115–122, 2007.

[21] M. A. Caldarone, *High-strength Concrete: A Practical Guide*, Taylor & Francis, 1st edition, 2009.

[22] M. K. Dash, S. K. Patro, and A. K. Rath, "Sustainable use of industrial-waste as partial replacement of fine aggregate for preparation of concrete: a review," *International Journal of Sustainable Built Environment*, vol. 5, no. 2, pp. 484–516, 2016.

[23] M.-A. Berube, J. Duchesne, and D. Chouinard, "Why the accelerated mortar bar method ASTM C 1260 is reliable for evaluating the effectiveness of supplementary cementing materials in suppressing expansion due to alkali-silica reactivity," *Cement, Concrete and Aggregates*, vol. 17, no. 1, pp. 26–33, 1995.

[24] M. D. A. Thomas and F. A. Innis, "Use of the accelerated mortar bar test for evaluating the efficacy of mineral admixtures for controlling expansion due to alkali-silica reaction," *Cement, Concrete and Aggregates*, vol. 21, no. 2, pp. 157–164, 1999.

[25] C. Gosselin, "Microstructure development of calcium aluminate cements accelerated by lithium sulphate," in *Proceedings of the Centenary Conference*, pp. 109–122, HIS BRE Press, Avignon, France, 2008.

[26] A. F. Bentivegna, *Multi-Scale Characterization, Implementation, and Monitoring of Calcium Aluminate Cement Based-Systems [Ph.D. thesis]*, University of Texas, Dissertation Austin, Tex, USA, 2011.

[27] S. Seraj, *Evaluating Natural Pozzolans for Use as an Alternative Supplementary Cementitious Materials in Concrete [Ph.D. thesis]*, University of Texas, Dissertation Austin, Tex, USA, 2014.

10

Strength and Stiffness of Stabilized Alluvial Silt under Frost Actions

Haibo Wang,[1] An Deng,[2] and Ping Yang[1]

[1] *Department of Civil Engineering, Nanjing Forestry University, Nanjing, Jiangsu 210037, China*
[2] *School of Civil, Environmental and Mining Engineering, University of Adelaide, Adelaide, SA, Australia*

Correspondence should be addressed to Haibo Wang; harbor006@foxmail.com

Academic Editor: Hossein Moayedi

The Yellow River alluvial silt was stabilized into pavement base materials for cold regions. The stabilizing additives were cement, fly ash, and lime, which were included in a range of combinations and dosages when mixed with the silt. Freeze-thaw cyclic impacts were conducted on the treated samples to assess materials performance of withstanding the frost actions. The tests were conducted on samples cured for 7 days to up to 180 days. Test results show that the cement-fly ash-treated samples outperform the other two stabilization categories with respect to material strength and stiffness developed under both normal and frost conditions. Under the normal conditions, the material unconfined compressive (UC) strength rises to 3.0 MPa on day 28 depending on the cement and fly ash dosage used. If subjected to frost actions, the fly ash inclusions warrant a residual UC strength value of 1.3 MPa and above. The antifrost performance of the cement-fly ash-treated samples is related to thermal buffer capacity of the fly ash particles. Water adsorption and material soundness results agree with the strength and stiffness development. An optimal dosage was 3–6% for the cement and 0.3 for cement to fly ash mass ratio.

1. Introduction

The Yellow River is the second longest river in China. As illustrated in Figure 1, the river originates in western China, flows toward the east, and empties into the Bohai Sea. When passing through the Loess Plateau situated in the midwest of China, the river erodes and thus carries downstream billions of tons of loess each year, which eventually deposits to the lower reaches of the river. The loess has deposited to a wide scope due to historical course changes and river floods, and the wide deposit scope is known as the Yellow River Flooding Area, which measures about 54,000 hectares. The loess deposits in this area are designated as the Yellow River alluvial silt.

The alluvial silt has been deemed as an important choice of fill material for infrastructures due to its vast volume and local accessibility. The infrastructures, however, are exposed to seasonal frost action with a ground surface temperature of down to −10°C from December to February and up to more than 30°C in July and August each year [1]. These frost actions have been damaging the fill material matrix and undermining the stability of the infrastructures, including crack propagation and excessive ground deformation or even instability failures [2–4]. A solution for this concern is to stabilize the silt using cementitious additives and examine the performance of the treated silt counteracting the frost action. Altun et al. [5] stabilized sandy and clayey silt of low plasticity by using Class C fly ash and cement and subjected the stabilized specimens to up to seven freezing and thawing cycles. The results showed that 30% fly ash and 2% cement are the optimum dosages with respect to material strength gain, and samples with 10% fly ash and 2% cement exhibit favorable antifrost performance. Also, on common silt soils but treated with geofiber and synthetic fluid, Hazirbaba and Güllü [6] found that these new additives help maintain the materials strength after seven freeze-thaw cycles. Similar hybrid additives were attempted by Güllü and Khudir [7] who stabilized low plasticity silt using jute fiber and steel fiber together with lime and assessed the effect of freeze-thaw cycles on material strength. These studies illustrate the likelihood of stabilizing the Yellow River alluvial silt to counteract the frost action.

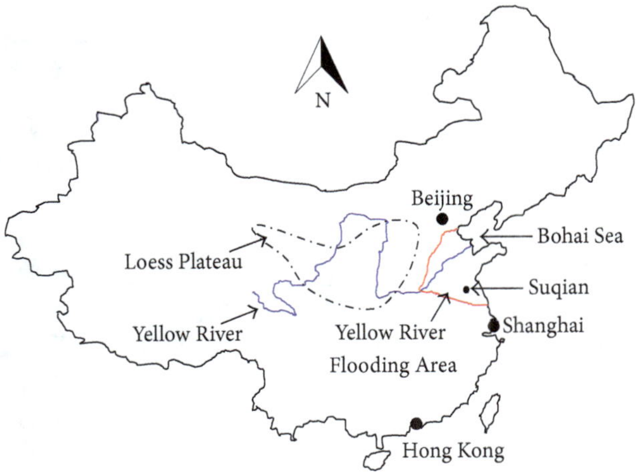

FIGURE 1: The Yellow River basin (not drawn to scale).

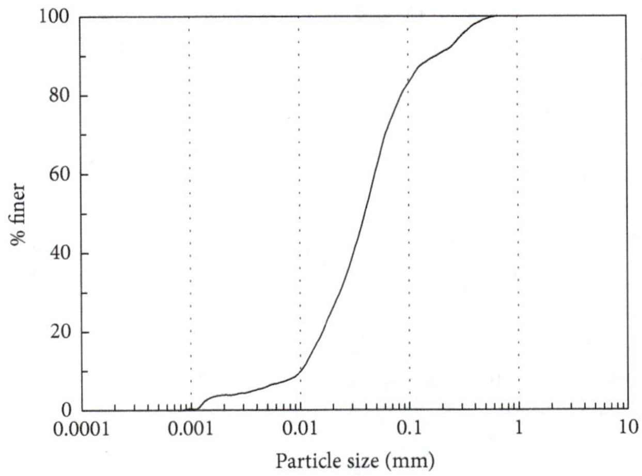

FIGURE 2: Grain size distribution of soil sample.

Despite the existence of studies on common silty soils, stabilizing the Yellow River alluvial silt is worth an investigation due to the silt unique characteristics. The alluvial silt has been subjected to the river weathering process and, separate from the common silty soils, exhibits special properties: round and uniform particles, low plasticity, zero to low cohesion, and hydraulic instability. The demonstrated region-specific properties likely influence the stabilization process and lead to a design different from other common silty soils. Previous studies have also shown that the stabilization and antifrost performance are dependent on the choices of the silty soils used [5, 7]. Similar dependence may occur to the Yellow River alluvial silt. To this end, this study adds value to the community with respect to shedding some light on how better this Yellow River alluvial silt can be stabilized under the seasonal frost environments.

In review of recent soil stabilization practices [8–14], the current study considered three common inorganic additives, Portland cement, Class F fly ash, and hydrated lime, as the stabilizers for the Yellow River alluvial silt. All of the three additives were included in a series of mass contents designed to examine the role of the additives in improving silt performance, aiming to work out a dosage for field applications. Specifically, take the cement as the major stabilizer and fly ash or lime as the minor, examining the likelihood of the minor stabilizer in improving the material performance. This is important as the hybrid scenarios likely excel in a difficult or severe environment such as the frost impact. In addition, the ingredients dosage spans a wide range so as to work out an optimum dosage. The results are supposed to be applicable to all climate environments, including regular benign conditions and severe seasonal impacts as examined in some of the previous studies. To attain this goal, this study subjected the treated soil samples to two artificial testing ambience settings: constant room temperature and freezing-thawing cyclic environments. Some critical test results including unconfined compressive (UC) strength, material mass loss, and water absorption observed under the respective testing ambience settings are presented and discussed.

2. Materials and Methods

2.1. Materials. The Yellow River alluvial silt was recovered from Suqian of Jiangsu Province, a midsize city situated in the south part of the Yellow River Flooding Area (Figure 1). Figure 2 shows the soil particle size distribution curve. Major soil index properties are summarized in Table 1. The particle size distribution was measured by a laser particle size analyzer Mastersizer 2000 which enables size analysis of both the coarse and the fine fractions of the silt. The vendor's manual specifies that a volume of sample flows across the focused beam of laser light and the intensity of light scattering is recorded, analyzed, and related to particles size. The Atterberg limits results for the alluvial silt sample are 22% for the plastic limit I_P and 33% for the liquid limit I_L. The silt is classified as ML sandy silt of low plasticity with 4.8% clay, according to the Unified Soil Classification System. Undisturbed native soil samples exhibit low bearing capacity with a California Bearing Ratio = 2.6% and UC strength = 0.11 MPa. Three cementitious additives were used, Portland cement, Class F fly ash, and hydrated lime, to stabilize the silt. The cement is rated as CEM I 32.5 R and has a density of 3,060 kg/m^3, specific area of 3,460 cm^2/g, initial setting time of 2.98 h, and final setting time of 5.67 h. The fly ash was collected from Xuzhou Thermal Power Plant. It has a density of 2,210 kg/m^3, specific area of 2,251 cm^2/g, and maximum dry density of 1,320 kg/m^3 obtained through the Proctor test. The Proctor test was conducted in accordance with China Standard for Soil Test Method [15]. The hydrated lime showed activity of 80%. Significant material composition differences are summarized in Table 2.

The additives were added in three combinations: cement alone, cement-fly ash combined, and cement-lime combined. All of the three combinations were blended together with corresponding optimum moisture contents which were obtained through the soil Proctor compaction tests. Each combination included a range of dosages with details shown in Table 3. The additive dosages are presented as percentage of dry soil mass. The dosages were designed to make the effect of additives content on the performance of the treated soil samples stand

Table 1: Soil index properties.

Property	Value
Specific gravity	2.70
Sand (%)	22.9
Silt (%)	72.3
Clay (%)	4.8
Liquid limit (%)	33
Plastic limit (%)	22
Plasticity index	11
Maximum dry density (kg/m^3)	1.78
Optimum moisture content (%)	15.7
USCS classification	ML

Table 2: Chemical compositions of materials.

Oxide	Percentage by mass (%)			
	Silt	Fly ash	Cement	Lime
SiO_2	61.95	50.11	21.40	—
Fe_2O_3	3.65	5.25	2.69	0.4
MgO	2.54	1.61	2.02	6.8
CaO	7.10	2.64	63.65	72.9
Na_2O	1.15	0.57	0.31	—
K_2O	2.01	0.45	0.67	—
Al_2O_3	12.42	27.2	4.58	0.8
SO_3	—	7.05	2.63	—
Loss on ignition	6.95	5.19	5.14	—

Note. —: not tested.

out. For instance, for the cement-treated silt, the cement content falls in the choices of 3%, 6%, and 8%; for the cement-fly ash-treated silt, the fly ash content increases from 7% to 15% with the cement content remaining unchanged at 3%, or from 6% to 30% with the cement content remaining unchanged at 6%. The chosen dosages were also designed into levels in line with those adopted in existing studies (e.g., [10, 12, 16]) and so help compare the test results.

2.2. Sample Preparation. Oven-dry silt and the additives were proportioned based on the designed dosages shown in Table 3. The dry mixtures were thoroughly blended in a mixer of 60 liters. The mixer capacity is about six times the volume of the intended ingredients of one batch and so offers decent room to facilitate motions of the ingredient particles and thus mixture uniformity. The weighed ingredients were fed into the mixer in dry states and in sequential equal portions. It is well acknowledged that the dry mixing process outperforms the wet mixing process with respect to achieving mixture uniformity when dealing with blending multiple ingredients. In addition, feeding the ingredients in portions avoids possible clods, clumps, or segregation. To confirm the mixture uniformity, the standard deviation of the UC test results was assessed. At least four samples from the same batch were tested and a value of 15% for the standard deviation was applied. If the value is above 15%, additional samples were tested until the criterion was satisfied. The criterion is a

Table 3: Additives used to treat silt.

Sample	Cement (%)	Fly ash (%)	Lime (%)	Water (%)
C3	3	0	0	16.0
C6	6	0	0	15.9
C8	8	0	0	15.9
C3FA7	3	7	0	15.4
C3FA10	3	10	0	15.3
C3FA15	3	15	0	15.0
C6FA6	6	6	0	16.4
C6FA10	6	10	0	15.6
C6FA20	6	20	0	15.0
C6FA30	6	30	0	14.9
C8FA20	8	20	0	15.8
C1.6L4	1.6	0	4	16.8
C2L4	2	0	4	16.9
C3L4	3	0	4	17.0
C3L7	3	0	7	17.3
C3L9	3	0	9	17.7

normal convention used to confirm consistency for UC test results of a geomaterial. The corresponding water content shown in Table 3 was included into the dry mixture to form a paste which was then subjected to further blending efforts. The complete blending process lasted about 30 minutes when no major clumps larger than 5 mm were observed. The fresh paste was then poured into a cylindrical steel mold of 50 mm diameter and 100 mm height. This sample dimension helps attain a uniform thermal profile within the sample when altering ambient temperature and therefore minimize the effect of transient freezing [17]. A static sample tamping process, as specified in standard ASTM D5102 [18], was adopted to achieve 96% degree of compaction. This degree of compaction meets the specifications stipulated for common civil infrastructures. Efforts such as use of multiple tamping devices were undertaken to compact the samples of a batch before the cement initial setting time. Calculate the mass and volume of the soil needed to achieve the 96% degree of compaction, feed the soil into the mold, and compress the soil by the plunger to the desirable volume. Once the sample was formed and extruded from the mold, it was then wrapped with a waterproof membrane and stored in a curing chamber with a constant temperature of 20 ± 1°C and relative humility of 96 ± 2% until a testing date.

2.3. UC Strength Test. The UC strength test was conducted on samples cured for 7, 28, and 180 days, respectively. The UC strength test was also conducted on cured samples after the designed freezing-thawing cycles. Both tests referred to the standard ASTM D5102 (2016). In a UC strength test, the sample was compressed at a rate of 1 mm/min. The rate of strain was chosen so that the time to failure was about 15 minutes. The UC strength was the peak stress obtained during the test. At least four replicate samples were tested for each examined dosage and, as mentioned before, a criterion of 15% standard deviation applies to the UC strength results before averaging them into the UC strength value for this dosage.

TABLE 4: Freezing-thawing program.

Sample	Freeze-thaw cycles		
	After 28-day curing	After 38-day curing	After 180-day curing
C3	1, 3		
C6	1, 3, 5, 7, 9		
C3FA15	1, 3	5, 7, 9	
C6FA10	1, 3	5, 7, 9	1, 3, 5, 7, 9
C6FA20	1, 3	5, 7, 9	
C6FA30	1, 3	5, 7, 9	
C2L4	1, 3, 5, 7, 9		
C3L4	1, 3, 5, 7, 9		
C3L7	1, 3, 5, 7		
C3L9	1, 3, 5		

2.4. Freezing and Thawing. The freezing and thawing processes were conducted in accordance with ASTM D560 [19]. Upon a curing date, the soil samples were unwrapped, weighed, inundated in water for 24 hours to saturate the samples, and reweighed. The saturated samples were then kept in a freezer at −15°C for 24 h to sustain freezing. The temperature values were determined to suit the local (i.e., Suqian city) climate conditions documented in the past ten years. Next, the sample was transferred to the same curing chamber for 2-3 hours until completely thawed and then hydrated in the curing chamber for another 21-22 hours. These steps formed one freeze-thaw cycle. Sufficient samples, at least 36 cylinders for one dosage, were prepared in order to account for the number of tests conducted under varying freeze-thaw cycles. Sample and test details are shown in Table 4. The shown test program considered three curing ages: 28 days, 38 days, and 180 days. After each curing age, the samples were subjected to a certain number of continuous freeze-thaw impacts and then UC strength tests. In the testing program, sample C6FA10 was subjected to all three curing ages. This is to examine the time-dependent performance of the fly ash which is known to possess a pozzolanic reaction in the long term. The 28-day and 38-day curing times were designed to examine strength development at early ages. Scaling occurs to stabilized soils when exposed to freeze-thaw impacts and this is assessed as a material property known as soundness. To assess the material soundness against the frost actions, all samples were weighed at the end of the designed cycles and before being subjected to the UC tests where possible.

3. Results and Discussion

3.1. Compaction. The Proctor compaction test results are shown in Figure 3. Values are shown in terms of additives and dosages used to stabilize the Yellow River alluvial silt. Figure 3(a) shows higher optimum moisture content (OMC) and lower maximum dry density (MDD) when stabilizing the silt with the cement. These results are in agreement with past studies such as that by Tripura and Singh [20]. The cement dependence of the compaction curves arises from noticeable water absorption capacity of the cement in relation to the silt. As a result, an extra volume of water is required to lubricate solid particles and attain the optimum compaction of the mixtures. The OMCs, MDDs, and the compaction curves, however, remain similar for the series of cement-treated silt samples. With the cement content of 3% to 8%, the values range from 15.7% to 16.1% for the OMCs and from 1,780 to 1,766 kg/m^3 for the MDDs.

The cement-lime-treated and cement-fly ash-treated samples show different trend lines. For the cement-lime-treated samples, the OMCs increase and the MDDs decrease with the increase of the lime content, if the cement content remains at 3%. This implies that the compaction is sensitive to the lime content. Similar sensitivity occurs to the cement-fly ash-treated samples. As what occurred in earlier studies [10, 16, 21], the decreases of MDDs occur due to the lower specific gravity of the fly ash and the lime—the replacement of soil particles in a given volume by particles of fly ash or lime of lower specific gravity compared to that of the soil particles. It is noteworthy that the cement-fly ash-treated samples show interesting trends for the OMCs. The OMC values increase for sample C6FA6 and decrease for the other series of samples. The trend lines differ from those obtained by Kumar et al. [10] which shows increasing OMCs (28–37%) with an increase of fly ash (0–20%). Their study used similar Class F fly ash but high plasticity black cotton soil with I_P = 49.65% and I_L = 68%. The values are higher than corresponding values of the Yellow River alluvial silt shown in Table 1. This implies that the soil plasticity likely influences the OMC values. The higher the plasticity, the higher the OMCs. This argument is acceptable as the additional moisture content works as a lubricant into the surface of additional clay particles.

An explanation for the "discrepancy" on the compaction curves among the three stabilization categories arises from the varying CaO contents in the additives. As shown in Table 2, a significant difference exists with respect to the CaO contents among the three additives: 2.64% for the fly ash, 63.65% for the cement, and 72.9% for the lime. In contact with water, the content of CaO determines the water volume needed to form calcium hydroxide, an ingredient for pozzolanic reactions. This means that the need of the water volume for the fly ash-based samples is not as significant as that for the cement- or the lime-based samples in order to enable thorough pozzolanic reactions. The need for water becomes less noticable when increasing the fly ash contents. Conversely, the need for the water volume reduces when increasing the fly ash fraction in the cement-fly ash-treated samples. In the meantime, the spherical nature of the fly ash particles facilitates packing of the mixture which further offsets the need for water. As a result, the more the fly ash content, the less the water volume for the cement-fly ash-treated samples, in relation to the cement- or cement-lime-treated samples, needed to compact the mixture.

3.2. UC Strength (Without Frost Actions). The UC strength results are shown in Figure 4. Values are shown for samples without the frost actions and categorized in terms of the curing times and the additives used. Varying strength levels are obtained for the silt, dependent on the additives content

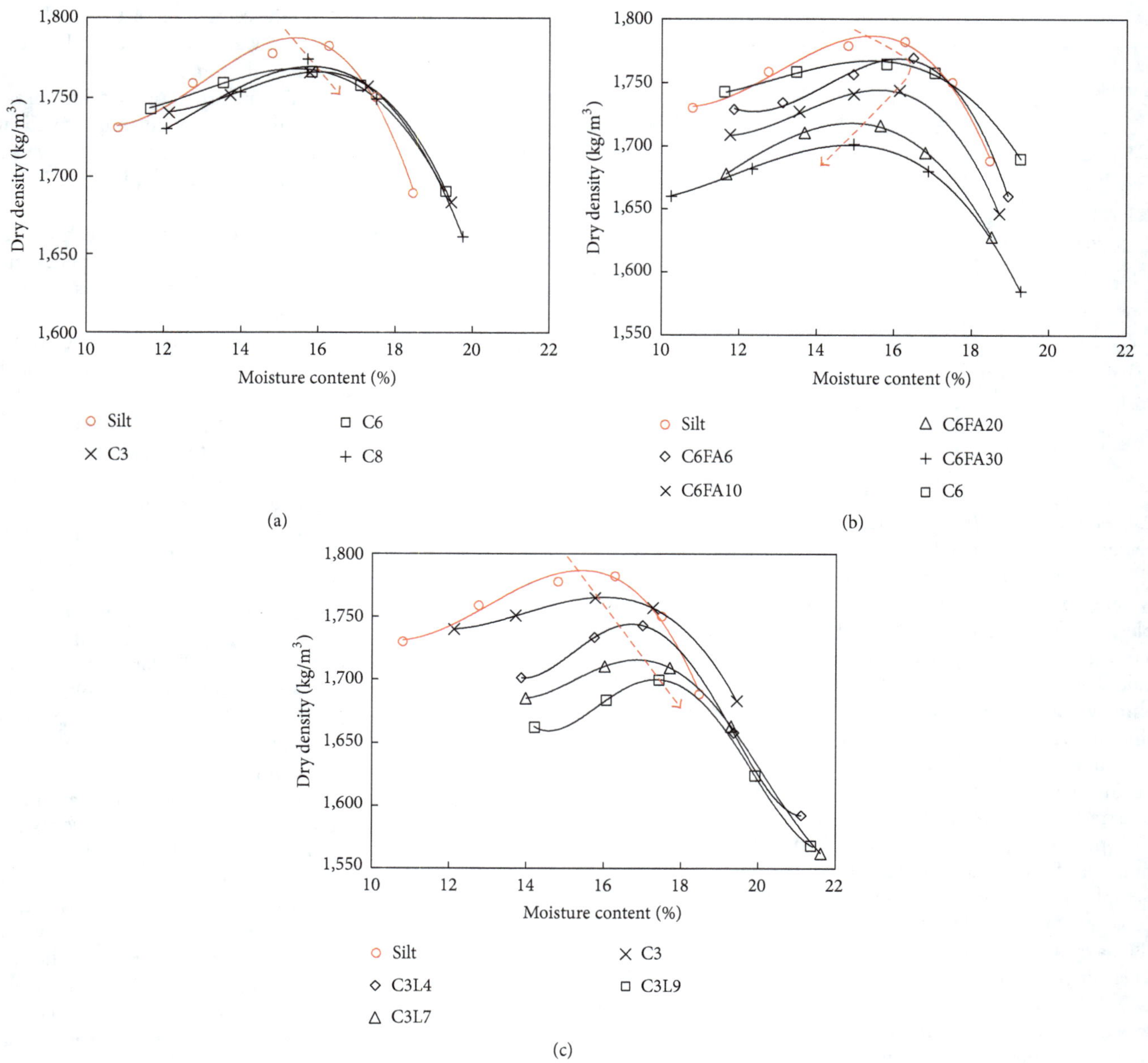

FIGURE 3: Proctor compaction test results: (a) cement-treated silt, (b) cement-fly ash-treated silt, and (c) cement-lime-treated silt.

used. The cement-treated samples (C3, C6, and C8) show UC strength increase linearly proportional to the increase of the cement content. The values fall in the ranges of 0.59 MPa to 1.48 MPa at 7 days and 0.68 MPa to 1.67 MPa at 28 days. This implies that moderate strength increases (14.2% on average) are gained from 7 days to 28 days, if the sample cement content remains the same. The strength develops due to the cement hydration creating a matrix of gel (i.e., the ettringite) to solidify the soil particles; and the gel grows over time. The strength values and the increase rates agree with the results for other cement-treated soils such as clay [22], marine sediments [23], and silty sand and clayey silt [24], where the cement was of similar contents.

Figure 4 shows that the 7-day UC strength of a cement-fly ash-treated sample is close to that of its dosage-corresponding cement-treated sample, such as C3FA7 versus C3 and C6FA10 versus C6. This agreement does not occur to the samples cured to 28 days when a cement-fly ash-treated sample shows UC strength significantly higher than that of its dosage-corresponding cement-treated sample. The reason is that, for the cement-fly ash-treated samples, significant strength increase (155.3% overall) is attained from the 7-day curing to the 28-day curing, where the cement and fly ash contents remain the same. The increase is merely 14.2% for the cement-treated samples. For the cement-fly ash-treated samples, the 28-day strength falls in the ranges of 1.46 MPa to 2.93 MPa. These comparisons suggest that the fly ash contributes to the strength development after the 7-day curing, which has been confirmed in similar fly ash stabilization studies [25, 26]. This strength dependence on the fly ash is associated with

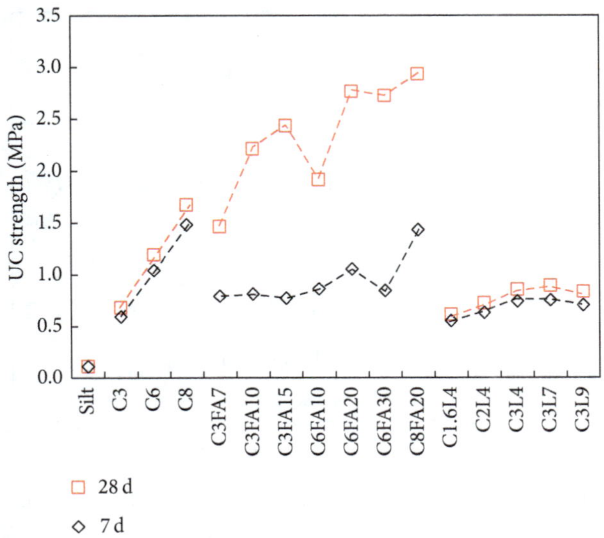

FIGURE 4: UC strength of stabilized samples without frost actions.

fly ash pozzolanic properties. The fly ash is a pozzolanic material, which when mixed with Portland cement and water reacts with calcium hydroxide released by the hydration of Portland cement to produce calcium silicate hydrates and calcium aluminate hydrates, improving the strength of the stabilized materials. These pozzolanic reactions are time-dependent processes and often remain noticeably active after 90 days of curing [27]. The effect of the fly ash on the strength development is also clear for samples with different fly ash contents. For instance, on the 28th day, the UC strength of the cement-fly ash-treated samples increased with the increase of the fly ash content where the cement content remained the same, such as 2.21 MPa for sample C3FA10 versus 1.46 MPa for sample C3FA7.

It is noteworthy that a strength "drop" appears in sample C6FA10 on day 28. The UC strength for this sample is 1.91 MPa. The strength for the neighboring samples is 2.43 MPa for sample C3FA15 and 2.76 for sample C6FA20. Although the differences are tolerable and the strength falls in the working range for backfill materials, it is of some value to research the possible details behind the strength inconsistency. There are no absolute answers. One possible explanation is the level of pozzolanic reactions versus the fly ash contents. The 28-day pozzolanic reactions are dependent on the fly ash contents due to fly ash's relatively slow hydration processes. The hydration processes rely on the calcium hydroxide released by the hydration of the cement to form additional calcium silicate hydrate and other cementitious compounds and then build up the strength. Seeing the increase of the cement content and the decrease of the fly ash content for sample C6FA10 in relation to sample C3FA15, the fly ash, as the source of pozzolan, is likely in short for the hydration processes and the pozzolanic reactions are not as thorough as sample C3FA15. Having increased the fly ash content by 10%, sample C6FA20 improves the ingredients fraction and thus the hydration processes and the strength. This argument is supported by comparing sample C6FA10 with sample C3FA10. The latter gives rise to a relatively less and probably appropriate level of calcium hydroxide through the cement hydration processes than the former does. The calcium hydroxide level is appropriate in that it enables the 10% fly ash pozzolanic reactions without leaving a significant excess of free calcium hydroxide, a less desirable hydration product due to its dissolvability. The fly ash to cement ratio is discussed subsequently.

No clear strength is gained from the increase of the lime content or the curing period for the cement-lime-treated samples. This implies that lime plays a less significant role in improving strength for the silt. Similar less significant strength development for lime-treated silt or clay was reported by Sakr et al. [28] and Le Runigo et al. [29]. Scholen [30] suggested that lime is sensitive to accessible water volume and the sensitivity causes insignificant strength development where the water volume such as the OMC is less than the volume desired for complete hydration. This suggestion is somehow in line with the results obtained in the current study. The chosen water contents for the cement-lime-treated samples were OMC = 16.8–17.7% (Table 3). The water contents were less than the water to lime ratio, that is, 0.3 or higher [31], desired to enable satisfactory pozzolanic reactions, and so involve less strength development. On a separate perspective, the lime-based samples have the potential to further develop strength given the access to moisture, which possibly hydrates the free lime, and create additional gel.

Normalization was conducted on additives dosage for the cement-fly ash-treated samples. Define mass ratio m as the fly ash content to the cement content. The optimum m value is where the UC strength approaches a peak value. The assessment for this optimum m value helps weigh the importance of the fly ash against the cement and thus suggests an optimal dosage where appropriate. Figure 5 shows the m values versus the UC strength. For the three cement contents 3%, 6%, and 8%, 7-day samples show an approximately flat development of the UC strength with the increase of m values; 28-day samples show an optimum value $m = 5.0$ if cement = 3% and $m = 3.3$ if cement = 6%. This reflects that the optimum m value is dependent on the cement content. The dependence of the optimum m value on the cement content indicates that a smaller fly ash to cement ratio is needed to achieve better strength development when the cement content increases. A range of $m = 5$ to 3 is plausible where the cement content falls in the range of 3% to 6%. This range is subject to improvement where additional test results are accessible.

3.3. UC Strength (Postfrost Actions). The UC strength results for all of the stabilized samples after frost actions are aggregated in Figure 6. Values are categorized in terms of the curing periods and the additives used. Overall, the cement-fly ash-treated samples show better postfrost performance than the cement-treated samples. The cement-lime-treated samples show performance similar to the cement-treated samples. This means that the fly ash improves material capacity counteracting the frost impact. Close UC strength values (0.1 to 0.2 MPa) were obtained between the cement- and cement-lime-treated samples after the 9th cycle (shown as converged ends for the bottom group of lines). After the same 9th cycle, the cement-fly ash-treated samples show varying strength

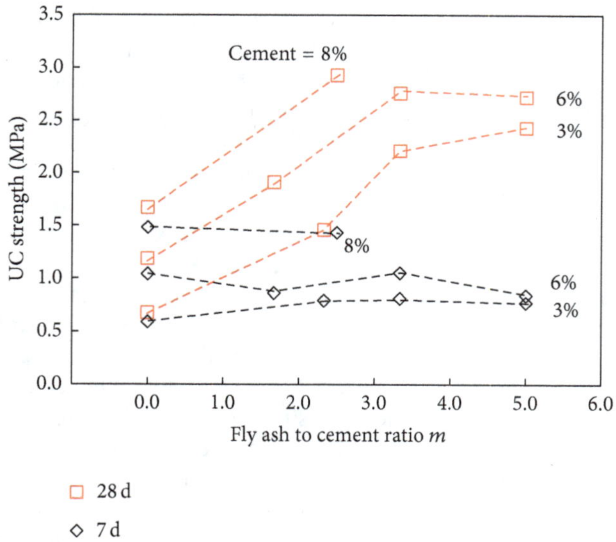

FIGURE 5: UC strength versus fly ash to cement mass ratio.

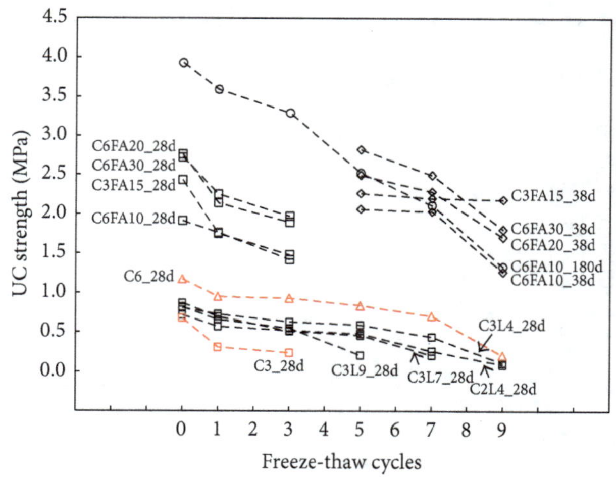

FIGURE 6: UC strength of stabilized samples after frost actions.

values (1.3 to 2.2 MPa) dependent on the fly ash contents. The higher the fly ash content, the higher the strength. This implies that the fly ash inclusion contributes to the material strength increase and counteracts the frost action. Similar results were reported by Altun et al. [5] for fly ash stabilized slit samples. The better antifrost performance of the cement-fly ash-treated samples arises partly from the hollow spherical nature of the fly ash particles. The fly ash consists of particles that resulted from the cooling of molten droplets of fused material contained within the exhaust gases leaving a combustion furnace where coal is burned to provide power. Fusion of silicate minerals results in the formation of hollow glassy spherical particles. The hollow spheres are elastic, light, and sometimes waterproof, which creates a foam matrix benefiting the stabilized silt counteracting the frost actions, as illustrated in Figure 7. The foam matrix helps buffer the volumetric increase that resulted from the ice lensing processes, which otherwise cracks or damages the silt matrix stabilized by other additives such as the cement. Similar concept of counteracting frost actions has been successfully applied to other foamed concrete products [32]. The cement or the lime, however, does not possess the hollow foamed structure, nor the buffering capacity when subjecting the samples to the freezing processes. To make it worse, lime-based geomaterials are expected to undergo a certain degree of swelling and soften when in contact with water. This means further deterioration in the thawing phase. As mentioned before, this observed heave is related to the additional generation of ettringite due to the existence of free lime.

In Figure 6, the UC strength drops with the increase of the freeze-thaw cycles for the cement-fly ash-treated samples. The drops are on average 16.8% after the 1st cycle, 13.4% from the 1st to the 3rd cycle, 8.1% from the 5th to the 7th cycle, and 25.8% from the 7th to the 9th cycle. These drops indicate that the strength losses become pronounced after the 1st freezing-thawing impact and worse after the 7th. These observations are in agreement with the test results reported by Qi et al. [33]. The cement-lime-treated samples show strength drops different from the cement-fly ash-treated samples. The differences lie in the drop percentage versus the cycle number. The drops are on average 17.4% after the 1st cycle, 16.6% from the 1st to the 3rd cycle, 21.4% from the 3rd to the 5th cycle, 41.9% from the 5th to the 7th cycle, and 71.2% from the 7th to the 9th cycle. The significant drops occur after the 5th cycle—weakening earlier and worse in magnitude than the cement-fly ash-treated samples. The accelerated material deterioration is likely associated with the larger pore space of the cement-lime samples. The cement-lime samples show an average void ratio of 0.57 versus 0.52 for the cement-fly ash-treated samples and 0.54 for the cement-treated sample. As suggested by Khoury and Zaman [34], the larger pore space is open to water access and thus vulnerable to ice lens evolutions which in turn induce cracking or fissuring due to icing pressures increased within the soil pores. As a result, additional pore spaces form and impair the soil matrix through further freezing-thawing iterations.

The curing processes help develop the UC strength and counteract strength drops caused by the frost actions where the fly ash is used. Despite being subjected to two more frost cycles, the 38-day cement-fly ash samples remain sound and give rise to UC strength higher than or equal to the strength of corresponding 28-day samples, as shown in Figure 6. This means that the additional 10-day curing time contributes to strength gain. This logic for the curing time versus the strength seems to not follow where a longer curing period is in place, for example, samples C6FA10_180 versus C6FA10_38. Both samples use the same dosage but were cured for different periods: 180 days for the former and 38 days for the latter. Both give rise to strength of similar level after the 7th freezing-thawing cycle, as well as the 9th cycle. This suggests two presumptions: (i) samples C6FA10_180 and C6FA10_38 attain similar UC strength before being exposed to the first freeze-thaw impact. Similar strength losses occur to the two sets of samples through the subsequent impacts. As a result, the UC strength values agree with each other after the 7th cycle, so do the strength values after the 9th cycle. Or (ii) before the 1st impact, sample C6FA10_180 attains strength higher than sample C6FA10_38 does. Despite starting with

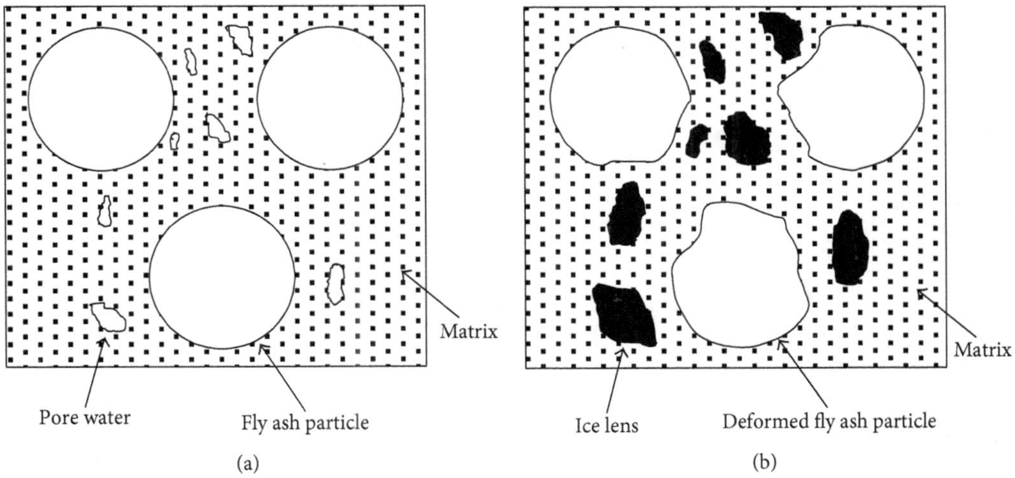

FIGURE 7: Schematic of foamed matrix: (a) before frost action and (b) after frost action.

the higher strength, sample C6FA10_180 deteriorates more severely at each impact and, after the 7th impact, ends up being close in strength to sample C6FA10_38. Presumption (ii) seems less plausible, as this presumption is inconsistent with the strength evolution of sample C6FA10_28. Sample C6FA10_28 gives rise to lower strength levels in relation to sample C6FA10_180 from the start and throughout corresponding impacts. Taking on presumption (i) means an insignificant strength having developed after the 38-day curing age.

Hori and Morihiro [35] and Yarbaşi et al. [36] linked material antifrost performance with material brittleness—the more brittle the material is, the more susceptible it is to cracking when frozen. This point somehow agrees with the shown test results: the higher the material strength (brittleness), the steeper the strength reduction curves, and vice versa. Although the relationship is noted between the material brittleness and the antifrost performance, no optimal dosage is clearly suggested from the curves of the UC strength versus the freeze-thaw cycles as shown in Figure 6. Sample C3FA15, when cured for 38 days, shows marginal strength loss throughout the 5th, 7th, and 9th cycle and is likely identified as a dosage of preference. If cured for 28 days, however, this sample shows less favorable performance in the first three cycles. In view of this, the preferred dosage is subject to further verification tests such as subjecting the 38-day samples to 1 and 3 frost cycles and assessing respective UC strength losses.

To examine ductile behavior of the stabilized silt, the stress-strain curves obtained in the UC strength tests are obtained and presented in Figure 8. The curves are categorized in terms of the freeze-thaw cycle number and drawn at the same scale in order to assess the effect of the frost actions across samples. The stress-strain curves are plotted for representative samples: C6FA10_28, C6FA10_38, C6FA10_180, C6_28, C3_28, and C3L4_28. These samples are chosen because of their comparable dosages such as C6FA10 versus C6 and C3L4 versus C3 which enable examination of the effects of the fly ash or lime on top of the cement. The dosage of 10% for the fly ash is chosen as this percentage is the least out of the series of cement-fly ash-treated samples and helps assess changes that occurred to the samples ductile behavior where a small content of fly ash is included. So does the 4% for the lime. The curves demonstrate distinct stress-strain characteristics among the samples. The fly ash inclusion clearly provides an increase in soil strength and ductility: higher peak strength and larger strain at failure. In addition, the ductility of the cement-fly ash-treated samples remains pronounced after respective frost actions. The cement- and the cement-lime-treated samples, however, show relatively less ductility. The demonstrated ductile behavior is in agreement with the aforementioned antifrost performance of the samples. Better ductility means a higher level of ability of the sample to deform under stress, or greater extent to which the sample can be plastically deformed to large strains without macroscopic fracturing or failure. To these ends, the fly ash outperforms the lime with respect to amending cemented samples and improving samples ductility and antifrost performance.

To gain further insights into the stress-strain characteristics, in particular material stiffness, secant modulus for each of the samples is plotted against corresponding peak UC strength, as shown in Figure 9. The secant modulus is the slope of the line from the origin to a point on a stress-strain curve where the stress is half the peak UC strength. The moduli are plotted in terms of the frost actions the samples are subjected to. In each plot, the moduli are categorized in terms of the additives used to treat the alluvial silt. It is shown that the cement-fly ash-treated samples are overall stiffer than the other two treatment categories when subjected to the same frost actions. For instance, after the 1st freeze-thaw cycle, the secant modulus ranges from 67 MPa to 102 MPa for the cement-fly ash-treated samples and 37 MPa to 41 MPa for the cement-lime-treated samples, depending on the dosages used. After the 9th cycle, the cement-fly ash-treated samples give rise to secant modulus even slightly higher than that of the respective intact samples of the other two treatment categories. It is noted that the secant moduli for all the three

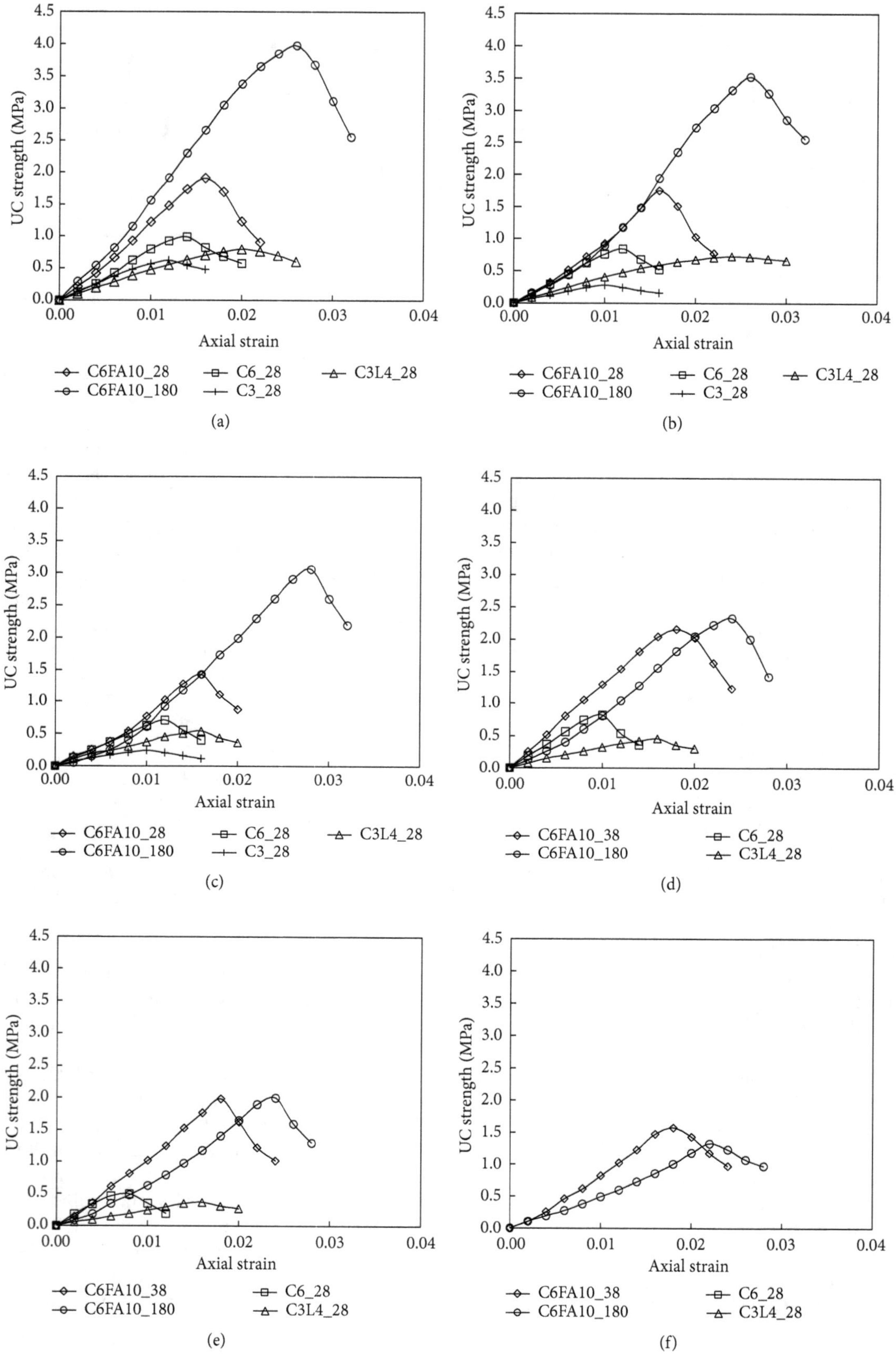

FIGURE 8: Stress-strain curves in UC strength tests drawn at the same scale for samples subjected to different freeze-thaw cycles: (a) no frost actions, (b) 1 cycle, (c) 3 cycles, (d) 5 cycles, (e) 7 cycles, and (f) 9 cycles.

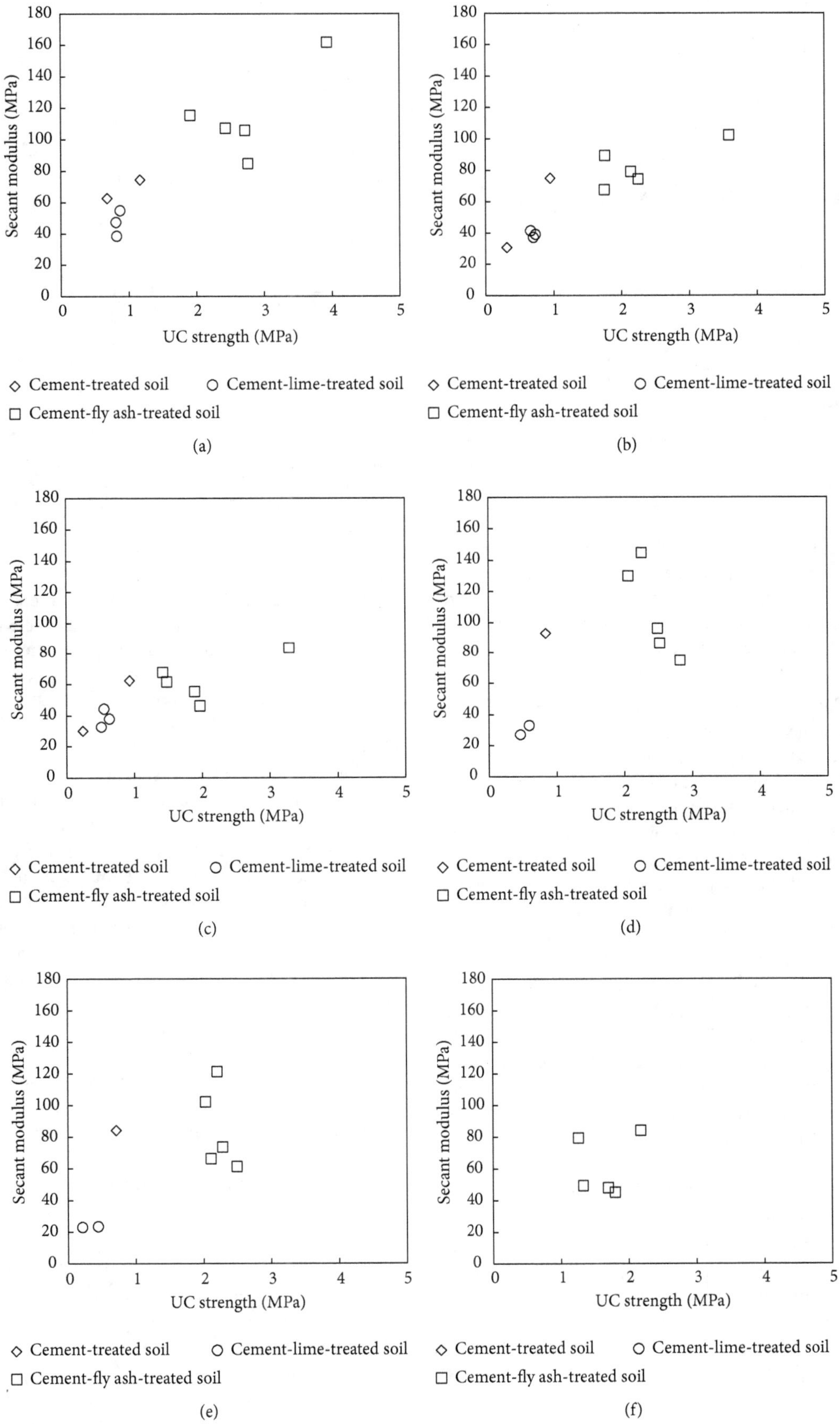

FIGURE 9: Secant modulus versus peak UC strength for samples subjected to different freeze-thaw cycles: (a) no frost actions, (b) 1 cycle, (c) 3 cycles, (d) 5 cycles, (e) 7 cycles, and (f) 9 cycles.

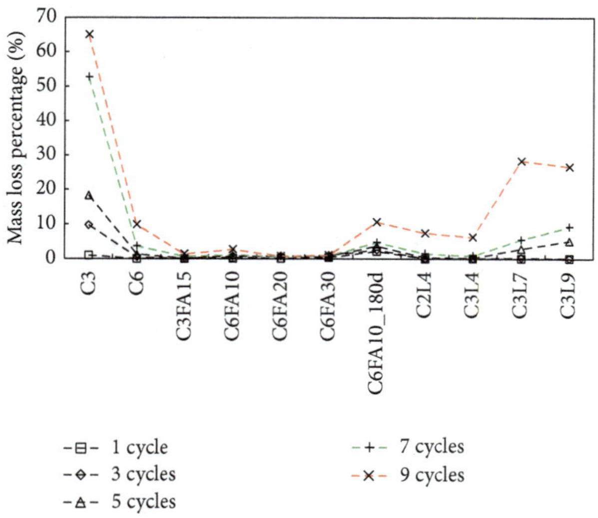

FIGURE 10: Mass loss percentage of stabilized samples after frost actions.

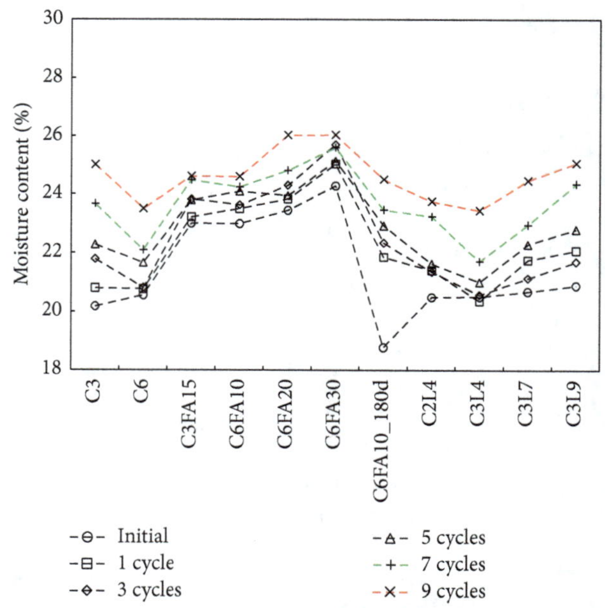

FIGURE 11: Moisture content of stabilized samples before and after frost actions.

treatment categories decrease with the increase of the freeze-thaw cycles. The first two decreases (from 0 to the 1st cycle and the 1st to the 3rd cycle) are relatively significant compared with the decreases for the rest. This means that the initial frost actions impair the materials severely.

3.4. Material Soundness and Moisture Absorption. Define mass loss percentage = (sample initial dry weight − sample dry weight after frost actions)/sample initial dry weight, to assess material soundness in response to the cyclic frost impact. The percentage results are shown in Figure 10. The cement-fly ash-treated samples exhibit noticeable less mass loss in contrast to the cement- and cement-lime-treated samples. For instance, after the 9th cycle, the average percentage of mass loss is 1.5% for the cement-fly ash-treated samples, 17.2% for the cement-lime-treated samples, and 37.4% for the cement-treated samples. These percentages are largely in agreement with results reported by Parsons and Milburn [37]. Their study showed, depending upon the cement dosage and soil types, mass loss of 2% to 7% for cement-treated soils, 7% to 19% for fly ash-treated soils, and 7% to 41% for lime-treated samples after 12 freeze-thaw cycles. A significant mass loss occurs to some samples, such as samples C3 of 18.26% loss of 5 cycles, C3L9 of 9.31% loss of 7 cycles, and C3L7 of 28.34% loss of 9 cycles. Due to the significant mass reductions, these samples hardly maintained respective integrity and were not subjected to the remaining UC strength tests.

It is of interest to compare the mass losses of sample C6FA10_180d with those of sample C6FA10_28d in order to assess the effect of the curing time on the material soundness. Despite being cured longer, sample C6FA10_180d underwent a higher percentage of losses and exhibited a trend of more mass loss if the frost actions continued. These observations are in line with the relation, which was suggested by Hori and Morihiro [35] and Yarbaşi et al. [36], that existed between the material brittleness, mass loss, and strength drop as shown in Figure 9. This note is subject to insights into material microstructure changes detected through conducting microscopical observations such as scanning electron microscopy. These microscopical observations help shed some light on how the additives influence antifrost performance of the alluvial silt. The observational devices, unfortunately, were not accessible in the first instance and the observations are left for the next stage of studies.

The mass losses and UC strength drops are further linked to the water volume absorbed into the materials matrix during the frost actions. As discussed above, the freezing phase tends to crack or fissure the solid samples through frost heaving processes—ice lensing. The subsequent thawing phase melts the ice lenses into liquid water. A certain fraction of the sample pore space is unoccupied by the water but intruded air due to the fact that ice is 8.3% lighter than water. The formed partially saturated conditions are supposed to introduce an additional volume of water into the created cracks and fissures. The resulting increases of the water content against the frost actions are shown in Figure 11. The water contents were worked out by deducting the material mass losses. It was assumed that the samples produced in one batch have the same initial water content. This assumption enables comparisons conducted on the water contents after the frost actions. The comparisons show the following:

(i) The water contents increase with the count of the freeze-thaw cycles across all series of samples. On average, an increase of 0.23% to 0.94% is observed per two cycles for the samples tested. The degree of increase varies among the sample categories: the least for the cement-fly ash-treated samples followed by the cement- and the cement-lime-treated samples. For instance, from 7 to 9 cycles, an average increase of water content is 0.63% for the cement-fly ash-treated samples, 1.12% for the cement-lime-treated samples, and 1.37% for the cement-treated samples.

(ii) The less noticeable changes in the water content of the cement-fly ash-treated samples agree with the less mass reduction (Figure 10) and the higher UC strength (Figure 6) of the samples.

(iii) Higher start water contents occur to the 28-day or 38-day cement-fly ash-treated samples in relation to the other two treatment categories. Seeing that the three categories adopt similar levels of water contents for fresh samples (Table 3), the fly ash is acknowledged to be effective in developing samples water absorption capacity during the curing periods. A longer curing period likely offsets this effectiveness as demonstrated by sample C6FA10_180d. This sample shows less start water content in relation to the other cement-fly ash-treated samples. This means that the fly ash has nearly completed pozzolanic reactions at 180 days and became less sensitive to water presence.

4. Conclusions

Stabilizing the Yellow River alluvial silt through using cement, fly ash, and lime to counteract frost actions was examined. The examination was implemented through conducting Proctor compaction, unconfined compressive strength, freeze-thaw cyclic impacts, and material soundness and water absorption tests on a series of samples prepared in varying dosages. The conclusions are made as follows:

(i) Including a certain mass of additives to the silt resulted in varying compaction curves, depending on the additives and the dosages used. The cement-treated samples showed curves independent of the cement contents, whereas dosage dependency was pronounced for the curves of the cement-fly ash- and the cement-lime-treated samples. All stabilized samples gave rise to maximum dry density less than that of the silt sample and optimum moisture content in the range of 14.9% to 17.7%. The cement-lime-treated samples attained optimum moisture content of up to 1% higher than the other two treatment categories.

(ii) The cement-fly ash additive was superior to the cement or cement-lime additives with respect to the development of the UC strength, secant modulus, and ductility for the Yellow River alluvial silt. The advantages were noticeable at 28 days. A range of 3 to 5 for the fly ash to cement mass ratio was plausible with respect to attaining desirable material UC strength. The cement-fly ash-treated samples are stiffer than the other two treatment categories when subjected to the same frost actions.

(iii) The first two frost actions impair the materials slightly more than the rest based on the strength development and secant moduli variation. This observation applies to all of the three stabilization categories; however, it is subject to additional test results where samples are comparable in terms of the dosage and curing time used.

(iv) The cement-fly ash additive outperformed the cement or cement-lime additives with regard to capacity of counteracting the frost actions. The preferred dosage was 3% for the cement and 15% for the fly ash.

(v) Moisture content builds up in response to the freeze-thaw cycles, on average 0.23% to 0.94% increase per two cycles for the samples tested. The fly ash is effective in developing samples water absorption capacity during the curing periods.

(vi) A correlation existed between the material strength, soundness, and water absorption capacity. The higher the strength, the more noticeable the strength drop in response to the frost actions, the less the mass loss, and the less the volume of water absorbed.

Competing Interests

The authors declare that there are no competing interests regarding the publication of this paper.

Acknowledgments

This study was supported by a grant from the National Natural Science Foundation of China (51478226) and the Priority Academic Program Development of Jiangsu Higher Education Institutions.

References

[1] T.-M. Gu, *Climate of North China Plain*, China Meteorological Press, Beijing, China, 1991.

[2] Z.-D. Zhu, S.-Y. Liu, and H.-J. Sun, "Study of stabilized silt in Xu-Su area," *Rock and Soil Mechanics*, vol. 25, no. 7, pp. 1155–1158, 2004.

[3] J.-H. Xiao, J.-K. Liu, L.-Y. Peng, and L.-H. Chen, "Effects of compactness and water Yellow-River alluvial silt content on its mechanical behaviors," *Rock and Soil Mechanics*, vol. 29, no. 2, pp. 409–414, 2008.

[4] X.-G. Song, H.-B. Zhang, S.-G. Wang, Z.-X. Jia, and Y.-H. Guan, "Hydrophilic characteristics and strength decay of silt roadbed in Yellow River alluvial plain," *Chinese Journal of Geotechnical Engineering*, vol. 32, no. 10, pp. 1594–1602, 2010.

[5] S. Altun, A. Sezer, and A. Erol, "The effects of additives and curing conditions on the mechanical behavior of a silty soil," *Cold Regions Science and Technology*, vol. 56, no. 2-3, pp. 135–140, 2009.

[6] K. Hazirbaba and H. Güllü, "California Bearing Ratio improvement and freeze-thaw performance of fine-grained soils treated with geofiber and synthetic fluid," *Cold Regions Science and Technology*, vol. 63, no. 1-2, pp. 50–60, 2010.

[7] H. Güllü and A. Khudir, "Effect of freeze-thaw cycles on unconfined compressive strength of fine-grained soil treated with jute fiber, steel fiber and lime," *Cold Regions Science and Technology*, vol. 106-107, pp. 55–65, 2014.

[8] T. M. Petry and D. N. Little, "Review of stabilization of clays and expansive soils in pavements and lightly loaded structures—history, practice, and future," *Journal of Materials in Civil Engineering*, vol. 14, no. 6, pp. 447–460, 2002.

[9] Z. Nalbantoğlu, "Effectiveness of class C fly ash as an expansive soil stabilizer," *Construction and Building Materials*, vol. 18, no. 6, pp. 377–381, 2004.

[10] A. Kumar, B. S. Walia, and A. Bajaj, "Influence of fly ash, lime, and polyester fibers on compaction and strength properties of expansive soil," *Journal of Materials in Civil Engineering*, vol. 19, no. 3, pp. 242–248, 2007.

[11] X. Kang, G.-C. Kang, and L. Ge, "Modified time of setting test for fly ash paste and fly ash-soil mixtures," *Journal of Materials in Civil Engineering*, vol. 25, no. 2, pp. 296–301, 2013.

[12] D. Ciancio, C. T. S. Beckett, and J. A. H. Carraro, "Optimum lime content identification for lime-stabilised rammed earth," *Construction and Building Materials*, vol. 53, pp. 59–65, 2014.

[13] N. Latifi, S. Horpibulsuk, C. L. Meehan, M. Z. A. Majid, and A. S. A. Rashid, "Xanthan gum biopolymer: an eco-friendly additive for stabilization of tropical organic peat," *Environmental Earth Sciences*, vol. 75, article 825, 2016.

[14] N. Latifi, C. L. Meehan, M. Z. A. Majid, and S. Horpibulsuk, "Strengthening montmorillonitic and kaolinitic clays using a calcium-based non-traditional additive: a micro-level study," *Applied Clay Science*, vol. 132-133, pp. 182–193, 2016.

[15] China Ministry of Construction, "Standard for soil test method," GB/T 50123-1999, China Ministry of Construction, 1999.

[16] A. Athanasopoulou, "Addition of lime and fly ash to improve highway subgrade soils," *Journal of Materials in Civil Engineering*, vol. 26, no. 4, pp. 773–775, 2014.

[17] J.-M. Konrad and M. Samson, "Influence of freezing temperature on hydraulic conductivity of silty clay," *Journal of Geotechnical and Geoenvironmental Engineering*, vol. 126, no. 2, pp. 180–187, 2000.

[18] ASTM, "Standard test method for unconfined compressive strength of compacted soil-lime mixtures," Tech. Rep. D5102, ASTM International, West Conshohocken, Pa, USA, 2009.

[19] ASTM, "Standard test methods for freezing and thawing compacted soil-cement mixtures," D560, ASTM International, West Conshohocken, Pa, USA, 2015.

[20] D. D. Tripura and K. D. Singh, "Characteristic properties of cement-stabilized rammed earth blocks," *Journal of Materials in Civil Engineering*, vol. 27, no. 7, Article ID 04014214, 2015.

[21] B. R. Phani Kumar and R. S. Sharma, "Effect of fly ash on engineering properties of expansive soils," *Journal of Geotechnical and Geoenvironmental Engineering*, vol. 130, no. 7, pp. 764–767, 2004.

[22] S. Horpibulsuk, R. Rachan, and A. Suddeepong, "Assessment of strength development in blended cement admixed Bangkok clay," *Construction and Building Materials*, vol. 25, no. 4, pp. 1521–1531, 2011.

[23] D. Wang, N. E. Abriak, and R. Zentar, "Strength and deformation properties of Dunkirk marine sediments solidified with cement, lime and fly ash," *Engineering Geology*, vol. 166, pp. 90–99, 2013.

[24] Y. Yi, M. Liska, and A. Al-Tabbaa, "Properties of two model soils stabilized with different blends and contents of GGBS, MgO, lime, and PC," *Journal of Materials in Civil Engineering*, vol. 26, no. 2, pp. 267–274, 2014.

[25] J. Prabakar, N. Dendorkar, and R. K. Morchhale, "Influence of fly ash on strength behavior of typical soils," *Construction and Building Materials*, vol. 18, no. 4, pp. 263–267, 2004.

[26] T. Wu, M. Chi, and R. Huang, "Characteristics of CFBC fly ash and properties of cement-based composites with CFBC fly ash and coal-fired fly ash," *Construction and Building Materials*, vol. 66, pp. 172–180, 2014.

[27] M. Chrysochoou, D. G. Grubb, K. L. Drengler, and N. E. Malasavage, "Stabilized dredged material. III: mineralogical perspective," *Journal of Geotechnical and Geoenvironmental Engineering*, vol. 136, no. 8, pp. 1037–1050, 2010.

[28] M. A. Sakr, M. A. Shahin, and Y. M. Metwally, "Utilization of lime for stabilizing soft clay soil of high organic content," *Geotechnical and Geological Engineering*, vol. 27, no. 1, pp. 105–113, 2009.

[29] B. Le Runigo, V. Ferber, Y.-J. Cui, O. Cuisinier, and D. Deneele, "Performance of lime-treated silty soil under long-term hydraulic conditions," *Engineering Geology*, vol. 118, no. 1-2, pp. 20–28, 2011.

[30] D. E. Scholen, "Stabilizer mechanisms in nonstandard stabilizers," in *Proceedings of the 6th International Conference on Low-Volume Roads*, pp. 252–260, TRB, Minneapolis, Minn, USA, 1995.

[31] S. Xu, J. Wang, and Y. Sun, "Effect of water binder ratio on the early hydration of natural hydraulic lime," *Materials and Structures*, vol. 48, no. 10, pp. 3431–3441, 2015.

[32] K. Natesaiyer, K. C. Hover, and K. A. Snyder, "Protected paste volume of air-entrained cement paste. Part II," *Journal of Materials in Civil Engineering*, vol. 5, no. 2, pp. 170–186, 1993.

[33] J. Qi, W. Ma, and C. Song, "Influence of freeze-thaw on engineering properties of a silty soil," *Cold Regions Science and Technology*, vol. 53, no. 3, pp. 397–404, 2008.

[34] N. N. Khoury and M. M. Zaman, "Environmental effects on durability of aggregates stabilized with cementitious materials," *Journal of Materials in Civil Engineering*, vol. 19, no. 1, pp. 41–48, 2007.

[35] M. Hori and H. Morihiro, "Micromechanical analysis on deterioratIon due to freezing and thawing in porous brittle materials," *International Journal of Engineering Science*, vol. 36, no. 4, pp. 511–522, 1998.

[36] N. Yarbaşi, E. Kalkan, and S. Akbulut, "Modification of the geotechnical properties, as influenced by freeze-thaw, of granular soils with waste additives," *Cold Regions Science and Technology*, vol. 48, no. 1, pp. 44–54, 2007.

[37] R. L. Parsons and J. P. Milburn, "Engineering performance of stabilized soils," *Transportation Research Record*, vol. 1837, pp. 20–29, 2003.

11

Experimental Investigation on the Behavior of Iron Powder-Reinforced Sand under Electromagnetic Field

Ying Lai,[1] Bin Zhu,[2] and Xiangtian Xu[2]

[1]Postdoctoral Fellow, MOE Key Laboratory of Soft Soils and Geoenvironmental Engineering, College of Civil Engineering and Architecture, Zhejiang University, 866 Yuhangtang Road, Hangzhou, Zhejiang, China
[2]Professor, MOE Key Laboratory of Soft Soils and Geoenvironmental Engineering, College of Civil Engineering and Architecture, Zhejiang University, 866 Yuhangtang Road, Hangzhou, Zhejiang, China
[3]Associate Professor, Institute of Transportation, Inner Mongolia University, Hohhot 010070, China

Correspondence should be addressed to Xiangtian Xu; jtxuxt@imu.edu.cn

Academic Editor: Guoqiang Xie

Applications of soil improvement have proliferated in recent years. To date, we have limited studies on the quantitative analyses of the autoadaptive material and specifically to model its stress-strain relationship. This paper explored an autoadaptive material, iron-powdered Ottawa sand, which was temporarily solidified by applying an electromagnetic field. A series of compression triaxial tests were carried out with various relative densities of specimens (60% and 80%), in four electromagnetic fields (0 A, 0.5 A, 1 A, and 2 A) and under three confining pressures (103 kPa, 206 kPa, and 310 kPa). The test results indicate that the strength of specimens increased while initial stiffness and brittleness reduced by adding iron powder. Moreover, the strength of the specimens increased by increasing the magnitude of the applied electromagnetic field. The behavior of the iron-powdered sand was described by using a revised Duncan–Chang model. The revised model was evaluated by comparing the simulated results with the corresponding test data. The comparison showed that the revised model can better capture the nonlinear stress-strain behavior of the specimens. With the application of the revised Duncan–Chang model, the standard error of the estimate between the experimental and predicted results is lowered down to 0.39 from 4.7. Future research is geared towards practical applications for temporary solidification of soil.

1. Introduction

Soil, the most abundant construction material in nature, is entirely inseparable from any constructions and has been used for engineering projects for thousands of years. However, natural soil conditions are not always adequate for required construction. Soil improvements have been intensively studied and successively applied over the past decades. For example, powder, sheets, and rods of polymeric or metal materials [1–6] were placed in soil to increase its yield stress and resistance. These processes of creating reinforced soil are soil strengthening on a macroscale. In comparison, soil stabilization is the process of soil strengthening on a microscale. The matrix of soil can be changed by physically or chemically binding the particles of soil and the reinforced material together [7]. Binding agents include cement or other chemical products [6, 8, 9–11]; binding processes involve heating or freezing [12, 13]. A possible disadvantage of a reinforced soil is that the strengthening effect depends on the orientation of inclusions. Specifically, a particular direction or surface can provide the maximum strengthening effect. Conversely, a specific direction or surface can weaken the strength of soil [4]. Moreover, the disadvantages of the present available soil stabilization techniques are that improvements in the stress resistance are accompanied by reductions in permeability and ductility.

Despite soil improvement techniques of reinforcement by including polymeric materials and stabilization by adding binding agents, an inventive method, an autoadaptive material has been successfully applied in other engineering disciplines. The autoadaptive materials, such as magneto- and electrostrictive solids and magneto- and electrorheological fluids, are the materials that can sense external disturbances and modify their characteristics intentionally to mitigate the effects of the external influence. This improvement characteristic and mitigation property offer an excellent potential application in earthquake engineering. For example, the magnetorheological fluid (MRF) dampers in semiactive vibration devices [14–16] are known to be well suited for suppressing the vibration of structures and systems.

Hryciw and Susila [17] were the first researchers to apply magnetorheological materials in geotechnical engineering. In their study, when a pair of permanent magnets was placed on top and bottom of a specimen, oil-based magnetizable particles transformed from a fluid state into a solid state. As a result, they observed a significant increase in the soil stiffness and strength and a simultaneous decrease in hydraulic conductivity. This innovative method of introducing autoadaptive media causes changes in the properties of soil and overcomes the shortcomings of the soil improvement methods mentioned above.

It is complex to accurately model composite materials consisting of natural (soil) and man-made materials (polymeric, chemical, and binding products). Generally, the constitutive model assumes soil is homogeneous. However, this assumption is not suitable for composite materials with a high degree of anisotropy. In spite of difficulties, some notable achievements have been made in developing a constitutive model and framework for composite materials, including the introduction of Mohr–Coulomb theory into reinforced soil by fiber [4, 18], the discrete framework accounting for potential weakness at the soil-inclusion interface [19], and a more complex simulation software developed in [20] for composite materials mixed with chemical products [21].

This research investigates the temporary reinforced effect of iron-powdered Ottawa sand 20–30 under controllable magnitudes of electromagnetic fields. It is interesting to note that the behavior of soil is different from that of metals. Specifically, the strength of soils decreases after peak strength during progressive straining; this phenomenon is termed as "strain softening." In comparison, metal materials exhibit "strain hardening"; an increase in strength occurs during progressive straining. To explore this, it is required to incorporate stress-strain theory with the behavior of Ottawa sand and iron powder under electromagnetic fields. One approach to study the strengthening effect of iron powder temporarily solidified under the electromagnetic fields is to analyze the stress-strain relationship and the strength criterion. Based on the results from triaxial compression tests, we present a stress-strain relationship for iron-powdered sand.

2. Test Conditions

The objective of this experiment was to evaluate the effect of different magnitudes of the electromagnetic field on the stress-strain relationship of iron-powdered sand. Undrained compression triaxial tests were performed with a triaxial cell apparatus capable of housing a specimen 3.56 cm in diameter and 7.11 cm in height. Axial compression loading is under the strain-controlled mode with a strain limit of 20% and strain rate of 0.3%/min. A detailed description of testing materials, sample preparation, and testing procedure is given below.

2.1. Test Materials. The study is undertaken on reconstituted specimens of Ottawa sand 20–30 with various amounts of iron powder. The specific gravity of Ottawa sand 20–30 is 2.65, e_{max} is 0.652, e_{min} is 0.478, D_{50} is 0.6, D_{50} is 0.76, and sphericity is 0.85.

The particle size of iron powder in the specimens needs to be determined by groutability ratio because of the groutability issue of granular soil. The precise groutability ratio is difficult to determine and has been extensively studied because this phenomenon is affected by the soil parameters such as the grain size of the soil and grout which mostly impacts the groutability, the fine contents, the relative density, and the grouting pressure. To determine the grain size of iron powder, we applied the empirical relationship of groutability developed in [22], which involved only the grain size of soil and grout. In Burwell's theory, the groutability index N was introduced to explain the grouting evenness:

$$N = \frac{D_{15}\,(\text{soil})}{d_{85}\,(\text{grout})}, \tag{1}$$

where N is the groutability of soil, D_{15} is the diameter through which 15% of the total soil mass passes, and d_{85} is the diameter through which 85% of the total grout mass passes. If N calculated by (1) is smaller than 11, the grout cannot be injected into the soil successfully. If N from (1) is larger than 25, the following equation should be checked additionally:

$$N = \frac{D_{10}\,(\text{soil})}{d_{95}\,(\text{grout})}, \tag{2}$$

where N is the groutability of soil, D_{10} is the diameter through which 10% of the total soil mass passes, and d_{95} is the diameter through which 95% of the total grout mass passes. If N, calculated by (2), is larger than 11, grouting is possible; otherwise, grout cannot be injected.

Based on these limitations of groutability provided by (1) and (2), we selected the iron powder by its particle size distribution, that is, d_{85} (grout) < 0.0228 mm and d_{95} (grout) <0.0472 mm.

2.2. Sample Preparation. The reconstituted specimens were compacted in a three-piece mold with a diameter of 3.56 cm and height of 7.11 cm. An adjusted method is developed for the sample preparation within a cavity formed by a rubber membrane that lined up with the three-piece split mold. Compaction is achieved by tamping each layer of the specimens by the Harvard Miniature with a circular base attached to its tip. The circular base has a slightly smaller diameter than that of the three-piece split mold, so the shear distortion on the membrane inside the three-piece split mold can be minimized while being pushed downward by the Harvard Miniature.

When preparing the specimens, we choose relative density as a control parameter because it remains constant even when adding iron powder. First, sand is used to determine the sample specifications, such as the weight of specimens and how many taps were needed to reach the target relative densities (60% and 80%). In order to find the specifications mentioned above, the maximum and minimum void ratios of Ottawa sand are determined according to the methods in [23, 24], which are 0.478 and 0.652, respectively. Then, the void ratio at target relative density can be calculated. Next, we calculated the weight of the sand specimens based on the void ratio at target relative densities (60% and 80%), the size of the specimen (3.56 cm of diameter and 7.11 cm of height), and the specific gravity of Ottawa sand. The amount of iron powder added to the sand is 10% of the total specimen weight.

The iron powder content is 10% by weight of the specimen, and iron powder is premixed with sand before specimen preparation. For the relative density at 60%, the sample is divided into 3 layers and each layer was tapped 3 times. The preliminary trial with Ottawa sand at 80% relative density using 3 layers yielded a poor distribution of specimens, so to produce more uniform specimens, the number of layers is increased to 5. At 80% relative density, the sample is divided into 5 layers and tapped 10 times for each layer. The soil for each layer is weighed, placed in separate containers, and kept sealed until placed into the three-piece split mold. The diameter and height of the compacted specimens are measured after the vacuum is applied and the three-piece mold was removed. The same compaction procedure is followed to obtain a comparable soil property index for both sand specimens and iron-powdered sand.

2.3. Triaxial Test Apparatus and Procedure. The GEOTAC triaxial apparatus made by Trautwein Soil Testing Equipment Company is employed for testing the new material. It consists of GeoJac^{TM} digital load actuators with a load capacity of 8896.44 N, load frame, and pressure chamber (Figure 1). The test procedure of triaxial compression test followed the procedure in [25–28] except for the following adaptions. First, because the test material is cohesionless as discussed in Sample Preparation, the soil compaction is done in the three-piece split mode which sits on the cell base of the triaxial apparatus and covers the rubber membrane wrapped around the lower platen fixed on the cell base. Then, the rubber membrane is filled with the sand and iron power mixture in layers, and each layer was compacted with Harvard Miniature as described in Sample Preparation. After sample compaction is finished, the magnetic coil is placed upon top of the three-piece mold, and the upper platen attached with the vacuum line goes through the magnetic coil and then is placed on top of the sample (Figure 1 (left)). Next, the upper platen is wrapped by the rubber membrane rolled up from the soil sample, vacuum is applied so that the sample is sealed, and the three-piece split mold is removed so the magnetic coil is around the sample (Figure 1 (middle)). Lastly, carefully place the pressure chamber on the cell base, align with the loading devicee with piston, and leave the electric wires extended from the magnetic coil to the precarved grooves on the cell base, which avoids the wires being pinched by the weight of the

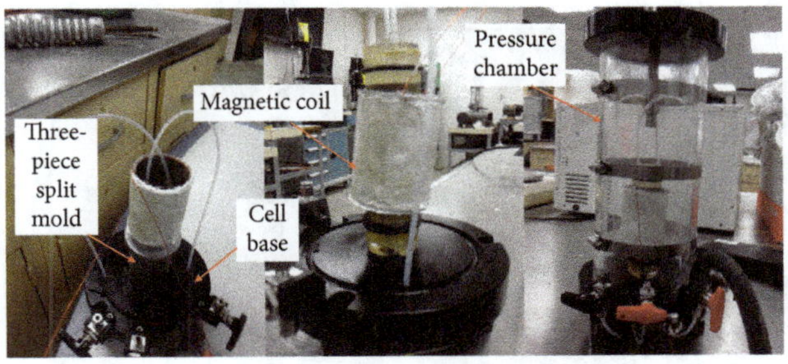

FIGURE 1: Triaxial apparatus with an electromagnetic coil attached.

pressure chamber and the force from applied confining pressure around the soil sample (Figure 1 (right)). These two pieces of wire from the magnetic coil are then connected to the power box capable of providing direct current and being manually controlled. Finally, the magnetic wire around the specimen was powered up before loading.

3. Experimental Results and Discussion

3.1. Stress-Strain Relationship. A series of triaxial tests are carried out at two relative densities (60% and 80%) under four currents (0 A, 0.5 A, 1.5 A, and 2 A) and three confining pressures (103 kPa, 206 kPa, and 310 kPa). Testing data of deviator stress from 30 triaxial compression tests are collected. The corresponding stress-strain curves under three confining pressures σ_3 (σ_3 = 103 kPa, 206 kPa, and 310 kPa) and at two relative densities (D_r = 60% and 80%) are shown in Figure 2. Apparently, the lowest peak stress is obtained from the clean sand specimens. The specimens consisting of iron powder and Ottawa sand under the magnetic fields yield similar curves, all of which are slightly above the curve from the clean sand. Therefore, the presence of the iron powder and the electromagnetic field increases the peak stress; adding 10% iron powder results in an increase of the peak stress, and applying a magnetic field leads to a further increase in peak stress. For instance, if the stress-strain curve is under the same confining pressure and at the same constant relative density, let us say σ_3 = 103 kPa and D_r = 60%, the peak stress increases with an addition of iron content (i.e., from 0% to 10%). As can be seen from Figure 2, the impact of electromagnetic field on the iron-powdered sand on the stress-strain relationship is evident. That is, the peak stress increases with the increase of current imposed through (I = 0 A, 0.5 A, 1 A, and 2 A). On the contrary, as can be seen from Figure 2, if one selects a stress-strain curve with the same current (i.e., I = 0.5 A) and same relative density (i.e., D_r = 60%), the peak stress increases with the increase of confining pressure (i.e., σ_3 = 103 kPa, 206 kPa, and 310 kPa). Similarly, another set of figures can be drawn for the case of the same confining pressure and the current, say σ_3 = 103 kPa and I = 0.5 A (Figure 2), and peak stress is found to increase with the increase of relative densities. Note that the strain-softening phenomenon is observed in the stress-strain curves in clean sand specimens and in iron-powdered sand.

The hyperbolic model developed in [29–37] has been applied extensively in geotechnical engineering to model nonlinear behavior of soil and rock. The model can reflect

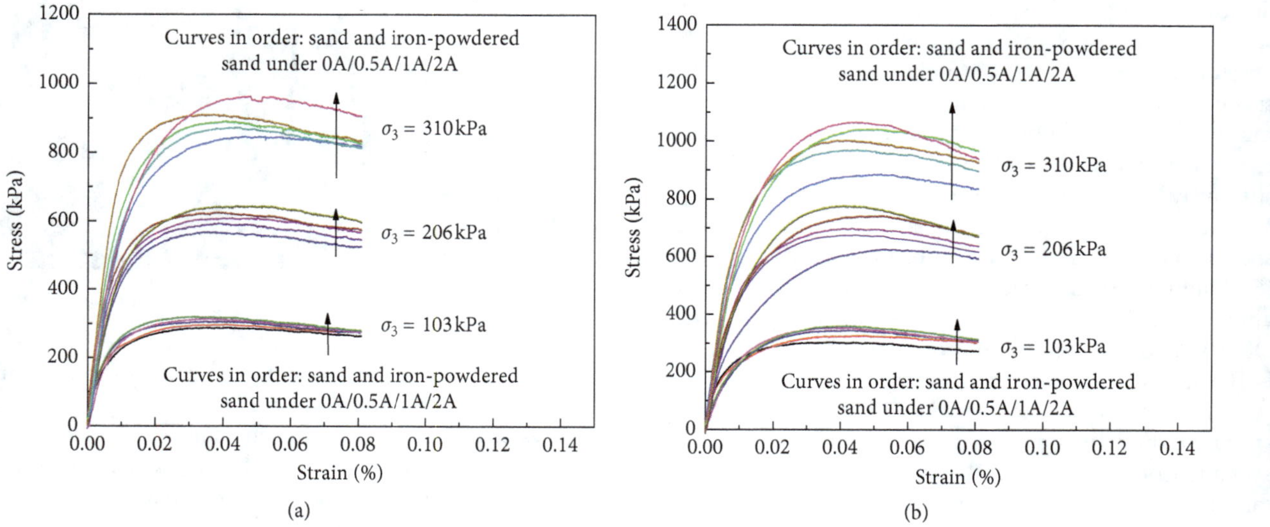

FIGURE 2: The stress-strain curves of sand specimens with different iron powder contents as well as under different confining pressures and magnitude of the electromagnetic field: specimens at (a) 60% relative density and (b) 80% relative density.

the elastic-to-plastic deformation of soil and rock and its parameters have explicit meaning:

$$\sigma_1 - \sigma_3 = \frac{\varepsilon_1}{d + e\varepsilon_1}, \tag{3}$$

where $\sigma_1$andσ_3 are the major and minor principle stress, respectively; ε_1 is the axial strain; and constants d and e are the fitting parameters which can be determined from the conventional compression or the extension triaxial test:

$$d = \frac{1}{E_i}, \qquad e = \frac{1}{(\sigma_1 - \sigma_3)_{\text{ult}}}. \tag{4}$$

Generally, the Duncan–Chang model can fit well for the stress-strain relationship with the strain-hardening phenomenon. However, the theoretical basis of the Duncan–Chang model is an elasticity model based on the generalized incremental Hook's law, which limits its application in geotechnical engineering. Thus, (6) is proposed to describe the soil strain-softening and the prediction of deviator stress from a known axial strain:

$$\sigma_1 - \sigma_3 = \frac{\varepsilon_1}{d + e\varepsilon_1} f\varepsilon_1, \tag{5}$$

where $\sigma_1 - \sigma_3$ is the deviator stress, ε_1 is the axial strain, and d, e, and f are the fitting parameters.

One more fitting parameter is chosen for the iron-powdered sand so that a better fit can be achieved between the predicted behavior and the strain-softening behavior observed in the triaxial compression tests. To show the capability of the proposed constitutive model by revising the Duncan–Chang model, we compare the observed and the predicted behaviors by the revised Duncan–Chang model and the Duncan–Chang model for each test. For the purpose of demonstration, Figures 3(a) and 3(b) show one set of comparison curves of the confining pressure under 103 kPa, the relative density at 60%, and the magnitude of current at 0 A and 2 A, respectively. As can be seen, the predicted results by the revised model agree perfectly with the observed response. This good agreement suggests that the revised model is capable of accurately predicting the behavior of Ottawa sand and the iron-powdered sand in which strain-softening is dominant.

3.1.1. Prefailure Behavior: Initial Stiffness. The elastic modulus for each testing condition is determined from the corresponding stress-strain curve and summarized in Table 1 and Figure 4. It is evident from Figure 4 that the confining pressure had a significant influence on the elastic modulus. As expected, the elastic modulus was found to increase with increasing confining pressure for both sand and iron-powdered sand under the electromagnetic field. In contrast to the pronounced impact from confining pressure to the elastic modulus, the impact of iron powder inclusion on elastic modulus depends on the relative density of specimens. First, the first three columns in Table 1 show the elastic modulus of specimens at 60% relative density decreases with the increase of iron content. However, for specimens at 80% relative density shown in the first three columns in Table 1, the elastic modulus is relatively independent of the iron content. Second, as shown in Figure 4(a) and Table 1, the elastic modulus increases with the increase of magnitude of current imposed through for specimens at 60% relative density under any confining pressure and magnitude of the electromagnetic field. However, at 80% relative density (Figure 4(b) and Table 1), the elastic modulus increases with the increase of magnitude of the electromagnetic field under any confining pressure except 310 kPa. When investigating the effect of relative density on elastic modulus, one can pick two data points from Figure 4 with the same testing condition (confining pressure and the magnitude of electromagnetic field) and find the specimens with higher relative

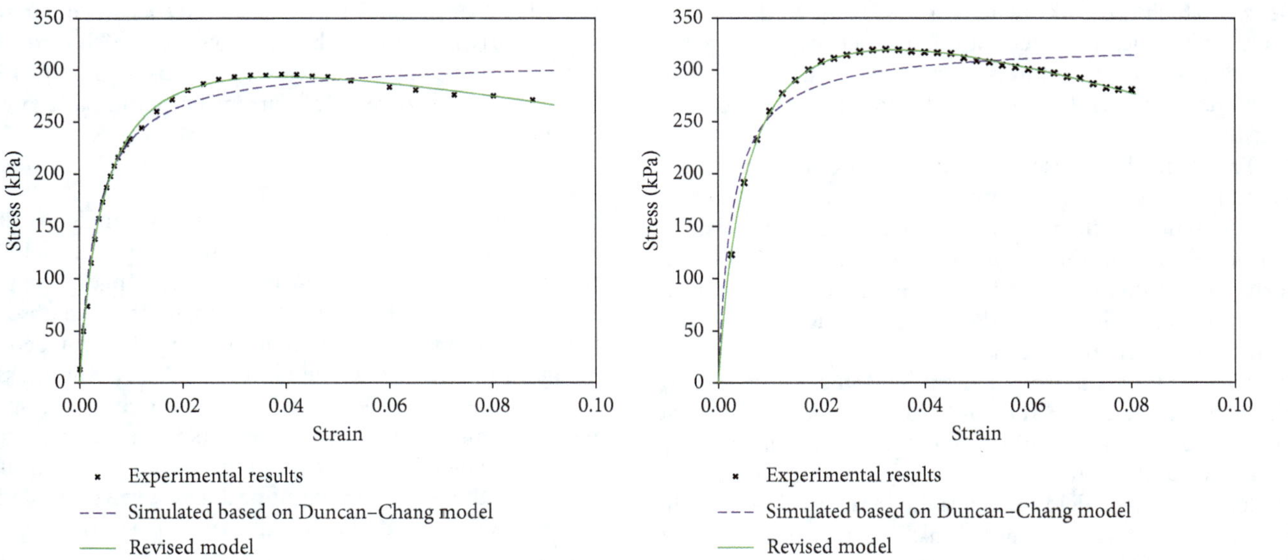

(a) (b)

Figure 3: Comparison between the Duncan–Chang model and the revised model: (a) iron-powdered sand at 60% relative density and under 103 kPa confining pressure; (b) iron-powdered sand at 60% relative density and under 2 A current and 103 kPa confining pressure.

Table 1: Elastic modulus.

σ_3(kPa)	Sand (kPa)	Iron-powdered sand (kPa)	Iron-powdered sand under 0.5 A (kPa)	Iron-powdered sand under 1 A (kPa)	Iron-powdered sand under 2 A (kPa)
Relative density at 60%					
103	30409.68	16464.47	22983.93	31533.01	38477.60
206	42855.95	33820.75	42773.88	50010.53	56840.49
310	61040.02	55637.26	62280.16	65811.87	76135.43
Relative density at 80%					
103	35069.21	33076.31	34522.15	37699.62	41039.19
206	48027.40	48448.40	54550.36	65218.19	68462.87
310	77618.35	80643.42	85292.63	66799.12	70504.34

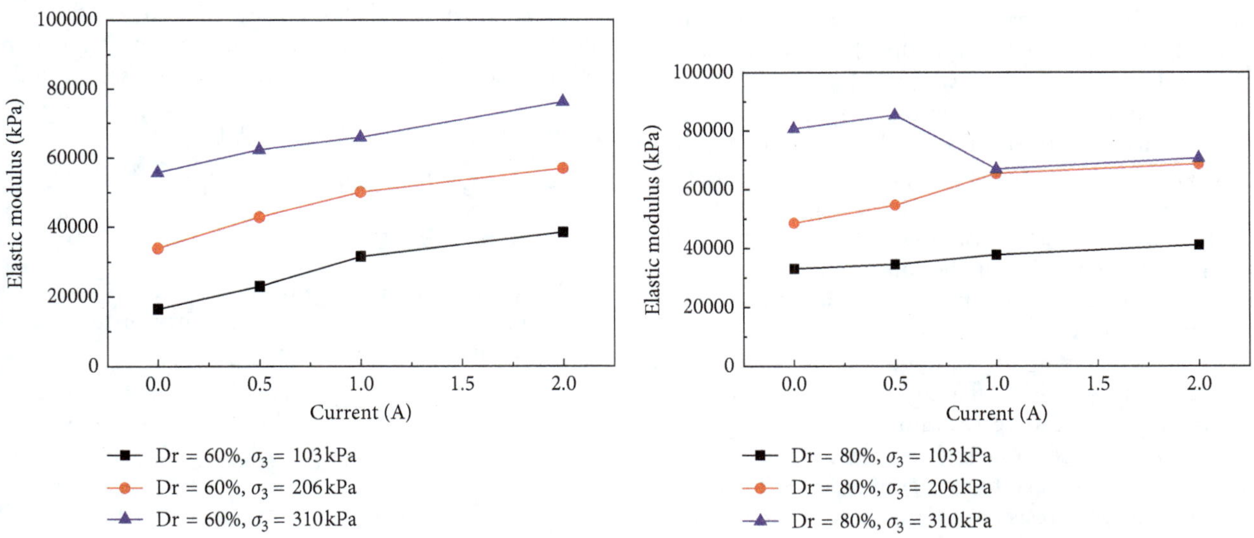

(a) (b)

Figure 4: Elastic modulus: iron-powdered sand at (a) 60% relative density and (b) 80% relative density under different magnitude of the electromagnetic field.

density which presents greater stiffness at the initial loading stage. In parallel, the effect of relative density as related to elastic modulus is captured from Figure 4 that the elastic modulus is moderately increased with increasing relative density.

From the above discussion, it concludes that the elastic modulus is a function of relative density, confining pressure, iron content, and the magnitude of electromagnetic field. It also implies that the ability of specimens to resist deformation at the initial loading stage arises from void ratio (which depends on relative density, confining pressure, and iron content) and the internal friction between soil particles (which depends on iron content and the magnitude of electromagnetic field). The confining pressure and relative density determine the initial void ratio of specimens with pure sand, and additionally, iron content and the magnitude of electromagnetic field influence the initial void ratio of the iron powder-reinforced specimens. The internal resistance for pure sand specimens is provided by interlock between sand particles and frictional forces from unsmooth surfaces. In comparison, the internal resistance for iron powder-reinforced specimens is additionally governed by the iron powder content and the magnitude of electromagnetic field. As discussed in Section 2.2, the iron content is 10% by weight of the sample, which implies that the specimen with relative density at 60% contains less iron powder compared with the specimen with relative density at 80%. At the sample preparation stage, we observe that the iron powder coating is around the sand particles. Therefore, at a relative density of 60%, most of the iron powder was coated around sand particles and little powder is filled into voids between particles to densify the specimens. Hence, the coating by iron powder reduced the internal friction of iron powder-reinforced specimens, and the elastic modulus decreased if iron powder is added (as shown in Figure 4(a)). However, compared with specimens with a relative density of 60%, specimens at 80% relative density contain smaller voids and more iron powder (iron powder is still 10% by weight of the specimen but is with a heavier weight since the specimen is heavier at higher relative density). Therefore, except coating around sand particles, iron powder filled most of the voids which densifies iron powder-reinforced specimens. Moreover, the densification effect from filling voids by iron powder dominates over the coating effect around particles; as a result, iron powder inclusion slightly increases the elastic modulus of specimens. When the electromagnetic field is applied by powering up the coil around the specimen, the elastic modulus generally increases with the increase of magnitude of current (Figure 4) except when the relative density is 80%, confining pressure is 310 kPa, and current is above 1 A. Figure 4(a) shows that the elastic modulus of 60% relative density sample increases with the current amplitude, which is reasonable since a higher magnitude of electromagnetic field is generated to strengthen the specimens and increase the internal resistance. However, at 80% relative density and under 310 kPa confining pressure, the elastic modulus has no continuous increase with the increase of magnitude of current. This is rational since the coating sand particles are compacted at a denser degree, which makes it easier for coating sand particles slide over each other with a higher driving force on the iron powder particles from the electromagnetic field. Therefore, the elastic modulus decreases with the increase of current amplitude at a higher confining pressure.

3.1.2. Peak Stress. The most impressive advantage of iron powder inclusion together with electromagnetic field reinforcement is the improvement of peak strength. The peak stress from each test is summarized in Table 2. An absolute measure to quantify this improvement is provided by strength ratio defined as the ratio of peak deviator stress of the reinforced specimen to that of the unreinforced specimen for similar conditions. The effects from the magnitude of electromagnetic field, confining pressure, and relative density on the improvement of peak stress are presented in Figure 5. As expected, the magnitude of the electromagnetic field has positive impact on strength improvement. Additionally, for all reinforced specimens, the improvement of strength is more obvious for specimens under 103 kPa to 206 kPa than those under 310 kPa, and similar trends are observed at all relative densities. This indicates that the benefit of iron powder reinforcement as well as the electromagnetic field is more significant at medium to low confining pressure. It is reasonable since the specimens under high confining pressure is already stiff and the effectiveness of iron powder together with the electromagnetic field is less remarkable. The strength improvement is more pronounced at 80% relative density. It verifies that iron powder coating around sand particles at 60% relative density reduces surface friction between sand particles in the microscopic view and reduces strength overall.

From the discussion above, we conclude that peak stress is a function of four variables: iron content, confining pressure, relative density, and current. The contribution of each variable could be evaluated by regressing the fitting parameters with the test variables. The fitting parameters d, e, and f (shown in Table 3) are assumed to be a function of the testing variables, that is, relative density, confining pressure, and current. The function is expressed as follows:

$$\begin{aligned} d &= \left(a_1 (\sigma_3)^{a_2}\right) \times \left(1 + m_d I_i + n_d I_i^2 + l_d I_i^3\right), \\ e &= \left(b_1 \sigma_3^2 + b_2 \sigma_3 + b_3\right) \times \left(1 + m_e I_i + n_e I_i^2 + l_e I_i^3\right), \\ f &= \left(c_1 \sigma_3^2 - c_2 \sigma_3 - c_3\right) \times \left(1 + m_f I_i + n_f I_i^2 + l_f I_i^3\right). \end{aligned} \tag{6}$$

Next, the determination of the function is as follows.

Taking the determination of the parameter d at 60% relative density as an example,

$$d = F(\sigma_3, I, D_r, \mathrm{Fe\%}), \tag{7}$$

which can be written as a piecewise function:

$$d = \begin{cases} f_0(\sigma_3), & \text{if Fe\%} = 0, \\ f_1(\sigma_3) \times f_2(I), & \text{if Fe\%} > 0. \end{cases} \tag{8}$$

First, $f_0(\sigma_3)$ is obtained by regressing the fitting parameter d and confining pressure values:

TABLE 2: Peak stress.

σ_3(kPa)	Sand (kPa)	Iron-powdered sand (kPa)	Iron-powdered sand under 0.5 A (kPa)	Iron-powdered sand under 1 A (kPa)	Iron-powdered sand under 2 A (kPa)
Relative density at 60%					
103	287.82	298.78	307.69	312.49	318.65
206	580.43	596.19	612.23	622.92	641.42
310	842.89	870.30	890.86	904.56	938.83
Relative density at 80%					
103	306.25	344.63	364.64	370.05	377.42
206	627.93	673.69	692.13	746.95	774.36
310	871.28	952.53	986.80	1027.91	1041.62

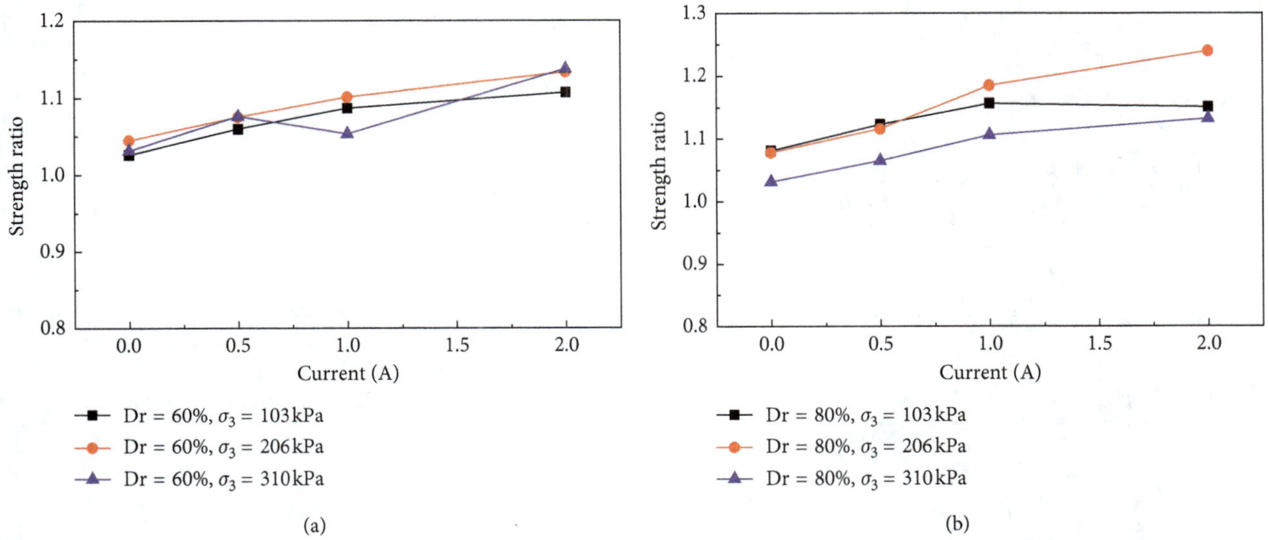

FIGURE 5: Variation of strength ratio with the magnitude of the electromagnetic field for different confining pressures and relative densities: specimens at (a) 60% relative density and (b) 80% relative density.

TABLE 3: The fitting parameter.

σ_3 (kPa)	Sand (kPa)	Iron-powdered sand (kPa)	Iron-powdered sand under 0.5 A (kPa)	Iron-powdered sand under 1 A (kPa)	Iron-powdered sand under 2 A (kPa)
d at 60% relative density					
103	61.77	54.18	59.03	67.01	64.64
206	138.25	130.01	148.96	129.76	159.71
310	189.05	205.85	216.72	196.42	243.50
d at 80% relative density					
103	58.96	67.91	86.09	91.19	93.30
206	169.67	171.83	168.42	190.48	237.16
310	194.13	219.43	239.95	286.44	309.82
e at 60% relative density					
103	0.0089	0.0051	0.0055	0.008	0.0062
206	0.0127	0.0095	0.0140	0.0077	0.0144
310	0.0124	0.0125	0.0112	0.0102	0.0163
e at 80% relative density					
103	0.030	0.010	0.014	0.0138	0.015
206	0.020	0.015	0.013	0.016	0.02
310	0.010	0.0111	0.012	0.019	0.019
f at 60% relative density					
103	−212.69	−138.50	−195.51	−263.63	−246.68
206	−546.91	−469.56	−537.34	−445.84	−591.54
310	−556.91	−750.52	−881.63	−679.92	−882.90
f at 80% relative density					
103	−556.90	−207.54	−368.15	−419.67	−411.32
206	−577.91	−689.71	−650.82	−758.04	−1153.73
310	−590.40	−770.31	−831.82	−1116.84	−1399.20

$$f_0(\sigma_3) = 3.8504\sigma_3^{1.0337}, \quad \text{if Fe\%} = 0. \tag{9}$$

Second, in order to determine $f_1(\sigma_3)$, by regressing data from column 1 and column 3 in Table 3, we obtain

$$d = 2.006\sigma_3^{1.2202}. \tag{10}$$

$f_2(I_i)$ can be expressed as follows:

$$f_2(I_i) = 1 + mI_i - nI_i^2 + lI_i^3, \tag{11}$$

$$\Pi = \sum_{i=1}^{j} [f_2(I_i) - y_i]^2 = \sum_{i=1}^{j} [1 + mI_i + nI_i^2 + lI_i^3 - y_i]^2. \tag{12}$$

In order to obtain m, n, and l, (15) is differentiated:

$$\frac{\partial\Pi}{\partial m} = 2\left[\mathrm{m}\sum_{i=1}^{j} I_i^2 + \mathrm{n}\sum_{i=1}^{j} I_i^3 + \mathrm{l}\sum_{i=1}^{j} I_i^4 - \sum_{i=1}^{j} I_i y_i + \sum_{i=1}^{j} I_i\right] = 0,$$

$$\frac{\partial\Pi}{\partial n} = 2\left[\mathrm{m}\sum_{i=1}^{j} I_i^3 + \mathrm{n}\sum_{i=1}^{j} I_i^4 + \mathrm{l}\sum_{i=1}^{j} I_i^5 - \sum_{i=1}^{j} I_i^2 y_i + \sum_{i=1}^{j} I_i^2\right] = 0,$$

$$\frac{\partial\Pi}{\partial l} = 2\left[\mathrm{m}\sum_{i=1}^{j} I_i^4 + \mathrm{n}\sum_{i=1}^{j} I_i^5 + \mathrm{l}\sum_{i=1}^{j} I_i^6 - \sum_{i=1}^{j} I_i^3 y_i + \sum_{i=1}^{j} I_i^3\right] = 0,$$

$$m = \frac{1}{\Delta}\cdot\begin{vmatrix} \left(\sum_{i=1}^{j} I_i y_i - \sum_{i=1}^{j} I_i\right) & \sum_{i=1}^{j} I_i^3 & \sum_{i=1}^{j} I_i^4 \\ \left(\sum_{i=1}^{j} I_i^2 y_i - \sum_{i=1}^{j} I_i^2\right) & \sum_{i=1}^{j} I_i^4 & \sum_{i=1}^{j} I_i^5 \\ \left(\sum_{i=1}^{j} I_i^3 y_i - \sum_{i=1}^{j} I_i^3\right) & \sum_{i=1}^{j} I_i^5 & \sum_{i=1}^{j} I_i^6 \end{vmatrix},$$

$$n = \frac{1}{\Delta}\cdot\begin{vmatrix} \sum_{i=1}^{j} I_i^2 & \left(\sum_{i=1}^{j} I_i y_i - \sum_{i=1}^{j} I_i\right) & \sum_{i=1}^{j} I_i^4 \\ \sum_{i=1}^{j} I_i^3 & \left(\sum_{i=1}^{j} I_i^2 y_i - \sum_{i=1}^{j} I_i^2\right) & \sum_{i=1}^{j} I_i^5 \\ \sum_{i=1}^{j} I_i^4 & \left(\sum_{i=1}^{j} I_i^3 y_i - \sum_{i=1}^{j} I_i^3\right) & \sum_{i=1}^{j} I_i^6 \end{vmatrix},$$

$$l = \frac{1}{\Delta}\cdot\begin{vmatrix} \sum_{i=1}^{j} I_i^2 & \sum_{i=1}^{j} I_i^3 & \left(\sum_{i=1}^{j} I_i y_i - \sum_{i=1}^{j} I_i\right) \\ \sum_{i=1}^{j} I_i^3 & \sum_{i=1}^{j} I_i^4 & \left(\sum_{i=1}^{j} I_i^2 y_i - \sum_{i=1}^{j} I_i^2\right) \\ \sum_{i=1}^{j} I_i^4 & \sum_{i=1}^{j} I_i^5 & \left(\sum_{i=1}^{j} I_i^3 y_i - \sum_{i=1}^{j} I_i^3\right) \end{vmatrix}, \tag{13}$$

where

$$\Delta = \begin{vmatrix} \sum_{i=1}^{j} I_i^2 & \sum_{i=1}^{j} I_i^3 & \sum_{i=1}^{j} I_i^4 \\ \sum_{i=1}^{j} I_i^3 & \sum_{i=1}^{j} I_i^4 & \sum_{i=1}^{j} I_i^5 \\ \sum_{i=1}^{j} I_i^4 & \sum_{i=1}^{j} I_i^5 & \sum_{i=1}^{j} I_i^6 \end{vmatrix}. \tag{14}$$

Based on the experimental results, we obtain the following parameters: $m = 0.30$, $n = -0.46$, $l = 0.17$, and $j = 12$, in this study.

Thus,

$$f_2(I_i) = 1 + 0.30I_i - 0.46I_i^2 + 0.17I_i^3,$$

$$d = \begin{cases} 3.85\sigma_3^{1.0337}, & \text{if Fe\%} = 0, \\ (2.006\sigma_3^{1.2202}) \\ \quad\times(1 + 0.30I_i - 0.46I_i^2 + 0.17I_i^3), & \text{if Fe\%} > 0. \end{cases} \tag{15}$$

The same process is applied to relate fitting parameters e and f at different test conditions.

At relative density 60%, the following can be obtained:

If no magnetic field is applied, then we have the following equations:

$$\begin{aligned} d &= 3.85\sigma_3^{1.0337}, \\ e &= \left(-9E-06\sigma_3^2 + 0.0007\sigma_3 + 0.001\right), \\ f &= \left(0.72\sigma_3^2 - 54.71\sigma_3 + 445.77\right). \end{aligned} \tag{16}$$

If a magnetic field is applied, we can obtain

$$\begin{aligned} d &= \left(2.006\sigma_3^{1.2202}\right) \times \left(1 + 0.30I_i - 0.46I_i^2 + 0.17I_i^3\right), \\ e &= \left(-3E-06\sigma_3^2 + 0.0004\sigma_3 - 0.0007\right) \\ &\quad \times\left(1 - 0.21I_i - 0.14I_i^2 + 0.13I_i^3\right), \\ f &= \left(0.11\sigma_3^2 - 27.08\sigma_3 + 242.64\right) \\ &\quad \times\left(1 + 0.72I_i - 0.97I_i^2 + 0.34I_i^3\right). \end{aligned} \tag{17}$$

At 80% relative density, the following can be obtained:

If no magnetic field is applied,

$$\begin{aligned} d &= 4.41\sigma_3^{1.0398}, \\ e &= \left(-5E-05\sigma_3^2 + 0.003\sigma_3 - 0.03\right), \\ f &= \left(0.71\sigma_3^2 - 53.81\sigma_3 + 400.71\right), \end{aligned} \tag{18}$$

If a magnetic field is applied,

$$\begin{aligned} d &= \left(3.64\sigma_3^{1.0969}\right) \times \left(1 + 0.316I_i + 0.076I_i^2 - 0.057I_i^3\right), \\ e &= \left(-2E-05\sigma_3^2 + 0.001\sigma_3 - 0.0042\right) \\ &\quad \times\left(1 - 1.200I_i + 2.154I_i^2 - 0.746I_i^3\right), \\ f &= \left(0.892\sigma_3^2 - 72.303\sigma_3 + 676.21\right) \\ &\quad \times\left(1 + 0.720I_i - 0.066I_i^2 - 0.021I_i^3\right). \end{aligned} \tag{19}$$

3.1.3. Postpeak Behavior. The postpeak behavior of the iron powder-reinforced autoadaptive material was examined in terms of brittleness (I_B) and energy absorption capacity (EAC).

(1) Brittleness. We can observe from stress-strain curves in Figure 2 that the residual strength of autoadaptive specimens remains higher than that of unreinforced specimens after peak strength, which suggests that the autoadaptive material would sustain greater load even at a larger deformation compared to unreinforced sand. To quantify this behavior, the brittleness index (I_B) defined in [38] is applied:

$$I_B = \frac{\sigma_{d,\text{peak}}}{\sigma_{d,\text{res}}} - 1, \quad (20)$$

where $\sigma_{d,\text{peak}}$ is the peak deviator stress and $\sigma_{d,\text{res}}$ is the residual deviator stress of the specimen. The ductility behavior increases as I_B decreases towards zero.

As expected, the brittleness is found to be greater for specimens with higher relative density (Figure 6 and Table 4). This implies that the relative density dominates the ductility behavior of sand and unreinforced sand over the magnitude of electromagnetic field. It is interesting to note that the inclusion of iron powder improved ductility of specimens; however, the existence of electromagnetic field decreased ductility (Figure 6). The peak and residual stress have been improved (Figure 2) by the inclusion of iron powder but with a greater improvement of residual stress; as a result, the brittleness index for the iron-powdered sand was reduced. This validates again that iron powder coating around the soil particles, which made soil particles easier to rearrange in the microscopic view and makes specimens easier to deform even at a relatively larger strain. Applying the electromagnetic field can strengthen the specimens and decreases the ductility of iron powder-reinforced specimens (Figure 6).

(2) Energy Absorption Capacity. As defined in [39], the energy absorption capacity is introduced to describe toughness of fiber-reinforced concrete under flexural loading condition. The energy absorption capacity (EAC) is defined by the area under the stress-strain curve up to a reference strain (8% is selected in this study).

The EAC of the autoadaptive specimen is from energy dissipated from the soil matrix deformation $(\text{EAC})_s$ and the energy dissipated through iron powder particles movement driven by electromagnetic force $(\text{EAC})_i$. Hence, the total energy absorption capacity (EAC) of the autoadaptive material can be expressed as follows:

$$\text{EAC} = (\text{EAC})_s + (\text{EAC})_i. \quad (21)$$

Table 5 and Figure 7 present the energy absorption capacity of each specimen. Figure 7 demonstrates the impact of the magnitude of electromagnetic field and confining pressure on energy absorption improvement. The general pattern interpreted in Table 5 and Figure 7 is quite similar to that previously demonstrated in Table 2 for the peak stress response. All of the testing factors noticeably affected the energy absorption capacity positively of both sand and iron-powdered sand under the electromagnetic field. As shown in the first three columns in Table 5, the effect of inclusion of iron powder is found to be more pronounced for the specimens under higher confining pressures. Figures 7(a) and 7(b) show the increase in magnitude of the electromagnetic field leads to a higher improvement of energy absorption capacity. If one picks one data point from Figures 7(a) and 7(b) for the same testing condition (the confining pressure and magnitude of current) except relative density, it can be noticed that the relative density has positive effect on the energy absorption.

Alternatively, the energy absorption capacity can also be evaluated by normalizing the EAC over the entire stress-strain curve, which is expressed as $[(\text{EAC})_\varepsilon/(\text{EAC})_{\varepsilon_{\text{ref}}}]$. The $(\text{EAC})_\varepsilon$ represents the area up to any axial strain ε on the stress-strain curve, and $(\text{EAC})_{\varepsilon_{\text{ref}}}$ is the area to a reference strain ε_{ref} (8% in this study). To make a distinction of effect from different variables (relative density, magnitude of the electromagnetic field, and confining pressure) on soil behavior, Figures 7(c) to 7(e) show three examples of normalized energy absorption against the axial strain with variation of parameters in each plot. First, the normalized energy absorption for iron powder-reinforced sand increases with the axial strain at an almost constant rate (Figure 7(c) to Figure 7(e)). The increase of relative density, confining pressure, and the magnitude of electromagnetic field leads to a lower energy absorption rate, which indicates soil behaves predominantly brittle. This agrees well with the deviator stress-strain curves shown in Figure 2 and brittleness index shown in Figure 6.

3.2. Shear Strength Parameters. To investigate the influence of the electromagnetic field on the friction angle and cohesive coefficient, we plot the Mohr–Coulomb failure envelopes at different test conditions. To precisely assess the friction angle and cohesive coefficient, we employ the method of linearizing triaxial test failure envelopes [40]. Generally, the friction angle and the cohesive coefficient are obtained by drawing a tangent line to a series of Mohr's circles. However, it is easier and more accurate to regress on the scattered points from the stress path than to draw a tangent line to a series of circles. Additionally, linear regression on the scattered points gives the statistical evaluation of the data fitting. Handy [40] pointed out that variability biases exist either by best fit by eye or by regression of the p-q diagram from the triaxial test data, which leads to an overestimation of the friction angle. The reason for these biases is that the variability is 45° to the y-axis if lateral stress σ_3 is constant. As can be seen in the Mohr–Coulomb envelope equation,

$$\tau = c + \sigma \times \tan \alpha. \quad (22)$$

The independent variable is σ, and it is assumed that the test variability is parallel to the y-axis. However, as shown in Figure 8, the variability of τ is not parallel to the y-axis, and it is a line inclined with the y-axis. This variability bias leads to the overestimation of the friction angle and will be involved when regressing p-q data points on the p-q plane.

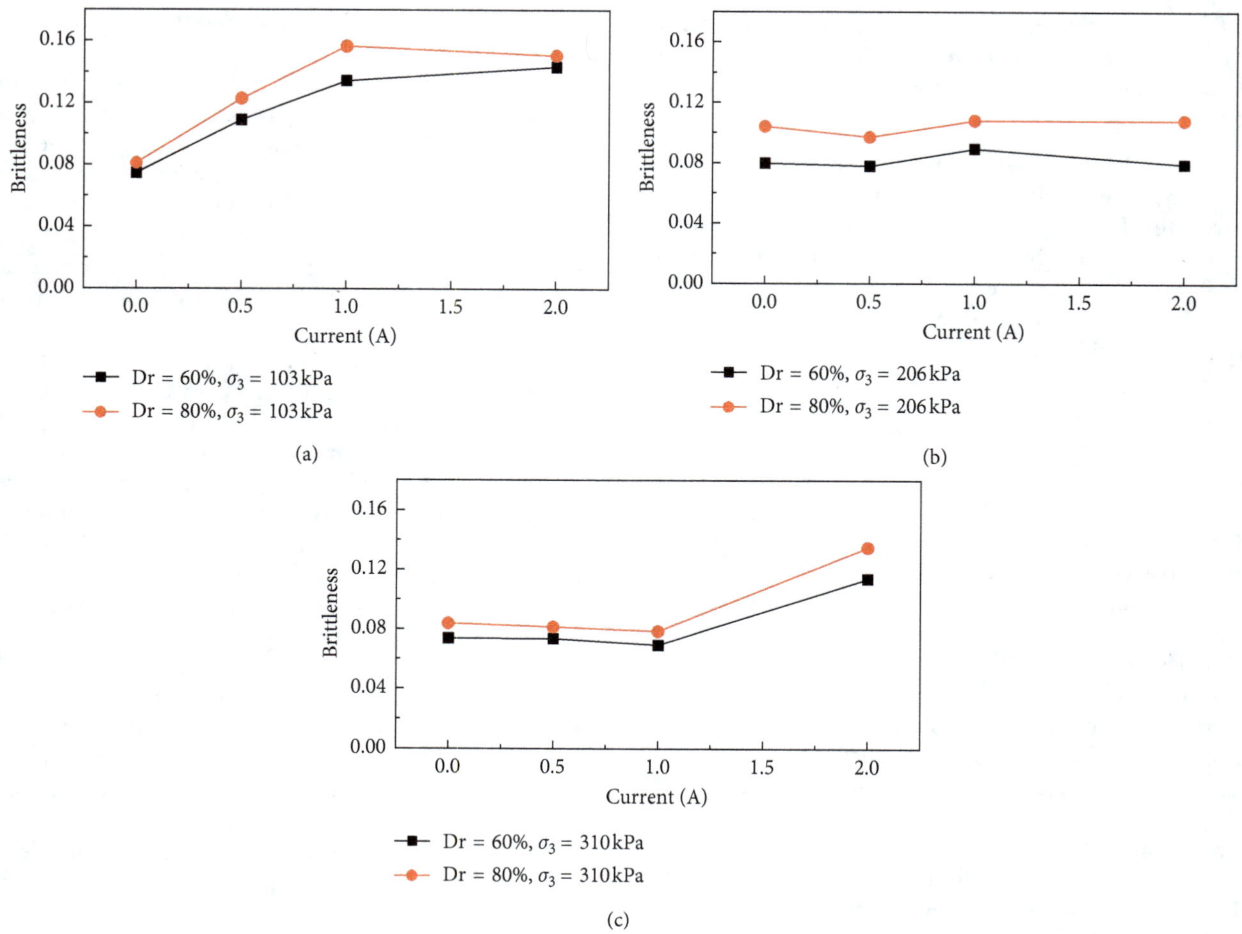

FIGURE 6: Brittleness index: effect of the magnitude of electromagnetic field on the brittleness index of specimens under (a) 103 kPa, (b) 206 kPa, and (c) 310 kPa.

TABLE 4: Brittleness.

σ_3 (kPa)	Sand (kPa)	Iron-powdered sand (kPa)	Iron-powdered sand under 0.5 A (kPa)	Iron-powdered sand under 1 A (kPa)	Iron-powdered sand under 2 A (kPa)
Relative density at 60%					
103	0.09	0.07	0.11	0.13	0.14
206	0.09	0.08	0.08	0.09	0.08
310	0.09	0.07	0.07	0.07	0.11
Relative density at 80%					
103	0.11	0.08	0.12	0.16	0.15
206	0.11	0.10	0.10	0.11	0.11
310	0.13	0.08	0.08	0.08	0.14

TABLE 5: Energy absorption capacity.

σ_3 (kPa)	Sand (kPa)	Iron-powdered sand (kPa)	Iron-powdered sand under 0.5 A (kPa)	Iron-powdered sand under 1 A (kPa)	Iron-powdered sand under 2 A (kPa)
Relative density at 60%					
103	3173.38	3253.88	3500.11	3518.13	3802.56
206	6355.85	6453.01	6432.52	6812.96	6773.75
310	8865.79	9128.71	9585.37	9443.23	9745.08
Relative density at 80%					
103	3541.93	3554.64	3788.64	3864.91	3819.25
206	6349.85	7497.74	7359.33	7959.01	7762.73
310	9117.75	10255.34	10546.79	10721.98	10820.73

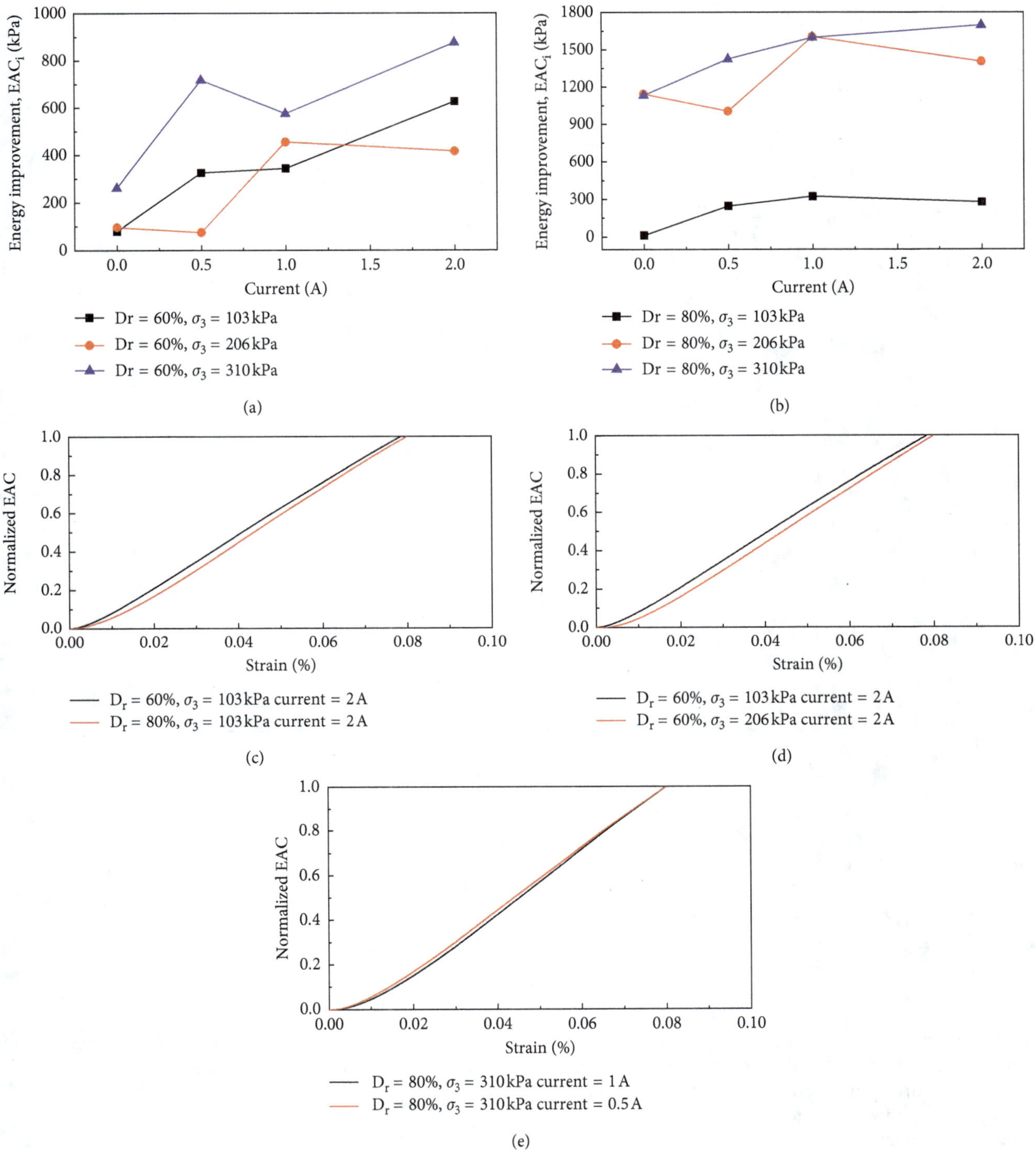

FIGURE 7: Energy absorption capacity: effect of the magnitude of electromagnetic field on energy absorption capacity improvement of specimens at (a) 60% relative density and (b) 80% relative density and effect of (c) relative density, (d) confining pressure, and (e) magnitude of the electromagnetic field on normalized energy absorption capacity.

This problem is simply solved by regressing q on p on the polar planes and rotating q and p axes by an angle θ (normally 45°). Consequently, the y-axis is parallel to the data variability direction as shown in Figure 9.

It is solved as follows: first, the abscissa of the center of the Mohr circle and the radius are calculated based on the minor principle stress and major principle stress from the triaxial tests:

$$p = (\sigma_1 + \sigma_3)/2,$$
$$q = (\sigma_1 - \sigma_3)/2. \tag{23}$$

Second, we convert the data set p-q from rectangular coordinates to polar coordinates, which are designated as θ and r. Third, 45° is added to θ, which gives us θ_r. Fourth, convert θ_r and r back to the rectangular plane, which are

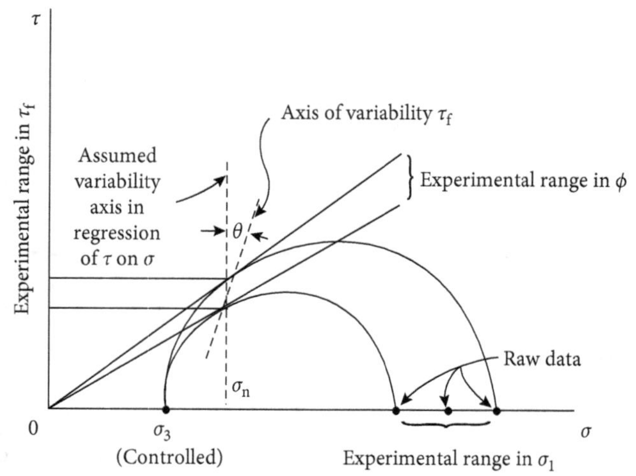

Figure 8: Incorrect regression of τ on σ [40].

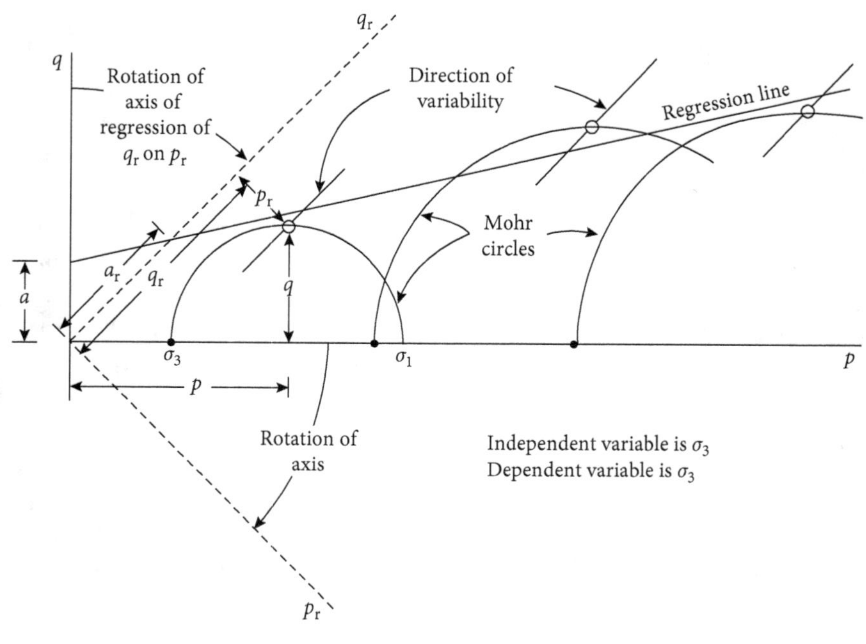

Figure 9: p-q diagram for q_r on p_r after rotation [40].

p_r-q_r data points. Last, perform regression on p_r-q_r data, and the slope and intercept of the regression line are termed as α and a_r, correspondingly. Finally, the following relationships are applied to relate α and a_r to friction angle ϕ and cohesive coefficient c:

$$\sin\phi = \tan\alpha, \quad c = a/\cos\phi, \tag{24}$$

where

$$a = a_r\left[\frac{\sin(90 - \Delta\theta - \alpha)}{\cos\alpha}\right]. \tag{25}$$

By following the procedure above, the friction angle and the cohesive coefficient can be obtained. The friction angle and the cohesive coefficient are listed in Table 6. The results of the nonrotated method compared with Handy's method are shown in Figure 10 and Table 6; Handy's method generally agrees with the corresponding test results compared with the nonrotated method.

4. Conclusions

The improvement of soil strength by inclusion of iron powder and magnetic field was evaluated in this study. This investigation included the performance on triaxial tests of Ottawa sand and iron-powdered sand under different magnetic fields and development of the stress-strain relationship. A revised Duncan–Chang model was proposed to simulate the strain-softening behavior of specimens. The following main conclusions and observations can be made from this investigation:

(1) The most impressive advantage of iron powder inclusion together with electromagnetic field reinforcement was the improvement of peak strength

TABLE 6: Friction angle and cohesive coefficient.

Type of the specimen	Friction angle, ϕ (°)		Cohesive coefficient, c (psi)	
	Nonrotated method	Rotated method	Nonrotated method	Rotated method
Sand at 60% D_r	35.082	35.080	0.004	0.001
Iron-powdered sand at 60% D_r	35.686	35.683	0.031	0.818
Iron-powdered sand at 60% D_r and 0.5 A	35.799	35.785	0.220	1.322
Iron-powdered sand at 60% D_r and 1 A	36.408	36.404	0.374	1.361
Iron-powdered sand at 60% D_r and 2 A	37.543	37.544	1.044	3.857
Sand at 80% D_r	37.413	37.415	0.005	−0.0001
Iron-powdered sand at 80% D_r	37.608	37.581	0.372	0.573
Iron-powdered sand at 80% D_r and 0.5 A	38.007	37.981	0.727	3.09
Iron-powdered sand at 80% D_r and 1 A	39.276	39.167	0.738	3.104
Iron-powdered sand at 80% D_r and 2 A	38.716	39.538	0.752	3.721

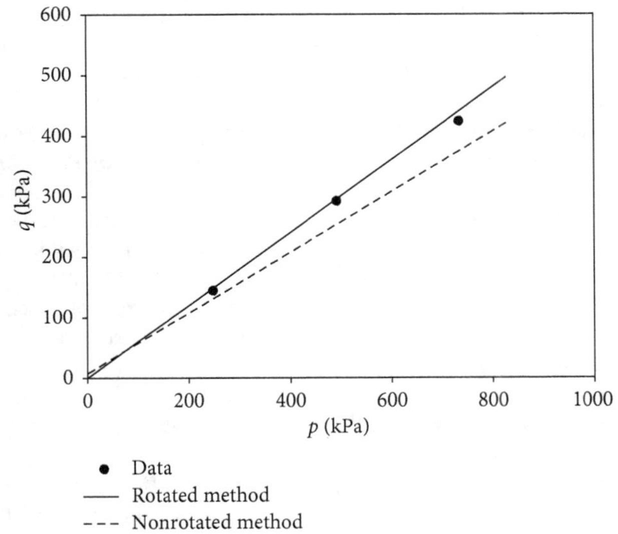

FIGURE 10: Sand with iron powder at 60% relative density under 2A current.

and shear strength. The positive impact was more significant at medium to low confining pressure. It is reasonable since the specimens under high confining pressure are already stiff and the effectiveness of iron powder together with the electromagnetic field is less remarkable.

(2) The addition of iron powder to pure sand increased the peak stress, energy absorption, and brittleness. The initial stiffness improved with the addition of iron powder for the specimens at 60% relative density; however, the initial stiffness of specimens at 80% relative density was insensitive to the iron powder inclusion. Moreover, the addition of iron powder slightly improved internal friction angle and most importantly introduced the cohesive coefficient to specimens which were zero for pure sand.

(3) By applying the electromagnetic field, the peak stress, energy absorption, friction angle, cohesive coefficient, and strength ratio, the initial stiffness improved with the increase of the electromagnetic field except for the specimens with 80% relative density and under confining pressure of 310 kPa. This is explained by that the specimens with 80% relative density and under 310 kPa confining pressure are already stiff, and therefore, the improvement is not apparent. In parallel, the brittleness of specimens increased with the increase of magnitude of the electromagnetic field which agrees well with the increase of stiffness.

(4) The initial stiffness of iron-powdered sand was determined by the confining pressure, relative density, iron powder content, and magnitude of the electromagnetic field. Among these controllable testing variables, the confining pressure and stiffness have major impact on the elastic modulus. The effect of the inclusion of iron powder on the elastic modulus was positive for specimens at 60% relative density and was less obvious on the specimens at 80% relative density. By applying the electromagnetic field, the initial stiffness improved except for the specimens that were already stiff (80% relative density and under 310 kPa confining pressure). It can be stated that the combination of the inclusion of iron powder and electromagnetic field has great potential in improvement of initial stiffness.

(5) At a relative density of 60%, most of the iron powder was coated around sand particles and few was filled into voids between particles to densify the specimens. Hence, the coating by iron powder reduced the surface friction of iron powder-reinforced specimens in the microscopic view and made the coating soil particles easier to rearrange. As a result, it reduced the initial stiffness and overall brittleness by making specimens easier to deform. However, compared with specimens with a relative density of 60%, specimens at 80% D_r contain smaller voids and more iron powder. Therefore, except coating around sand particles, iron powder filled most of the voids which densified iron powder-reinforced specimens. Moreover, the densification effect from filling voids by iron powder dominates over the coating effect around particles; as a result, the strength improvement was more pronounced at 80% relative density.

We conclude that this innovative method of autoadaptive media has the potential to be a valuable means of strengthening the soil by temporary solidification. This characteristic is applicable in earthquake engineering by sensing and intentionally mitigating the external disturbance. However, the development and evaluation of this investigation still require intensive laboratory tests. The challenge in practice is mixing the iron powder with sand uniformly in the field and effectively placing a magnetic field on-site.

Conflicts of Interest

The authors declare that they have no conflicts of interest.

Acknowledgments

This research was supported by the National Natural Science Foundation of China (Nos. 51809231 and 51808490) and "the Fundamental Research Funds for the Central Universities" (No. 2018FZA4016).

References

[1] D. H. Gray and T. Al-Refeai, "Behavior of fabric-versus fiber-reinforced sand," *Journal of Geotechnical Engineering*, vol. 112, no. 8, pp. 804–820, 1986.

[2] M. Hoshiya and J. N. Mandal, "Metallic powders in reinforced earth," *Journal of Geotechnical Engineering*, vol. 110, no. 10, pp. 1507–1511, 1984.

[3] H. A. D. Kirsten, "Equivalence of mesh- and fibre- reinforced shotcrete at large deflections," *Canadian Geotechnical Journal*, vol. 30, no. 3, pp. 418–440, 1993.

[4] M. H. Maher and D. H. Gray, "Static response of sands reinforced with randomly distributed fibers," *Journal of Geotechnical Engineering*, vol. 114, no. 11, pp. 1661–1677, 1990.

[5] S. Teixeira, B. Bueno, and J. Zornberg, "Pullout resistance of individual Longitudinal and transverse geogrid ribs," *Journal of Geotechnical and Geoenvironmental Engineering*, vol. 133, no. 1, pp. 37–50, 2007.

[6] C.-S. Tang, B. Shi, Y.-J. Cui, C. Liu, and K. Gu, "Desiccation cracking behavior of polypropylene fiber-reinforced clayey soil," *Canadian Geotechnical Journal*, vol. 49, no. 9, pp. 1088–1101, 2012.

[7] I. L. Casagrande, "Electro-osmosis in soils," *Géotechnique*, vol. 1, no. 3, pp. 159–177, 1949.

[8] Q. Tang, F. Gu, Y. F. Gao, T. Inui, and T. Katsumi, "Desorption characteristics of Cr(III), Mn(II) and Ni(II) in contaminated soil using citric acid and citric acid containing wastewater," *Soils and Foundations (JGS)*, vol. 58, pp. 50–64, 2018.

[9] N. D.. Quang and J. C. Chai, "Permeability of lime-and cement-treated clayey soils," *Canadian Geotechnical Journal*, vol. 52, no. 9, pp. 1221-1227, 2015.

[10] L. Yang and R. D. Woods, "Shear stiffness modeling of cemented clay," *Canadian Geotechnical Journal*, vol. 52, pp. 152–166, 2014.

[11] Y. Yi, M. Liska, C. Unluer, and A. Al-Tabbaa, "Carbonating magnesia for soil stabilization," *Canadian Geotechnical Journal*, vol. 50, no. 8, pp. 899–905, 2013.

[12] J. T. Holden, "Improved thermal computations for artificially frozen shaft excavations," *Journal of Geotechnical and Geoenvironmental Engineering*, vol. 123, no. 8, pp. 696–701, 1997.

[13] R. C. Joshi, G.. Achari, D. Horsfield, and T. S. Nagaraj, "Effect of heat treatment on strength of clays," *Journal of geotechnical engineering*, vol. 129, no. 6, pp. 1080–1088, 1994.

[14] S. B. Choi, "Vibration control of flexible structures using ER dampers," *Journal of Dynamic Systems, Measurement and Control Transaction of ASME*, vol. 121, pp. 134–138, 1999.

[15] S. J. Dyke, B. F. Spencer, M. K. Sain, and J. D. Calrson, "Anex perimental study of MR dampers on seismic protection," *Smart Materials and Structures*, vol. 7, pp. 693–703, 1998.

[16] T. Pranoto, K. Nagaya, and A. Hosoda, "Vibration suppression of plate using linear MR fluid passive damper," *Journal of Sound and Vibration*, vol. 276, pp. 919–932, 2004.

[17] R. D. Hryciw and E. Susila, "Temporary solidification of coarse grained soil by magnetorheological fluids," in *Proceedings of Soil dynamics symposium in honor of Professor Richard D. Woods, GSP*, vol. 134, pp. 1–14, 2005.

[18] S. Boussetta, M. Bouassida, and M. Zouabi, "Numerical behavior of reinforced soil by rigid inclusion," in *Proceedings of Geo-China International Conference*, pp. 225–232, 2016.

[19] C. Li and J. G. Zornberg, "Mobilization of reinforcement forces in fiber-reinforce soil," *Journal of Geotechnical and Geoenvironmental Engineering*, vol. 139, no. 1, pp. 107–115, 2012.

[20] COMSOL, *Comsol Multiphysics 4.3a (Computer software)*, COMSOL, Stockholm, Sweden.

[21] Q. Tang, F. Gu, Y. Zhang, Y. Zhang, and J. Mo, "Impact of biological clogging on the barrier performance of landfill liners," *Journal of Environmental Management*, vol. 222, pp. 44–53, 2018.

[22] E. B. Burwell, "Cement and clay grouting of foundations. Practice of the corps of engineering," ASCE Soil Mechanics and Foundations Division, vol. 84, pp. 1551/1 -1551/22, 1958.

[23] ASTM, *Standard Test Methods For Maximum Index Density And Unit Weight Of Soils Using A Vibratory Table. D4253*, ASTM International, West Conshohocken, PA, USA, 2006.

[24] ASTM, *Standard Test Methods For Minimum Index Density And Unit Weight Of Soils And Calculation Of Relative Density. D4254*, ASTM International, West Conshohocken, PA, USA, 2006.

[25] ASTM, *Standard Test Methods For Unconsolidated-Undrained Triaxial Compression Test On Cohesive Soils. D2850-15*, ASTM International, West Conshohocken, PA, USA, 2006.

[26] J. E. Bowels, *Engineering Properties Of Soils And Their Measurement*, McGraw-Hill book Company, New York, NY, USA, 1970.

[27] K. H. Head, *Manual of Soil Laboratory Testing*, Pentech Press, London, UK, 2nd edition, 1994.

[28] F. Tatsuoka, S. Shibuya, and R. Kuwano, *Advanced Laboratory Stress-Strain Testing of Geomaterials*, A. A. Balkema publishers, Tokyo, Japan, 2001.

[29] J. M. Duncan and C. Y. Chang, "Nonlinear analysis of stress and strain in soils," *Journal of the Soil Mechanics and Foundations Division, ASCE*, vol. 96, no. SM5, pp. 1629–1653, 1970.

[30] H. Altenbach, G. B. Stoychev, and K. N. Tushtev, "On elastoplastic deformation of grey cast iron," *International Journal of Plasticity*, vol. 17, no. 5, pp. 719–736, 2000.
[31] R. O. Davis and A. P. Selvadurai, *Elasticity and Geomecanics*, Cambridge University Press, USA, 1996.
[32] R. O. Davis and A. P. S. Selvadurai, *Plasiticity and Geomechanics*, Cambridge University Press, UK, 2002.
[33] P. V. Lade, "Elastic-plastic stress-strain theory for cohesionless soil with curved yield surfaces," *International journal of solids and structures*, vol. 13, no. 11, pp. 1019–1035, 1977.
[34] P. V. Lade and J. M. Duncan, "Elastoplastic stress-strain theory for cohesionless soil," *Journal of Geotechnical Engineering*, vol. 101, pp. 1037–1053, 1975.
[35] T. W. Lambe and R. V. Whitman, *Soil Mechanics*, John Wiley & Sons, Inc, New York, NY, USA, 1969.
[36] H. Matsuoka and T. Nakai, "Stress-deformation and strength characteristics of soil under three different principal stress," *Proceedings of the Japan Society of Civil Engineers*, vol. 1974, no. 232, pp. 59–70, 1974.
[37] K. H. Roscoe and A. N. Schofield, "Mechanical behavior of an idealized "wet clay"," in *Proceedings of 2nd European Conference on Soil Mechanics and Foundation Engineering*, vol. 1, pp. 47–54, Wiesbaden, Germany, 1963.
[38] N. C. Consoli, J. P. Montardo, P. D. Marques Prietto, and G. Savitri Pasa, "Engineering behavior of a sand reinforced with plastic waste," *Journal of Geotechnical and Geoenvironmental Engineering*, vol. 128, no. 6, pp. 462–472, 2002.
[39] ASTM, *Standard test method for flexural toughness and first-crack strength of fibre-reinforced concrete (Using beam with third-point loading). C1018*, ASTM international, West Conshohocken, PA, USA, 1997.
[40] R. L. Handy, "Linearizing tri-axial test failure envelopes," *Geotechnical Testing Journal*, vol. 4, no. 4, pp. 188–191, 1981.

12

The Effect of Polymer-Fiber Stabilization on the Unconfined Compressive Strength and Shear Strength of Sand

Jin Liu, Qiao Feng, Yong Wang, Yuxia Bai, Jihong Wei, and Zezhuo Song

School of Earth Sciences and Engineering, Hohai University, Nanjing 210098, China

Correspondence should be addressed to Jin Liu; jinliu920@163.com

Academic Editor: Belal F. Yousif

The mixed soil stabilizer of polyurethane organic polymer and polypropylene fiber was used to reinforce sand. The unconfined compressive test and direct shear test were carried out to evaluate the effects of polymer-fiber reinforced sand. The different contents of polymer and fiber were selected for the tests. The test results indicated that polymer-fiber mixture can improve strongly the strength of sand. The presence of polyurethane organic polymer enhances sand structural stability and the best composition of polymer and fiber was 4% and 0.2–0.3% of dry sand, respectively. Based on the test results and images of scanning electron microscope (SEM), the reinforcement mechanism was analyzed. The research results can be considered as the reference for sand reinforced engineering.

1. Introduction

Sand is usually easily destroyed due to its low strength and cohesion [1, 2]. In geotechnical engineering, most of natural sand needs to be reinforced to meet the engineering requirements. The physical and chemical reinforcement methods are mainly used in foundation, slope, and embankment. The fiber used as a physical reinforcing material has recently become a focus of intense interest [3–5]. Jiang et al. [6] introduced the engineering properties of soils reinforced by short discrete polypropylene fiber. Gao et al. [7] evaluated the unconfined compressive strength of fiber reinforced clay soil. Noorzad and Fardad [8] presented liquefaction resistance of sand reinforced with randomly distributed fiber under cyclic loading. The static liquefaction of fiber reinforced sand under monotonic loading was also studied by Ibraim et al. [9]. These results indicated that mixing of fibers with a soil mass might act as a spatial three-dimensional network to interlock soil particles to help the particles to form a unitary coherent matrix and restrict the displacement. The cohesive force of fiber reinforced soil is mainly generated by the interlock between adjacent sand particles. While the fiber is used for the loose sand, the cohesion of fiber reinforced sand is difficult to be improved strongly.

Recently, the chemical polymer materials have been used in sand reinforcement [10–15]. Liu et al. [12] evaluated effect of sand fixed with a new organic polymer sand-fixing agent. Ma et al. [13] presented sand reinforced with an eco-friendly polymer composite sand-fixing agent. Mahmoud and Attia [14] introduced the inverse emulsion polymerization for the synthesis of high molecular weight polyacrylamide and its application as sand stabilizer. Gong et al. [15] studied Kerqin sandy land of China reinforced using polymeric materials. The polymer can fill up the voids of sand and adsorbs ions on the surface of sand particle. The elastic and viscous membrane structure is formed through physicochemical bonds between molecules and sand particle to improve the sand strength and cohesion. The cohesion of polymer reinforced soil can be increased strongly. While the polymer materials are mixed with fiber to treat the sand, the cohesion of reinforced sand will be improved with the increment of connected force. Therefore, the polymer-fiber mixture can be considered as a new mixed material to reinforce the loose sand.

In this study, the polymer-fiber mixed with polyurethane organic polymer and polypropylene fiber was used as the soil stabilizer to reinforce sand. The unconfined compressive test and shear test were carried out to evaluate the effects of polymer-fiber reinforced sand. The different contents of

TABLE 1: Grain size distribution of sand.

Grain size (mm)	Weight percentage content (%)
2-1	0.2
1–0.5	17.1
0.5–0.25	48.9
0.25–0.1	31.7
0.1–0.075	2.1

polymer and fiber were selected for the tests. Based on the test results and images of scanning electron microscope (SEM), the reinforcement mechanism was analyzed and the best composition of polymer-fiber was obtained. The research results can be considered as the reference for sand reinforced engineering.

2. Materials

The sand used in this study was acquired from Nanjing, Jiangsu Province, China. Its property parameters are shown in Tables 1 and 2. The dominant component of the sand is the particle size between 0.5 and 0.1 mm (80%). Its specific gravity is approximately 2.65. As shown in Table 2, it has maximum dry density (ρ_{max}) of 1.66 g/cm^3, minimum dry density (ρ_{min}) of 1.34 g/cm^3, maximum void ratio (e_{max}) of 0.970, and minimum void ratio (e_{min}) of 0.590. Its particles have a mean grain size (D_{50}) of 0.30 mm, gradation coefficient (C_g) of 1.13, and uniformity coefficient (C_u) of 2.77.

The polypropylene fiber used in this study is shown in Figure 1(a). It has a length of 18 mm, diameter of 0.034 mm, specific gravity (G_f) of 0.91, tensile strength of 350 MPa, elasticity modulus of 3500 MPa, fusion point of 165°C, burning point of 590°C, and excellent dispersibility. The proportion of dry weight of sand is considered as the percentage of fiber and denoted as

$$P_f = \left(\frac{W_f}{W_s}\right) \times 100\%, \quad (1)$$

where W_f is the fiber weight and W_s is the dry sand weight.

Polyurethane organic polymer shown in Figure 1(b) was selected in this study. It has a main constituent of polyurethane resin and contains enormous amount of functional group –NCO. It is light yellow of oil liquid with a pH of 6-7, viscosity of 650–700 mPa·s, specific gravity of 1.18 g/cm^3, solid content of 85%, and coagulation time of 30–1800 s and holds water content larger than 40 times. The coagulation time decreases with the increase in the concentration. It is an environment-friendly product with no additional pollution. The proportion of dry weight of sand is considered as the percentage of polymer and denoted as

$$P_p = \left(\frac{W_p}{W_s}\right) \times 100\%, \quad (2)$$

where W_p is the polymer weight and W_s is the dry sand weight.

3. Experimental Methods

In this study, the laboratory tests of unconfined compression test and direct shear test were performed to evaluate the performance of polymer-fiber reinforced sand. In the two types of test, sand samples were first oven-dried. Four percentages of fiber (P_f) 0.1%, 0.2%, 0.3%, and 0.4% and four percentages of polymer (P_p) 1%, 2%, 3%, and 4% are considered for this investigation and the unreinforced one as a control. A moisture content of 10% was used for the mixing process of all test specimens.

In the preparation of all samples, the proposed content of fiber was first mixed with the dry sand in small increments by hand to obtain a uniform mixture. The proposed content of polymer and water were mixed to a polymer solution. The water was poured into a flask containing polymer along the flask wall gradually and the mixture was stirred continuously during the mixing process to obtain a uniform polymer dilution. After that the polymer solution was added into the fiber-sand mixture to obtain a final mixture. It is important to ensure all mixed fibers thoroughly. Finally, the mixtures were prepared for the samples of unconfined compression test and direct shear test.

3.1. Unconfined Compression Test. In unconfined compression test, samples were prepared with static compaction method based on ASTM standards. Four-layered compaction was adopted to keep the uniformity of test specimens with the diameter being 39.1 mm and height being 80 mm. After the specimen preparation, the reinforced specimens were kept in curing box with a temperature around 20°C and curing time of 48 hours. The unconfined compression tests were carried out at the loading rate of 2.4 mm/min until samples failed. Additionally, unconfined compression test was performed on specimen triplicates and average values were used.

3.2. Direct Shear Test. In direst shear test, the samples were prepared with static compaction method based on ASTM standards. Two-layered compaction was adopted to test specimens with 61.8 mm diameter and 20 mm height. After the specimen preparation, the reinforced specimens were kept in curing box with a temperature around 20°C and curing time of 48 hours. The direct shear tests were carried out at a strain rate of 0.02 mm/min under the normal pressures of 100, 200, 300, and 400 KPa in order to define the shear strength parameters (c and ψ). And the shear strength parameters were obtained on specimen triplicate and average values were used.

4. Test Results

The specimens with different percentages of fiber and polymer were tested in unconfined compression test and direct shear test. Their results are summarized in Tables 3 and 4. The detailed analysis of each test are given as follows.

4.1. Unconfined Compression Test. The results of samples in unconfined compression test are shown in Table 3. In this

TABLE 2: Property parameters of sand.

ρ_{max} (g/cm^3)	ρ_{min} (g/cm^3)	e_{max}	e_{min}	D_{50} (mm)	C_u	C_g	G_s
1.66	1.34	0.970	0.590	0.30	2.77	1.13	2.65

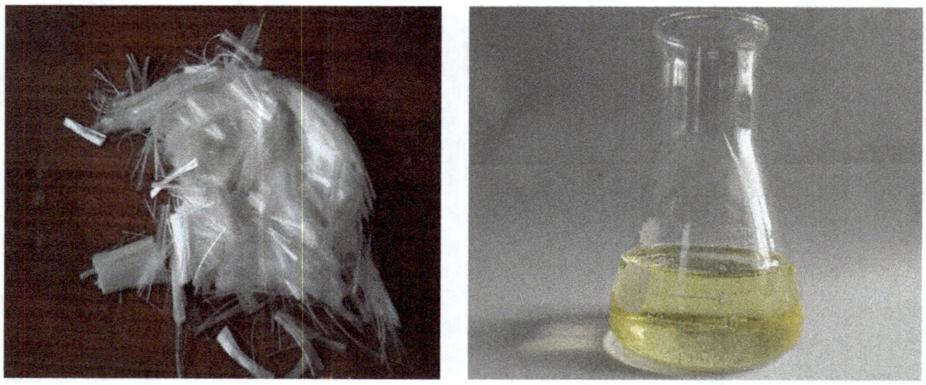

(a) (b)

FIGURE 1: The polymer-fiber material: (a) polypropylene fiber; (b) polyurethane organic polymer.

TABLE 3: Unconfined compressive strengths of sand samples.

Serial number	Dry density (g/cm^3)	Content of polymer (%)	Content of fiber (%)	UCS (kPa)	Percentage of strength improving (%)
S1	1.5	0	0/0.1/0.2/0.3/0.4	/	/
S2	1.5	1	0	22.99	0.00
S3	1.5	1	0.1	26.28	14.31
S4	1.5	1	0.2	34.22	48.85
S5	1.5	1	0.3	41.22	79.30
S6	1.5	1	0.4	38.02	65.38
S7	1.5	2	0	41.88	0.00
S8	1.5	2	0.1	52.57	25.53
S9	1.5	2	0.2	92.05	119.79
S10	1.5	2	0.3	131.45	213.87
S11	1.5	2	0.4	129.11	208.29
S12	1.5	3	0	52.69	0.00
S13	1.5	3	0.1	65.64	24.58
S14	1.5	3	0.2	116.85	121.77
S15	1.5	3	0.3	156.69	197.38
S16	1.5	3	0.4	131.91	150.35
S17	1.5	4	0	106.14	0.00
S18	1.5	4	0.1	118.71	11.84
S19	1.5	4	0.2	157.73	48.61
S20	1.5	4	0.3	163.23	53.79
S21	1.5	4	0.4	135.13	27.31

TABLE 4: Direct shear strengths of sand samples.

Serial number	Dry density (g/cm^3)	Content of polymer (%)	Content of fiber (%)	Cohesion (kPa)	Angle of internal friction (degree)
S22	1.5	0	0	5.21	30.12
S23	1.5	0	0.1	8.94	30.74
S24	1.5	0	0.2	13.14	35.00
S25	1.5	0	0.3	12.21	30.54
S26	1.5	0	0.4	10.4	27.31
S27	1.5	1	0	18.27	28.37
S28	1.5	1	0.1	20.31	29.6
S29	1.5	1	0.2	21.15	30.96
S30	1.5	1	0.3	15.32	27.19
S31	1.5	1	0.4	8.65	26.66
S32	1.5	2	0	25.43	27.7
S33	1.5	2	0.1	28.98	28.12
S34	1.5	2	0.2	32.15	30.54
S35	1.5	2	0.3	30.63	26.61
S36	1.5	2	0.4	17.31	25.84
S37	1.5	3	0	40.27	25.25
S38	1.5	3	0.1	49.38	27.7
S39	1.5	3	0.2	57.59	29.1
S40	1.5	3	0.3	43.62	26.57
S41	1.5	3	0.4	29.94	24.7
S42	1.5	4	0	70.19	25.36
S43	1.5	4	0.1	75.32	26.58
S44	1.5	4	0.2	80.36	27.9
S45	1.5	4	0.3	60.85	23.75
S46	1.5	4	0.4	47.56	22.66

FIGURE 2: Sample of unreinforced sand.

FIGURE 3: Sample of fiber reinforced sand.

study, the unreinforced and fiber reinforced samples are not able to form the required samples due to their poor cohesive force. The photos of unreinforced and fiber reinforced samples after pressure molding are shown in Figures 2 and 3, respectively. But all the polymer reinforced sand can form the required samples. The unconfined compression strength of samples reinforced with polymers 1%, 2%, 3%, and 4% is 22.99, 41.88, 52.69, and 106.14 kPa, respectively. It can be seen from Table 3 that the unconfined compression strength of samples is improved by the reinforced material of polymer-fiber. The strength improving is affected by the content of polymer and fiber.

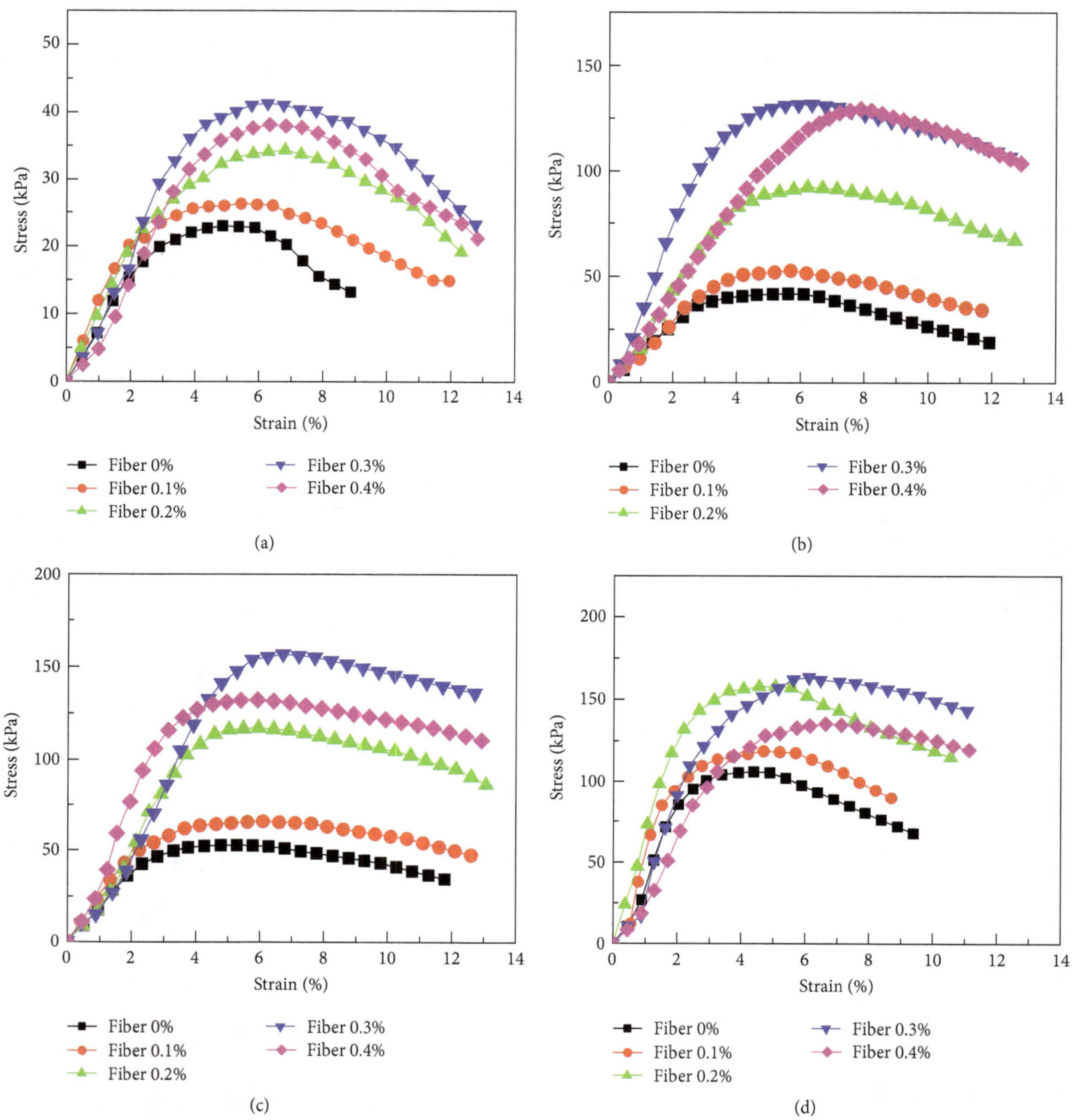

FIGURE 4: The strain-stress curves of unconfined compressive strengths tests. (a) Polymer 1%, (b) polymer 2%, (c) polymer 3%, and (d) polymer 4%.

The axial stress-strain curves of samples reinforced with different content of polymer-fiber are present in Figure 4. As seen in Figure 4, all the polymer-fiber reinforced samples tend to soften, with an obvious peak value, being mainly in a ductile failure state. The peak values are defined as the unconfined compressive strengths of sand samples. The axial stress increases with the increasing of the axial strain to reach the peak value and then decreases to a relative stable value. After the peak value, the axial stress decreased rate of samples reinforced with polymer 1% shown in Figure 4(a) is larger than the ones with polymers 2%, 3%, and 4% in Figures 4(b)–4(d). The samples with higher polymer content have the larger residual strength. With the same polymer content, the axial stress change paths are similar. The photos of samples before/after test are present in Figure 5. As shown in Figures 5(a) and 5(b), the samples with polymer 1% and polymer 1%-fiber 0.2% have enough cohesion to form intact required cylinders. After test, the failure shape of sample with polymer 1% is like a "petal" and the one of sample with polymer 1%-fiber 0.2% damage is along a main "sliding" surface. Based on all the failure shapes after tests, the samples only reinforced with polymer are mainly "petal" model and the ones reinforced with the mixture of polymer-fiber are mainly "sliding" model.

(a) (b)

FIGURE 5: Photos of samples before/after test. (a) Polymer 1%-fiber 0%; (b) polymer 1%-fiber 0.2%.

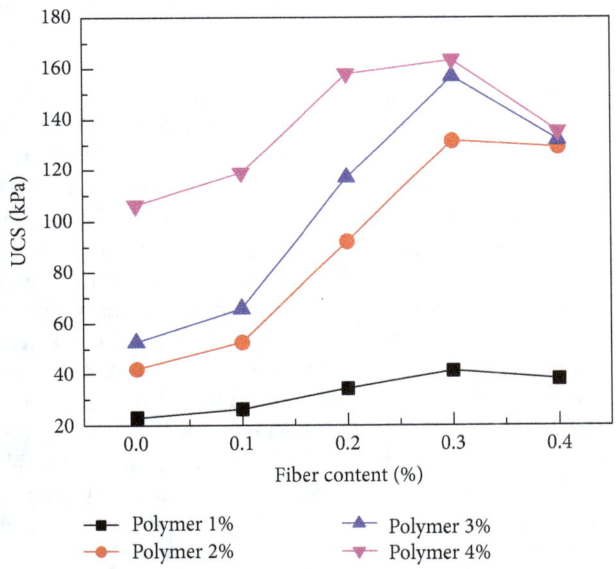

FIGURE 6: Unconfined compressive strength (UCS) of samples.

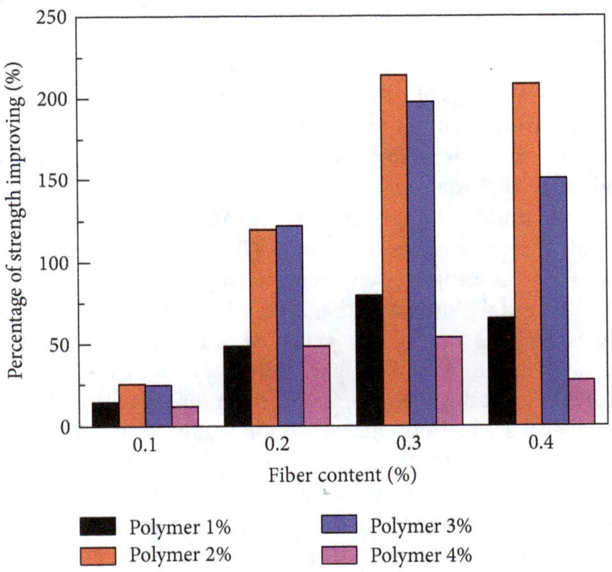

FIGURE 7: Percentage of strength improving of samples.

The unconfined compressive strength of reinforced samples is presented in Figure 6. As seen in Figure 6, the UCS of samples are affected by the different content of polymer-fiber. The UCS values of samples with the same polymer content increase with the increasing of fiber content to reach a higher value and then decrease. While the fiber content is 0.3%, four groups of samples with polymers 1%, 2%, 3%, and 4% reach the peak values 41.22 kPa, 131.45 kPa, 156.69 kPa, and 163.23 kPa, respectively. Meanwhile, the UCS of samples with the same fiber content increases with the increasing of polymer content. The highest value of sample with polymer 4%-fiber 0.3% is 163.23 kPa.

The samples reinforced with only polymer as a reference and the percentages of strength improving of samples with the different fiber content are shown in Figure 7. The UCS of samples reinforced with polymers 1%, 2%, 3%, and 4% are 22.99, 41.88, 52.69, and 106.14 kPa, respectively. As shown in Figure 7, the samples with polymers 2-3% have larger improved percentages. The improved percentages of samples reinforced with polymer 2%-fiber 0.3% and polymer 3%-fiber 0.3% are 213.87% and 197.38%, respectively.

4.2. Direct Shear Test. The results of samples in direct shear test are shown in Table 4. The unreinforced sample with a cohesion zero kPa and angle of internal friction of 30.12° is considered as a reference. The cohesion of all the reinforced samples is improved, especially for the one with higher polymer content. The sample with polymer 4%-fiber 0.2% has a higher cohesion value of 80.36 kPa. Angle of internal friction of samples is rarely affected. The samples reinforced with polymer-fiber mixture have slightly lower internal friction angle.

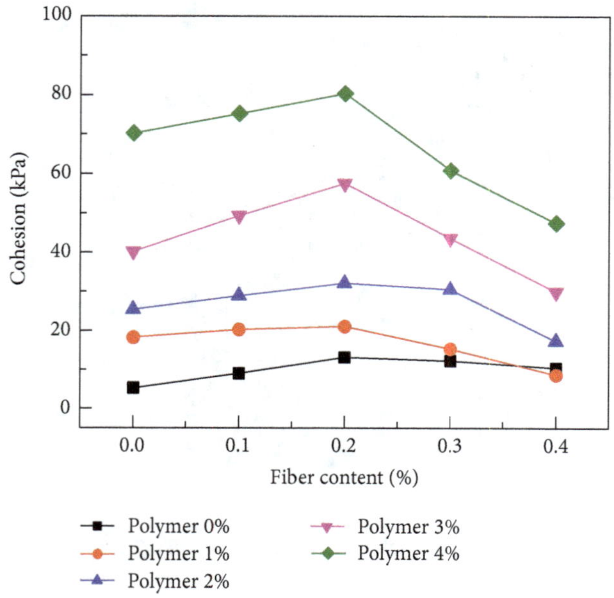

FIGURE 8: Cohesion of sand samples.

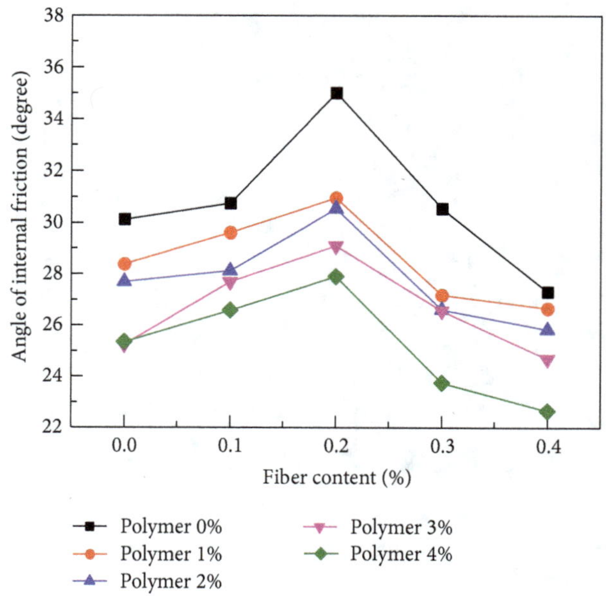

FIGURE 9: Angle of internal friction of sand samples.

The variations of cohesion with polymer-fiber content are shown in Figure 8. As seen in Figure 8, cohesion of samples increases with increase of fiber content to reach a high value and, after that, decreases with the fiber content increasing. With the same polymer content, the samples with fiber 0.2% have the highest cohesion values. While the fiber content is 0.2%, the cohesion of these samples with polymers 0%, 1%, 2%, 3%, and 4% are 13.14 kPa, 21.15 kPa, 32.15 kPa, 57.59 kPa, and 80.36 kPa, respectively. With the same fiber content, the cohesion of samples with higher polymer content also has the higher cohesion values. The sample reinforced with polymer 4%-fiber 0.2% has a highest cohesion value of 80.36 kPa.

The internal friction angle of samples treated with different content of polymer-fiber is present in Figure 9. It is observed that internal friction angle of samples is affected by content of polymer and fiber. The unreinforced sample has internal friction angle of 30.12°. With the same polymer content, the internal friction angle of samples increases with the increasing of fiber content to reach a high value and then decreases, and the samples reinforced with fiber 0.2% have the highest value. All the samples with polymer-fiber have slightly smaller internal friction angles than the one of unreinforced sample. Therefore, the sand stabilized by polymer has higher cohesion but has slightly smaller internal friction angle. The presence of a suitable fiber content can improve the cohesion and internal friction angle.

5. Reinforcement Mechanism

The unreinforced sand has low cohesion between the soil particles resulting in its loose structure (see Figure 2). The fiber reinforcement mechanism in sand mainly includes that the fiber and sand particles constrain each other to produce interface force. This interface force is mainly caused by the extrusion pressure and in the form of cohesion and friction. In the unconfined compression test, the sample has no confining pressure result that the fiber reinforced sand has no enough cohesion to form an intact sample (see Figure 3). In the direct shear test, the fiber reinforced sand with the confining pressure is easy to form a three-dimensional fiber-sand net. When the content of fiber is rather little, the space of fiber is large resulting in the fact that it is difficult to form the effective net between fiber and sand. When the content of fiber increases gradually, the fibers have smaller space to make the adjacent fiber intersect easily to form the effective fiber-sand net to improve the cohesion and friction. But while the content of fiber is rather more, lots of fiber filaments will gather in clusters inside the sand sample to make the nonuniform distribution of fibers. It is easy to form the weak area of stress due to the excess of fibers. So all the values of cohesion and internal friction angle of fiber reinforced samples increase with the fiber content increasing and then decrease (see Figures 8 and 9).

The polyurethane organic polymer has a large amount of polyurethane resin macromolecule and isocyanate group (–NCO). The sand reinforcement mechanism of polymer includes filling, chemical reaction, and enwrapping. The polymer solution fills up the voids and adsorbs ions on the surface of sand particles and other matter to form physicochemical bonds between molecules and other matters. The elastic and viscous membrane is formed through the bands of long-chain macromolecules to interlink the sand particles. This membrane structure can increase in bonding and interlocking forces between sand particles, and the UCS and cohesion of polymer reinforced samples are improved (see Figures 6, 8, and 9). The more the polymer content is, the more the bonds are produced to form an intact membrane enwraping the sand particles, so the UCS and cohesion of polymer reinforced sand increased with polymer content. But, while the sand particle surface is enwrapped with polymer, its surface roughness will be reduced resulting in the decrease slightly of internal friction angle of polymer

(a) (b)

Figure 10: SEM images of sample reinforced with polymer-fiber: (a) 30 times magnification and (b) 200 times magnification.

reinforced sand during shearing process. Meanwhile, the internal friction angle also decreases with the increasing with polymer content because of the better enwrapping on sand particle surface (see Table 4 and Figure 9).

The sand reinforced with the mixture of polymer-fiber and the reinforcement effect is improved stronger than the ones treated with only one material polymer or fiber. The SEM images of sand sample reinforced with polymer-fiber are presented in Figure 10. As seen in Figure 10(a), while the polymer-fiber is mixed with sand, the fiber as the framework is distributed in the sand, the polymer fills up the voids and adsorbs ions on the surface of sand particles and fiber. It is seen clearly from Figure 10(b) that the sand particle and fiber surface are enwrapped fully by the polymer to from a membrane structure. The loose sand and fiber are connected by the polymer membrane thus to shape firmly together as a whole, especially for the contact area of particles and fiber. So all sand samples reinforcing the mixture of polymer-fiber can be formed to the intact columnar sample under unconfined condition (see Figure 5).

While the sample is reinforced with polymer-fiber, the different content of polymer and fiber will affect the improvement effect. With the same polymer content, the framework structure is formed more completely with the increase of fiber content. After the completely framework structure formation, the excessive content of fiber will easily gather in clusters resulting in the fact that the polymer is very difficult to enwrap each fiber surface to form a whole stable structure. So the UCS increases with the increasing of fiber content; after the fiber content larger than 0.3%, the UCS decreases with the fiber content increasing. The cohesion and friction increase with fiber content until they reach 0.2%; they decrease with the fiber content increasing. With the same fiber content, there are more polymers to enwrap the sand and fiber surface to connect them together with the increasing of polymer content. So the UCS and cohesion increase with the increasing of polymer content for the samples with the same fiber content.

6. Conclusions

In order to study the unconfined compressive strength and shear strength of sand reinforced with polymer-fiber mixture, the unconfined compressive tests and direct shear tests on sand samples with different content of polymer and fiber were carried out. The test results and reinforcement mechanism were analyzed. Based on the results of the tests presented herein, the main conclusions can be summarized as follows:

(1) The UCS of sand can be improved strongly by the polymer-fiber mixture material. While the polymer content is a constant, the UCS first increases and then has a downward trend with increasing of fiber content. With the same fiber content, the UCS increases with the polymer content increasing. The sample reinforced with polymer 4%-fiber 0.3% has a highest UCS of 163.23 kPa.

(2) The cohesion of samples increases with increase of fiber content and then decreases with the fiber content increasing. With the same polymer content, the samples with fiber 0.2% have the highest cohesion values. With the same fiber content, the cohesion of samples with higher polymer content also has the higher cohesion values. The sample reinforced with polymer 4%-fiber 0.2% has a highest cohesion value of 80.36 kPa. All the samples with polymer-fiber have slightly smaller internal friction angles than the one of unreinforced sample.

(3) About the mechanism of polymer-fiber reinforced sand, the fiber as the framework is distributed in the sand; the polymer fills up the voids and adsorbs ions on the surface of sand particles and fiber. The sand and fiber are connected by the polymer membrane thus to shape firmly together as a whole to improve the strength and cohesion of sand. The research results can be applied as the reference for practical engineering, especially for reinforcement of foundation, embankment, and landfill.

Conflicts of Interest

The authors declare that there are no conflicts of interest regarding the publication of this paper.

Acknowledgments

This research was financially supported by the National Natural Science Foundation of China (Grant no. 41472241), Natural Science Foundation of Jiangsu Province, China (Grant no. BK20141415), and Water Conservancy Science and Technology Project of Jiangsu Province, China (Grant no. 2017010).

References

[1] T. Shogaki and K. Kaneda, "Feasible method, utilizing density changes, for estimating in situ dynamic strength and deformation properties of sand samples," *Soils and Foundations*, vol. 53, no. 1, pp. 64–76, 2013.

[2] Y. Li, J. Cui, T. Zhang, T. Okuro, and S. Drake, "Effectiveness of sand-fixing measures on desert land restoration in Kerqin Sandy Land, northern China," *Ecological Engineering*, vol. 35, no. 1, pp. 118–127, 2009.

[3] N. Unnikrishnan, K. Rajagopal, and N. R. Krishnaswamy, "Behaviour of reinforced clay under monotonic and cyclic loading," *Geotextiles and Geomembranes*, vol. 20, no. 2, pp. 117–133, 2002.

[4] C. Tang, B. Shi, W. Gao, F. Chen, and Y. Cai, "Strength and mechanical behavior of short polypropylene fiber reinforced and cement stabilized clayey soil," *Geotextiles and Geomembranes*, vol. 25, no. 3, pp. 194–202, 2007.

[5] J. Li and D. W. Ding, "Nonlinear elastic behavior of fiber-reinforced soil under cyclic loading," *Soil Dynamics and Earthquake Engineering*, vol. 22, no. 9, pp. 977–983, 2002.

[6] H. Jiang, Y. Cai, and J. Liu, "Engineering properties of soils reinforced by short discrete polypropylene fiber," *Journal of Materials in Civil Engineering*, vol. 22, no. 12, pp. 1315–1322, 2010.

[7] L. Gao, G. Hu, N. Xu, J. Fu, C. Xiang, and C. Yang, "Experimental study on unconfined compressive strength of basalt fiber reinforced clay soil," *Advances in Materials Science and Engineering*, vol. 2015, Article ID 561293, 8 pages, 2015.

[8] R. Noorzad and P. F. Amini, "Liquefaction resistance of Babolsar sand reinforced with randomly distributed fibers under cyclic loading," *Soil Dynamics and Earthquake Engineering*, vol. 66, pp. 281–292, 2014.

[9] E. Ibraim, A. Diambra, D. Muir Wood, and A. R. Russell, "Static liquefaction of fibre reinforced sand under monotonic loading," *Geotextiles and Geomembranes*, vol. 28, no. 4, pp. 374–385, 2010.

[10] G. Telysheva and G. Shulga, "Silicon-containing polycomplexes for protection against wind erosion of sandy soil," *Journal of Agricultural Engineering Research*, vol. 62, no. 4, pp. 221–227, 1995.

[11] J. Yang, F. Wang, L. Fang, and T. Tan, "The effects of aging tests on a novel chemical sand-fixing agent—polyaspartic acid," *Composites Science and Technology*, vol. 67, no. 10, pp. 2160–2164, 2007.

[12] J. Liu, B. Shi, Y. Lu et al., "Effectiveness of a new organic polymer sand-fixing agent on sand fixation," *Environmental Earth Sciences*, vol. 65, no. 3, pp. 589–595, 2012.

[13] G. Ma, F. Ran, E. Feng, Z. Dong, and Z. Lei, "Effectiveness of an eco-friendly polymer composite sand-fixing agent on sand fixation," *Water, Air, and Soil Pollution*, vol. 226, no. 7, article 2490, 2015.

[14] M. A. Mahmoud and N. F. Attia, "Inverse emulsion polymerization for the synthesis of high molecular weight polyacrylamide and its application as sand stabilizer," *International Journal of Polymer Science*, vol. 2015, Article ID 436583, pp. 1–10, 2015.

[15] W. Gong, Y. Zang, B. Liu et al., "Effect of using polymeric materials in ecological sand-fixing of Kerqin Sandy Land of China," *Journal of Applied Polymer Science*, vol. 133, no. 43, Article ID 44102, 2016.

13

Discretization of the Water Uptake Process of Na-Montmorillonite Undergoing Atmospheric Stress: XRD Modeling Approach

Walid Oueslati and Mahdi Meftah

UR 05/13-01 Physique des Matériaux Lamellaires et Nano-Matériaux Hybrides (PMLNMH), Faculté des Sciences de Bizerte, Université de Carthage, 7021 Zarzouna, Tunisia

Correspondence should be addressed to Walid Oueslati; walidoueslati@ymail.com

Academic Editor: Donatella Termini

This work focuses on the water-montmorillonite interaction under variable atmospheric conditions in order to characterize the interlamellar space (IS) configuration for possible use in the context of geological barrier for radioactive and industrial waste confinement. Atmospheric stress is achieved by applying, for Na-rich montmorillonite, a water sorption/desorption constraint, created at the laboratory scale. This hydrological disturbance allows the "demolition" of the clay history and to highlight the clay hydrous performance. The structural analysis is achieved using modeling of XRD profiles, which allowed us to determine the optimal structural parameters describing the IS configuration along the c^* axis. During the "in situ" XRD analysis, a sorption/desorption cycle is envisaged by variation of the relative humidity rate (%RH) from the saturated condition (94 %RH) towards extremely dry ones (2-3 %RH). Qualitatively, a new hydration hysteresis behavior of the "stressed" sample appears. Structural analysis achieved before and after perturbation allowed us to identify, respectively, the homogeneous hydration states, the hydrous transition domains, and the hydration heterogeneity degree. This latter parameter is characterized, quantitatively, by variable relative abundances of mixed-layer structure (MLS) population discerned over a wide explored RH range. Using the optimum structural parameter, the water molecule distribution versus the applied hydrous strain was quantified.

1. Introduction

The use of the smectite clay, like montmorillonite, as an canister envelope or as plugs or rock grout, in the context of geological barrier for industrial and/or radioactive hazardous waste confinement constitutes a green solution for severe environmental problems. In this regard, montmorillonite will be exposed to several environmental constraints (temperature cycle, extremely saturated conditions, high water pressure, etc.). Under these conditions, the clay-sealing ability is influenced, and two different scenarios can take place: (1) the crystal lattices may undergo permanent changes resulting in a largely reduced swelling and gel-forming capacity, and (2) the microstructure may become temporarily or permanently altered, which affects the cation exchange process [1–7].

The crystal structure of smectite consists of two sheets of principally tetrahedral Si with minor Al sandwiched to the top and bottom of a layer of octahedrally coordinated cations, usually Al but with some Mg and Fe [8, 9]. Furthermore, isomorphic substitution causes internal negative charges in the lamellae. The negative charges are compensated by adsorption of ions near the external surfaces of the clay minerals.

The smectite group of clay minerals with 2 : 1 layer structure is able to expand and contract its structures while maintaining its crystallographic integrity. Expansion takes place as water or some polar molecules enter the interlayer space [10].

The layers expand because the exchangeable cations attract more water than the relatively small layer charge [11]. The swelling is characterized by a stepwise expansion of the layer-to-layer distance with increasing water activity. This expansion has been described by the incorporation of 1, 2, and 3 planes of interlayer H_2O molecules, leading to the well-known 1W (d_{001} = 11.6–12.9 Å), 2W (d_{001} = 14.9–15.7 Å), and

3W (d_{001} = 18–19 Å) hydration states, in addition to the dehydrated one, 0W (d_{001} = 9.6–10.2 Å) [12].

The swelling of montmorillonite in various liquid systems has been investigated by several authors. These studies have focused on many parameters that come into play during the chemical clay-environment interaction. Indeed, to evaluate the mechanisms causing changes in the soil structure, Kaya and Fang [13] show that the physicochemical properties (i.e., cation exchange capacity (CEC), zeta potential (ζ), surface charge density, pore-size distribution, and Atterberg limits) of kaolinite, bentonite, and a local soil as the dielectric constant of the pore fluid change, such that as the dielectric constant of the pore fluid approaches that of the soil, the repulsive and attractive forces diminish. Smectite hydration and interlayer structure have been widely studied from both experimental and computational methods, mainly using X-ray diffraction (XRD) and some theoretical methods [14–31]. The studies [20, 32] show that hydration of swelling clays increases their basal spacing stepwise. Dazas et al. [31] investigate, under near-saturated conditions, the interlayer structure model of trihydrated low-charge smectite by X-ray diffraction and Monte Carlo modeling method. Oueslati et al. [29] characterize, using quantitative XRD analysis, the effect of the chemical composition of the soil solution on the structural change of Na-montmorillonite materials. On the contrary [30, 33], they investigate the effect of an environmental constraint, created in situ by varying the relative humidity rate, on the cation exchange performance and the structural changes of Na-montmorillonite in contact with Cd^{2+}, Co^{2+}, Zn^{2+}, Ni^{2+}, and Ba^{2+} chloride solutions.

Within this scope, and in extension with earlier works, this study is aimed at quantifying using the XRD modeling approach, the effect of an in situ applied hydrous perturbation on the hydration behavior and the structural changes of the most used host material in the earlier studies (i.e., the Na-exchanged montmorillonite). The structural parameters were determined by quantitative XRD analysis using an indirect method based on the comparison of the experimental 00l reflections with those calculated from structural models. This investigation allowed us to determine, respectively, the nature, amounts, position, and organization of exchangeable cations in the interlamellar space, along the c^* axis.

2. Materials and Methods

2.1. Starting Materials. The dioctahedral smectite used in this work is a reference montmorillonite sample (SWy-2) from the Source Clays Repository of The Clay Minerals Society [34]. The host material is characterized by a half-cell structural formula as follows [35]:

$$\left(Si_{3.923}Al_{0.077}\right)\left(Al_{1.459}Ti^{4+}_{0.018}Fe^{3+}_{0.039}Fe^{2+}_{0.045}Mg^{2+}_{0.382}\right) O_{10}(OH)_2\left(Ca^{2+}_{0.177}Na^{+}_{0.027}\right) \quad (1)$$

This clay mineral represents a major octahedral charge and exhibits an extremely limited substitution in the tetrahedral sheets where the cation exchange capacity (CEC) is 101 meq/100 g [36]. In order to guarantee a maximum dispersion, a preliminary treatment for the host materials is performed. The Na-rich montmorillonite suspension is prepared according to a classical protocol [37]. The obtained sample is referred to as SWy-2-Na.

2.2. Hydrous Perturbation. In order to achieve an atmospheric constraint (i.e., desorption process), a starting material suspension is placed in a sand bath at 50°C for 24 h to ensure a complete drying process. This sample undergoes a second humectation route (sorption process) by excessive addition of water followed by a moderate agitation. These two steps are repeated 50 times to ensure the destruction of the sample hydration history which is a principal parameter affecting its probable long use in the context of geological barrier.

2.3. XRD Measurements. For all samples, oriented slides were prepared by drying at room temperature an aqueous clay suspension on glass slides. XRD patterns were then recorded using a Bruker D8 diffractometer equipped with an MHG Messtechnik humidity controller coupled to an Anton Paar CHC+ chamber. Experimental XRD patterns were registered every 3 %RH scale at the fixed relative humidity condition values. Intensities were measured with a SolXE Si(Li) solid-state detector (Baltic Scientific Instruments) for 4 s per 0.04° 2θ step over the 2–32° 2θ Cu-Kα angular range. The divergence slit, the two Soller slits, the antiscatter slits, and the resolution slits were 0.3°, 2.3°, 0.3°, and 0.1°, respectively.

The samples were kept at 23°C in the CHC+ chamber during the whole data collection. The samples were kept also under a constant flow of mixed dry/saturated air to maintain the desired relative humidity (RH) after an initial equilibration. RH was continuously monitored with a hygrometer located close to the sample. To carry out the adsorption cycle, the RH value extended from 4% to the almost saturated condition (94 %RH). The reverse cycle (i.e., desorption) is accomplished by varying the moisture values in the opposite direction starting the dehydration process from 94 %RH and decreasing towards extremely dry ones (4 %RH).

2.3.1. Semiquantitative XRD Investigation. The diffractometer installation is monitored by DIFFRACPlus software (Bruker AXS GmbH, Karlsruhe, Germany) which allowed the calculation of qualitative parameters such as the basal spacing d_{001} from the first-order (001) Bragg reflections; the FWHM (full width at the half maximum intensity for the 001 reflection) and the rationality of the reflection position (ξ) related to the 00l reflections [38]. The combination of semiquantitative parameters with the profile geometry description (i.e., symmetric and asymmetric X-ray peaks) provides preliminary information about the hydration state evolution all over the explored RH range.

2.3.2. Quantitative XRD Investigation. The XRD modeling method is widely used to quantify hydration properties of smectite as a function of relative humidity [12, 13, 24–30, 33, 39, 40]. This indirect method consists of adjusting the experimental patterns by fitting positions and profiles of the 00l reflections over the explored angular range. Theoretical XRD patterns are calculated using the algorithms developed

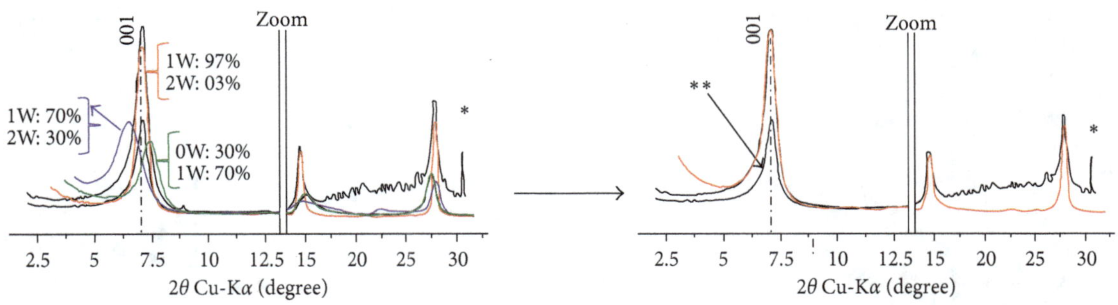

FIGURE 1: Example of an experimental XRD profile modeled using three MLSs containing (97%-1W, 3%-2W), (30%-0W, 70%-1W), and (70%-1W, 30%-2W), respectively. *Halite (NaCl) complex, **the 001 line is recorded twice time in order to check the sample balance with its environment.

in [41] and detailed later in [42]. The abundances of the diverse layer types (W_i), the mode of stacking of the different kinds of layers, and the mean number of layers per coherent scattering domain (CSD) are determined through XRD profile modeling. Within a CSD, the stacking of layers is described by a set of junction probability parameters (P_{ij}). The relationships between these probabilities and the abundances W_i of the different types of layers are given in [41]. All XRD profiles are simulated following the fitting strategy detailed in [24, 25], where the authors reproduce the experimental XRD pattern using various contributions to obtain a good fit, when it is not obtained with a unique periodic structure. Each contribution contains different layer types in variable proportions. However, the use of two or more mixed-layer structures to fit all features of experimental XRD patterns does not imply the actual presence of two or more populations of particles in the sample as discussed below. As a consequence, layers in the same hydration state as those present in the different mixed-layer structures must have identical properties at a given RH value. Each given layer type was thus assigned a unique chemical composition, a unique layer thickness value, and a unique set of atomic coordinates for all mixed-layer structures at a given RH. A detailed example of an experimental XRD profile modeled using three MLSs containing (97%-1W, 3%-2W), (30%-0W, 70%-1W), and (70%-1W, 30%-2W), respectively, is reported in Figure 1.

3. Results

3.1. Unstressed Sample

3.1.1. Qualitative XRD Analysis

(1) Sorption Process. For the starting materials, the evolution of the experimental XRD patterns versus %RH, reported in the Figure S1, shows a significant fluctuation of the position and intensity of the investigated $00l$ reflection along the hydration process. An additional reflection attributed to the Halite crystallization is present in up to 60 %RH. The qualitative investigation of the d_{001} value evolution, the FWHM, and the ξ parameters versus %RH (Figure 2) indicates heterogeneous hydration behavior for the transitional hydration state domain. This character is confirmed by the obtained high value of the cited parameters [29]. Based on this criterion, three principal transitional hydration domains along the sorption process can be reached and situated, respectively, between 10 and 34 %RH for 0W-1W, 49–67 %RH for 1W-2W, and 91–94 %RH for 2W-3W. For the three defined RH transition ranges, the XRD patterns are characterized by an asymmetric 001 reflection geometry accompanied by the corresponding high FWHM and ξ values (i.e., at 22 %RH, FWHM = 2.48° (2θ) and ξ = 1.84 Å; at 55 %RH, FWHM = 1.45° (2θ) and ξ = 1.58 Å; at 91 %RH, FWHM = 0.867° (2θ) and ξ = 1.13 Å), indicating probably an interstratified hydration character with more one layer type. Indeed, the d_{001} value shifts gradually from 0W (9.98 Å) towards 3W (18.86 Å) layer-type phase for the extreme %RH condition. The greatest clear 1W → 2W transition is observed from 52 %RH to 61 %RH. The 2W → 3W transition is observed from 85 %RH to 94 %RH.

(2) Desorption Process. The desorption sequence is achieved by decreasing the RH rate. The d_{001} basal spacing shifts from 18.86 Å, characteristic of a 3W hydration state (3W), to 9.96 Å, characteristic of a dehydrated state (0W). This evolution presents characteristic levels of almost homogeneous hydration states of hydration (3W, 2W, and 1W). Furthermore, there are transition regions 3W → 2W, 2W → 1W, and 1W → 0W, which are characterized by experimental profiles with an asymmetric diffraction line shape and an irrationality of the reflection positions, indicating probably a possible interstratified hydration state (Figure S2 and Figure 2). By comparing the widths (as a function of %RH) of the transition domains along the sorption and desorption processes (Figure 2), an offset is shown for 2W → 1W and 1W → 0W transitions.

Indeed, the 2W → 1W transition starts from 52 %RH and ends at 34 %RH, which is the RH value where the same transition ends during the sorption process. Similarly for the 1W → 0W transition, which starts from 13 %RH and ends at 4 %RH, there is a large offset to the hydration process whose 1W → 0W transition starts at 13 %RH and ends at 34 %RH.

This hydrous behavior shows the existence of hydration hysteresis reported in Figure 2. Also, during the desorption process, the absence of the pure dehydrated phase (0W), despite having reached the limit of 4 %RH, is noted.

In order to decode the insertion/disengagement mechanism of IS water molecules, this result can be explained by a freedom degree decrease for all the constituents inside the IS, which complicate the water molecule release process.

This fact can be interpreted by a transformation of the free water from the surrounding environment into structural

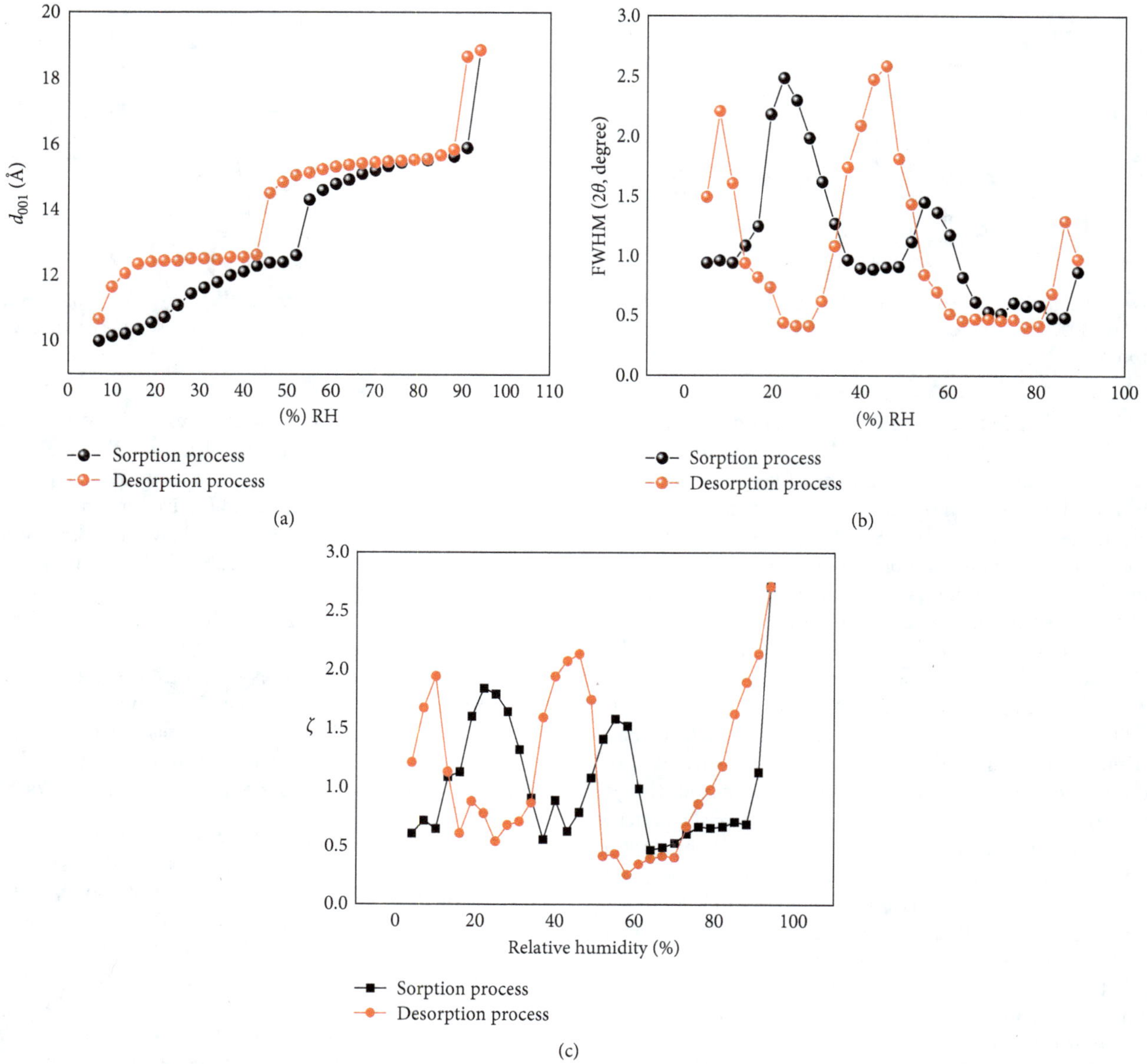

FIGURE 2: Evolution of the d_{001} (a), FWHM (b), and ξ parameters (c) versus %RH in the case of the unstressed sample along the sorption and desorption processes. The ξ parameter, which accounts for the departure from rationality of the 00l reflection series, is calculated as the standard deviation of the $l \times d_{00l}$ values calculated for the X_i measurable reflections ($X_i = 3$ in this case) over the 2θ Cu-Kα angular range (degree). Blue and yellow circles indicate interstratified hydration character (I) and homogeneous hydration character (H).

water, incorporated in the internal and external pore surfaces, which complicates its loss by dehydration.

The investigation of FWHM and ξ parameter shows the difference between the two processes for the transitions 2W → 1W and 1W → 0W (Figure 2). This problem does not arise for the 3W → 2W transition if we want to compare it with the 2W → 3W transition along the hydration process. Indeed, in this case, the high %RH value induces an elevated amount of water present in the IS; thereafter, the existence of a large fraction of water considered as "free water" will facilitate the process of dehydration.

3.1.2. Quantitative XRD Analysis

(1) Sorption Process. The hydration behavior obtained from the qualitative XRD analysis is refined by the quantitative analysis in order to quantify the clay "particle" composition.

The obtained best agreements between experimental and calculated intensity are reported in Figure S1, where a selection in the presentation was made in order to view areas with large fluctuations in the basal spacing value. All structural parameters used to fit experimental pattern are reported in Table 1 and Figure 3.

The distribution of water molecules in the theoretical model respects the results published in the literature [24, 29]. Indeed, for the 1W hydration state, one water sheet is located in the middle of the IS at the same position as the exchangeable cation. For the 2W hydration state, two water sheets are distributed on both sides of the middle of the IS. For the 3W hydration state, the exchangeable cation "bathes" in 3 water sheets superimposed along the c^* axis.

TABLE 1: Some structural parameters obtained from the quantitative XRD analysis in the case of the unstressed sample along the sorption process.

%RH	Layer stacking mode*	*M*
4	R0 — R0	7
7	R0 — R0	7
10	R0 — R0	8
13	R0 — R1 ($P_{AA}=0.80$) R1 ($P_{AA}=0.85$)	8
16	R0 — R1 ($P_{AA}=0.85$) R1 ($P_{AA}=0.70$)	8
19	R0 — R1 ($P_{AA}=0.85$) R1 ($P_{AA}=0.70$)	8
22	R1 ($P_{AA}=0.75$) R1 ($P_{AA}=0.70$) R0	8
25	R1 ($P_{AA}=0.75$)	9
28	R1 ($P_{AA}=0.75$)	9
31	R1 ($P_{AA}=0.75$)	7
34	R1 ($P_{AA}=0.70$)	6
37	R1 ($P_{AA}=0.70$) R1 ($P_{AA}=0.70$)	6
40	R1 ($P_{AA}=0.70$)	5
43	R1 ($P_{AA}=0.75$)	5
46	R1 ($P_{AA}=0.87$) R1 ($P_{AA}=0.70$) R0	5
49	R1 ($P_{AA}=0.87$) R1 ($P_{AA}=0.70$) R0	5
52	R1 ($P_{AA}=0.85$) R1 ($P_{AA}=0.96$) R1 ($P_{AA}=0.70$) R1 ($P_{AA}=0.70$)	5
55	R1 ($P_{AA}=0.70$) R1 ($P_{AA}=0.90$) R0	5
58	R1 ($P_{AA}=0.70$) R1 ($P_{AA}=0.90$) R1 ($P_{AA}=0.75$)	6
61	R1 ($P_{AA}=0.75$) R1 ($P_{AA}=0.88$)	6
64	R1 ($P_{AA}=0.75$)	5
67	R1 ($P_{AA}=0.80$)	5
70	R1 ($P_{AA}=0.85$)	5
73	R1 ($P_{AA}=0.85$)	5
76	— R0 R0	5
79	— R0 R0	5

TABLE 1: Continued.

%RH	Layer stacking mode*	*M*
82	— R0 R0	5
85	— R0 R0	7
88	R0 R0 R0	7
91	R0 R0 R0	7
94	R0 R0	6

Note: 3W, 2W, 1W, and 0W are attributed to the layer hydration state. n_{H_2O}: the number of H_2O molecules per half unit cell is fixed to 2.5, 5.6, and 6.1, respectively, for 1W, 2W, and 3W. Z_{Na}: the position of exchangeable cations per half unit cell calculated along the c^* axis is fixed to 8.30 Å, 9.50 Å, 11.25 Å, and 12.60 Å, respectively, for 0W, 1W, 2W, and 3W hydration state. n_{Na}: the number of H_2O molecules per half unit cell is fixed to 0.33, indicating full saturation of the cation exchange capacity (CEC) of the minerals. *R, Reichweite factor; R0 and R1 describe the MLS with random interstratifications or with partial segregation; M: the average number of layers per particle; P_{AA}: layer succession law.

During the hydration process and for RH up to 43%, the experimental profile is reproduced by interstratification between the 0W and 1W phases with variable proportions, indicating the absence of a mono-homogeneous phase at 0W for humidity below 13 %RH (Figure 3).

On the contrary, an introduction of an R1 stacking mode type is noted at 13 %RH, which is consistent with the variation of the FWHM and ξ parameters. The first appearance of the 2W phase begins at 46 %RH and maintains a heterogeneous hydration state between 3 layer types (0W, 1W, and 2W) up to 58 %RH, where the structure is reproduced using a random distribution between the three mentioned phases. From 61 %RH, the contribution of the 0W phase disappears, and the experimental profile is completely reproduced by a combination of the 1W and 2W phases up to 76 %RH, where a minority phase related to the 3W hydration state is introduced. This combination continues up to 91 %RH, and then the 1W phase contribution disappears and the structure is completely modeled by 2W and 3W layer types. It should be noted that, at 94 % RH, the experimental diffractogram is reproduced by a majority 3W phase and the sample retains its interstratified character, which is in contradiction with the qualitative description mentioning a homogeneous phase. In contrast to the qualitative analysis, the quantitative analysis allowed us to show the existence of structural heterogeneities related to the coexistence of populations of several layers within the clay particle over the whole explored RH range.

(2) Desorption Process. For the desorption sequence and by decreasing humidity values up to 64 %RH, the experimental profile is modeled by a distribution of the two phases,

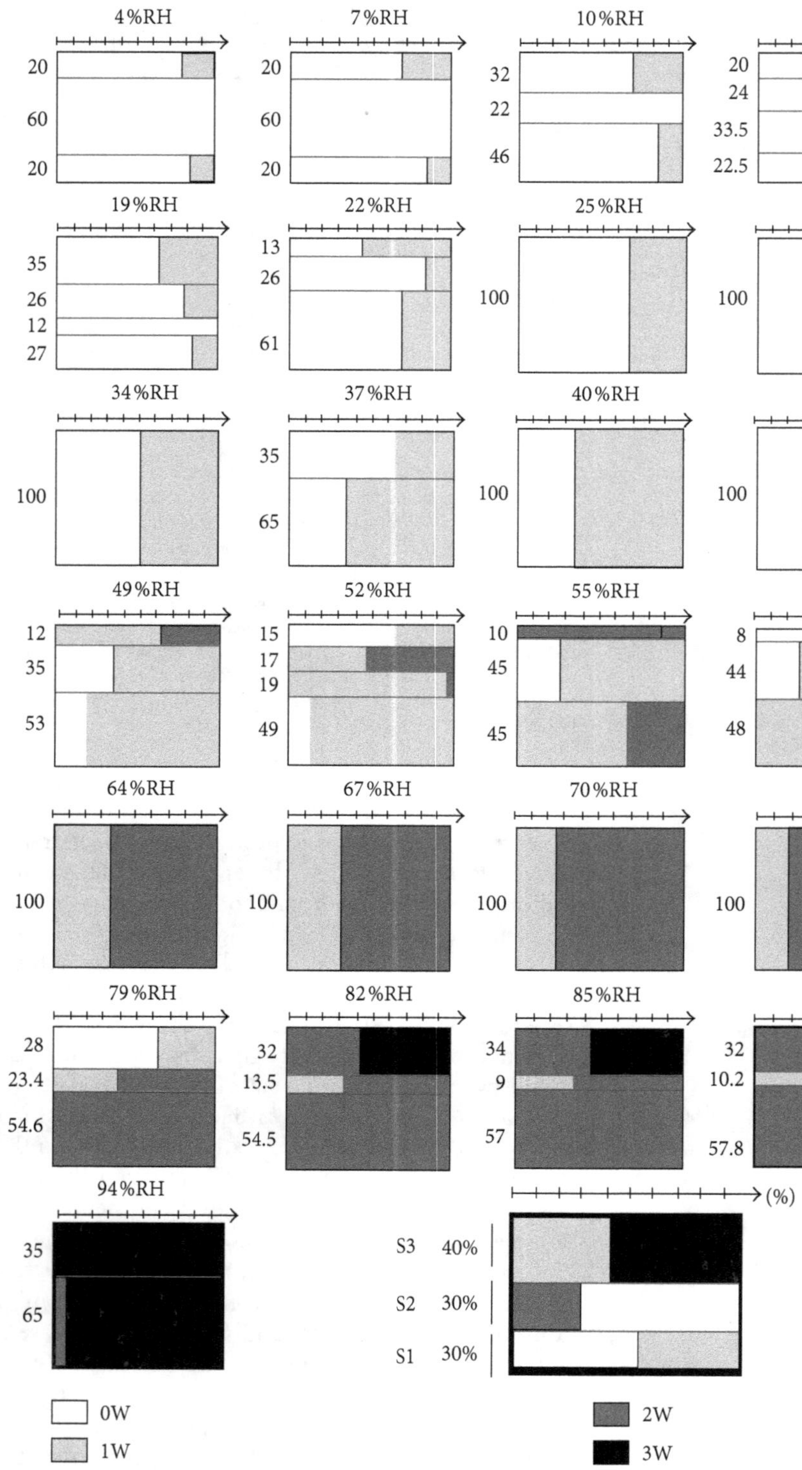

Figure 3: Schematic representation of the structure model used to fit the experimental XRD pattern in the case of the unstressed sample along the sorption process. Relative proportions, expressed in wt.%, of the three elementary mixed-layer structure (MLS) contributions are plotted on the *y*-axis, whereas their compositions (relative proportions of the different layer types) are plotted on the *x*-axis.

respectively, 3W and 2W. The basal distances and the proportions of the phase contributions used indicate a transition to the 2W hydration state. From 61 %RH and up to 58 % RH, the experimental profile is fully reproduced using segregated three phases mixed randomly. From 55 %RH and up to 34 %RH, the 3W phase disappears, which means a high retention of the water content, despite the relatively low RH and the absence of a sudden transition. At this humidity range, the structure is reproduced by a 2W/1W mixed structure. The first appearance of the 0W phase is obtained at 31 %RH, and its contribution in the theoretical model continues to increase as a function of the decrease in the RH rate. It should be noted that, for humidity ranging from 40 to 16 %RH, layers are stacked with random

Table 2: Some structural parameters obtained from the quantitative XRD analysis in the case of the unstressed sample along the desorption process.

%RH	Layer stacking mode	M
94	R0 R0	6
91	— —	3
88	R0 R1 ($P_{AA}=0.80$)	4
85	R1 ($P_{AA}=0.80$) R0	4
82	— R1 ($P_{AA}=0.80$)	6
79	— R1 ($P_{AA}=0.80$)	6
76	R1 ($P_{AA}=0.86$) R0	6
73	— R1 ($P_{AA}=0.86$)	6
70	— R1 ($P_{AA}=0.92$)	6
67	— R1 ($P_{AA}=0.86$)	6
64	— R1 ($P_{AA}=0.96$)	7
61	— R1 ($P_{AA}=0.96$) R0	7
58	R0 R1 ($P_{AA}=0.96$) R0	7
55	R0 R1 ($P_{AA}=0.96$)	7
52	R1 ($P_{AA}=0.85$)	6
49	R1 ($P_{AA}=0.88$)	6
46	R1 ($P_{AA}=0.88$) R0	6
43	R1 ($P_{AA}=0.88$) —	7
40	R0 R0	7
37	R0 R0	7
34	R0 R0	7
31	R0 R0 R0	7
28	— R0 R0	8
25	— R0 R0	8
22	— R0 R0	8
19	— R0 R0	8
16	R0 R0	8
13	R1 ($P_{AA}=0.75$) R0	8
10	R1 ($P_{AA}=0.50$) R1 ($P_{AA}=0.55$)	8
7	R1 ($P_{AA}=0.80$)	7
4	R1 ($P_{AA}=0.89$)	8

distribution, and from 13 %RH, a partial tendency to segregation governs the stacking mode (Table 2 and Figure 4).

3.2. Stressed Sample

3.2.1. Qualitative XRD Description

(1) Sorption Process. From a macro point of view, the hydration behavior similar to that of the unstressed sample is observed in the case of the sample having undergone 50 drying/wetting cycles. Indeed, the d_{001} basal spacing value goes from 9.86 Å at 4 %RH relative to the 0W phase to 18.47 at 94 %RH, indicating a highly hydrated specimen (3W) (Figure S3). But the difference in behavior between the two hydration samples lies in the mechanism and the path followed during the swelling process. Indeed, for the stressed sample, the absence of homogeneous hydration levels characterized by homogeneous phases at 1W and 2W is noted. The evolution of experimental XRD patterns shows a 0W–2W transition over a wide range of humidity that starts at 16 %RH and ends at 70 %RH. This behavior is not observed in the case of the unstressed sample. The second difference is the absence of a homogeneous 3W phase for the maximum used RH values and the shift of the RH range for the homogeneous 2W phase. However, the hydration evolution versus %RH can be described by three main steps: (i) a logic starting from the 0W hydration state, (ii) a transition from 0W to 2W passing through an ephemeral 1W state, and (iii) an incomplete transition from 2W to 3W for high RH rates.

The exploitation of the calculated values of FWHM and ξ, reported in Figure 5, confirms the interstratified character observed during the 0W–2W and 2W–3W transitions. Indeed, a maximum of the FWHM value and ξ parameters is obtained for two ranges of humidity, respectively, that is, from 16% to 70 %RH and from 88% to 94 %RH. The homogeneous 0W and 2W hydration phases are characterized by low values of the abovementioned two parameters.

(2) Desorption Process. The evolution of experimental XRD patterns along the desorption process is shown in Figure S4. Along the explored RH range, a transition from an interstratified 3W hydration state to an interstratified 0W phase, characterized by an asymmetric 00l reflection with irrational positions, is observed (Figure 5). The followed dehydration way differs from that adopted during the hydration process indicating the appearance of a hydration hysteresis phenomenon. Depending on the RH value, during the dehydration

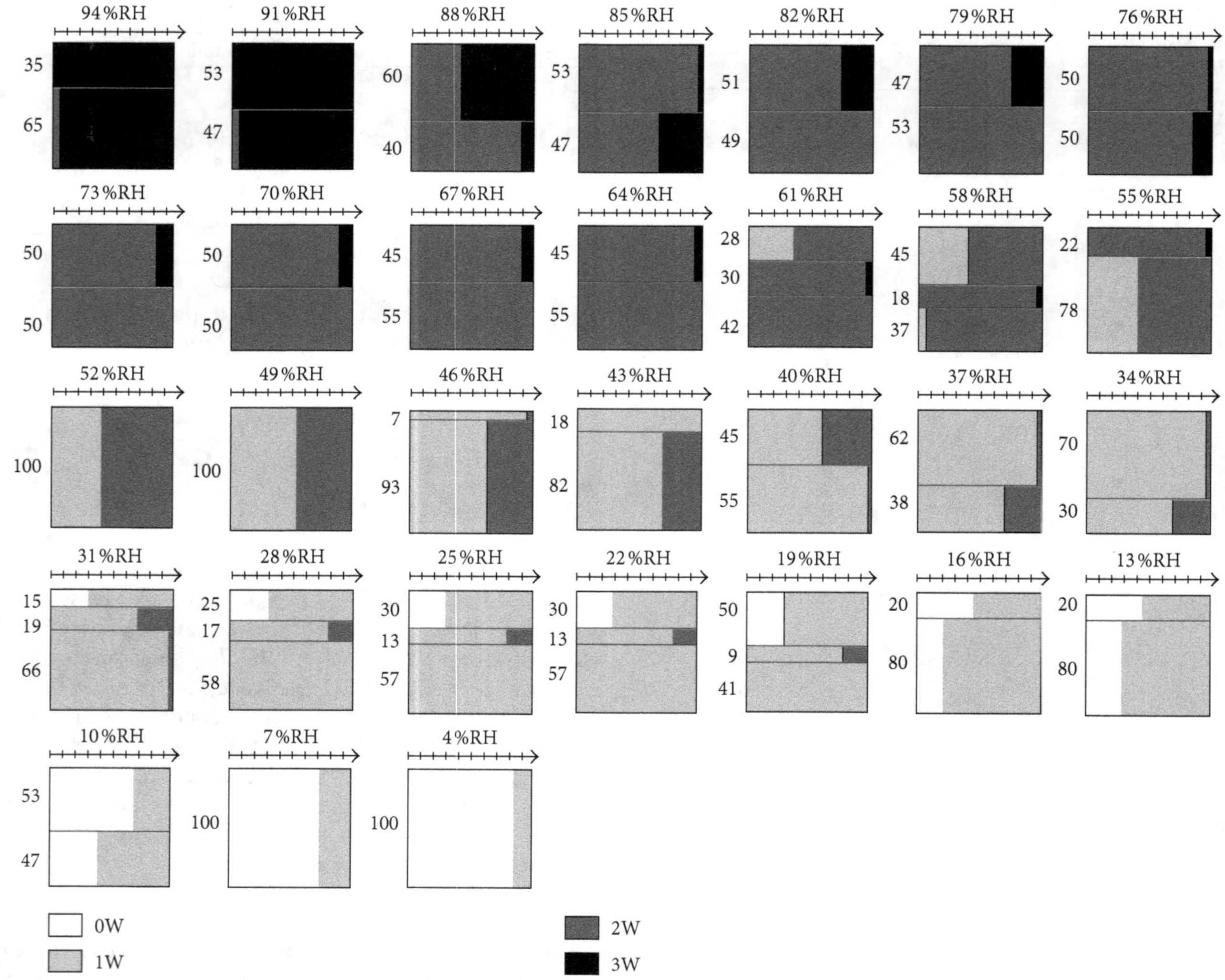

FIGURE 4: Schematic representation of the structure model used to fit the experimental XRD pattern in the case of the unstressed sample along the desorption process.

sequence, two hydration levels related, respectively, to the homogeneous phases 1W and 2W appear. The investigation of the FWHM and ξ parameters (Figure 5) allowed us to identify the interstratified type of the obtained phases during the dehydration process.

Indeed, from 91 %RH and up to 82 %RH, a fast 3W-2W transition with the absence of a purely trihydrated phase is observed. The homogeneous 2W phase begins at 79% and ends at 64 %RH. By comparing this result to that observed in the case of the "unstressed" sample, a decrease in the width of the RH range over which it is spreading is noted. This RH shift is probably attributed to an instability in the IS configuration or the easy release of the IS water molecule, due to RH value fluctuations. Indeed, from 61 %RH and up to 31 %RH, the evolution of the d_{001} basal spacing value indicates a slow progressive 2W-1W transition that can be interpreted as a small loss of interlayer water content. The 1W homogeneous state is observed over a very short RH range compared to the "unstressed" sample whose 1W level started at 31 %RH and reached 16 %RH. This result is consistent with the above, especially since a 1W-0W transition starts a bit early (from 23 %RH), which is not the case for the unstressed sample. The evolution of the different parameters from the qualitative analysis as a function of the %RH variation is reported in Figure 5. Several differences appear between the hydration and dehydration mechanisms obtained for the "stressed" sample. Whatever be the orientation of the hydration sequence and based on the low values of FWHM and ξ, only the homogeneous 2W state exists and a short and near-homogeneity is observed for the 1W state. On the contrary, an absence of the homogeneous 3W state, obtained in the case of the unstressed sample, is noted. The hysteresis effect (Figure 5) is present during both cycles, indicating a swelling behavior depending on the environmental conditions. The observed fluctuations, in terms of %RH width, the absence of homogeneous hydration domains, and characteristic phases of extreme RH conditions (4% and 94 %RH), confirm the major interstratified character observed for the "stressed" sample during the sorption/desorption sequence.

The evolution of the two parameters FWHM and ξ made it possible to determine a limit of homogeneity and/or heterogeneity of hydration and to confirm the geometric anomalies observed on the experimental XRD profiles.

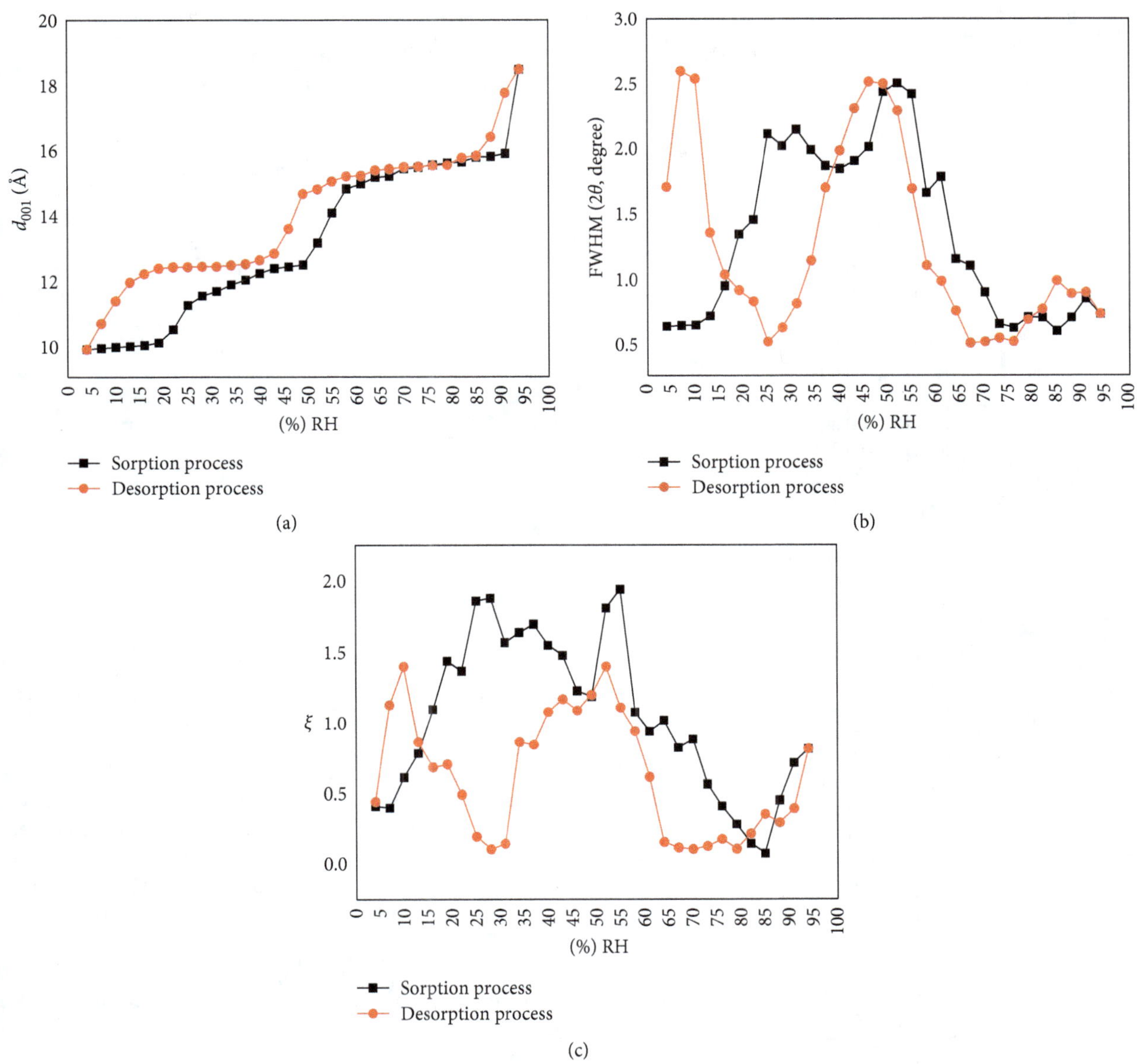

FIGURE 5: Evolution of the d_{001} (a), FWHM (b), and ξ parameters (c) versus %RH in the case of the stressed sample.

In general and for both sorption and desorption mechanisms, the "stressed" sample has more hydration heterogeneities than the "unstressed" sample. The interpretation of this result can be completed only after a quantitative analysis which makes it possible to quantify the abundances of the phases present in the structure and the contents of the IS.

3.2.2. Quantitative XRD Analysis

(1) Sorption Process. The obtained results deduced from the qualitative analysis are refined by quantitative analysis in order to quantify the composition of the clay "particle." All structural parameters used to model experimental XRD patterns are shown in Figure S3 and are summarized in Table 3 and Figure 6. Contrary to the results obtained during the qualitative analysis, concerning the homogeneous 0W and 2W hydration phases, the quantitative analysis shows the absence of any hydration homogeneities along the sorption process. In fact, for a humidity ranging from 4% to 37 %RH, the experimental profiles are reproduced by a 0W/1W mixture with variable abundances. The investigation of junction probabilities and the succession law between layers indicates a major segregation trend (R1). Also, qualitatively, the 2W transition starts at 52 %RH, which is not really the case since the introduction of a 2W layer "population" is done quantitatively at 40 %RH and the system (i.e., crystalline particle) is totally reproduced by 3 layer types (i.e., 0W, 1W, and 2W). This configuration does not exclude the 0W phase (which is in complete disagreement with the qualitative description), the contribution of which persists in the diffracted intensity, even for very high values of humidity. The three layer types are stacked with a segregation tendency.

At 34 %RH, the decrease of the average number of layers per half unit cell is due to hydration transition domain. This

TABLE 3: Some structural parameters obtained from the quantitative XRD analysis in the case of the stressed sample along the sorption process.

%RH	Layer stacking mode	M
4	R0 — R0	7
7	R1 ($P_{AA}=0.90$) — R1 ($P_{AA}=0.80$)	9
10	R1 ($P_{AA}=0.90$) — R1 ($P_{AA}=0.80$)	10
13	R1 ($P_{AA}=0.90$) — R1 ($P_{AA}=0.80$)	11
16	R1 ($P_{AA}=0.90$) — R1 ($P_{AA}=0.80$)	11
19	R0 R1 ($P_{AA}=0.90$)	11
22	R1 ($P_{AA}=0.80$) R1 ($P_{AA}=0.90$) R1 ($P_{AA}=0.70$)	11
25	R1 ($P_{AA}=0.90$) R1 ($P_{AA}=0.70$)	10
28	R1 ($P_{AA}=0.90$) R1 ($P_{AA}=0.70$)	8
31	R1 ($P_{AA}=0.70$)	8
34	R1 ($P_{AA}=0.70$)	7
37	R1 ($P_{AA}=0.70$)	7
40	R1 ($P_{AA}=0.70$) R0	7
43	R1 ($P_{AA}=0.75$) R1 ($P_{AA}=0.70$) R0	6
46	R1 ($P_{AA}=0.70$) R1 ($P_{AA}=0.80$)	6
49	R1 ($P_{AA}=0.70$) R1 ($P_{AA}=0.90$) R1 ($P_{AA}=0.57$)	6
52	R1 ($P_{AA}=0.70$) R1 ($P_{AA}=0.90$) R1 ($P_{AA}=0.57$)	6
55	R1 ($P_{AA}=0.70$) R1 ($P_{AA}=0.90$) R1 ($P_{AA}=0.57$)	7
58	R1 ($P_{AA}=0.70$) R1 ($P_{AA}=0.90$) R1 ($P_{AA}=0.60$)	6
61	R1 ($P_{AA}=0.85$) R1 ($P_{AA}=0.75$) R1 ($P_{AA}=0.65$)	6
64	R1 ($P_{AA}=0.85$) R1 ($P_{AA}=0.75$) R1 ($P_{AA}=0.68$)	6
67	R1 ($P_{AA}=0.85$) R1 ($P_{AA}=0.75$) R1 ($P_{AA}=0.90$)	6
70	R1 ($P_{AA}=0.85$) R1 ($P_{AA}=0.75$) R1 ($P_{AA}=0.90$)	6

TABLE 3: Continued.

%RH	Layer stacking mode	M
73	R1 ($P_{AA}=0.80$) R1 ($P_{AA}=0.65$) R1 ($P_{AA}=0.92$) R1 ($P_{AA}=0.95$)	6
76	— R1 ($P_{AA}=0.92$) R1 ($P_{AA}=0.85$)	6
79	R0 — R1 ($P_{AA}=0.85$)	6
82	R1 ($P_{AA}=0.95$) R0 R1 ($P_{AA}=0.85$)	7
85	R0 R0 R1 ($P_{AA}=0.85$)	7
88	R0 R0 R1 ($P_{AA}=0.85$)	7
91	R1 ($P_{AA}=0.93$) R0 R0	6
94	R1 ($P_{AA}=0.85$) R1 ($P_{AA}=0.97$)	6

decrease in the M parameter is quantitatively reflected by diversity and a maximum number of MLSs present in the proposed theoretical model. The elevated heterogeneity degree is confirmed from 61 %RH, where the experimental profile is totally reproduced by 4 theoretical distributions (0W, 1W, 2W, and 3W). This combination of "population" is up to 76 %RH.

This result is probably due to a hard exfoliation mechanism of the IS under the effect of the constraint created by the drying/wetting stress.

From 79 %RH and up to 91 %RH, the 0W phase disappears and the system is reproduced by 3 phases only. At 94 %RH, the 1W phase disappears and the homogeneous 3W hydration state is never reached; however, the experimental profile is reproduced by a distribution between two phases (2W and 3W) stacked with a segregation tendency.

Physically, the existence of these hydration heterogeneities, in the quantitative analysis, agrees with the results of the qualitative analysis, indicating an increase in the values of FWHM and ξ on the same humidity range.

(2) Desorption Process. All structural parameters resulting from the quantitative analysis are summarized in Table 4 and Figure 7. In addition to the absence of the homogeneous hydration state over the entire explored RH range and contrary to the results of the qualitative analysis, the water molecule release is done in a progressive and continuous manner in the absence of any specific level. Indeed, from 94% up to 73 %RH, the experimental profiles are reproduced based on an arrangement of 2W and 3W phases stacked with a segregation tendency in the majority of cases. This result

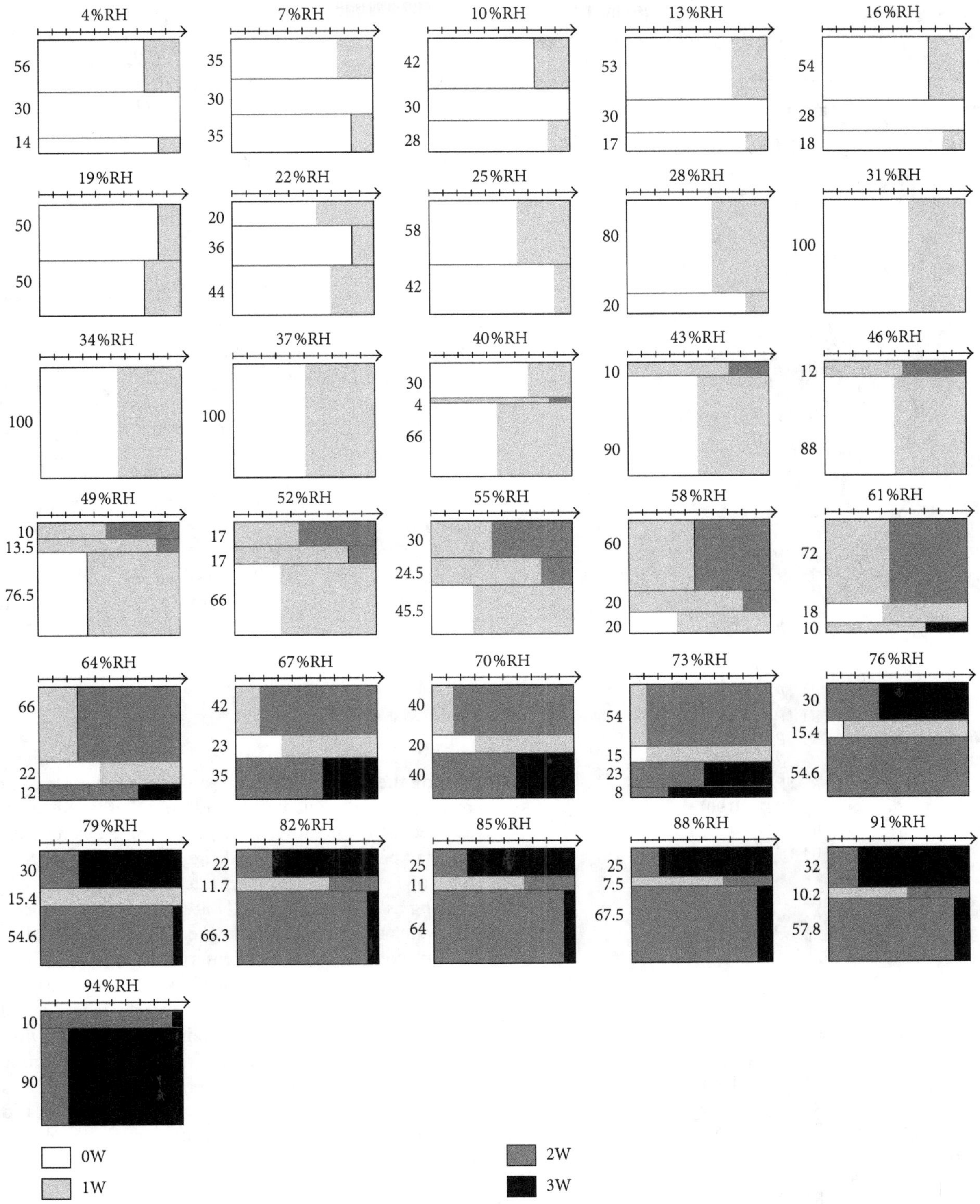

FIGURE 6: Schematic representation of the MLS used to fit the experimental XRD pattern in the case of the stressed sample along the sorption process.

contradicts with the qualitative analysis which assumes a homogeneous 2W phase starting at 79 %RH and ending at 64 %RH. From 70 %RH, the 1W phase is introduced in the MLS, and up to 52 %RH, the structure is fully reproduced by three layer types. At 49 %RH, the transition 3W-2W is achieved and a disappearance of the 3W phase is noted. The system is modeled using random distribution of 1W and 2W phases up to 46 %RH. The 0W phase is introduced, with minor proportions, in the MLS from 43 %RH and continues its progression in terms of abundance up to 4 %RH. At 25 %RH, the total disappearance of the 2W phase indicates the end of the 2W-1W transition.

The proposed MLS raises the ambiguity of the 3W-2W, 2W-1W, and 1W-0W transition zone limits, observed

Table 4: Some structural parameters obtained from the quantitative XRD analysis in the case of the stressed sample along the desorption process.

%RH	Layer stacking mode	M
94	R1 ($P_{AA}=0.86$) R1 ($P_{AA}=0.90$)	6
91	R1 ($P_{AA}=0.90$) R1 ($P_{AA}=0.90$)	6
88	R1 ($P_{AA}=0.75$) R1 ($P_{AA}=0.90$)	5
85	R0 R1 ($P_{AA}=0.86$)	5
82	R0 R1 ($P_{AA}=0.86$)	5
79	— R1 ($P_{AA}=0.75$)	5
76	— R1 ($P_{AA}=0.75$)	5
73	— R1 ($P_{AA}=0.75$)	5
70	R0 R1 ($P_{AA}=0.75$)	5
67	R0 R1 ($P_{AA}=0.75$)	5
64	R0 R0 R1 ($P_{AA}=0.75$)	5
61	R0 R0 R1 ($P_{AA}=0.75$)	5
58	R1 ($P_{AA}=0.95$) R1 ($P_{AA}=0.90$) R1 ($P_{AA}=0.75$)	5
55	R1 ($P_{AA}=0.95$) R1 ($P_{AA}=0.90$) R1 ($P_{AA}=0.75$)	5
52	R1 ($P_{AA}=0.75$) R0 R1 ($P_{AA}=0.75$)	5
49	R1 ($P_{AA}=0.75$) —	5
46	R1 ($P_{AA}=0.70$) —	5
43	R1 ($P_{AA}=0.75$) R1 ($P_{AA}=0.92$)	5
40	R1 ($P_{AA}=0.75$) R1 ($P_{AA}=0.92$)	6
37	R1 ($P_{AA}=0.75$) R1 ($P_{AA}=0.90$)	5
34	R1 ($P_{AA}=0.75$) R1 ($P_{AA}=0.90$)	5
31	R1 ($P_{AA}=0.87$) R1 ($P_{AA}=0.90$)	5
28	R1 ($P_{AA}=0.87$) R0	5
25	R1 ($P_{AA}=0.79$) R1 ($P_{AA}=0.90$)	5
22	R1 ($P_{AA}=0.79$) R1 ($P_{AA}=0.90$)	6
19	R1 ($P_{AA}=0.79$) R1 ($P_{AA}=0.85$)	6
16	R1 ($P_{AA}=0.90$) R1 ($P_{AA}=0.87$)	6
13	R1 ($P_{AA}=0.95$) R1 ($P_{AA}=0.87$)	5
10	R0 — R0	5
7	R1 ($P_{AA}=0.55$) — R0 ($P_{AA}=0.75$)	7
4	— R1 ($P_{AA}=0.75$)	7

during the qualitative analysis, and precisely determines the boundaries of these transitions, despite the existing overlaps. Indeed, transition zones are characterized by asymmetry in the shape of characteristic reflections and irrational positions, and modeling has allowed a more in-depth description of hydration heterogeneities linked to these zones. The absence of a clean transition zone has been shown whatever being the RH value.

During the desorption process, the 2W phase persists up to 25 %RH, and the 1W phase endures even at 4 %RH, indicating probably the hard water molecule release mechanism for the "stressed" sample. This behavior results from the effect of the applied constraint.

4. Discussion

4.1. Water Molecule Distribution and IS Configuration. During the sorption/desorption cycle, the correlation of the results from the qualitative and quantitative XRD analyses allowed us to quantify, with precision, the composition of the clay particle studied and the contents of the IS of each type of layer. The observed fluctuations of the d_{001} basal spacing as a function of %RH are related to the variation of the water molecule amounts present in the IS.

The coexistence of different hydration states, which is an omnipresent character regardless of the constraint, is probably due to a complex insertion/release water molecule mechanism. This process remains unexplored with conventional analysis methods. The results, obtained from XRD modeling, can provide answers on the evolution of the interlamellar water abundances and their organization inducing an adequate description of the evolution of the layer thickness, according to the variation of the surrounding humidity. The evolution of the theoretical water molecules (n_{H_2O}) abundance (Figure 8), resulting from the different layer-type populations versus %RH, for the unstressed and stressed samples takes the form of a hydration hysteresis. This behavior is consistent with the results obtained by Oueslati et al. [30], who demonstrate the hysteresis material response, after external solicitation. The obtained hysteresis can be divided into three main sections, respectively, I, II, and III, as a function of the value of %RH. In fact, section I starts at 4 %RH and finishes at 37 %RH, section II is spread over the 40%–82 %RH range, and finally, section III starts at 85 %RH and ends at 94 %RH.

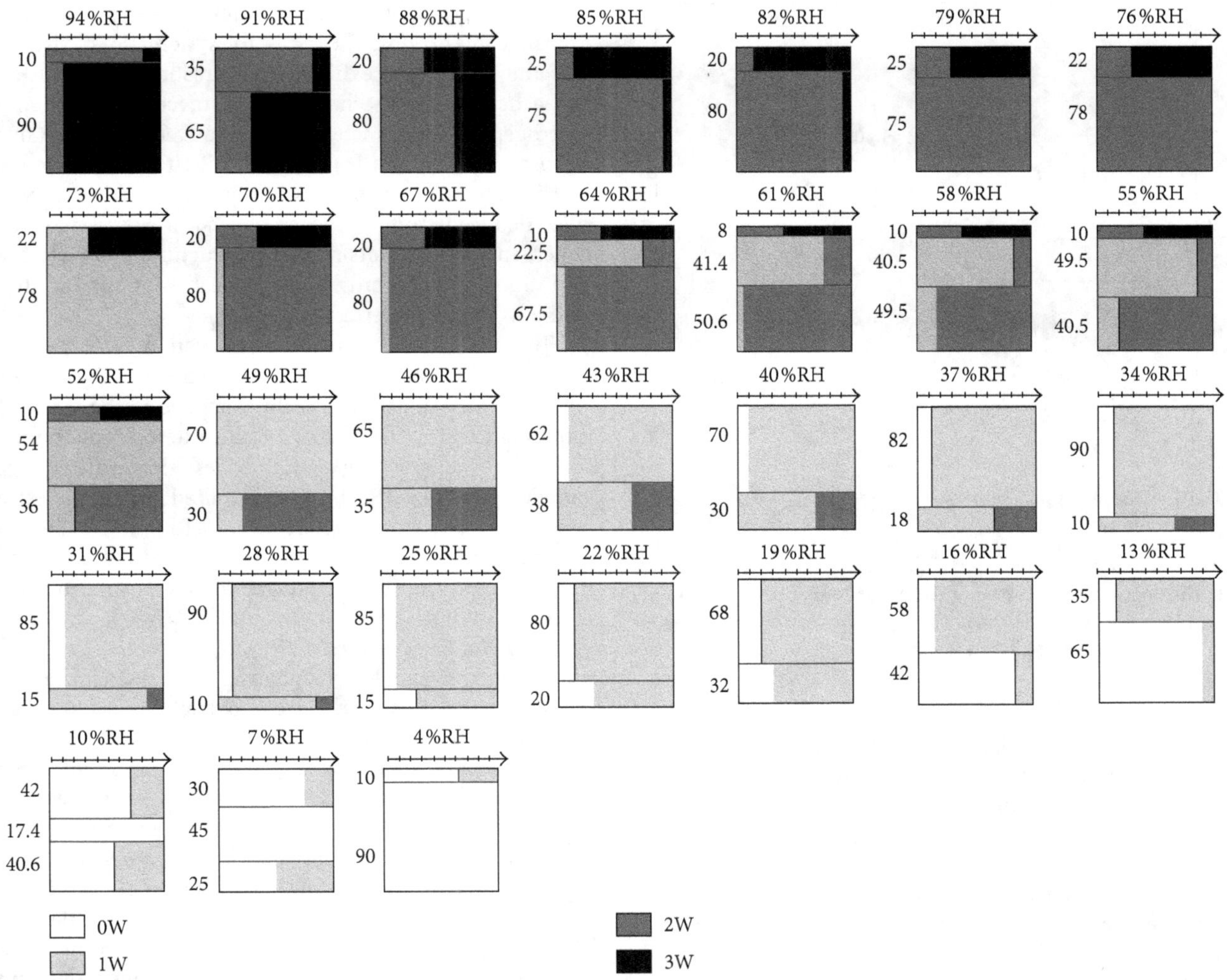

FIGURE 7: Schematic representation of the MLS used to fit the experimental XRD pattern in the case of the stressed sample along the desorption process.

4.1.1. Section I. For the low %RH value and until 37 %RH, a same evolution of the water molecule amount, for the unstressed and stressed samples, is noted and the variation of the environmental constraint (i.e., 50 drying/wetting cycles) does not affect the water retention mechanism. Quantitatively, this domain is totally described, at maximum, by 1W phases which are characterized by a relatively small layer thickness and a quite limited interlayer water amount. This condition imposes a restricted geometrical configuration of the IS (exchangeable cation + one water sheet) inducing a minimum freedom degree.

4.1.2. Section II. Within this section, there is a maximum water amount fluctuation and the sample has a high sensitivity to the applied atmospheric stress. Indeed, the decrease in the hysteresis thickness, for the stressed sample, indicates a significant variation in the water loss process as a function of %RH during the desorption sequence. This section can be divided in two specific phases, mentioned, respectively, (1) and (2), in blue in Figure 8. The difference between the two phases appears essentially on the curves relating to the desorption sequence.

For the phase (1), the starting sample is characterized by a hydration level with an average of $n_{H_2O} > 6$, whereas the stressed sample has a pseudolevel at $n_{H_2O} = 5.2$. In addition, the bearing relative to the starting sample is spread over a wide humidity range from 82 %RH to 55 %RH (~27%), which is not the case for the stressed sample with a fairly narrow range from 82 %RH to 67 %RH (~15%).

For the phase (2), there is a gradual and continuous decay for both samples but still keeping the offset in the number of water molecules. This behavior can be interpreted by the hypothesis of the transformation of a fraction of the "free" IS water into "structural" water, under the effect of the drying/wetting constraint, inducing a reduction in the number of vacant sites on the internal surfaces of the tetrahedral sheet and necessarily a decrease in the number of inserted water molecules. At the macroscopic scale, this means that, in the case of the "unstressed" sample, a water retention capacity is higher than that of the "stressed" one.

4.1.3. Section III. From 85% towards 94 %RH, a maximum water molecule abundance is reached and the same hydration/dehydration way is observed for the two studied

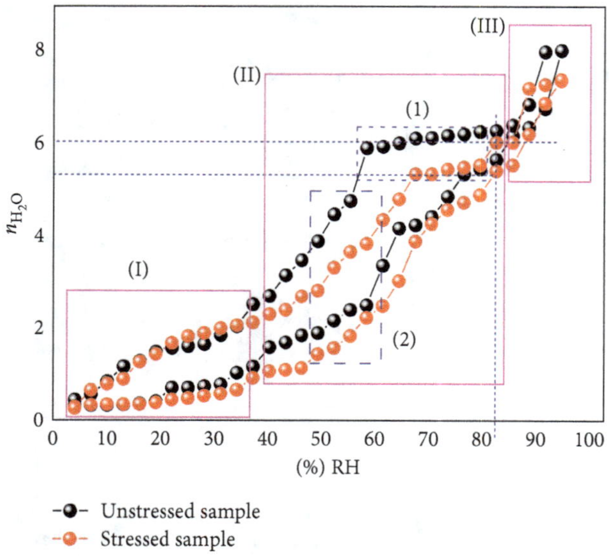

FIGURE 8: Evolution of the water molecule amounts (n_{H_2O}) versus %RH. The bottom black curve represents the sorption process of the unstressed sample. The red curve at the bottom represents the sorption process of the stressed sample. For (1) and (2) in blue, see text.

samples. This can be explained by the same principle as for Section I. Indeed, at the 3W layers hydration state, the interlamellar space is completely saturated by the geometric arrangement of the water sheets and the exchangeable cations. The stereographic configuration and the condition imposed by the physical interactions that govern the nature and arrangement of the chemical species present a constraint which limits the hydration behavior changes despite the external applied atmospheric constraint.

4.2. Hydration State Discretization. The ability of smectites to incorporate interlayer H_2O molecules and the subsequent change in d_{00l} have been studied for several decades. These observations suggested several possible models where crystalline swelling is controlled by the balance between the repulsive force owing to 2 : 1 layer interactions and the attractive forces between hydrated interlayer cations and the negatively charged surface of siloxane layers [43–46]. Few studies have taken into account the coexistence of different hydration states using the XRD profile modeling approach (with 0 to 3 planes of interlayer H_2O molecules) [47–50]. More recently, the studies [51, 52] refined this approach by fitting the 00*l* reflections over a large angular range and showed that randomly interstratified structures, each containing different layer types, coexisted in their montmorillonite samples. Ferrage et al. [24] investigate the distribution of interlayer water in bihydrated smectite from X-ray diffraction profile modeling of 00*l* reflections where they propose a new configuration for the interlayer structure of bihydrated smectite. They focused on the montmorillonite hydration properties by modeling experimental X-ray diffraction patterns recorded under controlled relative humidity (RH) conditions. Indeed, the humidity rate varies with an increment of ~20%, and it is an understandable view that the purpose of the work was to study the hydration heterogeneities in the case of several exchangeable cations. They show the systematic coexistence of smectite layer types exhibiting contrasting hydration states and a heterogeneous character for smectite hydration. This heterogeneity is characterized qualitatively using the fluctuation of FWHM and (ξ) parameters. This rule is used after that in [29, 30] to define hydration heterogeneities. Several rules deduced from the work of [25] are respected in this paper, and some results related to structural parameters such as position and amounts of exchangeable cations and the use of the MLS are in perfect concordance.

This work focuses on the discretization of the water retention mechanism, using the XRD profile modeling approach, when an external hydrous strain is applied in the case of Na-montmorillonite. The choice of the external environmental constraint (i.e., 50 drying/wetting cycles), the relative humidity variation steps (~3 %RH), and the realized limit of the extreme conditions (i.e., from 4 %RH to 94 %RH) make this work a fundamental study in order to understand the interlamellar water amount fluctuation. Indeed, the localization of the hydration transition region is accurately defined, and the hydration heterogeneities, related to the presence of more than one layer type, can be quantitatively explained. Among the strong keys of this work is the overall characterization of the abundance limits, of the dehydrated phases (0W) and the extremely hydrated phases (3W), along the sorption/desorption sequence. Indeed, after 50 drying/wetting cycles, the dehydrated and the extremely hydrated phases are more present in the stressed sample than the host material (Figure 9).

This configuration is probably due to a chemical and electrostatic equilibrium change in the interlamellar space. This change is explained by a possible adjust in the internal surface of the interlayer space following sorption/desorption cycles. Subsequent to the excessive repetition of the insertion/release water process in the starting materials, a water molecule residue has been incorporated in the internal structure which subsequently reduces the vacant site proportion inside the interlamellar space and affects its initial "function" after that (i.e., the cation exchange capacity (CEC)). By examining the evolution of the 1W and 2W abundances (Figure 9) versus %RH, a similar variation is observed for the stressed and the unstressed samples.

It remains to note that the 1W phase never reached the maximum abundance in the case of the stressed and unstressed samples. However, the 1W phase is the layer type mostly present along all the relative humidity scales. This result is interpreted by the transitional character of the 1W hydration state explained by a structural interlamellar space change with a permanent presence of the interlamellar water molecules regardless of %RH rates. The 1W phase can be considered as a transitional hydration state support. In terms of heterogeneities, a trend towards homogeneity is noted for both hysteresis terminals, whereas a heterogeneity tendency is noted for the transitional areas.

5. Conclusion

The XRD profile modeling of the 00*l* reflections allowed us to characterize the structural changes along the c^* axis of the

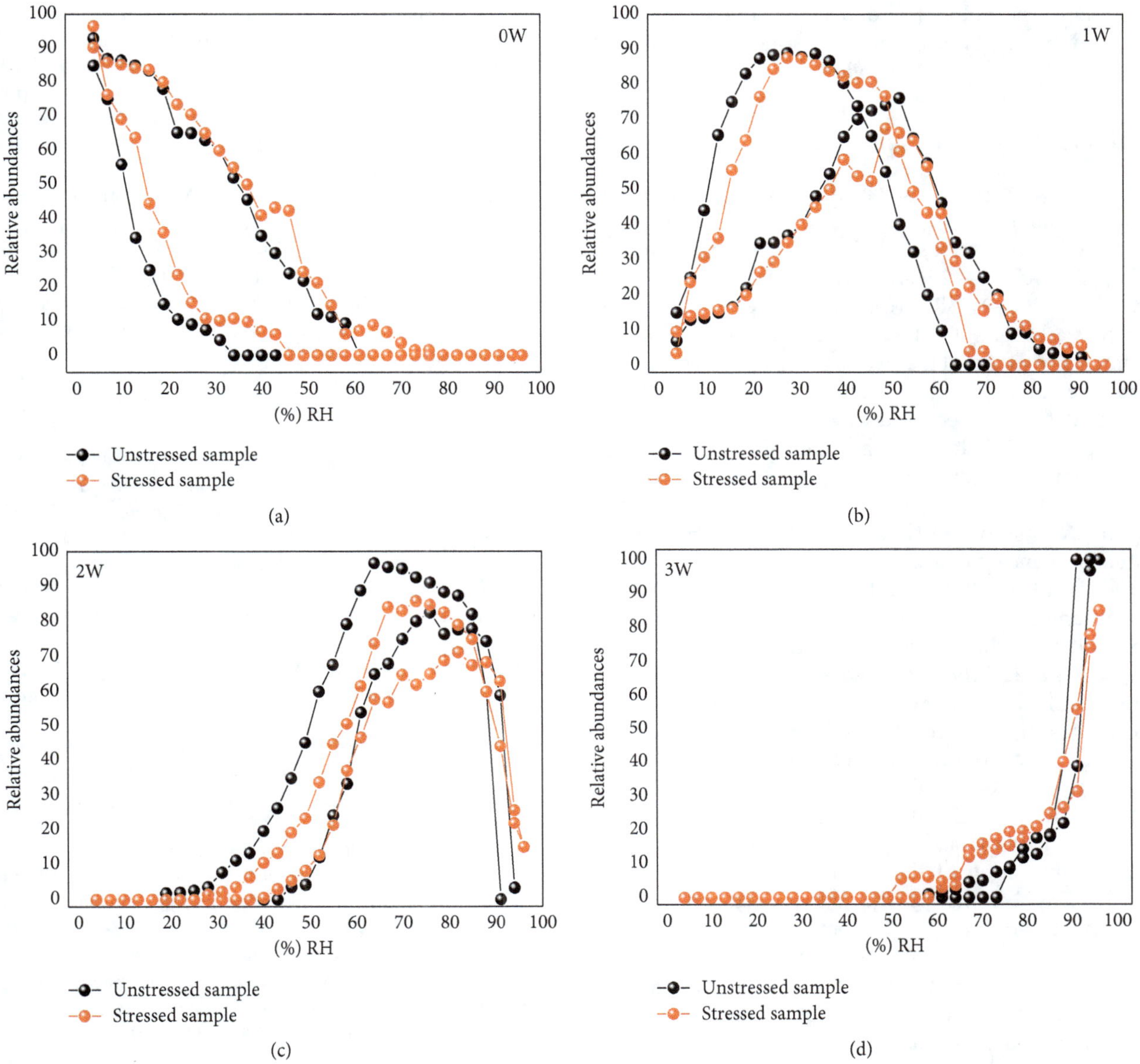

FIGURE 9: Evolution of the individual relative abundance of each hydration state as a function of the relative humidity (%RH) in the case of unstressed and stressed samples.

SWy-2-Na sample before and after application of the atmospheric stress. The obtained results are as follows:

(i) A new hydration behavior of the "stressed" sample described by fluctuations in the hydration hysteresis as a function of the RH rates.

(ii) The application of the sorption/desorption process by "in situ" variation of the %RH, before and after applying the constraint, made it possible to identify the homogeneous hydration states and the hydration transition zones, which are characterized by a high heterogeneity degree.

(iii) Structural heterogeneities are characterized by an elevated MLS number. Indeed, two possible configurations of IS contents, 0W/1W/2W and 1W/2W/3W, with variable relative abundances, are discerned over a wide explored RH range.

(iv) Iteration of the insertion/release phenomenon of the water molecules, during drying/wetting cycles, induces an increase in the relative abundance of the 0W and 3W population at the expense of the abundances of the 1W and 2W phases. This result explains the limits observed during the sorption/desorption process in the case of the stressed sample.

Conflicts of Interest

The authors declare that they have no conflicts of interest.

Acknowledgments

The authors specially thank Bruno Lanson (ISTerre, Grenoble, France) for his availability, his fruitful discussions on the XRD modeling approach, and his advice on the work presentation. The authors also thank Nathaniel Findling (ISTerre, Grenoble, France) for the assistance during XRD and sorption/desorption data collection.

Supplementary Materials

Figure S1: best agreement obtained between experimental (black) and modeled (red) XRD patterns in the case of the SWy-2-Na sample versus %RH. The sorption sequence by increasing %RH: 4–25 %RH (a), 28–49 %RH (b), 52–73 %RH (c), and 76–94 %RH (d). For all XRD patterns, the intensity is arbitrary. The theoretical decomposition of the experimental patterns (blue, red, and green) is on the left. The obtained best agreement is on the right. For each %RH value, the 001 line is recorded twice (in black) in order to check the sample balance with its environment. *Halite (NaCl) complex. Figure S2: best agreement obtained between experimental (black) and modeled (red) XRD patterns in the case of the SWy-2-Na sample versus %RH. The desorption sequence by decreasing %RH: 94–70 %RH (a), 67–46 %RH (b), 43–25 %RH (c), and 22–4 %RH (d). For all XRD patterns, the intensity is arbitrary. The theoretical decomposition of the experimental patterns (blue, red, and green) is on the left. The final best agreement is on the right. *Halite (NaCl) complex. Figure S3: best agreement obtained between experimental (black) and modeled (red) XRD patterns in the case of the stressed sample versus %RH. The sorption sequence obtained by increasing %RH: 4–25 %RH (a), 28–46 %RH (b), 49–70 %RH (c), and 73–94 %RH (d). For all XRD patterns, the intensity is arbitrary. The theoretical decomposition of the experimental patterns (blue, red, and green) is on the left. The obtained best agreement is on the right. *Halite (NaCl) complex. Figure S4: best agreement obtained between experimental (black) and modeled (red) XRD patterns in the case of the stressed sample versus %RH. The desorption sequence obtained by increasing %RH: 4–25 %RH (a), 28–46 %RH (b), 49–70 %RH (c), and 73–94 %RH (d). For all XRD patterns, the intensity is arbitrary. The theoretical decomposition of the experimental patterns (blue, red, and green) is on the left. The obtained best agreement is on the right. *Halite (NaCl) complex. (*Supplementary Materials*)

References

[1] H.-J. Herbert, J. Kasbohm, H. Sprenger, A. M. Fernández, and C. Reichelt, "Swelling pressures of MX-80 bentonite in solutions of different ionic strength," *Physics and Chemistry of the Earth, Parts A/B/C*, vol. 33, no. S1, pp. S327–S342, 2008.

[2] T. Higashihara, K. Kinoshita, Y. Akagi, S. Sato, and T. Kozaki, "Transport number of sodium ions in water-saturated, compacted Na-montmorillonite," *Physics and Chemistry of the Earth, Parts A/B/C*, vol. 33, no. S1, pp. S142–S148, 2008.

[3] M. V. Villara, M. Sánchez, and A. Gens, "Behaviour of a bentonite barrier in the laboratory: experimental results up to 8 years and numerical simulation," *Physics and Chemistry of the Earth, Parts A/B/C*, vol. 33, no. S1, pp. S476–S485, 2008.

[4] T. Kozaki, J. Liua, and S. Sato, "Diffusion mechanism of sodium ions in compacted montmorillonite under different NaCl concentration," *Physics and Chemistry of the Earth, Parts A/B/C*, vol. 33, no. 14-16, pp. 957–961, 2008.

[5] T. Kozaki, T. Sawaguchi, A. Fujishima, and S. Sato, "Effect of exchangeable cations on apparent diffusion of Ca2+ ions in Na- and Ca-montmorillonite mixtures," *Physics and Chemistry of the Earth, Parts A/B/C*,vol. 35, no. 6–8, pp. 254–258, 2010.

[6] O. Karnland, M. Birgersson, and M. Hedström, "Selectivity coefficient for Ca/Na ion exchange in highly compacted bentonite," *Physics and Chemistry of the Earth, Parts A/B/C*, vol. 36, no. 17–18, pp. 1554–1558, 2011.

[7] K. K. Norrfors, M. Bouby, S. Heck et al., "Montmorillonite colloids: I. Characterization and stability of dispersions with different size fractions," *Applied Clay Science*, vol. 114, pp. 179–189, 2015.

[8] W. F. Bradley, R. E. Grim, and G. F. Clark, "A study of the behavior of montmorillonite on wetting," *Zeitschrift Kristallographie*, vol. 97, pp. 260–270, 1937.

[9] J. Méring and R. Glaeser, "Sur le rôle de la valence des cations échangeables dans la montmorillonite," *Bulletin de la Société Francaise de Minéralogie et Cristallographie*, vol. 77, pp. 519–530, 1954.

[10] S. Dultz, J.-H. An, and B. Riebe, "Organic cation exchanged montmorillonite and vermiculite as adsorbents for Cr(VI): effect of layer charge on adsorption properties," *Applied Clay Science*, vol. 67-68, pp. 125–133, 2012.

[11] D. M. Moore and R. C. Reynolds Jr., *X-Ray Diffraction and the Identification and Analysis of Clay Minerals*, Oxford University Press, New York, NY, USA, 1997.

[12] B. Dazas, B. Lanson, J. Breu, J.-L. Robert, M. Pelletier, and E. Ferrage, "Smectite fluorination and its impact on interlayer water content and structure: a way to fine tune the hydrophilicity of clay surfaces?," *Microporous and Mesoporous Materials*, vol. 181, pp. 233–247, 2013.

[13] A. Kaya and H. Y. Fang, "The effects of organic fluids on physicochemical parameters of fine-grained soils," *Canadian Geotechnical Journal*, vol. 37, no. 5, pp. 943–950, 2000.

[14] M. H. Fu, Z. Z. Zhang, and P. F. Low, "Changes in the properties of a montmorillonite-water system during the adsorption and desorption of water: hysteresis," *Clays and Clay Minerals*, vol. 38, pp. 485–492, 1990.

[15] E. S. Boek, P. V. Coveney, and N. T. Skipper, "Monte Carlo molecular modeling studies of hydrated Li-, Na-, and K-smectites: understanding the role of potassium as a clay swelling inhibitor," *Journal of the American Chemical Society*, vol. 117, no. 50, pp. 12608–12617, 1995.

[16] E. S. Boek, P. V. Coveney, and N. T. Skipper, "Molecular modeling of clay hydration: a study of hysteresis loops in the swelling curves of sodium montmorillonites," *Langmuir*, vol. 11, no. 12, pp. 4629–4631, 1995.

[17] F.-R. C. Chang, N. T. Skipper, and G. Sposito, "Computer simulation of interlayer molecular structure in sodium montmorillonite hydrates," *Langmuir*, vol. 11, no. 7, pp. 2734–2741, 1995.

[18] F.-R. C. Chang, N. T. Skipper, and G. Sposito, "Monte Carlo and molecular dynamics simulations of interfacial structure in lithium-montmorillonite hydrates," *Langmuir*, vol. 13, no. 7, pp. 2074–2082, 1997.

[19] N. T. Skipper, F. R. C. Chang, and G. Sposito, "Monte Carlo simulation of interlayer molecular structure in swelling clay minerals. 1. Methodology," *Clays and Clay Minerals*, vol. 43, no. 3, pp. 285–293, 1995.

[20] S. Karaborni, B. Smit, W. Heidug, J. Urai, and E. van Oort, "The swelling of clays: molecular simulations of the hydration of montmorillonite," *Science*, vol. 271, no. 5252, pp. 1102–1104, 1996.

[21] Q. H. Zeng, A. B. Yu, G. Q. Lu, and R. K. Standish, "Molecular dynamics simulation of the structural and dynamic properties of dioctadecyldimethyl ammoniums in organoclays," *Journal of Physical Chemistry B*, vol. 108, no. 28, pp. 10025–10033, 2004.

[22] A. Meleshyn and C. Bunnenberg, "Swelling of Na/Mg-montmorillonites and hydration of interlayer cations: a Monte Carlo study," *Journal of Chemical Physics*, vol. 123, no. 7, article 074706, 2005.

[23] D. E. Smith, Y. Wang, A. Chaturvedi, and H. D. Whitley, "Molecular simulations of the pressure, temperature, and chemical potential dependencies of clay swelling," *Journal of Physical Chemistry B*, vol. 110, no. 40, pp. 20046–20054, 2006.

[24] E. Ferrage, B. Lanson, B. A. Sakharov, and V. A. Drits, "Investigation of smectite hydration properties by modeling of X-ray diffraction profiles. Part 1. Montmorillonite hydration properties," *American Mineralogist*, vol. 90, no. 8-9, pp. 1358–1374, 2005.

[25] E. Ferrage, B. Lanson, N. Malikova, A. Plançon, B. A. Sakharov, and V. A. Drits, "New insights on the distribution of interlayer H_2O molecules in bi-hydrated smectite from X-ray diffraction profile modeling of 00l reflections," *Chemistry of Materials*, vol. 17, no. 13, pp. 3499–3512, 2005.

[26] E. Ferrage, B. Lanson, B. A. Sakharov, N. Geoffroy, E. Jacquot, and V. A. Drits, "Investigation of smectite hydration properties by modeling of X-ray diffraction profiles. Part 2. Influence of layer charge and charge location," *American Mineralogist*, vol. 92, no. 10, pp. 1731–1743, 2007.

[27] W. Oueslati, M. S. Karmous, H. Ben Rhaiem, B. Lanson, and A. Ben Haj Amara, "Effect of interlayer cation and relative humidity on the hydration properties of a dioctahedral smectite," *Zeitschrift für Kristallographie Supplements*, vol. S26, pp. 417–422, 2007.

[28] W. Oueslati, H. B. Rhaiem, B. Lanson, and A. B. H. Amara, "Selectivity of Na–montmorillonite in relation with the concentration of bivalent cation (Cu^{2+}, Ca^{2+}, Ni^{2+}) by quantitative analysis of XRD patterns," *Applied Clay Science*, vol. 43, no. 2, pp. 224–227, 2009.

[29] W. Oueslati, H. Ben Rhaiem, and A. Ben Haj Amara, "XRD investigations of hydrated homoionic montmorillonite saturated by several heavy metal cations," *Desalination*, vol. 271, no. 1–3, pp. 139–149, 2011.

[30] W. Oueslati, H. Ben Rhaiem, and A. Ben Haj Amara, "Effect of relative humidity constraint on the metal exchanged montmorillonite performance: an XRD profile modeling approach," *Applied Surface Science*, vol. 261, pp. 396–404, 2012.

[31] B. Dazas, E. Ferrage, A. Delville, and B. Lanson, "Interlayer structure model of tri-hydrated low-charge smectite by X-ray diffraction and Monte Carlo modeling in the Grand Canonical ensemble," *American Mineralogist*, vol. 99, no. 8-9, pp. 1724–1735, 2014.

[32] T. Sato, T. Watanabe, and R. Otsuka, "Effects of layer charge, charge location, and energy change on expansion properties of dioctahedral smectites," *Clays and Clay Minerals*, vol. 40, no. 1, pp. 103–113, 1992.

[33] W. Oueslati, M. Ammar, and N. Chorfi, "Quantitative XRD analysis of the structural changes of Ba-exchanged montmorillonite: effect of an in situ hydrous perturbation," *Minerals*, vol. 5, no. 3, pp. 507–526, 2015.

[34] W. F. Moll, "Baseline studies of the Clay Minerals Society Source Clays: geological origin," *Clays and Clay Minerals*, vol. 49, no. 5, pp. 374–380, 2001.

[35] A. R. Mermut and A. F. Cano, "Baseline studies of the Clay Minerals Society Source Clays: chemical analyses of major elements," *Clays and Clay Minerals*, vol. 49, no. 5, pp. 381–386, 2001.

[36] J. L. Post, B. L. Cupp, and F. T. Madsen, "Beidellite and associated clays from the DeLamar Mine and Florida Mountain area, Idaho," *Clays and Clay Minerals*, vol. 45, no. 2, pp. 240–250, 1997.

[37] D. Tessier, "Experimental study of the clay materials organization. Hydration, swelling and structure along drying and re-wetting," Thesis, University of Paris VII, Publication INRA Versailles, Paris, France, 1984.

[38] S. W. Bailey, "Nomenclature for regular interstratifications," *American Mineralogist*, vol. 67, pp. 394–398, 1982.

[39] M. Ammar, W. Oueslati, H. B. Rhaiem, and A. B. H. Amara, "Quantitative XRD analysis of the dehydration–hydration performance of (Na^+, Cs^+) exchanged smectite," *Desalination and Water Treatment*, vol. 52, no. 22–24, pp. 4314–4333, 2014.

[40] M. Ammar, W. Oueslati, H. B. Rhaiem, and A. B. H. Amara, "Effect of the hydration sequence orientation on the structural properties of Hg exchanged montmorillonite: quantitative XRD analysis," *Journal of Environmental Chemical Engineering*, vol. 2, no. 3, pp. 1604–1611, 2014.

[41] V. A. Drits and C. Tchoubar, *X-ray Diffraction by Disordered Lamellar Structures: Theory and Applications to Microdivided Silicates and Carbons*, Springer, Berlin, Germany, 1990.

[42] B. A. Sakharov and B. Lanson, "X-ray identification of mixed-layer structures. Modelling of diffraction effects," in *Handbook of Clay Science*, F. Bergaya and G. Lagaly, Eds., vol. 5B, pp. 51–135, Elsevier, Amsterdam, Netherlands, 2013.

[43] K. Norrish, "The swelling of montmorillonite," *Discussions of the Faraday Society*, vol. 18, pp. 120–133, 1954.

[44] H. van Olphen, "Thermodynamics of interlayer adsorption of water in clays," *Journal of Colloid Science*, vol. 20, no. 8, pp. 822–837, 1965.

[45] D. A. Laird, "Influence of layer charge on swelling of smectites," *Applied Clay Science*, vol. 34, no. 1–4, pp. 74–87, 2006.

[46] B. Lanson, "Modelling of X-ray diffraction profiles: investigation of defective lamellar structure crystal chemistry,"*EMU Notes in Mineralogy*, Chapter 4, vol. 11, pp. 151–202, Mineralogical Society Great Britain & Ireland, Ireland, 2011.

[47] T. Iwasaki and T. Watanabe, "Distribution of Ca and Na ions in dioctahedral smectites and interstratified dioctahedral mica/smectites," *Clays and Clay Minerals*, vol. 36, no. 1, pp. 73–82, 1988.

[48] I. Bérend, J. M. Cases, M. François et al., "Mechanism of adsorption and desorption of water vapour by homoionic montmorillonites: 2. The Li^+, Na^+, K^+, Rb^+ and Cs^+ exchanged forms," *Clays and Clay Minerals*, vol. 43, no. 3, pp. 324–336, 1995.

[49] J. M. Cases, I. Bérend, M. François, J. P. Uriot, L. J. Michot, and F. Thomas, "Mechanism of adsorption and desorption of water vapour by homoionic montmorillonite: The Mg^{2+}, Ca^{2+}, Sr^{2+} and Ba^{2+} exchanged forms," *Clays and Clay Minerals*, vol. 45, no. 1, pp. 8–22, 1997.

[50] J. Cuadros, "Interlayer cation effects on the hydration state of smectite," *American Journal of Science*, vol. 297, no. 8, pp. 829–841, 1997.

[51] L. Calarge, B. Lanson, A. Meunier, and M. L. Formoso, "The smectitic minerals in a bentonite deposit from Melo (Uruguay)," *Clay Minerals*, vol. 38, no. 1, pp. 25–34, 2003.

[52] A. Meunier, B. Lanson, and B. Velde, "Composition variation of illite-vermiculite-smectite mixed-layer minerals in a bentonite bed from Charente (France)," *Clay Minerals*, vol. 39, no. 3, pp. 187–196, 2004.

14

Dynamic Behavior of Clay-Aggregate Mixtures

Kang Fei and Jinxin Xu

Institute of Geotechnical Engineering, Yangzhou University, Yangzhou 225127, China

Correspondence should be addressed to Kang Fei; kfei@yzu.edu.cn

Academic Editor: Philip Eisenlohr

Clay-aggregate mixtures are frequently used in engineering practice. To improve the understanding of the effects of coarse particles on the dynamic behavior of clay-aggregate mixtures, series of stress controlled cyclic triaxial tests were performed on clay specimens with various glass bead contents. The results show that the initial shear modulus of clay-aggregate mixtures increased with the increase of the coarse aggregate content and the confining stress. At the confining stress of 100, 200, and 400 kPa, the addition of 32% coarse particles caused an increase in the initial shear modulus of 117%, 110%, and 67%, respectively. Moreover, the normalized shear modulus decreased and damping ratio increased with the coarse aggregate content, and the influence of the confining stress on the strain-dependent dynamic properties was negligible. The specimen with a higher coarse aggregate content was observed to have larger cyclic shear strength and smaller excess pore water pressure, but the effects of the coarse aggregate content became less pronounced under large cyclic stresses.

1. Introduction

Clay-aggregate mixtures refer to clays mixed with sand, gravel, or cobble particles. These types of composite soils are frequently used in engineering practice. For example, to improve the strength and stiffness of the core of earth dams, a certain amount of gravels is usually added to the clay. A complete understanding of the dynamic behavior of such composite materials is very important for engineering applications.

In the past, studies of the dynamic behavior of soil mixtures have concentrated mostly on the effects of clay content on the cyclic shear strength of clay-sand mixtures. Thevanayagam et al. [1] reported that the increasing fines content resulted in a decrease in the resistance to liquefaction. Chien et al. [2] stated that the liquefaction resistance decreased with the fines content under the constant relative density. However, Kuwano et al. [3] concluded that the liquefaction resistance increased with the clay content based on the results of cyclic triaxial tests. Ovando-Shelley and Pérez [4] also suggested that the addition of clay increased the liquefaction potential of kaolin-sand mixtures. The influences of the clay plasticity on the strength increment were studied by Ishihara [5], Beroya et al. [6], and Park and Kim [7]. Many other researchers observed that the cyclic shear strength decreased with the fines content at first. Beyond a transition fines content, the strength increased with a further increase in fines content [8–11].

The conflicting pattern observed in the cyclic shear strength with the clay content is a reflection of the roles of clay and coarse particles in the load transfer [12, 13]. Cabalar and Mustafa [14, 15] pointed out that the mechanical behavior of sand-clay mixtures depended on the relative concentration of the sand and clay samples. When the clay content is low, the coarse particles are in contact with others, and the formed coarse particle structure dominates the dynamic behavior, whereas the clay fraction acts only to fill the voids between coarse particles. For this case, the intergranular void ratio was suggested to use to describe the behavior of the soil mixture [16–19]. In contrast, when the clay content is high, coarse particles are floated in the clay matrix and the overall behavior is controlled by the clay fraction. The mechanical behavior of the soil mixture is expected to be similar to that of the pure clay with the same interfine void ratio [20]. However, Yagiz [21], Zhao et al. [22], and Fei [23] pointed out that the coarse particles still have a significant influence on the strength of clay-aggregate mixtures, even if the coarse aggregate content is not high enough to form a stable skeleton.

The strain-dependent dynamic deformation characteristics of clay-aggregate mixtures have also attracted the attention of researchers. On the basis of the results of dynamic hollow cylinder tests, Yamada et al. [24, 25] have suggested empirical relationships to estimate the strain-dependent shear modulus and damping ratio of clay-sand mixtures. In the proposed method, the effects of sand content were accounted by the equivalent plasticity index. But, for clay-gravel mixtures, the plasticity index cannot be obtained by conventional laboratory tests due to the existence of coarse particles. Shafiee and Ghate [26] observed that both the initial shear modulus and the damping ratio increase with the increase in coarse aggregate content, while the opposite trend was observed for the normalized shear modulus at different strain levels. However, a different pattern in dynamic properties with the coarse aggregate content was observed by Meidani et al. [27]. They concluded that the samples with higher gravel content have higher normalized shear modulus and lower damping ratio. One possible reason is that the gravels were in contact with others under the tested gravel content, and the formed gravel structure dominated the dynamic behavior.

As reviewed, the dynamic behavior of clay-aggregate mixtures has not been comprehensively studied yet, especially for the soil mixtures with moderate coarse aggregate contents. For example, the published experimental data of composite clay cores are very limited. To obtain a full understanding of the effects of the coarse particles on the dynamic characteristics of clay-aggregate mixtures, stress controlled cyclic triaxial tests were carried out in this study. On the basis of the test results, the effects of the coarse aggregate content on the strain-dependent shear modulus, the damping ratio, the dynamic pore water pressure, and the cyclic strength were analyzed.

2. Experimental Program

2.1. Test Materials. The clay used in the investigation was kaolinite. Its liquid limit and plastic limit were 43 and 21, respectively. The specific gravity of the clay was 2.62. For all the specimens, the dry density of the clay was kept constant at 1.55 g/cm^3; in other words, the interfine void ratio was constant. The sampling water content of the clay was 21%.

Glass beads were used as the coarse particles added to the clay. The specific gravity of the glass beads was 2.54, which is similar to that of natural gravels. The diameter of the glass beads is 1.4 cm. The main purpose of using the uniform-sized glass beads was to eliminate the effects of shape and graduation of coarse particles. The volume of the glass beads relative to the total volume of the mixture P was selected as 0, 8%, 16%, 24%, and 32%, respectively. These coarse aggregate contents cover the normal range used in the composite clay core of dams.

2.2. Specimen Preparation and Test Procedure. The cyclic triaxial tests were performed by GDS dynamic triaxial system. The specimens were 10 cm in diameter and 20 cm in height. All the test specimens were prepared by a moist tamping technique. First, the glass beads were mixed with the dry clay based on the designated coarse aggregate content. Then, the required amount of water was added to the soil mixture, and the soil was mixed thoroughly to ensure homogeneous distribution of coarse particles. Subsequently, the mixed soil was kept inside a plastic bag for 3 days to moisture homogenization. Afterwards, the soil was compacted using a 2.5 kg rammer to the desired dry density of the clay. The number of rammer blows was determined by trial and error. After the compaction, the specimens were saturated by vacuum-pumping for 24 h.

Two series of cyclic triaxial tests were performed in this study. The first series was used to determine the strain-dependent modulus and the damping ratio of clay-aggregate mixtures. In these tests, the specimens were first isotropically consolidated to an initial effective confining stress σ_0' of 100, 200, and 400 kPa. Then, the specimens were subjected to sinusoidal undrained cyclic loading at a frequency of 0.1 Hz and 10 cycles were applied in each loading stage. The small loading frequency was chosen to ensure the equalization of the pore water pressure in the sample. At the end of each loading step, the drainage valve was opened to dissipate the excess pore water pressure. The second series was used to determine the cyclic strength of clay-aggregate mixtures. For this purpose, undrained cyclic stresses with different amplitudes were applied after consolidation. In this series of tests, σ_0' was fixed at 100 kPa, and only the specimens with 0, 16%, and 32% coarse particles were tested. The loading frequency was 0.1 Hz, and the cyclic stress ratio CSR (half of the cyclic deviator stress divided by the effective confining stress $\sigma_d/2\sigma_0'$) ranged from 0.1 to 0.3. The failure condition of the sample was defined as a double amplitude strain of 5%.

3. Test Results

3.1. Initial Shear Modulus. By connecting the extreme points of the cyclic shear stress-shear strain hysteresis loops under different loading stages, the skeleton curves at various coarse aggregate contents can be obtained. The results of the specimens with 0% and 32% coarse aggregates are shown in Figure 1, where τ is the shear stress and γ is the shear strain. As observed, the nonlinear characteristics of the backbone curves are obvious and the hyperbolic model can be used to describe the stress-strain response. Similar results were also found for other coarse aggregate contents. By using the hyperbolic model, the initial shear modulus G_0 (the tangential slope of the backbone curve when the strain is zero) can be determined. The obtained relationships between the initial shear modulus and the coarse aggregate content at various initial confining stresses are shown in Figure 2.

It is generally believed that the mechanical behavior of clay-aggregate mixtures is mainly governed by the clay matrix up to a transition coarse aggregate content [28]. This transition coarse aggregate content is usually estimated by the maximum possible void ratio of the coarse grains. Based on the research of McGeary [29], the transition content is 53.36% for the case of one-size spheres packed in a simple cubic pattern. This value is significantly larger than those adopted in this study. It implies that the initial shear modulus of the tested clay-aggregate mixtures should be close to that

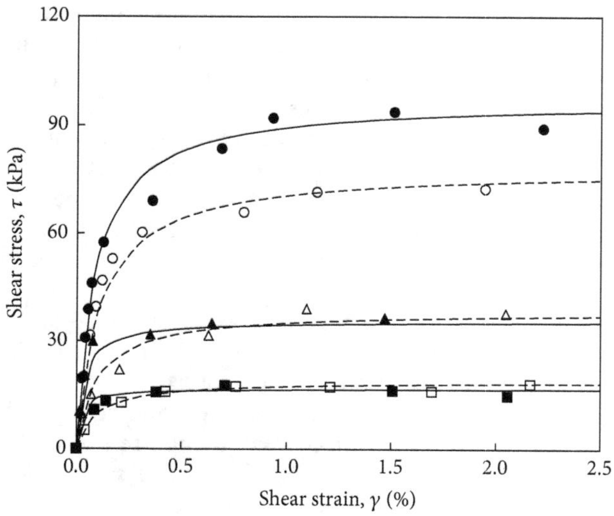

FIGURE 1: Backbone curve of clay-aggregate mixture.

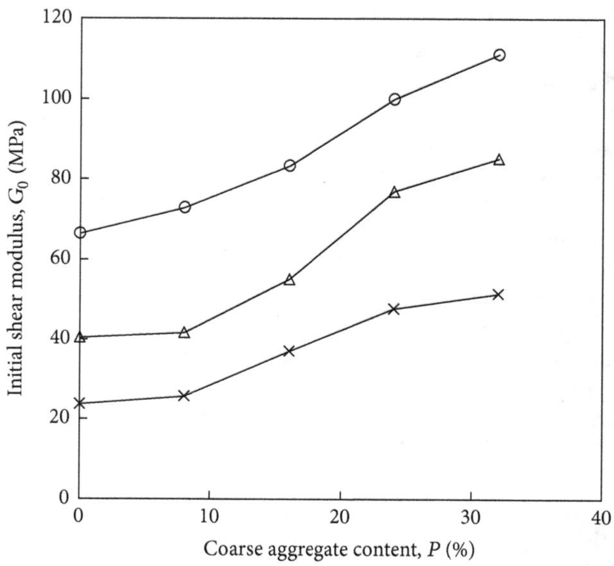

FIGURE 2: Relationship between initial shear modulus and coarse aggregate content.

of pure clay, and the effect of the coarse aggregate content is negligible. However, Figure 2 shows that the coarse aggregate content still has a remarkable effect on the initial shear modulus, even though the dry density of the clay fraction was kept constant. At the confining stress of 100, 200, and 400 kPa, the addition of 32% coarse particles caused an increase in the initial shear modulus of 117%, 110%, and 67%, respectively. This phenomenon can be explained from the point of the view of force chains. For the specimens with relatively low coarse aggregate contents, the coarse particles were far away from each other; the dynamic behavior was controlled by the clay. When more coarse particles were added to the host clay, the coarse particles made active contacts with the clay grains and the force chain consisted of clay and coarse particles together. In addition, the coarse particles also introduced kinematic constrains for deformation of the surrounded clay. Therefore, the larger coarse aggregate content led to a stronger force chain and a higher initial shear modulus.

Within the adopted coarse aggregate contents in this study, both coarse particles and clay grains participate actively in the force chain, so the global void ratio is expected to be an index of active contacts. The variations of the initial shear modulus versus the global void ratio at different confining stress are shown in Figure 3. As can be seen, there is a good agreement and a definite trend between the initial shear modulus and the global void ratio. It is therefore possible to estimate the initial shear modulus in terms of the global void ratio and the effective confining stress in the same way as that for clays [30]. Thus, the initial shear modulus

TABLE 1: List of values of A, B, and n for various data sources.

	A	B	n
This study	400	1.60	0.60
Yamada et al. [24]	49	5.13	0.50
Meidani et al. [27]	380	2.00	0.59

of clay-aggregate mixtures can be expressed in the following form:

$$G_0 = A\frac{(B-e)^2}{1+e}p_a\left(\frac{\sigma_0'}{p_a}\right)^n, \quad (1)$$

where p_a is atmospheric pressure; e is global void ratio. A, B, and n are material constants to be determined experimentally. Based on the test results, the values of A, B, and n were taken as 400, 1.60, and 0.60, respectively. The predicted initial shear moduli using (1) were represented by lines in Figure 3. It is observed that there is a good correlation between the predicted and the measured values. It should be noted that because the dry density of the clay matrix of all specimens remained constant, the addition of coarse particles reduced the global void ratio. Hence, the effect of coarse aggregate content is represented by the global void ratio in (1).

To further verify the reliability of the proposed approach, the test data from literature are compared with the predicted results in Figure 4. The material constants used in the prediction are shown in Table 1. In the study of Yamada et al. [24], a series of cyclic hollow cylindrical torsional shear tests were carried out on clay-sand mixtures. The mean size of the sand particles was 1 mm. In the cyclic triaxial tests conducted by Meidani et al. [27], the coarse particles added to the clay were gravels with an average size of 7.1 mm. In

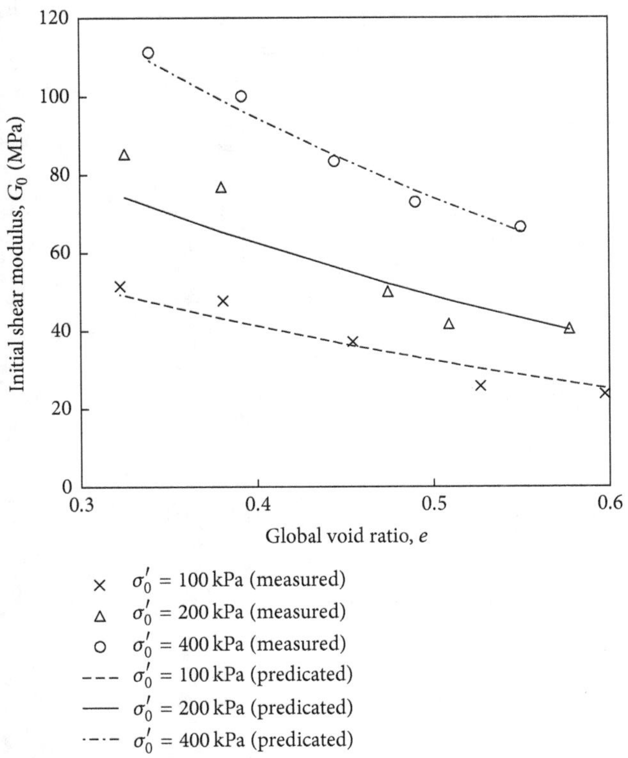

FIGURE 3: Comparison between computed and measured initial shear modulus.

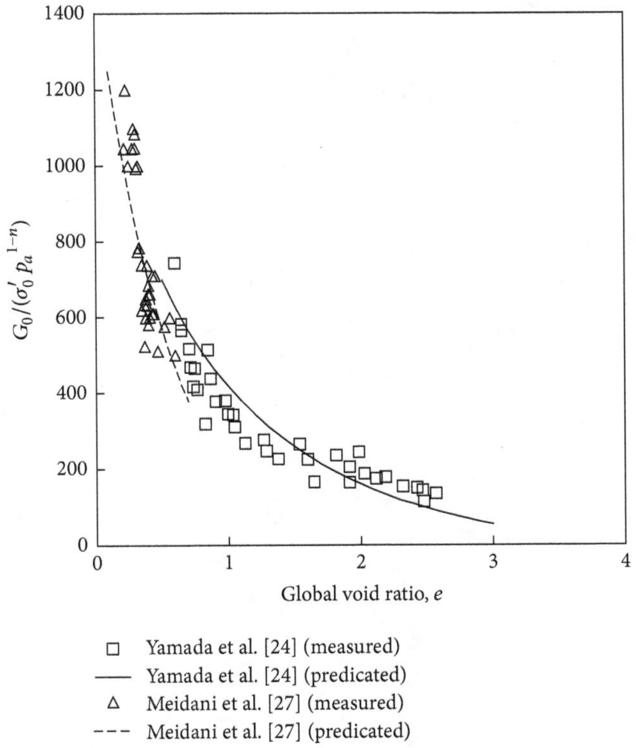

FIGURE 4: Comparison between computed initial shear modulus and literature data.

order to eliminate the influence of confining pressure, the initial shear modulus was normalized by $\sigma_0'^n \cdot p_a^{1-n}$ in the figure. Though the coarse particle sizes were different, there is a reasonably good correlation between the experimental and the calculated values. Equation (1) can be used to estimate the initial shear modulus of clay-aggregate mixtures; the effect of coarse aggregate content is reflected indirectly by the global void ratio.

3.2. Strain-Dependent Normalized Shear Modulus. The relationships between the normalized shear modulus G/G_0 and the shear strain amplitude γ at different confining stresses and coarse aggregate contents are shown in Figure 5. For each coarse aggregate content, the data points are observed to be basically distributed in a narrow belt zone; the influence of the confining stress on the normalized shear modulus of clay-aggregate mixtures is negligible. It is also observed that the coarse aggregate content has a significant effect on the normalized shear modulus. The rate of reduction in shear modulus with strain becomes faster as the coarse aggregate content increases. The similar observation has also been made by Yamada et al. [25]. One reason for this finding is that the clay-aggregate skeleton will collapse gradually due to the slippage between the clay and the coarse particles. Therefore, the effect of coarse particles in increasing the shear modulus will disappear gradually with the development of the shear strain; the shear modulus will be dominated by the clay fraction. Because the initial shear modulus increases with the increase in the coarse aggregate content, the samples with high coarse aggregate content have lower values of G/G_0 at a specific strain level.

From the above results, the following hyperbolic model was used to describe the relationship between the normalized shear modulus and the shear strain of clay-aggregate mixtures:

$$\frac{G}{G_0} = \frac{1}{1 + k_1 \left(1 + P/\left(a + bP\right)\right) \gamma}, \tag{2}$$

where k_1, a, and b are test parameters; P is the volume content of aggregates in decimal representation. The value of k_1 can be determined by the reciprocal of the shear strain at $G/G_0 = 0.5$ for the pure clay specimen. Parameters a and b are used to represent the effects of the coarse aggregate content. In this study, k_1, a, and b are obtained as 980, 0.24, and 0.37, respectively. The predicted $G/G_0 \sim \gamma$ relationships are also presented in Figure 5 as solid lines; the computed values are found to agree well with the test values.

3.3. Strain-Dependent Damping Ratio. The relationships between the damping ration D and the shear strain γ are shown in Figure 6. The results show that both the shear strain level and the coarse aggregate content have remarkable effects on the damping ratio, while the effect of the confining stress is negligible. It is observed that the damping ratio of the clay-aggregate mixture increased with increasing shear strain until a steady value D_{max} was reached. This observation is similar to the general trend of increasing equivalent damping ratio for pure clays.

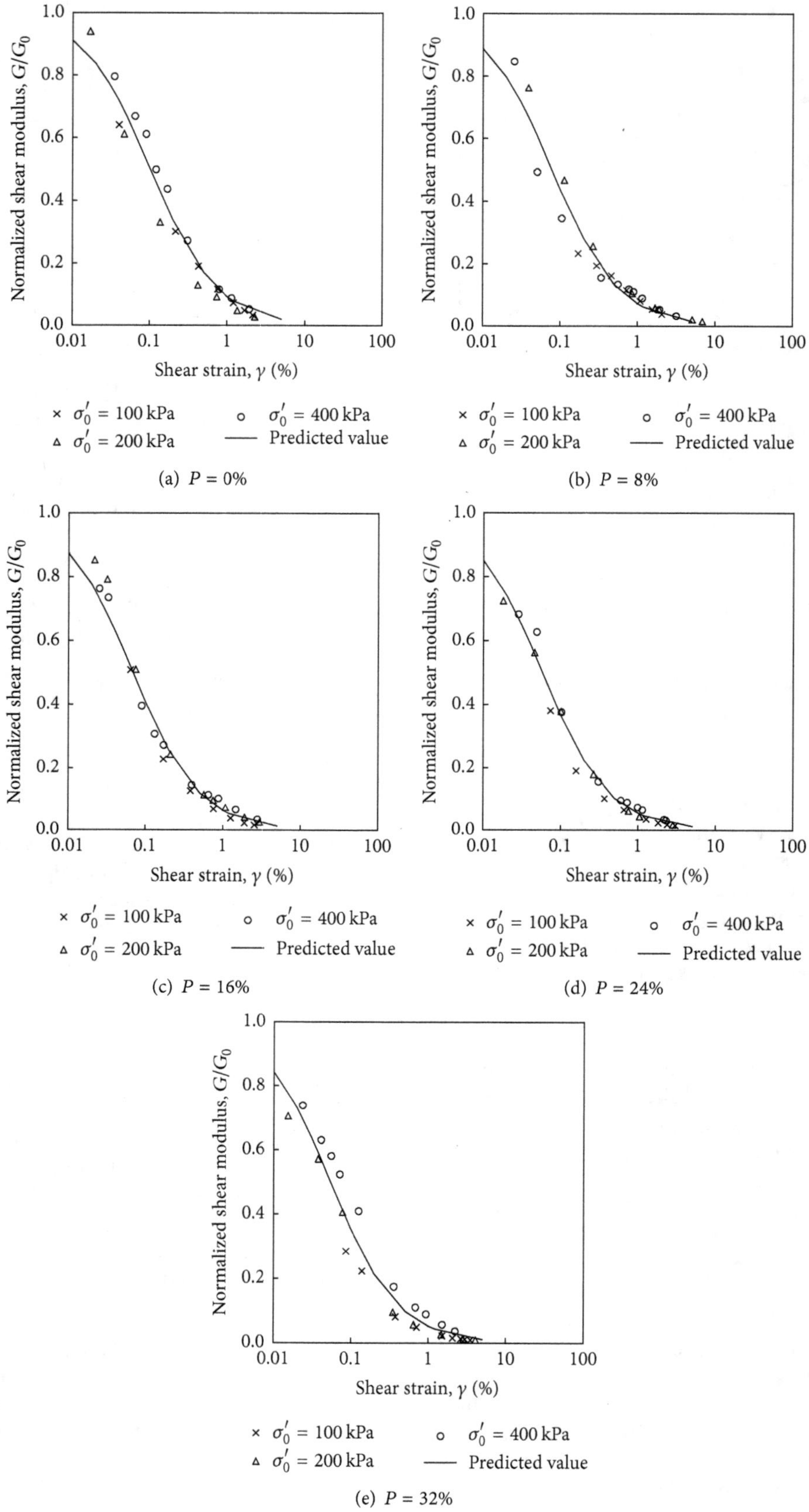

FIGURE 5: Relationship between normalized shear modulus and shear strain.

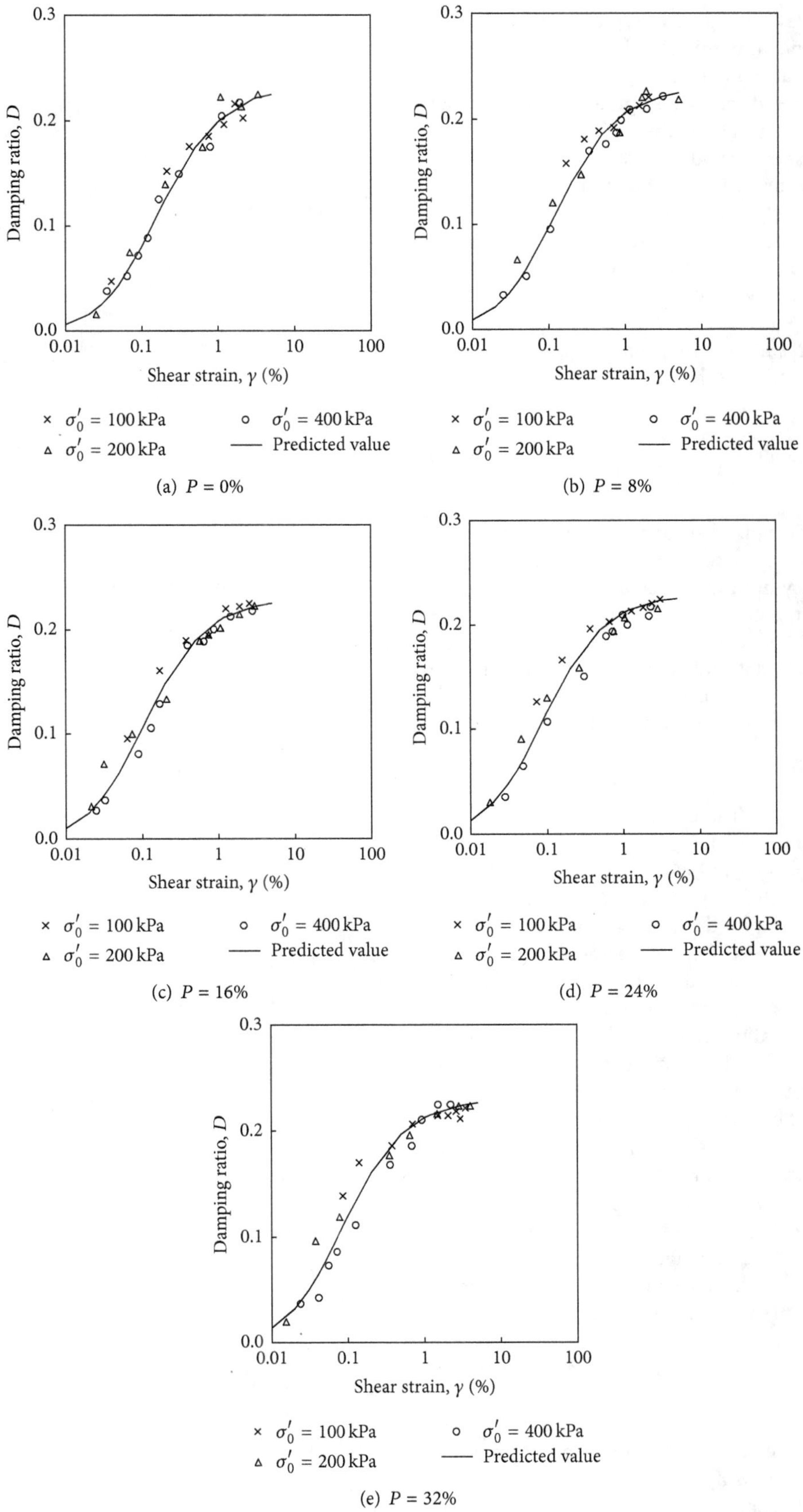

FIGURE 6: Relationship between normalized shear modulus and shear strain.

In addition to the material damping, the friction and the slippage at the clay-aggregate interface may contribute to the energy loss. Since the number of interfaces increased with the coarse aggregate content, samples with larger coarse aggregate content showed larger damping ratios. However, when the shear strain was large enough ($\gamma > 1\%$), the clay matrix has been fully sheared and the effect of coarse particles was negligible. The values of D_{max} of all samples are found to be approximately 0.23, which is close to the measured value of the pure clay. It should be noted that mixtures with extremely high coarse aggregate content may have a different trend, since the coarse particles will control the dynamic behavior in those cases.

Based on the test results, the following empirical relationship was proposed to estimate the damping ratio:

$$D = D_{max}\left(1 - \frac{G}{G_0}\right)^{\beta}, \tag{3}$$

where G/G_0 is the normalized shear modulus computed by (2); D_{max} is the maximum damping ratio, which can be determined by the asymptotic value of the damping ratio curve; β is the test parameter. In this study, D_{max} and β were obtained as 0.23 and 1.5, respectively. The fitted curves are compared with the test data in Figure 6. The effect of the coarse particles on damping ratio curves is represented well by the normalized shear modulus at different coarse aggregate contents.

3.4. Cyclic Shear Strength. The cyclic axial strain (ε_d) time histories for the specimens with the coarse aggregate content of 0 and 32% (CSR = 0.15 and $\sigma_0' = 100$ kPa) are shown in Figure 7. As noted, the addition of coarse particles led to decreasing strain amplitude at a given loading cycle number. But the strain development characteristic did not change significantly. The induced strain amplitudes of both specimens were found to constantly increase with the loading cycles and a progressive failure type was observed. Therefore, it is concluded that the clay specimens mixed with the adopted coarse aggregate contents are unlikely to liquefy.

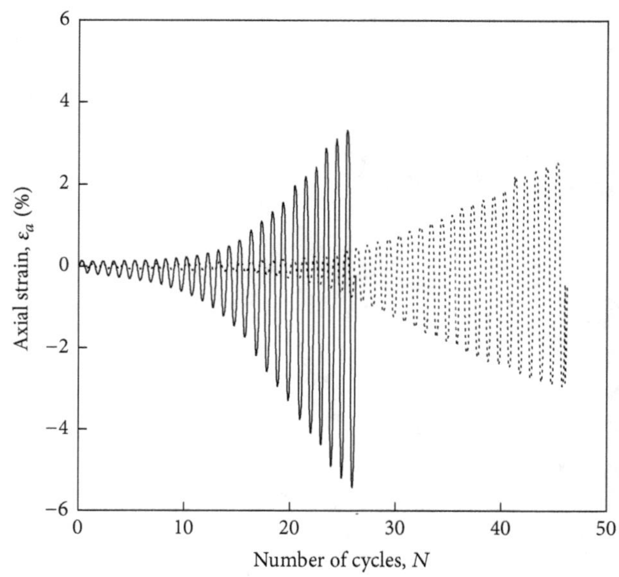

Figure 7: Time history of axial strain (CSR = 0.15; $\sigma_0' = 100$ kPa).

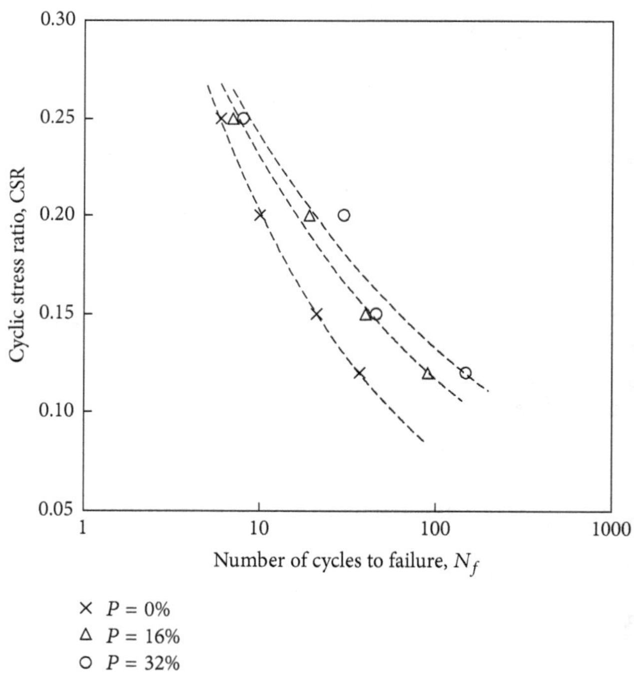

Figure 8: Relationship between cyclic stress ratio and number of cycles at failure.

The relationships between the cyclic stress ratio (CSR) and the number of cycles corresponding to a double amplitude axial strain of 5% (N_f) are shown in Figure 8. It is found that the value of N_f increased significantly with the coarse aggregate content, especially at small values of CSR. For larger CSR values, the slippages at the clay-aggregate interfaces were more easy to develop. Hence, the effects of the coarse aggregate content on the resistance to deformation became less pronounced.

Usually, the cyclic shear strength is defined as the value of CSR corresponding to a specific failure number of loading cycles [31]. In this study, $N_f = 20$ is used to determine the cyclic shear strength, which corresponds to an earthquake magnitude of 7.5. Relationships between the cyclic shear strength CSR_{20} and the coarse aggregate content are plotted in Figure 9. It is clear that CSR_{20} increased rapidly with increasing coarse aggregate contents, though the dry density of the clay fraction was kept constant. The values of CSR_{20} were 0.15, 0.19, and 0.21 for the coarse aggregate content of 0%, 16%, and 32%, respectively.

3.5. Excess Pore Water Pressure. Figure 10 presents the time histories of the excess pore water pressure ratio r_u (defined as the ratio of the excess pore water pressure to the initial effective confining stress) of the specimens with the coarse aggregate content of 0 and 32% (CSR = 0.15 and σ_0' =

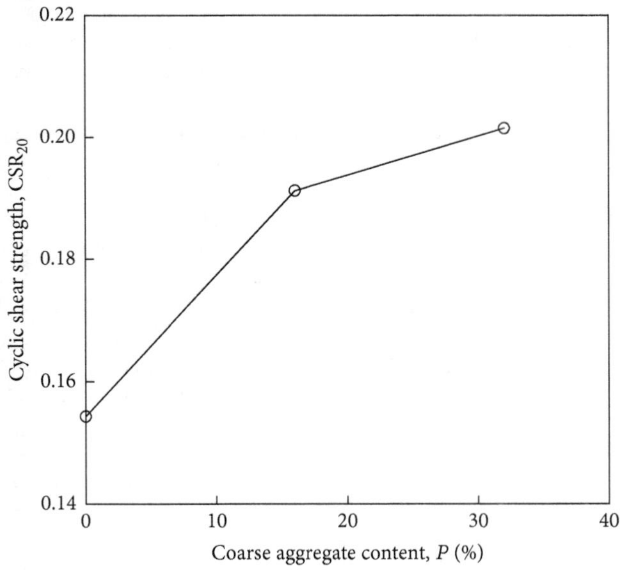

Figure 9: Relationship between cyclic shear strength and coarse aggregate content.

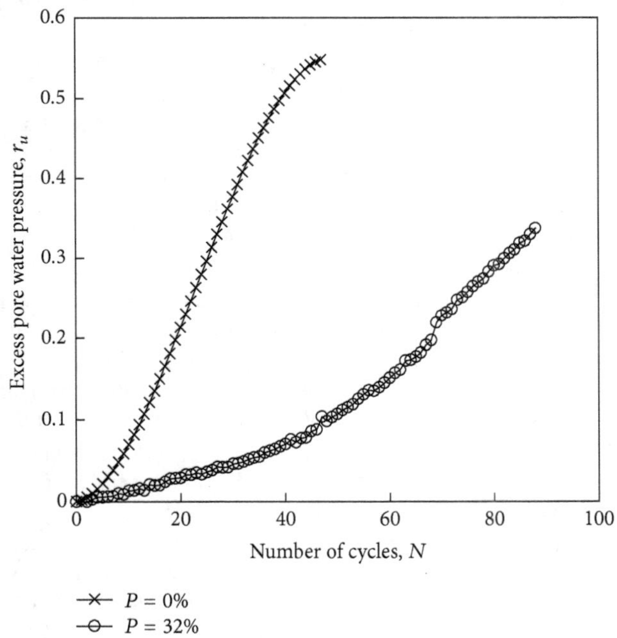

Figure 10: Time history of excess pore water pressure ratio (CSR = 0.15; $\sigma_0' = 100\,\text{kPa}$).

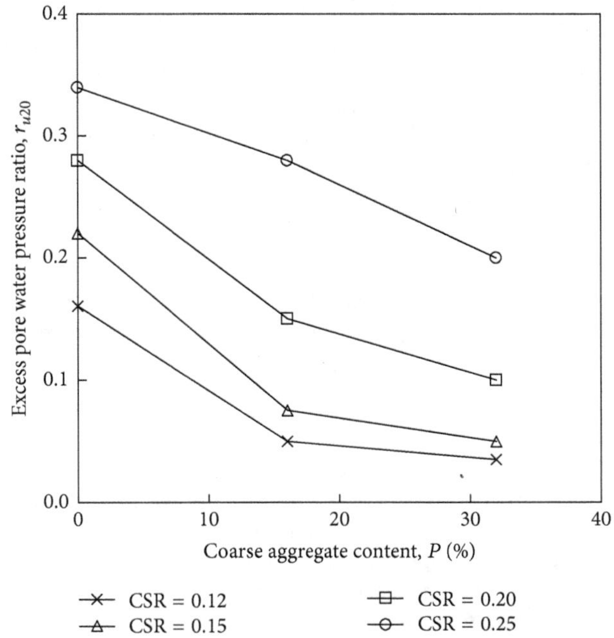

Figure 11: Relationship between excess pore water pressure ratio and coarse aggregate content.

100 kPa). The excess pore pressures at the end of each cycle were given in the figure, such that the cyclic pore pressures due to variations of the mean stress are not included. As can be seen, the pore water pressures generated during the cyclic loading were not sufficient to trigger liquefaction for either the pure clay or the clay-aggregate mixture. For each coarse aggregate content, the rates of pore water pressure generation are approximately constant, which is consistent with the progressive-type failure behavior. Similar trends were also observed for the other specimens. It is also noted that the pore water pressure generated more slowly for the specimen mixed with coarse particles. It suggests that the soils with high coarse aggregate contents tend to be less contractive, causing a slower rate of excess pore water pressure generation.

The effects of the coarse aggregate content on the excess pore water pressure ratio are clearly shown in Figure 11, where the test results are presented in terms of r_{u20} (pore water pressure ratio for the 20th cycle) against the coarse aggregate content. It is obvious that the influence of the coarse aggregate content on r_{u20} also depends on the applied cyclic shear stress. At low cyclic shear stress (CSR $\leq$ 0.2), an increase of the coarse aggregate content led to a significant decrease in the generated excess pore water pressure. At large cyclic shear stress (CSR = 0.25), the soil skeleton consisting of clay and coarse particles was susceptible to collapse after a few loading cycles; then the pore water pressure generation was dominated by the clay fraction. Therefore, the same variation in the coarse aggregate content resulted in a small reduction in r_{u20}.

4. Conclusions

To improve the understanding of the dynamic behavior of clay-aggregate mixtures, series of stress controlled cyclic triaxial tests were performed on clay specimens mixed with different amounts of coarse particles. The main findings were as follows:

(1) The initial shear modulus increased with the coarse aggregate content, even though the dry density of the clay fraction was kept constant. The initial shear modulus of clay-aggregate mixtures can be estimated in terms of the global void ratio and the confining stress.

(2) The normalized shear modulus decreased and damping ratio increased with the increase in the coarse aggregate content. However, the effect of coarse particles was negligible at high shear stain levels. The empirical relationships to determine the strain-dependent normalized shear modulus and damping ratio were proposed.

(3) Adding coarse particles to a clay matrix led to increasing of the cyclic shear strength. At small cyclic stress ratios, the number of loading cycles to failure increased significantly with the coarse aggregate content. For large cyclic stress ratios, the effects of the coarse aggregate content on the resistance to deformation became less pronounced.

(4) The clays with high coarse aggregate contents tended to be less contractive, causing a slower rate of excess pore water pressure generation during the undrained cyclic loading. The decrease in the excess pore water pressure is especially significant at low cyclic shear stress levels.

Competing Interests

The authors declare that there is no conflict of interests regarding the publication of this paper.

Acknowledgments

The authors acknowledge support from the Natural Science Foundation of Jiangsu Province (no. BK20141279) and the Qing Lan Project by the Jiangsu Province Government.

References

[1] S. Thevanayagam, K. Ravishankar, and S. Mohan, “Effects of fines on monotonic undrained shear strength of sandy soils,” *Geotechnical Testing Journal*, vol. 20, no. 4, pp. 394–406, 1997.

[2] L.-K. Chien, Y.-N. Oh, and C.-H. Chang, “Effects of fines content on liquefaction strength and dynamic settlement of reclaimed soil,” *Canadian Geotechnical Journal*, vol. 39, no. 1, pp. 254–265, 2002.

[3] K. Kuwano, H. Nakazawa, K. Sugihara, and H. Yabe, “Undrained cyclic shear strength of sand containing fines,” in *Proceedings of the 11th Japan National Conference on Geotechnical Engineering*, vol. 1, pp. 993–994, 1996.

[4] E. Ovando-Shelley and B. E. Pérez, “Undrained behaviour of clayey sands in load controlled triaxial tests,” *Geotechnique*, vol. 47, no. 1, pp. 97–111, 1997.

[5] K. Ishihara, “Liquefaction and flow failure during earthquakes,” *Geotechnique*, vol. 43, no. 3, pp. 351–415, 1993.

[6] M. A. A. Beroya, A. Aydin, and R. Katzenbach, “Insight into the effects of clay mineralogy on the cyclic behavior of silt-clay mixtures,” *Engineering Geology*, vol. 106, no. 3-4, pp. 154–162, 2009.

[7] S.-S. Park and Y.-S. Kim, “Liquefaction resistance of sands containing plastic fines with different plasticity,” *Journal of Geotechnical and Geoenvironmental Engineering*, vol. 139, no. 5, pp. 825–830, 2013.

[8] V. C. Xenaki and G. A. Athanasopoulos, “Liquefaction resistance of sand-silt mixtures: an experimental investigation of the effect of fines,” *Soil Dynamics and Earthquake Engineering*, vol. 23, no. 3, pp. 1–12, 2003.

[9] I. B. Gratchev, K. Sassa, V. I. Osipov, and V. N. Sokolov, “The liquefaction of clayey soils under cyclic loading,” *Engineering Geology*, vol. 86, no. 1, pp. 70–84, 2006.

[10] M. Derakhshandi, E. M. Rathje, K. Hazirbaba, and S. M. Mirhosseini, “The effect of plastic fines on the pore pressure generation characteristics of saturated sands,” *Soil Dynamics and Earthquake Engineering*, vol. 28, no. 5, pp. 376–386, 2008.

[11] U. Kim, D. Kim, and L. Zhuang, “Influence of fines content on the undrained cyclic shear strength of sand-clay mixtures,” *Soil Dynamics and Earthquake Engineering*, vol. 83, pp. 124–134, 2016.

[12] S. Thevanayagam, T. Shenthan, S. Mohan, and J. Liang, “Undrained fragility of clean sands, silty sands, and sandy silts,” *Journal of Geotechnical and Geoenvironmental Engineering*, vol. 128, no. 10, pp. 849–859, 2002.

[13] J. F. Peters and E. S. Berney, “Percolation threshold of sand-clay binary mixtures,” *Journal of Geotechnical and Geoenvironmental Engineering*, vol. 136, no. 2, pp. 310–318, 2010.

[14] A. F. Cabalar and W. S. Mustafa, “Behaviour of sand–clay mixtures for road pavement subgrade,” *International Journal of Pavement Engineering*, pp. 1–13, 2015.

[15] A. F. Cabalar and W. S. Mustafa, “Fall cone tests on clay-sand mixtures,” *Engineering Geology*, vol. 192, pp. 154–165, 2015.

[16] R. Salgado, P. Bandini, and A. Karim, “Shear strength and stiffness of silty sand,” *Journal of Geotechnical and Geoenvironmental Engineering*, vol. 126, no. 5, pp. 451–462, 2000.

[17] A. F. Cabalar, “The effects of fines on the behaviour of a sand mixture,” *Geotechnical and Geological Engineering*, vol. 29, no. 1, pp. 91–100, 2011.

[18] A. F. Cabalar and R. A. Hasan, “Compressional behaviour of various size/shape sand-clay mixtures with different pore fluids,” *Engineering Geology*, vol. 164, pp. 36–49, 2013.

[19] J. Chu and W. K. Leong, “Effect of fines on instability behaviour of loose sand,” *Geotechnique*, vol. 52, no. 10, pp. 751–755, 2002.

[20] S. Thevanayagam, “Intergrain contact density indices for granular mixes-I: framework,” *Earthquake Engineering and Engineering Vibration*, vol. 6, no. 2, pp. 123–134, 2007.

[21] S. Yagiz, “Brief note on the influence of shape and percentage of gravel on the shear strength of sand and gravel mixtures,” *Bulletin of Engineering Geology and the Environment*, vol. 60, no. 4, pp. 321–323, 2001.

[22] H. F. Zhao, L. M. Zhang, and D. S. Chang, “Behavior of coarse widely graded soils under low confining pressures,” *Journal of Geotechnical and Geoenvironmental Engineering*, vol. 139, no. 1, pp. 35–48, 2013.

[23] K. Fei, “Experimental study of the mechanical behavior of clay-aggregate mixtures,” *Engineering Geology*, vol. 210, pp. 1–9, 2016.

[24] S. Yamada, M. Hyodo, R. P. Orense, and S. V. Dinesh, “Initial shear modulus of remolded sand-clay mixtures,” *Journal of Geotechnical and Geoenvironmental Engineering*, vol. 134, no. 7, pp. 960–971, 2008.

[25] S. Yamada, M. Hyodo, R. P. Orense, S. V. Dinesh, and T. Hyodo, “Strain-dependent dynamic properties of remolded sand-clay mixtures,” *Journal of Geotechnical and Geoenvironmental Engineering*, vol. 134, no. 7, pp. 972–981, 2008.

[26] A. Shafiee and R. Ghate, “Shear modulus and damping ratio in aggregate-clay mixtures: an experimental study versus ANNs

prediction," *Journal of Applied Sciences*, vol. 8, no. 18, pp. 3068–3082, 2008.

[27] M. Meidani, A. Shafiee, G. Habibagahi et al., "Granule shape effect on the shear modulus and damping ratio of mixed gravel and clay," *Iranian Journal of Science and Technology, Transaction B: Engineering*, vol. 32, no. 5, pp. 501–518, 2008.

[28] M. M. Monkul and G. Ozden, "Compressional behavior of clayey sand and transition fines content," *Engineering Geology*, vol. 89, no. 3-4, pp. 195–205, 2007.

[29] R. K. McGeary, "Mechanical packing of spherical particles," *Journal of the American Ceramic Society*, vol. 44, no. 10, pp. 513–522, 1961.

[30] K. Ishihara, *Soil Behaviour in Earthquake Geotechnics*, Clarendon Press, Oxford University Press, 1996.

[31] B. Seed and K. L. Lee, "Liquefaction of saturated sands during cyclic loading," *Journal of Soil Mechanics & Foundations*, vol. 92, no. 6, pp. 105–134, 1966.

15

Experimental Study on Sedimentary Rock's Dynamic Characteristics under Creep State Using a New Type of Testing Equipment

Wanpeng Huang,[1,2] Wenbin Xing,[2] Shaojie Chen,[1,2] Yang Liu,[3] and Kai Wu[2]

[1]*State Key Laboratory of Mining Disaster Prevention and Control, Shandong University of Science and Technology, Qingdao 266590, China*
[2]*College of Mining and Safety Engineering, Shandong University of Science and Technology, Qingdao 266590, China*
[3]*College of Mechanical and Electronic Engineering, Shandong University of Science and Technology, Qingdao 266590, China*

Correspondence should be addressed to Wanpeng Huang; hwp20033@sdust.edu.cn and Wenbin Xing; 913231434@qq.com

Academic Editor: Michael J. Schütze

Under high stress, rock in the creep state is vulnerable to external impact load, causing the irreversible perturbation deformation of rock or even sudden failure. To explore the dynamic characteristics of sedimentary rock in the creep state, a siltstone specimen was experimentally studied using a new type of experimental system for rock creep perturbation. Compared to the currently available equipment for studying rock dynamics, this new type of experimental system provides a long-term and stable high static stress to maintain the creep state of rock specimen. This is independent of power supply because it provides static stress by gravity load. Moreover, the equipment provides a dynamic impact load through the free fall of impact weights. This study shows that the perturbation deformation of sedimentary rock increased in two stages: decay phase and sustained development phase. When the static stress reached up to ~85% of the rock's ultimate strength and the axial strain reached up to 80% of the ultimate failure strain, the rock became sensitive to impact load. This static stress level is basically the same as its long-term strength. With increasing impact strength, the increment curve of the rock creep perturbation deformation was transformed from the decay phase to the sustained development phase in advance, making the rock sensitive to external perturbation at a low static loading stress level.

1. Introduction

With increasing mining depth, coal mines face a higher in situ stress field environment. The concentrated stress caused by roadway excavation will make the force condition of the surrounding rock reach or exceed the long-term strength of rock. The rock mass shows a continuous creep state. The stability of deep roadway's surrounding rock in the creep state is very poor. Moreover, its creep deformation often lasts for several months or even years and also is easily affected by external dynamic perturbation loading. Such perturbations include the mining activities near the coal seams, blasting vibration in underground, strata fracture of roof, and impact pressure caused by rock burst [1–3]. The external perturbation will cause a sharp increase in the creep deformation amount of surrounding rock during creep and even sudden failure, resulting in a large deformable instability of deep roadway and other dynamic disaster phenomena. Therefore, studies on the dynamic response characteristics of a rock medium under impact perturbation loading in the rheological state have important theoretical and practical significance for the safe mining of deep coal mines.

Currently, experimental studies on rock dynamics are mostly based on impact experiments under the rock uniaxial or triaxial static compression using devices such as split-Hopkinson pressure bar (SHPB), light gas gun, and drop hammer [4–6]. In the SHPB system, the rock specimen is clamped by two stress transmission bars. The static stress on specimen is created by a hydraulic system, and the dynamic load is created by a striker launcher to impact the stress transmission bars. In 1968, Kumar first tested the dynamic strength of a rock using SHPB [7]. Demirdag et al. studied the dynamic mechanical behavior of some carbonate rocks using a classical servohydraulic testing machine and an SHPB testing system [8]. Moreover, some modifications to SHPB were carried out, allowing it to be more adaptive for certain rock dynamic experiments. For example, Li et al. studied the dynamic characteristics of granite subjected to an intermediate loading rate and proposed a new testing technique involving the modification of SHPB so that the test specimen is subjected to coupled axial static prestress, axial impact loading, and optional confining pressure [9–13]. The static stress system of a drop weight machine is similar to SHPB. The main difference is the loading method of dynamic stress. It uses a heavy hammer that drops from a certain height to impact the piston plate. The shock wave created by the impact is transmitted to a specimen. The drop weight machine was usually used in the dynamic mechanical experiments of rock or concrete. Okubo et al. conducted large-scale penetration tests of rock samples using a wedge-type drop hammer in 1997 [14]. In the recent years, drop weight machine has been used to study the dynamic fracture characteristic of rock and relationship between the impact energy, rock strength, and so forth [15–17].

According to the above-mentioned literature analysis, the static stress loading system of the existing experimental equipment for rock dynamics is similar to the equipment used in rock experiments such as MTS. It uses a hydraulic system to provide static stress. This loading mode is stable but suffers from the fact that the stress system must be preserved using a power source. Therefore, this equipment is only suitable for the short-time experiments of rock dynamics. The rock impact experiment in the creep state belongs to the category of long-term dynamic experiment. The static loading system of the equipment should be able to provide a long-term (e.g., a few days, months, or even years) and stable high static stress to maintain the stable creep state of rocks. At the same time, the experimental system can also provide a dynamic impact load to a rock specimen to study the dynamic response of rock under creep condition. According to literature, currently, a few studies have been reported on the dynamic mechanical characteristics of rock creep disturbance, and the experimental equipment needed for this type of study is limited.

To conduct this type of studies, we independently developed a new type of experimental system for rock creep disturbance. Further, a creep perturbation experiment of hard siltstone specimens was also carried out using the new experimental system. A preliminary study on the creep perturbation deformation law of rock specimens was conducted, and the rock's antiperturbation strength (the static stress level at which the rock becomes sensitive to an impact load) was investigated, thus laying a theoretical foundation for the stability control of deep rock engineering under perturbation effect.

Based on the preliminary study, we define the mechanical phenomenon as "rock creep disturbance effect"; that is, the rock can produce the corresponding creep strain increment after the external impact load is applied. It belongs to the category of rock rheological mechanics and rock dynamics and reflects the creep deformation rule of the rock under the creep condition under the influence of an external impact load. This is a new research direction in rock creep.

2. Experimental Equipment

To study the dynamic characteristics of a rock in the rheological state, a new type of equipment known as a testing system for rock creep perturbation effect (RRTS) was developed. This new equipment was able to apply a long constant axial pressure to a rock specimen by gravity loads that increased to two grades. The equipment, independent of power supply, can produce long-term and stable loads on samples. It produced perturbation loads by the shock of steel rings to the upper bearing plate. It is suitable for long-term static creep tests and RRTS experiments under uniaxial and conventional triaxial conditions [18]. The testing equipment is shown in Figure 1. It consists of two parts: the testing machine and data monitoring system.

2.1. Force-Amplification Structure and Principle of Stress Loading System. The stress applied to a rock specimen originates from the gravity load of circular ring-shaped weight made of cast iron. The weights were 1 kg, 3 kg, 5 kg, and 10 kg. The equipment's two grades of force amplification can increase the original gravity load by 60–100 times. The first grade of force amplification is achieved by a mechanical gear transmission structure, and the second grade of force amplification is achieved by a hydraulic transmission structure.

2.1.1. First Grade of Mechanical Force Amplification. This structure (Figure 2) includes a transmitted rotary table, shaft, and transmitted gear. The edge of the rotary table had chains, and the weights were hung on the chains to impose gravity load on the rotary table. Then, the gravity load was delivered to the transmitted gear connected to a transmitted rack. Because of the size differences in the rotary table and gear, the gravity load achieves first-grade amplification.

According to moment equilibrium conditions, the force in the first grade is amplified as follows:

$$F_1 = \frac{R_0}{R_1} W_0, \tag{1}$$

where F_1 is the amplified force applied on the small cylinder, W_0 is the gravity load and also the total weight of weights, and R_0 and R_1 are the radii of the transmitted rotary table and gear, respectively.

2.1.2. Second Grade of Hydraulic Force Amplification. This structure (Figure 3) includes a transmitted rack, a small cylinder, a large cylinder, and an oil-transferring pipeline. Force F_1 is directly pressed on the pressure axis of the small cylinder using the transmitted rack. Then, the force is transferred to the large cylinder through the oil-transferring pipeline. The piston on the top of the large cylinder is pushed upwards to provide load to the rock specimen. Because

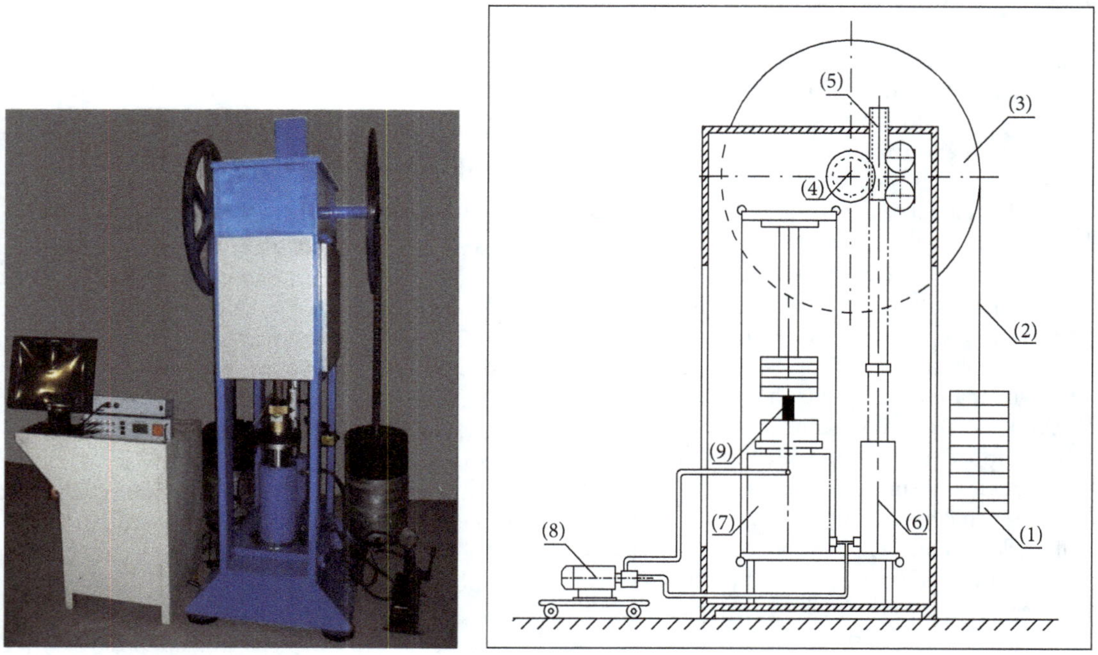

(a) Picture of machine

(b) Overall structure

FIGURE 1: Structure of testing equipment.

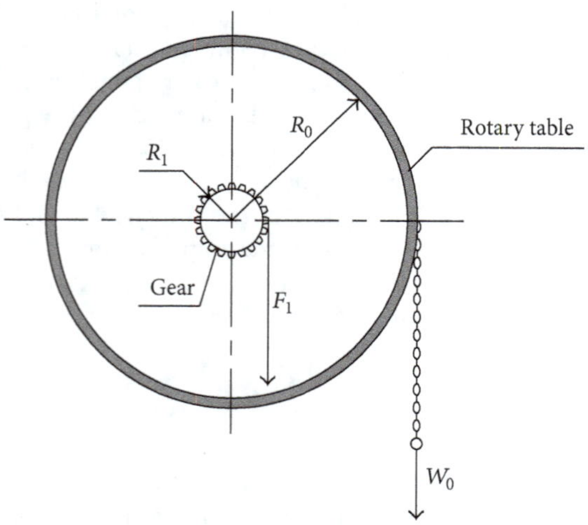

FIGURE 2: Schematic diagram of mechanical force amplification.

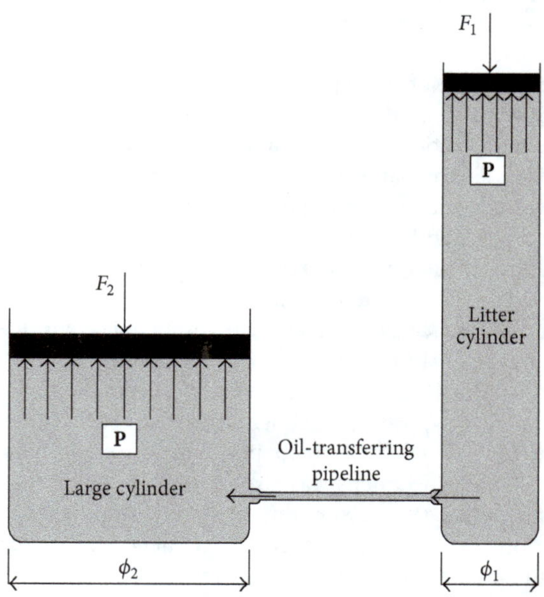

FIGURE 3: Schematic diagram of hydraulic force amplification.

of the size differences of the cylinders, the load achieves second-grade amplification. According to static equilibrium conditions, the force amplified in the second grade can be expressed as follows:

$$F_2 = \left(\frac{\phi_2}{\phi_1}\right)^2 \cdot F_1, \quad (2)$$

where ϕ_1 and ϕ_2 are the diameters of the large and small cylinders, respectively, and F_2 is the force applied by the large cylinder and also the stress load on the rock specimen.

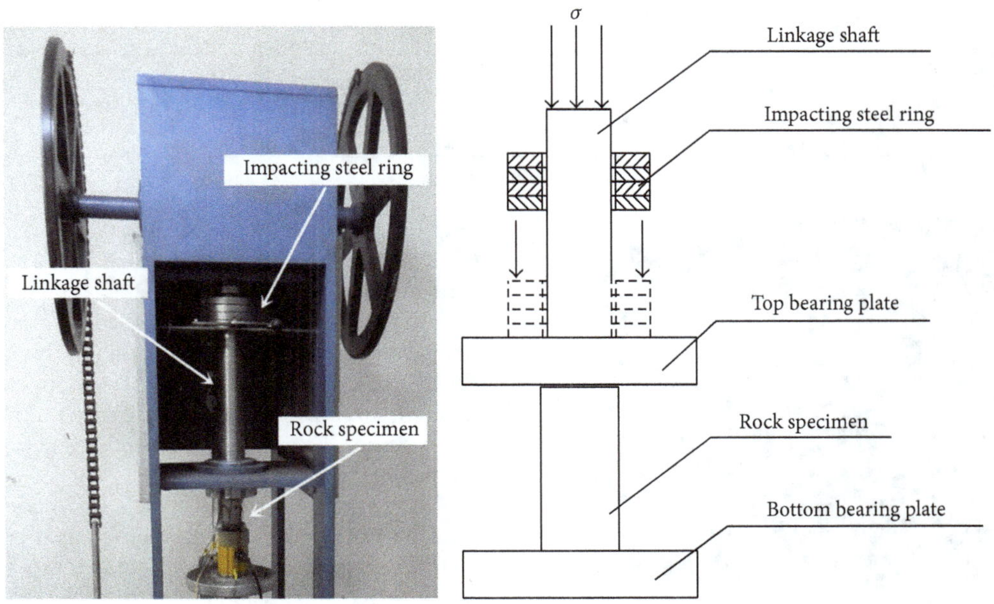

FIGURE 4: Application mode structure of perturbation loading.

Combining formulas (1) and (2), the stress load through the two grades of force amplification can be expressed as follows:

$$F_2 = \frac{R_0}{R_1}\left(\frac{\phi_2}{\phi_1}\right)^2 W_0. \tag{3}$$

2.2. Loading Mode of Impact Perturbation. The load infliction of the external perturbation loading occurs through the free fall of the impact weights. The weights impact the top bearing plate located on the upper part of the equipment as shown in Figure 4. Then, the impact load can be transferred to creep rock specimen.

In the concrete method, steel-ring weights are present around the linkage of equipment's top bearing plate. Each steel ring has three different weights of 3 kg, 5 kg, and 10 kg. Therefore, different numbers of steel rings can be combined to different impact blocks. The impact block falls down from different relative heights of 100–400 mm, thus exerting various perturbation loads on the rock specimen. After the rock specimen's deformation is in the creep stage at a certain level of static stress, the steel rings are lifted up to a certain relative height and then allowed to fall freely. When the steel rings fall on the top bearing plate, the impact load generated is transferred to the rock specimen (Figure 4).

According to the impulse and energy theorem, if the energy loss is ignored, the geopotential energy of impact weights can be completely transformed by impact momentum to the rock specimen. Thus, the impact momentum of perturbed load imposed on rock specimen ΔW can be calculated as follows:

$$\Delta W = F \cdot \Delta t = m \cdot v = m\sqrt{2gh}, \tag{4}$$

where F is the average impact force imposed on the specimen, N; Δt is the action time period of impact force, s; m is the weight of the steel ring, kg; v is the impact velocity when the steel ring drops to the bearing plate; g is the acceleration of gravity: $g = 9.8\ \mathrm{m/s^2}$; and h is the falling height, m. According to theoretical calculation, the impact momenta imposed on the rock specimens of 5 kg and 10 kg in weight were 7 N·s and 14 N·s, respectively.

3. Process and Method of Experiment

3.1. Rock Specimens. Gray-black siltstones collected from −1200 m horizontal strata in the Xinwen coalfield in Shandong Province were selected as the rock specimen in the experiment. After the processing, the specimens were screened for ultrasonic testing (UT). According to the UT results, the longitudinal wave velocity of siltstone specimens was 3.48 km/s in average, and the dynamic modulus was 30.8 GPa in average. The crosswise and axial strain gages were stuck on specimens to measure the transverse and axial strains of rock specimens in the loading procedure. During the experiment, the JC-II load transducer was padded under the rock specimen to record the actual capacity of the load, as shown in Figure 5.

3.2. Confirmation of Long-Term Strength. This is an impact perturbation experiment in the dynamic state under creep state of the rock. According to previous studies, in the long-term rock creep experiments, the creep process can be divided into three creep stages (deceleration, stable, and accelerated creep stage). Also, the sensitivity of rock to external disturbances at different creep stages is different. The threshold stress of the rock entering the stable creep stage is defined as the rock's long-term strength. To study the dynamic response characteristics of rock to external disturbance loads in different creep stages, first the long-term strength was determined by conducting a static creep

TABLE 1: Perturbation parameters of two impact weights.

Perturbation weight (kg)	Force peak (N)	Effective impact time (s)	Total action time (s)
5	450	0.015	0.028
10	960	0.013	0.03

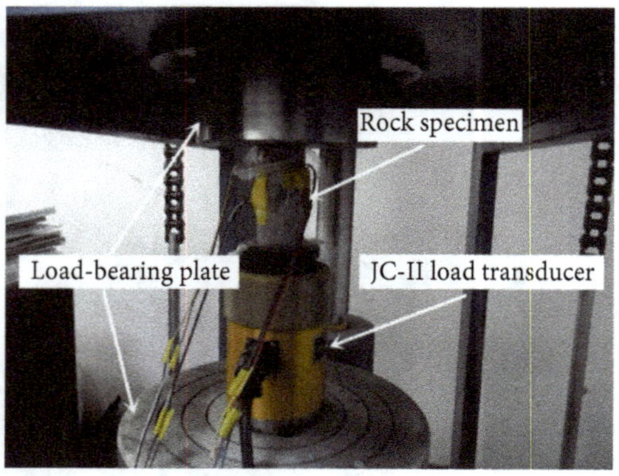

FIGURE 5: JC-II load transducer.

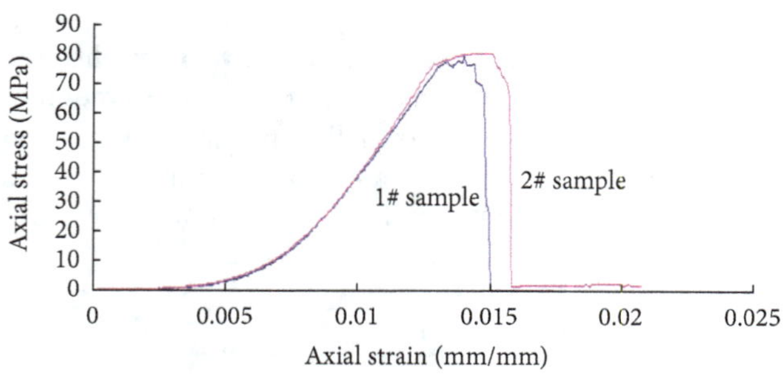

FIGURE 6: Stress-strain curve of sedimentary rock under uniaxial press.

experiment to estimate the capacity of static loading stress when the rocks are in the stable creep state. Then, based on the standard of long-term strength of rocks, each level of static loading stress that impacts loading was exerted.

Before the static creep test of rock, the conventional uniaxial compression experiments (short-term compression experiment and the mechanical parameters obtained are not related to time) of two rock samples were carried out using an MTS servo-controlled rock mechanics system. Furthermore, the stress-strain curves obtained for sedimentary rock are shown in Figure 6. According to the analysis of testing dates, the ultimate strength and strain of two samples are close. The ultimate strengths (the maximum stress in the case of rock failure in a short-term compression experiment) of 1# and 2# samples were 81.17 MPa and 82.88 MPa, respectively. The ultimate strength of sedimentary rock was about 82.03 MPa. When the ultimate strength of a rock is determined, the stress rating in the creep experiment can be determined by taking the ultimate strength as the extreme.

According to a previous experimental study on rock mechanics, under the normal uniaxial compression condition, the range of the statistics of the specific value of long-term and ultimate strengths of soft rock was about 60–80%; for a medium and hard rock, the specific value was generally between 85 and 95% [2, 19–23].

First, the uniaxial compressive creep test of the three specimens was carried out before starting the impact perturbation experiment. The static stress in the experiment was pushed in different levels by considering the rock's ultimate strength as the extreme. The different stages of creep curves of 3# and 4# are shown in Figure 7.

The long-term strength was confirmed using the creep transition method by fitting the obtained creep curve. This method considers the static stress level, where the rock sample's creep rate is about 0 as the rock's long-term strength [24]. As shown in Figure 7, according to the analysis of fitting to rock's creep curve, the creep rates of both samples tend to be 0 at the first two static stress levels. After the third stress level, the curve started to upturn. Furthermore, the fitting results show that the creep rate is above 0 at these stages. The result of third sample shows a similar trend. According to comprehensive analysis, the sedimentary rock's long-term strength is about 69.7–71.3 MPa. Average is 70.5 MPa. The specific rate of long-term and ultimate strength was 86%.

3.3. Experimental Method for Rock Creep Perturbation Effect. Before starting the experiment, the impact capacities of selected weights of 5 kg and 10 kg were measured with a dropping height of 10 cm. The vibration oscillogram of impact force (diagram *F-t*) obtained from the test is shown in Figure 8. According to the testing results, the total action time of impact wave was approximately 0.250.3 s, and the vibration wave demonstrated good integrity and transitivity. The time interval between two adjacent impacts is about 120 s. With the short duration of each impact wave, it can be considered that the next impact wave will not influence the above impact wave, and each impact is independent. The tested impact force peak of 5 kg weight imposed on the rock sample was about 450 N, and the converted amplitude of stress was about 0.23 MPa. The effective impact time Δt (the effective time-width of impact wave) was about 0.015 s. The impact force peak of 10 kg weight imposed on the rock sample was about 960 N, and the converted amplitude of stress was about 0.49 MPa. The effective impact time Δt was about 0.013 s. Considering the force peak as the average action force during the effective impact time Δt, the impact momentum of each impact was about 6.75 N·s (5 kg weight) and 12.48 N·s (10 kg weight). This result is quite similar to the theoretical calculation. The perturbation parameters of the combination of two impact weights are shown in Table 1.

The rock creep perturbation experiment was carried out under different static loading stress levels, and 80% of the long-term strength of the specimen was the initial static loading stress level. The experiment was divided into six

TABLE 2: Actual static stressin each level.

	Stress level 1	Stress level 2	Stress level 3	Stress level 4	Stress level 5	Stress level 6
Load application (MPa)	56.5	59.8	64.1	66.8	70	79.6
Ratio to long-term strength (%)	80.1%	84.8%	90.9%	94.7%	99.3%	112.9%
Ratio to ultimate strength (%)	68.9%	72.9%	78.2%	81.5%	85.4%	97.1%
Single impact load (MPa)	0.23/0.49	0.23/0.49	0.23/0.49	0.23/0.49	0.23/0.49	0.23/0.49

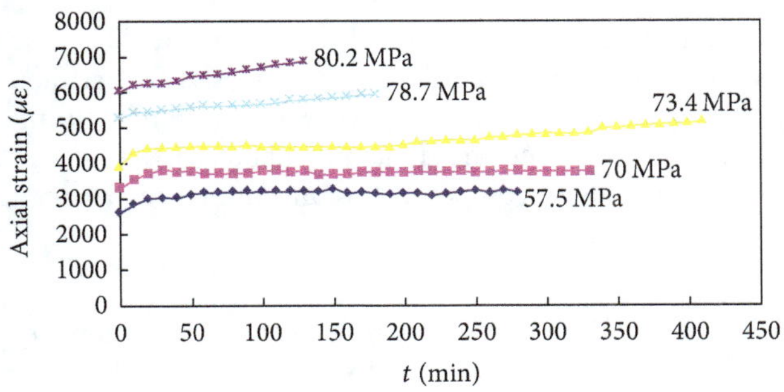

(a) 3# sample

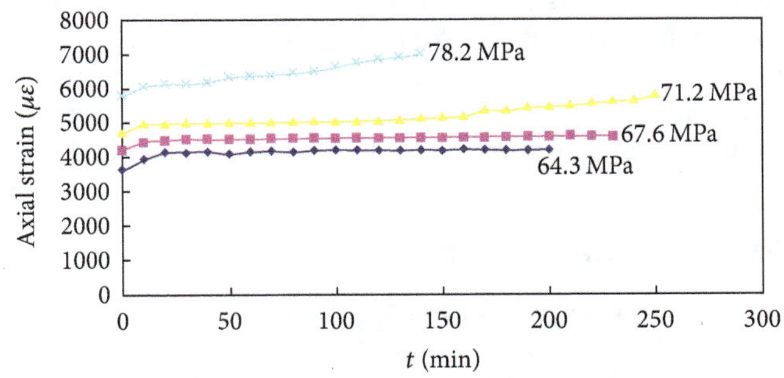

(b) 4# sample

FIGURE 7: Different stages of creep curve of 3# and 4# samples.

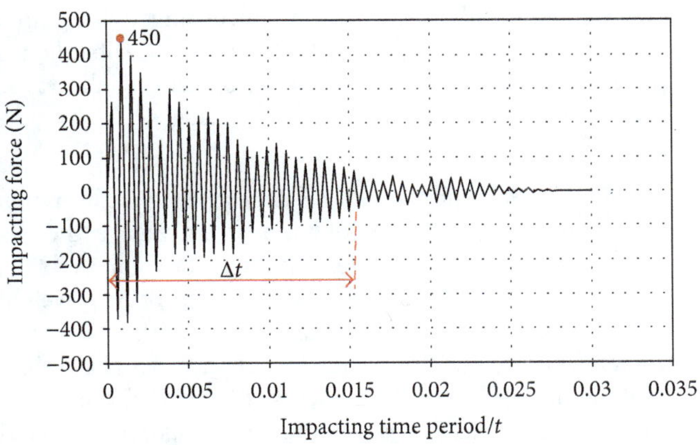

(a) Impact weight of 5 kg

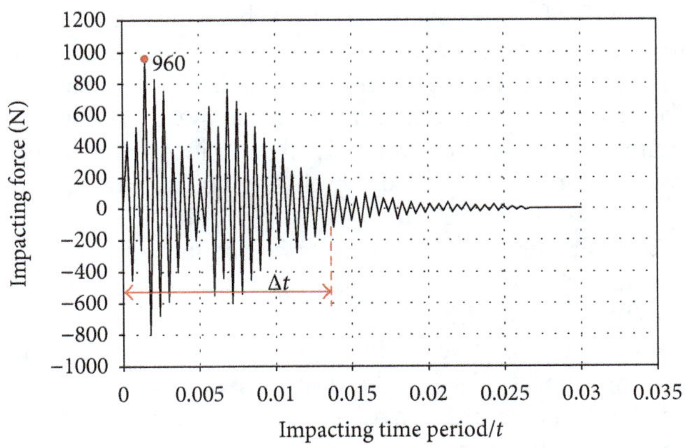

(b) Impact weight of 10 kg

FIGURE 8: Vibration oscillogram of impact force (force-time diagram F-t).

levels. Taking 2# specimen in the first group as an example, the imposed static loading stress of each level is shown in Table 2. According to previous tests, the creep deforming rate was unstable in the initial stage of every level of static load. If the static stress was maintained for 24 h or longer, the rock sample's creep rate became more stable. Therefore, the loading of static stress in each level lasted for about 48 h. After the creep deformation of the rock specimen was stable, the impact perturbation experiment was started. The selected impact weights were divided into two types, that is, 5 kg and 10 kg. The specimens were used to test the perturbation deformation regularity of hard siltstone specimens under different impact loadings. Then, the dynamic response of the specimens in the creep perturbation state was analyzed and summarized.

During the experiment, the continuous impact of several times was taken as one perturbation. After completing the impact for each time, the related strain amount of the specimens was recorded. The experiment was designed to impose two times of perturbation under each static stress level. Two perturbations were carried out at a time interval of 12 h, and a total of 12 perturbations were carried out during the entire experiment (under the sixth stress level, the specimen was destroyed when the first perturbation was imposed; therefore, in fact a total of 11 perturbations were conducted). The experiment was started with multistage loading. First, the loading pressure reached the first level of static loading stress. Under this level of stress, the specimen creeps for 48 h. After reaching a stable state of the amount of creep deformation, the first perturbation experiment was conducted (including several times of impact). Then, the homologous accumulative total deformations of perturbation were recorded. The day after the first perturbation, the second impact perturbation was conducted under the same static stress level, and the homologous accumulative total deformation of perturbation was recorded. After the perturbation experiment under first static stress level, the stress of the load reached the second static stress level. The third and fourth perturbation experiments were conducted after the creep was stable. In this manner, the perturbation experiments of the third and fourth static stress levels were conducted, and the accumulative total perturbation deformation was recorded.

4. Experimental Results

Three rock specimens were tested in each group of experiments. The results show a similar regularity. Therefore, the experimental methods and process control are accurate. Moreover, the results also prove the regulation, that is, slight divergence in the creep perturbation deformation on sedimentary rock specimen. According to the experimental data obtained for typical number 2 rock specimen in the two groups of experiments, the perturbation did not obviously affect the lateral creep deformation of rock. Therefore, only the relationship between axial cumulative disturbance deformation and impact numbers was analyzed. The initial dependent variables and cumulative disturbance deformation through various levels of static stress are shown in Table 3; ε_0 is specimen's ultimate strain.

In the first group of experiments with an impact weight of 5 kg, the rock specimen's ultimate strain was 6945 $\mu\varepsilon$ before the rock failure at the last stress level. This is regarded as the specimen's ultimate strain ε_0. In the second group with a loading impact of 10 kg, the ultimate strain was 6782 $\mu\varepsilon$. This should be regarded as the specimen's ultimate strain ε_0 in this group. When the impact loading was 5 kg, the increment curves between perturbation impact times N and axial cumulative disturbance deformation under various stages of static stress level are shown in Figure 9. To analyze more clearly, the above curves were integrated to obtain the entire curve shown in Figure 10. Using the same method, the integrated curve of 10 kg impact loading was obtained as shown in Figure 11.

5. Discussion

5.1. Characteristic of New Equipment. The application of new equipment showed high stability and accuracy in the experimental process. First, the static stress applied to rock specimen up to 56.5–79.6 MPa originating from the gravity load could be enlarged by the equipment's two grades of force-amplification system. That made the specimen remain in a stable creep state. Further, in the 48-hour loading process of each stress level, the static stress remained very stable with almost no reduction according to the results of JC-II load transducer. Thus, the experimental purpose of loading a long-term stable static stress was achieved. During the experiment, the static stress could be increased conveniently from a lower stress level to a higher stress level by increasing the iron weights. Second, in the dynamic impact process of experiment, each impact caused a certain amount of disturbance deformation to the rock specimen in the creep state. Further, the disturbance deformation obviously showed some regularity (Figure 7). This indicates that the impact energy well propagated into the interior of rock specimen and disturbed its creep deformation. Finally, during the experiment, the new equipment could be easily operated because of its simple structure. According to the analysis of experimental process and results, this new type of equipment is more suitable for studying the dynamic characteristics of a rock in the creep state compared to conventional experimental equipment.

5.2. Characteristic Analysis of Rock Creep Perturbation Deformation. The experiment results show that, under a high stress, external perturbation loading was applied to sedimentary rock during the flow deformation, also leading to an irreversible increase in perturbation deformation. This is similar to the perturbation deformation of a soft rock except that the sedimentary rock has a higher level of static stress, sensitive to outside perturbation. According to the curve analysis of the experimental data, the curve of siltstone test specimen cumulative disturbance deformation follows the same regulation, occurring through two stages.

5.2.1. Decay Stage of Cumulative Disturbance Deformation. Rock axial strain is <80% of limit strain before the disturbance at the first four stress levels (56.5 MPa, 59.8 MPa, 64.1 MPa, and 66.8 MPa). With the impact of 5 kg weight, the dynamic load of each impact was 0.23 MPa; the impact momentum was 6.75 N·s. With the impact of 10 kg weight, the dynamic load of each impact was 0.49 MPa; the impact momentum was 12.48 N·s. As shown in Figures 10 and 11, at static stress level 1 (56.5 MPa), when impacting with 5 kg weight, each impact caused the corresponding disturbance deformation in the first seven impacts. The axial cumulative disturbance deformation increased rapidly with a linear growth. To the 8th impact, the cumulative impact loads reached 1.84 MPa, and the cumulative impact momenta reached 54 N·s. The cumulative disturbance deformation started to decline and nearly approached zero; that is, several posterior impacts did not increase the axial deformation any longer, and the rock started to impact hardening. With the impact of 10 kg, even if the impact loads and momentum increased, the disturbance deformation still no longer increased to the seventh impact (the cumulative impact loads reached 3.43 MPa, and the cumulative impact momenta reached 87.36 N·s). Then, under the static stress levels 2, 3, and 4 (59.8 MPa, 64.1 MPa, and 66.8 MPa), the variation in the cumulative disturbance deformation showed a similar regularity. This proves that even when increasing the impact strength, the cumulative disturbance deformation did not

TABLE 3: Experimental data of rock specimen's cumulative disturbance deformation.

Impact loading	Strain ($\mu\varepsilon$)	Perturbation at stress level 1		Perturbation at stress level 2		Perturbation at stress level 3		Perturbation at stress level 4		Perturbation at stress level 5		Perturbation at stress level 6
		1st	2nd	1st	2nd	1st	2nd	1st	2nd	1st	2nd	—
5 kg	ε_r	3810	3928	4185	4298	4564	4700	4780	4934	5487	5866	6625
	$\Delta\varepsilon$	97	99	102	111	116	135	131	133	175	163	304
	$\varepsilon_r/\varepsilon_0$	54.9%	56.6%	60.3%	61.9%	65.7%	67.7%	68.8%	71%	79%	84.5%	95.4%
10 kg	ε_r	3750	3890	4105	4250	4412	4604	4830	5085	5409	5799	6267
	$\Delta\varepsilon$	108	113	128	130	142	150	173	184	238	252	348
	$\varepsilon_r/\varepsilon_0$	55.3%	57.4%	60.5%	62.7%	65.1%	67.9%	71.2%	75.0%	79.8%	85.5%	92.4%

ε_r: rock's creep value before the perturbation; $\Delta\varepsilon$: cumulative disturbance deformation.

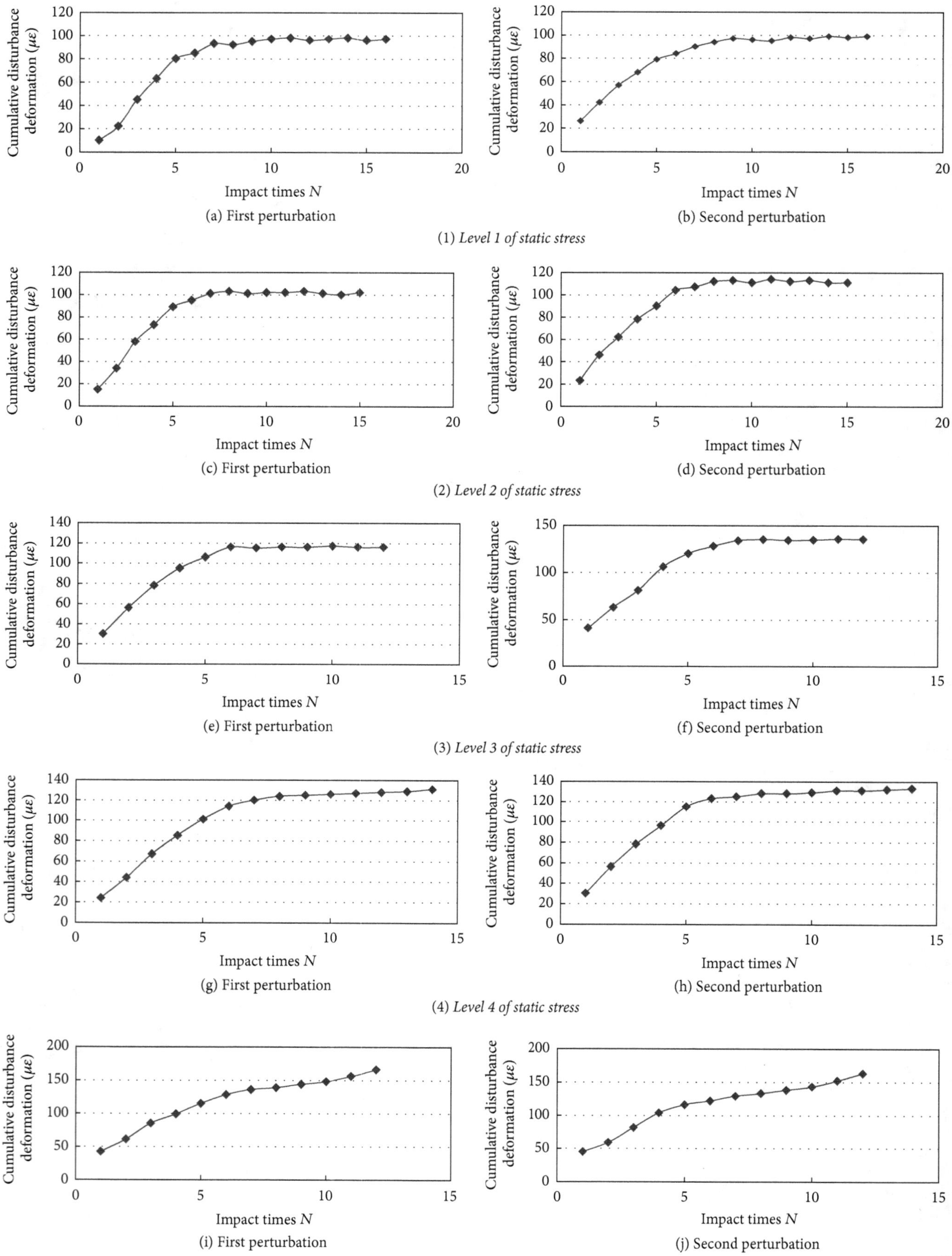

(a) First perturbation
(b) Second perturbation
(1) *Level 1 of static stress*

(c) First perturbation
(d) Second perturbation
(2) *Level 2 of static stress*

(e) First perturbation
(f) Second perturbation
(3) *Level 3 of static stress*

(g) First perturbation
(h) Second perturbation
(4) *Level 4 of static stress*

(i) First perturbation
(j) Second perturbation
(5) *Level 5 of static stress*

FIGURE 9: Continued.

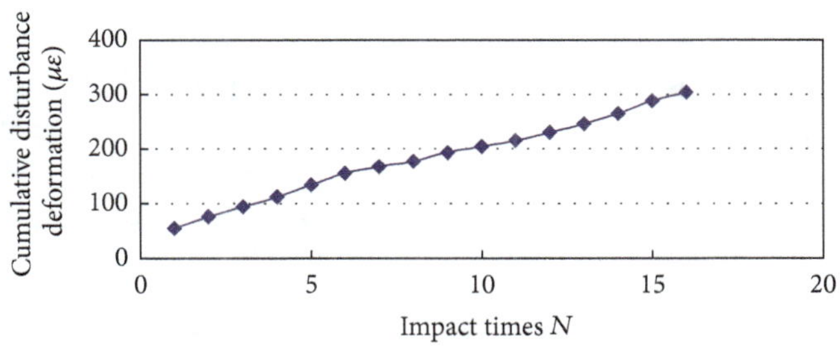

(6) *Level 6 of static stress*

FIGURE 9: Relationship curves between specimen cumulative disturbance deformation and impact times under every stress level when the impact loading was 5 kg.

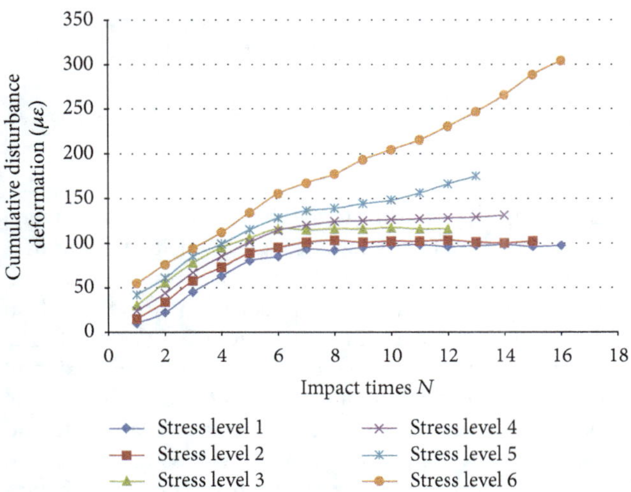

FIGURE 10: Integrated curve when the impact loading was 5 kg.

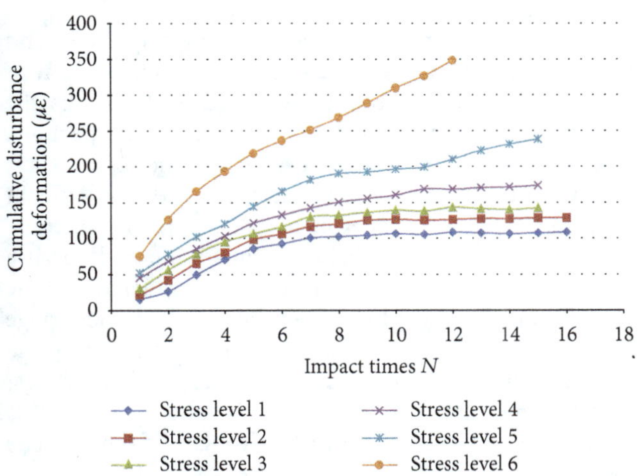

FIGURE 11: Integrated curve when the impact loading was 10 kg.

increase continuously under the low static stress level at the first creep stage; that is, the rocks are insensitive to the external perturbation loading.

Cumulative disturbance deformation is directly proportional to the rock specimen's static stress level. Under the same perturbation strength, the perturbation deformation increased with the increase in the static stress level.

5.2.2. Sustainable Developing Stage of Cumulative Disturbance Deformation. At the fifth static stress level (70 MPa), the applied loading reached 85.4% of the rock's ultimate strength and 99.3% of the long-term strength. According to the previous creep experiment, it can be concluded that the rock specimens entered a stable creep stage. The rock's axial total strain was 80% of the ultimate failure strain before the perturbation. After the perturbation, with the increase in impact times, the cumulative disturbance deformation decreased. In the two groups of experiment, the cumulative disturbance deformation first decreased and then increased. This proves that, with the gradual increase in impact time, the rock's internal injury started to aggravate and transformed from the impact-hardening stage to the injury-failure stage. Moreover, the rock became sensitive to the impact load. When the loading reached the sixth stress level (79.6 MPa), perturbation impact was carried out after two days' creep. The cumulative disturbance deformation generally increased gradually under the linear tendency. At that time, the rock's internal injury further increased, and the development of fracture showed an irreversible growth. The specimens were destroyed after the 16th (cumulative impact load: 3.68 MPa) and 13th (cumulative impact load: 6.37 MPa) impact tests in the two experiments.

5.3. Effect of Impact Loading Strength on Specimens. A comparative analysis of Figures 6 and 7 shows that the higher the impact strength, the higher the perturbation deformation under the same static stress level impacted by different impact weights. This indicates that, under the creep condition, particularly in the stable creep stage, as the external impact strength increased, the damage inside the rock worsened with heavy perturbation deformation. In the experiment of siltstone specimens impacted by 10 kg impact weight under the 4th static stress level, in terms of data analysis, the increment of rock creep perturbation deformation did not remain unchanged but increased slightly. However, it was small and basically remained as a single digit. In the later stage, the curve started to shift upwards with a continuous increasing tendency. In the two experiments, the comparative enlarged curves of the perturbation deformation increment

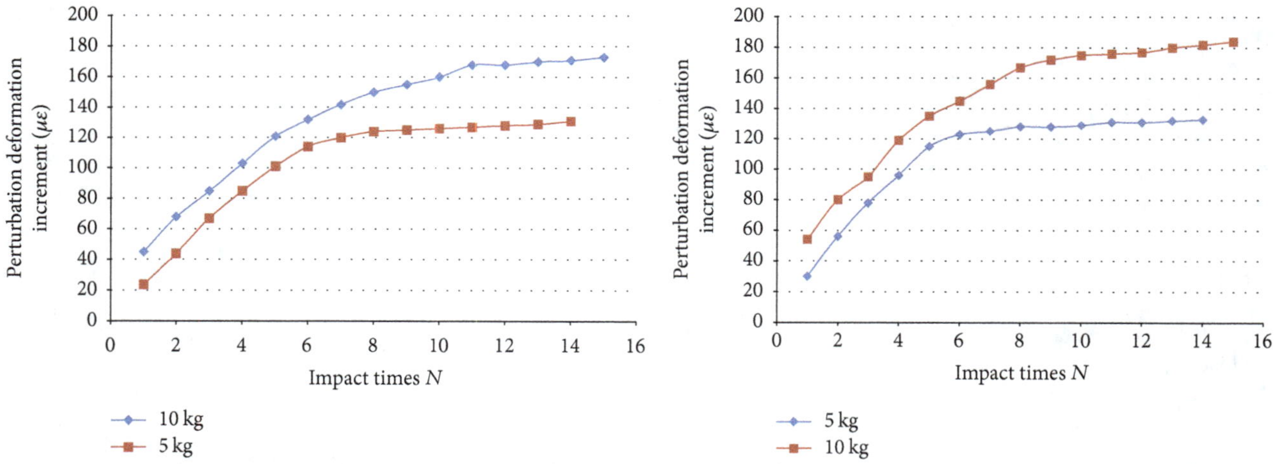

(a) Contrast curve of first perturbation

(b) Contrast curve of second perturbation

FIGURE 12: Comparative curves of perturbation deformation increment under the 4th static loading stress.

under the 4th static stress level are shown in Figure 12. By comparison, the differences between the two experiments are clearly illustrated. In the experiment of 5 kg impact strength, the curve is horizontal with the typical characteristics at the decay stage, whereas under 10 kg impact strength, the curve slightly increased in the early transitional period from the decay stage to the continuous improvement stage. In the later period, although the creep perturbation deformation slightly increased, it still shows a trend of slow development.

Thus, with the increase in impact strength, the time of the increment line of rock creep perturbation transition from the decay stage to the continuous improving stage becomes short. This causes the following problems: the rock becomes sensitive to external perturbation under a low static stress level, and the damage deformation inside the rock is irreversible. Because of the old experimental specimen and lack of number of impact loadings, the differences in rock perturbation deformation are not obvious. Hence, the above experiments are enough for qualitative studies on rock perturbation response to different impact strengths but are not enough for quantitative studies on rock perturbation response. More experiments should be carried out, and the relationship could be studied using statistics.

5.4. Analysis of Rock Antiperturbation Strength. In terms of the analysis of siltstone specimens' dynamic characteristics under impact loading, the reactions of the creep deformation of specimens to the outer perturbation load are different at different static stress levels. Two areas exist: sensitive area and nonsensitive area. When the static stress is less than a certain value, the effects of outer perturbation load on rock creep deformation are not obvious. As the impact time increases, the increment of perturbation deformation will gradually decrease and eventually remain stable along with rock impact hardening; when the static stress is greater than a certain value, the rock creep is sensitive to outer perturbation. Moreover, with the increase in impact times, the increment of perturbation deformation will increase, and the rock is finally damaged. According to the above characteristics, when the rock creep deformation becomes sensitive to impact loading, the threshold value of the static stress is defined as the antiperturbation strength of rock σ_r, which is the critical stress of sensitive area and nonsensitive area as well. The antiperturbation strength of rock is shown in the variation curve of perturbation deformation increment, that is, the stress value of the inflection point at the decay stage and continuously developing stage. When (stress on rock) $\sigma < \sigma_r$, the increment curve of rock perturbation deformation shows a sustained downward trend, and it is not sensitive to perturbation. When $\sigma > \sigma_r$, the increment curve of rock perturbation deformation shows a sustainable development status, but it is sensitive to perturbation.

Both experiments tested three siltstone specimens, and the static stress exerted on specimens is shown in Table 4.

According to the analysis of the experimental results, the perturbation deformation increment curve of rock specimen obviously changed from stress level 5. This indicates a transition from the decay phase to the continued development phase. In 10 kg impact experiments, the experimental curve at stress level 4 shows a slight upward trend, and because the strain is in the single digit level, it is considered that the rock is still in the perturbation deformation increment stage of decay under this stress level. This indicates that the change should start in stress level 5. The actual stress level 5 in the test is between 69.2 and 70.4 MPa, which is about 98.16–99.86% of the long-term strength and 84.39–85.85% of the ultimate strength.

According to the experimental analysis, the change in hard siltstone test specimen influenced by external perturbation is not obvious when it is below the long-term strength (flow in the first stage). When the static stress level is close to or even surpasses the long-term strength, the rock specimen in the steady creep state starts to respond apparently under the effect of external perturbation, and the cumulative disturbance deformation shows a sustained development. When the load stress increases further, the impact load would even

TABLE 4: Static loading stress level of different rock specimens.

Impact loading	Rock specimen	Stress level 1	Stress level 2	Stress level 3	Stress level 4	Stress level 5	Stress level 6
5 kg	1# specimen	56.2	58.8	64.5	66.3	70.2	79.5
	2# specimen	56.5	59.8	64.1	66.8	70	79.6
	3# specimen	56.6	59.6	64.8	66.0	69.7	79.8
10 kg	1# specimen	56.3	59.7	64.7	66.4	69.2	79.0
	2# specimen	56.8	59.4	64.0	66.5	70.4	79.4
	3# specimen	56.5	59.5	64.2	66.5	68.8	79.6

cause sudden crack damage, similar to the dynamic disaster damage in an underground excavation space in engineering such as mine rockburst and coal wall sheet.

By comprehensive analysis, the antiperturbation and long-term strengths of the hard siltstone specimens are basically the same, $\sigma_r = \sigma_s$, consistent with the results of previous studies on the moderate intensity of a red sandstone.

5.5. Practical Significance of Rock's Antiperturbation Strength. Deep underground rock engineering suffers from complex mechanical conditions in a high in situ stress field. In this field, the surrounding rock mass often demonstrates continuous strong creep characteristics and causes dynamic disasters such as a large deformation and sudden instability under the influence of impact loading. The parameter of rock's antiperturbation strength in this paper is an important index to characterize the influence of external perturbation on the surrounding rock mass. For deep underground rock engineering, the analysis of the perturbation intensity of rock determined in the experiments helps to analyze and determine whether the surrounding rock is in a flow perturbation sensitive area or nonsensitive area as well as the dynamic response to the external perturbation load. Then, the basis for the rational supporting design of deep rock mass engineering can be provided. This has a great significance in engineering practice.

6. Conclusions

We conclude the following:

(1) To analyze sedimentary rock's dynamic characteristics in the creep state, a new type of rock creep perturbation experiment system was developed. This equipment has two grades of force-amplification structure which can increase the original gravity load by 60–100 times. It can provide a longtime and constant high static stress. The load infliction of external perturbation loading occurs through the free fall of the impact weights. This new equipment is suitable for carrying out long-term static creep tests and dynamic disturbance tests of rock samples in the creep state.

(2) When a siltstone was influenced by certain external perturbation in the creep process, its perturbation deformation increased irreversibly. The incremental curve of the perturbation deformation generally had two stages: a decay phase and a sustained development phase. When the static stress was up to ~85% of the rock's ultimate strength and the axial strain was up to 80% of the ultimate failure strain, the rock became sensitive to impact load and tended to failure.

(3) An increase in the impact strength will aggravate the internal damage in rock mass, induced by a large deformation by the perturbation of the rock. With the increase in impact strength, the cumulative disturbance deformation curve of the rock will be transformed from the decay phase to the sustained development phase in advance, causing the rock to be sensitive to external perturbation in the lower static stress level and decreasing the antiperturbation strength of the rock.

(4) According to the actual measured results, the antiperturbation and long-term strengths of the hard siltstone specimens were basically equivalent: $\sigma_r = \sigma_s$. Further, the experimental study of the antiperturbation strength of rock is of great practical importance in the supporting design of the stability of deep underground rock engineering in the creep perturbation state.

Conflicts of Interest

The authors declare that they have no conflicts of interest.

Acknowledgments

This study was supported by the National Natural Science Foundation of China (nos. 51304127, 51774195, and 51474134), China Postdoctoral Science Foundation (no. 2016M590646), Qingdao Postdoctoral Applied Research Foundation (no. 2016121), Shandong Provincial Natural Science Fund for Distinguished Young Scholars (JQ201612), and Taishan Scholar Talent Team Support Plan for Advantaged and Unique Discipline Areas.

References

[1] Y. F. Gao, Q. Z. Fan, X. H. Cui, and Z. L. Fu, *Experimental Study on The Perturbation Effect of Rock Rheology*, Science Press, Beijing, China, 2007.

[2] C. X. Liu, J. D. He, and M. Z. Zhang, "Long-Term strength test and application of hard fine sandstone in deep site," *Journal of Mining and Safety Engineering*, vol. 27, no. 4, pp. 581–584, 2010.

[3] L. Jiang, A. Sainoki, H. S. Mitri, N. Ma, H. Liu, and Z. Hao, "Influence of fracture-induced weakening on coal mine gateroad stability," *International Journal of Rock Mechanics and Mining Sciences*, vol. 88, pp. 307–317, 2016.

[4] Q. B. Zhang and J. Zhao, "A review of dynamic experimental techniques and mechanical behaviour of rock materials," *Rock Mechanics and Rock Engineering*, vol. 47, no. 4, pp. 1411–1478, 2014.

[5] J. A. Hudson and J. P. Jarrison, *Engineering Rock Mechanics*, Elsevier Science, Oxford Ltd, 1997.

[6] J. Zhao, H. B. Li, M. B. Wu, and T. J. Li, "Dynamic uniaxial compression tests on granites," *International Journal of Rock Mechanics and Mining Sciences*, vol. 36, no. 2, pp. 273–277, 1999.

[7] A. Kumar, "The effect of stress rate and temperature on the strength of basalt and granite," *Geophysics*, vol. 33, no. 3, pp. 501–510, 1968.

[8] S. Demirdag, K. Tufekci, R. Kayacan, H. Yavuz, and R. Altindag, "Dynamic mechanical behavior of some carbonate rocks," *International Journal of Rock Mechanics & Mining Sciences*, vol. 47, no. 2, pp. 307–312, 2010.

[9] X. B. Li, T. S. Lok, and J. Zhao, "Dynamic characteristics of granite subjected to intermediate loading rate," *Rock Mechanics and Rock Engineering*, vol. 38, no. 1, pp. 21–39, 2005.

[10] X. Li, Z. Zhou, T.-S. Lok, L. Hong, and T. Yin, "Innovative testing technique of rock subjected to coupled static and dynamic loads," *International Journal of Rock Mechanics and Mining Sciences*, vol. 45, no. 5, pp. 739–748, 2008.

[11] Z. L. Zhou, X. B. Li, Z. Ye, and K. W. Liu, "Obtaining constitutive relationship for rate-dependent rock in SHPB Tests," *Rock Mechanics and Rock Engineering*, vol. 43, no. 6, pp. 697–706, 2010.

[12] Z. Zhou, X. Li, Y. Zou, Y. Jiang, and G. Li, "Dynamic Brazilian tests of granite under coupled static and dynamic loads," *Rock Mechanics and Rock Engineering*, vol. 47, no. 2, pp. 495–505, 2014.

[13] J. Zhao and H. B. Li, "Experimental determination of dynamic tensile properties of a granite," *International Journal of Rock Mechanics and Mining Sciences*, vol. 37, no. 5, pp. 861–866, 2000.

[14] S. Okubo, K. Fukui, and J. Kawakami, "Technical note large-scale penetration test using a drop hammer," *Rock Mechanics and Rock Engineering*, vol. 30, no. 2, pp. 113–118, 1997.

[15] D. J. Reddish, L. R. Stace, P. Vanichkobchinda, and D. N. Whittles, "Numerical simulation of the dynamic impact breakage testing of rock," *International Journal of Rock Mechanics and Mining Sciences*, vol. 42, no. 2, pp. 167–176, 2005.

[16] D. N. Whittles, S. Kingman, I. Lowndes, and K. Jackson, "Laboratory and numerical investigation into the characteristics of rock fragmentation," *Minerals Engineering*, vol. 19, no. 14, pp. 1418–1429, 2006.

[17] D. Wu, Y.-C. Liu, Z.-X. Zheng, and S. Wang, "Impact Energy Absorption Behavior of Cemented Coal Gangue-Fly Ash Backfill," *Geotechnical and Geological Engineering*, vol. 34, no. 2, pp. 471–480, 2016.

[18] Y. Gao, W. Huang, G. Qu, B. Wang, X. Cui, and Q. Fan, "Perturbation effect of rock rheology under uniaxial compression," *Journal of Central South University*, vol. 24, no. 7, pp. 1684–1695, 2017.

[19] M.-R. Shen and H.-J. Chen, "Testing study of long-term strength characteristics of red sandstone," *Rock and Soil Mechanics*, vol. 32, no. 11, pp. 3301–3305, 2011.

[20] L. Q. Li, W. Y. Xu, and W. Wang, "Estimation of long-term strength for Xiangjiaba sandstone based on creep tests," *Engineering Mechanics*, vol. 27, no. 11, pp. 127–136, 2010.

[21] Q. S. Liu, W. Gao, and L. Yuan, *The Deep Rock Roadways Stability Control Theory and Supporting Technology Application in Coal Mine*, Science Press, Beijing, China, 2010.

[22] X. Cui and Z. Fu, "Experimental study on rheology properties and long-term strength of rocks," *Chinese Journal of Rock Mechanics and Engineering*, vol. 25, no. 5, pp. 1021–1024, 2006.

[23] Y. J. Yang, T. Peng, and G. G. Li, "Study on long term strength of piliar rockmass at yungang grottoes," *Chinese Journal of Rock Mechanics and Engineering*, vol. 28, no. S2, pp. 3042–3048, 2009.

[24] B. Damjanac and C. Fairhurst, "Evidence for a long-term strength threshold in crystalline rock," *Rock Mechanics and Rock Engineering*, vol. 43, no. 5, pp. 513–531, 2010.

16

Microstructures, Mineral Compositions and Mechanical Properties of Red-Layers in Southern China

Huizhi Zhang[1,2] **and Jifeng Liu**[1,2]

[1]*Key Laboratory of Engineering Materials and Structural Reinforcement, Sanming University, Fujian Province University, Fujian, China*
[2]*Department of Architecture and Civil Engineering, Sanming University, Fujian, China*

Correspondence should be addressed to Huizhi Zhang; zhz888_1981@163.com

Academic Editor: José A. Correia

In order to reveal the water-induced damage mechanism and mechanical properties of red-layers in Southern China, typical red-layer samples are experimentally studied by scanning electron microscope (SEM), energy spectrum analysis, XRD, and mechanical tests. The results show that some red-layer samples contain a large number of smectite and illite-smectite mixed-layer clay minerals and their microstructures are mainly stacked microaggregates in which mutual agglomeration is formed by flake and tabular clay particles. Widespread distribution micropores and fractures in microaggregates provide the necessary moisture migration channels. The various cemented materials produce a kind of irreversible structural strength, and this strength is vulnerable for the aqueous medium conditions. The influence of microstructures and mineral compositions on red-layer mechanical properties is discussed. The results could provide the reference for further research about the water-induced damage mechanism of red-layers in Southern China.

1. Introduction

There are widely distributed red-layers in Southern China, for example, in Yunnan, Guangdong, Guizhou, and Sichuan provinces. The red-layers in those areas are brownish red or brownish yellow clay soil or soft rock, which is the laterized product of carbonate rock. Mainly including expansive clay minerals, soluble minerals, or organic minerals, the red-layers have many special engineering geological properties in comparison with common clay. In general, most red-layers have high water content, high plasticity, large pore ration, and low density, the physical and mechanical properties of red-layers are easily affected by water [1, 3]. Engineering construction disturbance or water intrusion will induce damage in microstructures or/and change in mineral compositions, which could cause the red-layers rapidly expanse and collapse. Such complex properties bring a serious impact on the safety and stability of major engineering construction in those regions [4–9]; for instance, squeezed muck wear formation ahead of shield cutters seriously affected Guangzhou Metro Lines construction speed (Figure 1, [10, 13]).

The physical and mechanical characteristics of red-layers or clay soil have been widely studied. Mendes and Toll [14] revealed that the critical state stress ratio is not affected by the initial water content, and the volumetric behavior differences lead to sample property differences when samples are subjected to wetting and drying after compaction. Chavali et al. [15] investigated the alkali-induced swelling in nonswelling kaolinitic clays, and the results showed that the magnitude of alkali-induced swelling in the clays is attributed to dispersion of clay particles and formation of new mineral. Harjupatana et al. [16] developed a noninvasive method for simultaneous measurement of the 3D displacement field and the water content distribution of a wet solid material and applied it in validating models involving transport of water and the resulting swelling deformation of bentonite. Aziz et al. [17] revealed that the observed soil

Figure 1: Squeezed muck wear in shield cutters, Guangzhou Metro Line.

behavior could be critical for embankments constructed with rock-fill materials. Wong et al. [18] reviewed the research of strength and stiffness reduction in rocks attributed to an increase in water content in recent years. Ma et al. [8] conducted a series of constant water content triaxial tests on statically compacted specimens, the results showed that Yunnan red clay failure envelope is affected by its saturation degree, and a new semiempirical shear strength model in terms of degree of saturation could be proposed. Zhao et al.'s [6] experimental results showed that the damage form of the remodeled red clay changed from the ballooning excessive damage to shear failure with the reduction of water content. Zhang et al. [7] proposed and verified a X-ray computed tomography (CT) method to nondestructively measure the water content of red clay cylinders, and the test results showed that the rainfall, ground water level, and degree of compaction are factors that could significantly affect the moisture migration distance and migration rate. Hu et al. [9] investigated the effect of water content on the stable swelling capacities, swelling process, and characteristics of red bed mud-stone, the results revealed that there were three phases exiting in the confined free sweeling test, and vertical loads greatly restrained the confined swelling. However, limited studies have been conducted on the relationship of red-layer mechanical properties and its microstructure and composition changes. Some studies showed that changes in mineral compositions and microstructures yielded macroscopic degradation of mechanical parameters [3, 19–21]; therefore, the relationships of mechanical behaviors and microstructure and mineral compositions of red-layers should be fully investigated.

In this study, for investigating mechanical behaviors of red-layers influenced by microstructures and mineral compositions, a series of experimental tests are conducted on red-layer specimens which are sampled in Guangzhou Metro Line.

2. Sample Preparation

The typical red-layer specimens are sampled at Tiyu West road station, Zhujiang new town station, Wushan road station, Kecun road station, and Tianhe road station of Guangzhou Metro Line, separately. The details of specimens are given in Table 1. For other details of the samples, see Liu and Zhang [22]. The samples' lithologies are shown in Figure 2.

The specimens for all mineral XRD tests are prepared as follows: (1) the core of the rock or soil sample block is dried for 48 h in an electric oven below 60°C, and then the specimens are cooled to room temperature. (2) 1-2 g of the samples is crushed to less than 1 mm particle size by a grinding machine, and then the particles are ground to the whole particle size of less than 40 μm. (3) The aluminum sample frame is closely placed on the flat glass plate, and the sample powder is spooned into the 20 mm × 18 mm framework and is vertically compressed for molding. The lower side is adopted for the test surface.

The sample preparation for XRD tests of the relative clay mineral content is as follows: (1) the air-dried sample is crushed to a particle size of less than 1 mm, the crushed power is placed into the beaker, and then the sample is soaked with distilled water and dispersed by the ultrasonic treatment. (2) The suspension with a particle size of less than 2 μm is drawn and centrifuged for accelerating the clay mineral settlement. (3) The centrifuged sample is then dried for 48 h in an electric drying oven below 60°C, and the dried sample is ground to finer particles by an agate mortar. (4) 40 mg of the dried sample particle is placed into 10 ml tubes and 0.7 ml of distilled water is added into the tubes; after complete stirring, the clay mineral is fully dispersed by the ultrasonic treatment, the suspension is quickly placed on glass slides and air-dried, and then a natural directional sheet (N) is prepared. (5) After the experiment, the glycol vapor is used to make a glycol saturated tablet (NG) by heating a natural directional sheet (N) at 40–50°C constant temperature for 7 h, and then the sheet is cooled to room temperature. (6) At the end of the experiment, the glycol saturated tablet (NG) kept at 550°C constant temperature for 2 h, and then the specimen is cooled to room temperature to make a heating sheet (550°C).

There is a time interval between the sampling and the test. For more realistic viewing of microstructures of the red-layers, it is necessary to remake the specimen. The specimens for SEM analysis are prepared as follows: (1) After removing surface, the samples are immersed in pure anhydrous alcohol for 30 minutes, and then the pure anhydrous alcohol is replaced with anhydrous alcohol to ensure the purity of the alcohol before the test. The purpose of using anhydrous alcohol is to ensure original microstructures of the specimen, because anhydrous alcohol is easily evaporated and carries away the residual moisture in the samples. (2) The samples are air-dried and then are cohered on the table, and heavy metal membrane are sprayed on surfaces to avoid discharge phenomenon. (3) 30 mm thickness heavy metal membrane is sprayed on the morphology analysis samples. The energy spectrum analysis samples are without metal spray coating to avoid the metal film effects on test precision.

3. Experimental Test

The typical tests for analyzing the microphysical property of red-layers are SEM analysis which include micromorphology

Table 1: Microexperimental soil samples of typical red-layers.

Number	Sampling site	Sampling depth (m)	Corings	Total sampling	Lithology	Disturbed or not
1	Tiyu west road station	16	3	12	Moderately weathered red clay	Disturbed
2	Zhujiang new city station	17.2–17.8	3	12	Red clay	Disturbed
3	Wushan road station	12	3	12	Moderately to strongly weathered red clay	Undisturbed

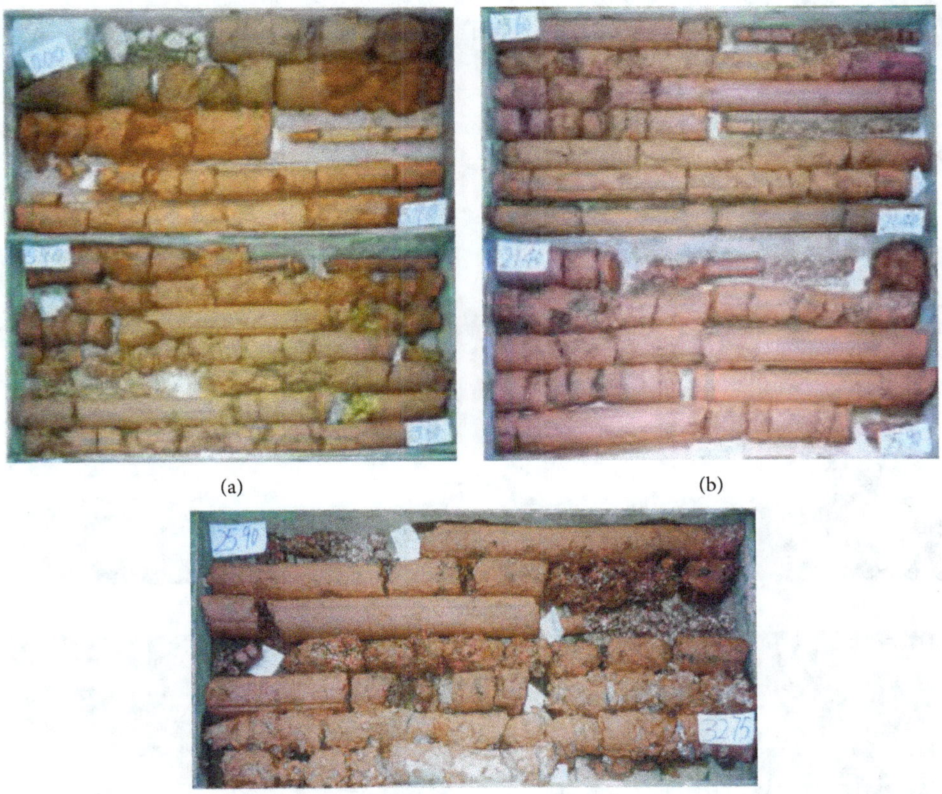

(a) (b) (c)

Figure 2: Corings of Wushan road station in Guangzhou Metro Line 3.

scanning, energy spectrum analysis, and XRD which include clay mineral and total mineral analysis. Samples are prepared in accordance with the test requirements. The used instruments are LEO-435VP scanning electron microscope, FEI Quanta200 field emission scanning electron microscope, and D/MAX-2500 X diffraction instrument.

4. Results

4.1. Analysis of Scanning Electron Microscope and Energy Spectrum Test Results. The mineral compositions are expressed as follows: C = chlorite; I = illite; S = smectite; Na-F = albite; I/S = illite and smectite mixed layer; K = kaolinite; K-F = kaolinite and potash feldspar mixed layer; Q = quartz crystal; I/C = illite and chlorite mixed layer; C/S = chlorite and smectite mixed layer.

The SEM images of Tiyu west road station red clay samples are shown in Figure 3, which shows that the soil samples are relatively loose and the larger pores between the particles are about 100–200 μm with the great connectivity. The samples contained a lot of honeycomb and flaky illite and smectite mixed layer, a small amount of needle-like chlorite and flake-flocculent chlorite and smectite mixed layer, and authigenic polyhedral quartz and strip albite crystal existing in particles.

The SEM images of Zhujiang new city station red clay samples are shown in Figure 4. It could be known from Figure 4 that the soil samples are relatively loose, and the larger pores between the particles are with the diameter of 30–100 μm with the good connectivity. The samples contain a large number of flake-flocculent illite and smectite mixed layer, illite and chlorite mixed layer, layered illite, and flake kaolinite. The local parts of the samples contain leached surface potash feldspar crystals. Under the magnification of 9440–20,000 times, a large number of flake aggregates could be observed.

The SEM images of Wushan road station red clay samples are shown in Figure 5. One could see from Figure 5 that the red

FIGURE 3: SEM images of red clay samples in Tiyu west road station. (a) ×4140 honeycomb I/S with a small amount of needle-like C. (b) ×10,000 flake-flocculent I/S.

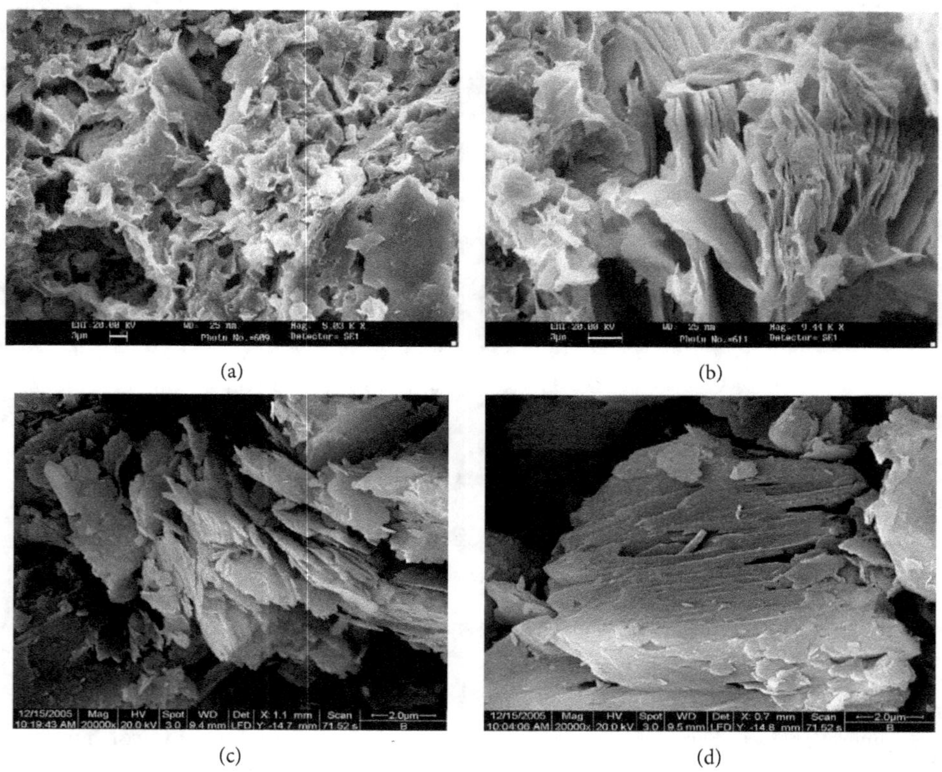

FIGURE 4: SEM images of red clay samples in Zhujiang new city station. (a) ×5030 flake-flocculent I/C and quartz crystal. (b) ×9440 flake I and K and flake-flocculent I/S. (c) ×20,000 leaf shaped and laminated C. (d) ×20,000 leached K-F crystal, containing fibrous sepiolite.

clay samples are loose, the larger pores between the particles are with the diameter of 30–50 μm, and the pores have the good connectivity. The samples contain a large number of honeycomb, flake, and flake-filamentous illite, smectite mixed layer, and curly flake smectite. And a small amount of needle-like chlorite and filamentous illite with the dissolved albite crystals are observed as well. Figure 5(b) shows the lamellar stack dimers under the magnification of 40,000 times, which is the important mesostructure characteristic of the complex red-layer by test observation. It can be estimated from Figure 5(c) that the thickness of the polymer monolithic stack is approximately 10 Å(1 nm), and these large amount of pores and a laminated polymers provide favorable conditions for water intrusion and damage of red-layers.

The energy spectrum analysis of red clay samples are carried out by the LINKISIS-type energy spectrometer; for simplicity of consideration, only the scanning images of Tiyu west road station red clay samples are given (Figure 6). Tiyu west road station red clay samples contain a large number of illite and smectite mixed layer, chlorite and smectite mixed layer, massive albite crystals, and filamentous chlorite. There are massive authigenic quartz crystals spreading between the particles. Zhujiang new city station red clay samples contain a lot of orderly flake, disorderly

(a) (b) (c)

Figure 5: SEM images of red clay samples in Wushan road station. (a) ×10000 filament I/S. (b) ×40,000 the typically flake I/S. (c) ×40,000 flake I/S, curled lamellar S.

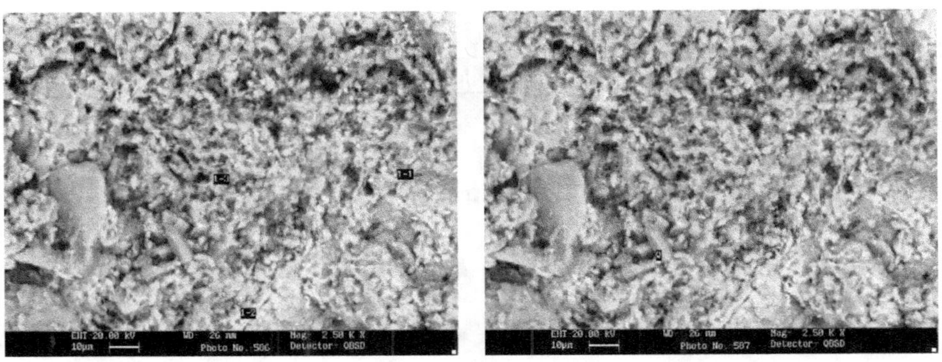

(a) (b)

Figure 6: Energy spectrum pictures of red clay samples in Tiyu west road station. (a) ×2500 1-1 I/S; 1-2 I/S; 1–3 C/S. (b) ×2500 1–4 authigenic quartz crystal.

flake-filamentous illite, and smectite mixed layer. There are bulk albite crystals existing between the local layers. There is a trend that a large number of kaolinite transform to illite. Wushan road station red samples contain a large number of aflocculent illite, smectite mixed layer, and flake chlorite. The local massive albite crystals and quartz crystals also exist in the samples.

4.2. Analysis of X-Ray Diffraction Results. The three clay minerals and three nonclay minerals samples are tested by the XRD. The results are analyzed in accordance with "X-ray diffraction analysis of the relative content of clay minerals in sedimentary rocks" and "X-ray diffraction quantitative analysis of the clay mineral aggregate and common nonclay minerals in sedimentary rocks" [23, 24].

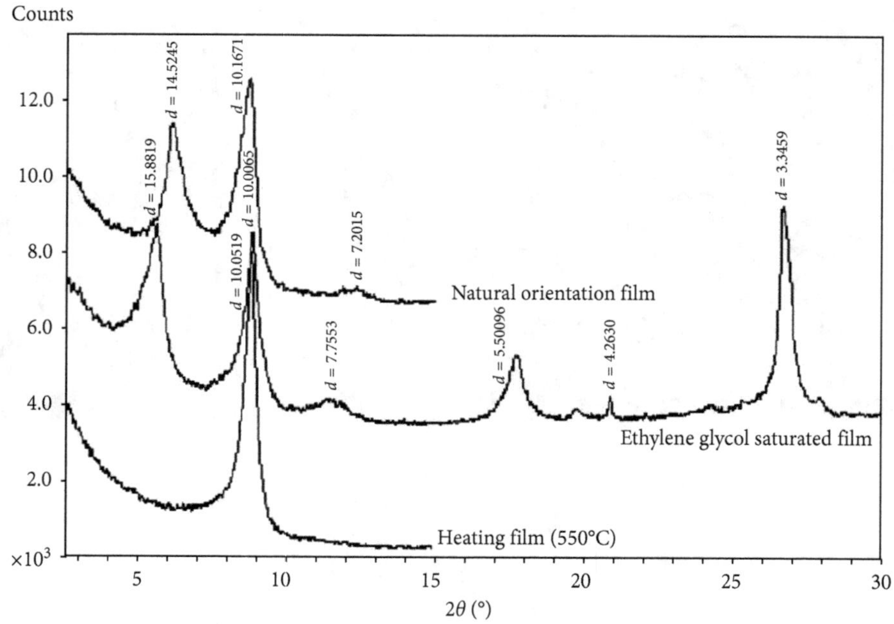

FIGURE 7: X-ray diffraction pattern of red clay in Tiyu west road station.

TABLE 2: XRD results of the remaining samples.

Number	Sampling site	Sampling depth (m)	Lithology	Clay mineral relative content (%)						Mixed layer ratio (%S)	
				S	I/S	I	Kao	C	C/S	I/S	C/S
KD01	Tiyu west road station	16	Moderately weathered red clay	54	—	46	—	—	—	—	—
KD02	Zhujiang new city station	17.2–17.8	Red clay	64	—	36	—	—	—	—	—
KD03	Wushan road station	12	Moderately to strongly weathered red clay	2	46	42	4	6	—	15	—
KD04	Zhujiang new city station	18.1–19.0	Sandstone	—	47	44	3	6	—	15	—
KD05	Kecun station	27.6	Red sandstone	3	75	20	1	1	—	40/25	—
KD06	Tianhe station	16	Shield excavated soil	—	—	56	38	6	—	—	—

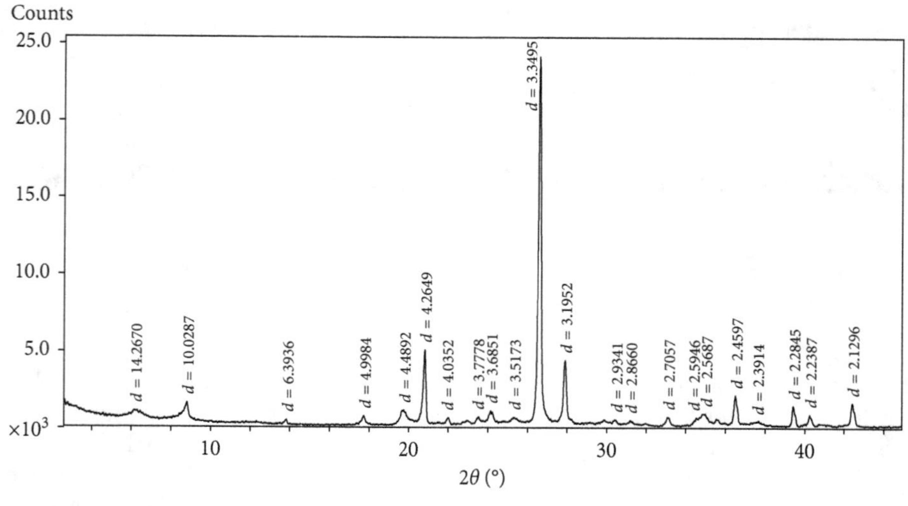

FIGURE 8: Total mineral X-ray diffraction pattern of red clay samples in Zhujiang new city station.

TABLE 3: All mineral XRD test results.

Number	Sampling site	Sampling depth (m)	Lithology	Mineral type and content (%)						Total clay minerals (%)
				Quartz	Feldspar	Plagioclase	Calcite	Hematite	Barite	
PKD01	Tiyu west road station	16	Moderately weathered red clay	47.2	—	14.5	—	2.8	—	35.5
PKD02	Zhujiang new city station	17.2–17.8	Red clay	45.7	—	10.9	—	3.5	—	39.9
PKD03	Wushan road station	12	Moderately to strongly weathered red clay	33.6	—	14.8	12.2	2.8	—	36.6
PKD04	Zhujiang new city station	18.1–19.0	Sandstone	27.7	—	5.4	21.1	5.7	—	40.1
PKD05	Kecun station	27.6	Red sandstone	44.1	8	6.1	7.4	1.4	—	33.0
PKD06	Tianhe station	16	Shield excavated soil	32.9	15.6	6.5	—		—	45.0

For the clay mineral XRD test, each soil sample has three piece directional diffraction and three diffraction patterns named "natural directional sheet" (N), "ethylene glycol" (NG), and "saturated tablet 550°C heating plate" [25, 27]. The clay mineral compositions can be identified according to the diffraction characteristics of three diffraction patterns such as diffraction peak or diffraction angle and peak value. The relative content can be determined by diffraction intensity. If the pattern is not very standard characteristic, or the diffraction patterns are between the two components, it can be identified as a mixed layer of illite and smectite, or chlorite and smectite mixed layer.

For simplicity of consideration, only diffraction pattern of Tiyu west road station red clay samples is shown in Figure 7. It is revealed that the soil samples mainly contain smectite and illite, and the relative content of smectite and illite is 54% and 46%, respectively. Other clay mineral components in the diffraction are not found. The XRD results of the remaining samples are shown in Table 2.

The XRD results of typical nonclay minerals of Zhujiang new city station samples are shown in Figure 8. It is shown that the content of clay minerals is 39.9%, and the main nonclay minerals are 45.7% quartz and 10.9% plagioclase, respectively. The hematite content is 3.5%, and other nonclay minerals are not found. The test results are shown in Table 3.

5. Conclusion and Discussion

The typical Southern China red-layers rock and soil samples are experimentally studied by SEM, energy spectrum analysis, XRD, and mechanical tests, and then the results are analyzed. Through this research, we can draw the following conclusions:

(1) The local area of red-layers in Southern China contains a large amount of smectite, illite, and smectite mixed-layer clay minerals. The microstructures are mainly layered or stacked microaggregates which are formed by the agglomeration of flaky and flat clay particles. The penetrated or partially penetrated cracks separate the lamellar structures. The stacked microaggregates are the main structural basic units which make up the microstructures and determine the swelling and shrinkage properties of the complex red-layers. The large amount of fine-grained expansion clay minerals is the main material basis for determining the complex red-clays properties. It is also an important internal factor which impacts its engineering properties such as water swelling, easily sliming, and strong viscosity [21, 28–30].

(2) The cohesion of various cemented materials in the stacked microaggregates enhances the structural connection between microaggregates of complex red soils, and the cohesion results in a certain structural strength. This structural strength is also a kind of irreversible bond strength which is formed by the physical and chemical connection. While disturbed by construction or intruded by water, or the solution medium conditions in the rock or soil changed, the physical and chemical properties of cemented material will be very different, the structural strength of microaggregates will be lost, the total strength of the complex red-layers will decrease, and the rate of expansion will increase.

Influence of water content on the red-layer specimens is a complex phenomenon. The water-induced changes in microstructures and mineral compositions yield degradation of physical and mechanical parameters. Therefore, it is important to analyze the effects of water content on the degradation mechanisms of the specimens, as those conditions are considered during the aforementioned experiments.

The typical red-layers samples have layered or honeycomb microstructures and easily water swelling and weakening minerals such as chlorite and smectite. The pores between the particles have good connectivity which provides the water invasion tunnels. With the lower water content, the particles and layers make good contact with each other and the contact surface is flat and smooth, forming good microstructures which induce relatively good physical and mechanical properties.

During the process of water content increasing, the water infiltrates inside the specimen. The chemical reactions such as dissolution and corrosion occur in the mineral particles, and the chemical migration and dissolution occur at the cemented or contact surfaces between mineral particles, which can make the mineral compositions and microstructures change and the macro physical and mechanical degradation of the specimens.

Since the red-layers specimens comprises many minerals which can easily expand in the water, under water invasion the cracks would likely propagate by the stress concentrated at the crack endpoints [31]. It is favorable for the formation of infiltration channels and water molecules seepage within the specimens. The reactions between minerals and aqueous solution are accelerated, which increase the change in microstructures and promote the extension and coalescence of cracks and microtunnels. These factors are the mechanism of water content effect on the red-layer properties like shear strength, swelling, creep, and so on.

(3) The swelling behavior of microaggregates affects water migration, water-induced damage, and rheological properties. The test results show that the complex red-layers comprise a large amount of lamellar micro-aggregates formed by flaky and flat clay particles. The thickness of the monolithic slice is about 10 Å, the length of the monolith is 1-2 μm and the width is 2–5 μm, and the lamellar microfracture width is 0.1–0.5 μm. There are the electrostatic repulsion, Van der Waals force, and hydration force existing between the stacking layers. The microaggregates structure arrangement state and the distance between the two relatively independent slices play a major role in the red clay hydration (volumetric chemical potential) at certain water content. The strong viscosity of red-layers under certain water content is due to its surface to surface microstructures in layered microaggregates.

Conflicts of Interest

The authors declare that they have no conflicts of interest.

Acknowledgments

The support of Fujian Province Natural Science Fund (Grant no. 2016J01743), Sanming University Science Research Development Fund (Grant no. JAT160454), and Fujian Province Young Teacher Education Research Project Fund (Grant no. B201603) is gratefully acknowledged.

References

[1] L. Song, *Research of Micro-and Meso-Softening Mechanism of Red Beds Soft Rock with Water*, Southwest Jiaotong University, Chengdu, Sichuan, China, 2014.

[2] Q. T. Bi, G. P. Jiang, and S. Y. Ding, "Water content influence on the shearing strength of red clay," *Earth and Environment*, vol. 33, no. 3, pp. 144–147, 2005.

[3] S. B. Li, J. Chen, and G. D. Yi, "Experimental study on the relationship between micro-characteristics and compressive strength of the red bed rock," *Geotechnical Investigation and Surveying*, vol. 4, no. 3, pp. 1–5, 2013.

[4] Q. L. Cui, H. N. Wu, S. L. Shen, and Y. S. Xu, "Geological difficulties of socket diaphragm walls in weathered granite in Shenzhen, China," *Bulletin of Engineering Geology and the Environment*, vol. 75, no. 1, pp. 263–273, 2016.

[5] H. J. Dong, X. F. Tan, Y. Zhao, S. H. Zhai, and Q. Yang, "Characteristic analysis of wear of cutters and excavation parameters of shield tunneling machine in red beds in Guangzhou," *China Harbour Engineering*, vol. 4, pp. 11–14, 2013.

[6] R. Zhao, S. Y. Zuo, W. Song, S. H. Li, and C. H. Huang, "Experiment and mechanism analysis of water contents on triaxial shear strength of the remodeled red clay of Guiyang," *Hydrogeology and Engineering Geology*, vol. 42, no. 5, pp. 90–95, 2015.

[7] J. H. Zhang, Q. P. Jiang, Y. Q. Zhang, L. L. Dai, and H. X. Wu, "Nondestructive measurement of water content and moisture migration of unsaturated red clays in South China," *Advanced in Materials Science and Engineering*, vol. 2015, Article ID 542538, 7 pages, 2015.

[8] S. K. Ma, M. S. Huang, P. Hu, and C. Yang, "Soil-water characteristics and shear strength in constant water content triaxial tests on Yunan red clay," *Journal of Central South University*, vol. 20, no. 5, pp. 1412–1419, 2013.

[9] W. J. Hu, Y. Ding, Z. Y. Xia et al., "Experimental study on confined swelling characteristic of redbed mudstone in Chongqing," *Journal of Disaster Prevention and Mitigation Engineering*, vol. 35, no. 5, pp. 607–611, 2015.

[10] W. B. Shi, Y. L. Lian, and X. L. Wang, "Discussion on the relationship between shear strength and water content of guiyang red clay," *Subgrade Engineering*, no. 4, pp. 80–85, 2011.

[11] Y. H. Wang, Q. X. Zhang, and P. C. Tian, "Effects of water content variation on the deformation and shear strength of laterite clay," *Geotechnical Investigation & Surveying*, vol. 37, no. 7, pp. 10–13, 2009.

[12] J. X. Liu and Y. T. Liu, "Study on engineering characteristics and centrifugal test of red bed," *Subgrade Engineering*, no. 2, pp. 78–80, 2010.

[13] Y. Xu, C. Y. Zhou, Z. Liu, D. L. Su, and Z. C. Du, "Model tests for failure mechanism of typical soft rock slopes of red beds under rainfall in South China," *Chinese Journal of Rock Mechanics and Engineering*, vol. 35, no. 3, pp. 549–557, 2016.

[14] J. Mendes and D. G. Toll, "Influence of initial water content on the mechanical behavior of unsaturated sandy clay soil," *International Journal of Geomechanics*, vol. 16, no. 6, pp. 1–18, 2016.

[15] R. V. P. Chavali, S. K. Vindula, H. P. Reddy, A. Babu, and R. J. Pillai, "Swelling behavior of kaolinitic clays contaminated with alkali solutions: a micro-level study," *Applied Clay Science*, vol. 135, pp. 575–582, 2017.

[16] T. Harjupatana, J. Alaraudanjoki, and M. Kataja, "X-ray tomographic method for measuring three-dimensional deformation and water content distribution in swelling clays," *Applied Clay Science*, vol. 114, pp. 386–394, 2015.

[17] M. Aziz, I. Towhata, S. Yamada, M. U. Qureshi, and K. Kawano, "Water-induced granular decomposition and its effects on geotechnical properties of crushed soft rocks," *Natural Hazards and Earth System Science*, vol. 10, no. 6, pp. 1229–1238, 2010.

[18] L. N. Y. Wong, V. Maruvanchery, and G. Liu, "Water effects on rock strength and stiffness degradation," *Acta Geotechnica*, vol. 11, pp. 713–737, 2016.

[19] H. F. Deng, M. L. Zhou, J. L. Li, X. S. Sun, and Y. L. Huang, "Creep degradation mechanism by water-rock interaction in the red-layer soft rock," *Arabian Journal of Geosciences*, vol. 9, no. 12, pp. 601–613, 2016.

[20] M. C. Wend, L. S. Tsai, Y. M. Hsieh, and F. S. Jeng, "An associated elastic-visco-plastic constitutive model for sandstone involving shear-induced volumetric deformation," *International Journal of Rock Mechanics and Mining Sciences*, vol. 47, no. 5, pp. 1263–1273, 2010.

[21] Y. B. He and H. Liu, "Creep characteristics of red layer and tunnel surrounding rock displacement prediction based on PSO-SVM," *Highway Engineering*, vol. 41, no. 2, pp. 75–79, 2016.

[22] J. F. Liu and H. Z. Zhang, "Water content influence on properties of red-layers in Guangzhou metro line, China," *Advances in Materials Science and Engineering*, vol. 2017, Article ID 4808909, 12 pages, 2017.

[23] The People's Republic of China Oil and Gas Industry Standard, *X ray diffraction analysis method of the relative content of clay minerals in sedimentary rock*, 1995.

[24] The People's Republic of China Oil and Gas Industry Standard, *X ray diffraction quantitative analysis method of the clay mineral aggregate and common non-clay minerals in sedimentary rocks*, 1995.

[25] X. Y. Zhao and Y. Y. Zhang, *Analysis of Clay Minerals and Clay Minerals*, Ocean Press, Beijing, China, 1990.

[26] J. S. Yu, X. R Lei, J. H. Zhang et al., *X-Ray Powder Mineral Identification Handbook (Map)*, Huazhong University of Science and Technology Press, Wuhan, China, 2011.

[27] C. H. Jiang, *Polycrystalline X-Ray Diffraction Technology and Application*, Chemical Industry Press, Beijing, China, 2014.

[28] Y. C. Wang and J. Chen, "Preliminary research on the relationship between microscopic structure and shear strength of red-bed soft rocks," *Geoscience*, vol. 27, no. 3, pp. 738–742, 2013.

[29] Z. K. Ding and A. Z. Lu, "Parameter identification of non-stationary creep equation of rocks," *Rock and Soil Mechanics*, vol. 25, no. 1, pp. 37–40, 2004.

[30] C. X. Chen, H. F. Lu, C. H. Yuan et al., "Experimental research on deformation properties of red-bed soft rock," *Chinese Journal of Rock Mechanics and Engineering*, vol. 29, no. 2, pp. 261–269, 2010.

[31] L. S. Tang, P. Zhang, and Y. Wang, "Propagation of the complex cracks in rocks II. Under water pressure and chemical damage," *Acta Scientiarum Naturalium Universitatis Sunyatseni*, vol. 42, no. 1, pp. 90–94, 2003.

17

Chemical, Mineralogical and Refractory Characterization of Kaolin in the Regions of Huayacocotla-Alumbres, Mexico

L. M. Romero-Guerrero,[1] R. Moreno-Tovar,[1] A. Arenas-Flores,[1] Y. Marmolejo Santillán,[2] and F. Pérez-Moreno[1]

[1]*Área Académica de Ciencias de la Tierra y Materiales, Universidad Autónoma del Estado de Hidalgo, Carr. Pachuca-Tulancingo Km. 4.5 Col. Carboneras, 42184 Mineral de la Reforma, Hidalgo, Mexico*
[2]*Área Académica de Química, Universidad Autónoma del Estado de Hidalgo, Carr. Pachuca-Tulancingo Km. 4.5 Col. Carboneras, 42184 Mineral de la Reforma, Hidalgo, Mexico*

Correspondence should be addressed to F. Pérez-Moreno; fpmoreno@uaeh.edu.mx

Academic Editor: Fernando Rubio-Marcos

In the present work, the chemical, mineralogical, refractory, and microstructural characterizations of kaolinites from the Huayacocotla-Alumbres region, which is between Veracruz and Hidalgo border, by X-ray diffraction (XRD), polarization optical microscopy (POM), scanning electron microscopy (SEM), refractoriness proof (pyrometric cone equivalent), and thermogravimetric analysis (TGA) were carried out. The analysis by POM showed that the kaolinization degree in this region is variable due to the presence of primary minerals, such as plagioclase, feldspar, and quartz. Additionally, hydrothermal alteration of the epithermal type was determined by oxidation of sulfides (pyrite and galena) and chlorite association. With the X-ray diffraction technique, andalusite and kaolinite were identified as the majority phases in Huayacocotla and quartz was identified as the majority phase in Alumbres. The minority phases, such as dickite, kaolinite, and cristobalite, were observed in both zones. The SEM technique was useful in the determination of the morphology of kaolinite and impurities of Na, Mg, K, and Fe of the complex clay illite-andalusite-dickite group. Thermogravimetric analysis was useful to discover the decomposition temperature and reveal the significant difference between 400 and 800°C, which showcases the greatest mass loss due to dehydration and carbonates decomposition. The mullite phase was detected at approximately 1000°C in the kaolin samples. The refractoriness tests were important to determine the stability temperature of kaolin, which is between 1300 and 1600°C. This stability temperature makes it feasible to use the kaolin as a refractory material for both low and high temperatures. The variables that affect the kaolin stability temperature were determined by principal components with the XLSTAT free program.

1. Introduction

Kaolin is a natural clay; it is an alteration product of feldspathic rocks whose main component is kaolinite [1]. Kaolin has wide application in industry due to its whiteness, inertness to chemical agents, electric insulator properties, mold ability, and ease of extrusion; also, it resists high temperatures, is nontoxic, and has high refractoriness. In Mexico, there are several kaolin deposits, and the most important ones are located in the States of Guanajuato, Hidalgo, and Veracruz [2].

Some characteristic problems of kaolin are the presence of minerals, such as iron oxide (hematite) and iron sulfides (pyrite), which act as coloring agents during the burning of ceramic pieces [3] and that the bending strength of ceramic tile is diminished when the pore size is reduced at a temperature higher than 1000°C [4]. Application techniques for clay purification that are used in pharmaceutical formulations are expensive because the clay used should have a high specific area, high adsorption capacity, high chemical inertness, and favorable rheological properties [5, 6].

Kaolin is an important raw material employed in ceramic manufacture; some kaolinite clay types have been used to prepare ceramics based on mullite because it has high-stability temperature, high refractoriness, high resistance to corrosive agents, and other desirable properties [7, 8]. The mineralogical compositions of clays are important aspects to evaluate since they directly influence the physical and mechanical qualities of fabricated ceramic products [7–10].

The kaolin exploitation has acquired great economic importance in recent years because its consumption has been increased in the manufacturing of products, such as paper, paints, inks, and plastics [1].

This study aims to assess the chemical, mineralogical, and refractory characteristics of kaolin from the Huayacocotla-Alumbres region in order to determine the material quality and propose possible application for it.

2. Materials and Methods

2.1. Study Area. The studied area, shown in Figure 1, is located in the east-central portion of the Mexican Republic, located to the northeast of Hidalgo state and northwest of Veracruz state. The area of Huayacocotla, Ver., is located between the coordinates 20°24′13″ N; 98°26′56″ W and 20°25′04″ N; 98°25′57″ W and the area of Alumbres, Hgo., is located between 20°40′18″ N; 98°41′15″ W and 20°40′22″ N; 98°41′10″ W, respectively.

The samples of kaolin were collected randomly on surface outcrop of mine that has been exploited in two zones that are shown in Figure 2.

The first zone is in Huayacocotla, Veracruz, localized among 20°24′13″ N; 98°26′56″ W and 20°25′04″ N; 98°25′57″ W, specifically in the mines of Rosa de Castilla, La Paloma, and Puerto Lobo; the access to this zone is leaving from Pachuca city in the direction to Tuxpan town by 130 Federal road and after passing the city of Tulancingo, 51 statal road must be taken to reach the mines in Agua Blanca, Huayacocotla, Ver.

The second zone is in Alumbres, Hidalgo, that is localized among 20°40′18″ N; 98°41′15″ W and 20°40′22″ N; 98°41′10″ W in Lote la Luz and Alumbres mines; the access to this zone is leaving from Pachuca city in the direction to Tampico town by the 105 Federal road and the towns Atotonilco el Grande, San Agustín Metzquititlan, and Zacualtipán must be traversed before reaching Alumbres, Hgo.

In the first zone, twenty-six samples were taken, and in the second, fifteen; these samples were chosen on the basis of their color, texture, and particle size, among other characteristics with an approximated weight of 1.5 kg each. They were placed in a plastic bag, sealed, and labeled with the location name, the date, and the georeferenced sampling point.

2.2. Analysis. The samples were dried at environmental temperature (30°C) without oven-drying over five days to remove the enough humidity to work with them. They were mixed and quartered to obtain a representative sample that was milled in an agate mortar to diminish the particle size; then, they were sieved on a set of 100, 200, 325, and 400 mesh to obtain subsamples with ranges of particle size 74–150; 45–74 and 38–45 μm is preferred for quantitative and qualitative analysis by spectrophotometric techniques [11, 12].

The chemical analysis was performed by triplicate; a 0.2 g clay sample of 38–45 μm of particle size was digested with an acid mixture of 4 ml of $HCl : HNO_3$ (3 : 1) and 7 ml of HF in a microwave oven, brand CEM mod. MARSLX. The sample temperature and pressure were raised up to 170°C and 170 psi in 25 min. These conditions were maintained for 15 min. Then, the samples were cooled, and 35 ml of H_3BO_3 (4%) was added to the containers to make a second digestion. The temperature and pressure were maintained at the same values as in the first digestion for 15 min; then, the volumetric flasks were filled with distilled water up to the 50 ml mark [13, 14]. The analysis was performed with an (ICP) spectrophotometer, PerkinElmer, model Optima 8300 XL, employing a multielemental-certified high-purity standard to determine the chemical composition of the principal elements of the oxide base in the laboratory of ICP and AA at the Autonomous University of Hidalgo State [15].

The detection limits for the analyzed samples were Mg (0.4), Mn (0.012), Na (0.04), Al (0.16), Ti (0.04), Ca (0.4), K (0.4), Fe (0.08), and Si (4) (mg L^{-1}).

The concentration conversion from mg·L^{-1}, provided by the ICP analysis, to the oxide percentage was carried out by using the periodic table to determine the molecular weight of the oxide compounds and employing the following equation:

$$\%O_m = \frac{C_L \times V_{af} \times F_d \times F_g}{W_s} \times \frac{1\ \mathrm{g}}{1000\ \mathrm{mg}} \times 100, \qquad (1)$$

where O_m = metal oxide; C_L = concentration obtained by ICP (mg L^{-1}); V_{af} = flask volume (L); F_d = dilution factor; W_s = sample weight (g); and F_g = gravimetric factor = molecular weight of oxide (g)/atomic weight of metallic element (g).

The samples of 5 g with a particle size range of 74–150 μm were prepared for analysis by polarizing optical microscopy (POM), and the minimum quantity of the samples was placed in the sample holder (<1 g) with two resin drops in xylene and was covered with an object to obtain a uniform surface for subsequent identification with an Olympus Microscope model BX41 with the objectives 10X and 20X in transmitted light with parallel Nichols (PN) [16, 17].

The scanning electron microscopy (SEM) was performed employing a microscope JEOL, model JSM-6300, at 30 kV equipped with an EDS detector to take photomicrographs and to do chemical microanalysis. The samples of 0.2 g with a particle size range of 45–74 μm were placed in the metal sample holders using a double-side graphite adhesive tape; the samples were covered with a thin gold foil for 10 sec to achieve greater conductivity before their analysis.

The analysis by X-ray diffraction (XRD) was made on the clay samples of 5 g with a particle size range of 38–45 μm. The samples were placed in the sample holders to be analyzed in an INEL diffractometer, model EQUINOX 2000,

FIGURE 1: Kaoline area of Huayacocotla, Ver.,-Alumbres, Hgo., Mexico.

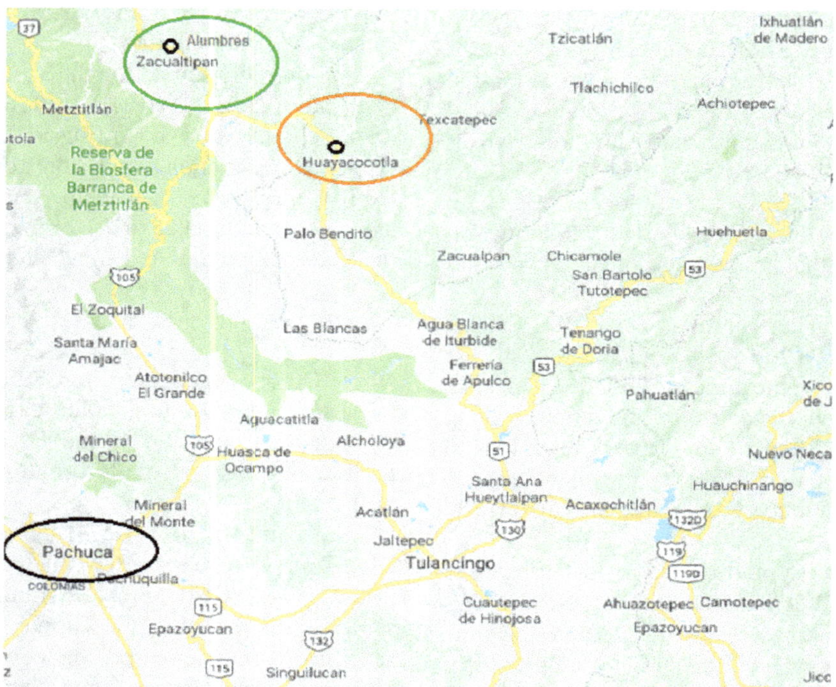

FIGURE 2: Access to the study zones of kaolin leaving from Pachuca, Hidalgo, Mexico.

with a simultaneous detector 2θ of 5–110° and Co $K\alpha$ (1.789010 Å) cathode at 220 V for 30 min for the identification of the main mineral phases.

The refractoriness test was performed by employing a pyrometric cone equivalent (PCE) test, according to the C-24 ASTM standard [18]. The Arabian rubber and the kaolin with a particle size range of 38–45 μm were mixed to develop the pasta for manufacturing the test cones. The amount of Arabian rubber used was varied depending on the kaolin features. This pasta was placed in the cone mold and dried for a few minutes before removing it from its mold for its synthesis in the temperature range of 450°C up to 1600°C with a heating rate of 2.5°C min^{-1} in a Nabertherm furnace. Based on the SiO_2, Al_2O_3, and FeO content, five samples from Huayacocotla (H-RMC-1, H-RMC-4, H-RMC-5, H-RMC-15, and H-RMC-21), three samples from Alumbres (A-RMC-25, A-RMC-26, and A-RMC-29), and three Orton cones were used for this test.

The thermogravimetric analysis (TGA) was done with an equipment brand *Mettler Toledo* model *TGA/SDTA851*^{e}. Nine kaolin samples were selected at random from the two areas of the study; four of them corresponded to the Huayacocotla area (H-RMC-1, H-RMC-4, H-RMC-12, and H-RMC-21) and five to the Alumbres area (A-RMC-25,

A-RMC-26, A-RMC-29, A-RMC-38, and A-RMC-44). The samples of 3 ± 0.5 g with a particle size range of 74–150 μm were settled in 70 μL alumina crucibles and were heated from 50°C up to 1100°C with a heating rate of 10°C min^{-1} in a N_2 atmosphere with a 100 ml min^{-1} flow rate.

The statistical principal component analysis was performed using the XLSTAT Software. V. 18.07 [19] for determination of factors that affect kaolin stability temperature; this was done with the ICP and PCE data.

3. Results and Discussion

3.1. Chemical Analysis by ICP. The oxides' average percentage of principal elements are shown in Tables 1 and 2, and the minimum and maximum values and the standard deviation of the samples' total number in both zones are presented.

The values obtained for SiO_2 were heterogeneous in the two areas. In Huayacocotla, the minimum percentage is 21.7% and the maximum is 84.1% with the average of 65.3 ± 16.3%. In the Alumbres area, the minimum value is 41.6% and the maximum is 87.8%, with an average of 69.3 ± 14.6%.

The Al_2O_3 concentration in the Huayacocotla samples is heterogeneous, with a minimum value of 6.3%, a maximum value of 30.7%, and an average of 19.3 ± 6.6%. In the Alumbres area, the behavior is similar, with a minimum value of 7.8%, a maximum value of 36.0%, and an average of 19.3 ± 7.8%.

Additionally, FeO is present in the heterogeneous concentration in both areas. In Huayacocotla, the minimum value is 0.1% and the maximum value is 6.2%. In Alumbres, the values are 0.1% and 19.0%, the minimum and maximum, respectively. The last area shows that the FeO percentage is higher than that in the first area; this is also reflected by the material coloration.

TiO_2 has been evaluated, and it has a relatively low concentration in the two areas; Huayacocotla presented 0.1% and 0.7% as the minimum and maximum values, respectively, and 0.4 ± 0.4% as the average. Alumbres presented 0.1% and 1.3% as the minimum and maximum values, respectively, and 0.5 ± 0.4% as the average. TiO_2 can be associated with presence of rutile, which is the most common mineral of titanium dioxide; however, polymorphs rarely include brookite and anatase, both form unique and distinctive crystals on the earth, and this is typical in felsic rocks [20].

The CaO concentration is heterogeneous, the minimum value recorded in Huayacocotla is <0.1%, the maximum value is 0.6%, and the average value is 0.2 ± 0.2%. Alumbres has the minimum value of 0.1%, maximum value of 0.4%, and the average value of 0.2 ± 0.1%.

The Na_2O concentration in Huayacocotla is higher than that in Alumbres. The first has a minimum value of 0.1%, a maximum value of 1.6%, and an average value of 0.6 ± 0.5%, while the Alumbres region has a minimum value of 0.1%, a maximum value of 0.4%, and an average value of 0.2 ± 0.1%.

Concentrations of CaO and Na_2O in both areas can mean reflection of altered calco-sodium plagioclase that is represented by clay in this region. Additionally, the percentages of MgO and MnO in the same areas are <0.01%; thus, these results are omitted from Tables 1 and 2. The oxide percentages of elements in the kaolin samples were similar to that in potassium feldspar (K-Feld) reported by Walsh and Howie [15].

3.2. Polarizing Optical Microscopy (POM). The analysis by POM allowed the observance of clay samples from Huayacocotla with a habit crystalline acicular that is weakly associated with minerals of high relief. Some of the clay samples present intergrowths with opaque minerals; these samples are appreciable crystalline forms of K-Feld, quartz, and plagioclase with iron oxides development on their cleavage, as shown in Figure 3(a).

Figure 3(b) shows minerals as high-relief impurities with crystalline habits from anhedral to subhedral; these are represented by quartz, plagioclase, and K-Feld. The last two minerals are associated with opaque mineral intergrowths, and their cleavage is also observed with some spheroidal minerals, which possibly correspond to chlorite.

Kaolin samples from Alumbres contain significant amounts of opaque minerals with a crystalline cubic system, such as pyrite, some coarse grits intergrowth with iron oxides on strongly oxidized clay, and high developments of metallic pyrite minerals, as is shown in Figure 4(a); therefore, this represents a hydrothermal alteration area [21].

Sometimes, pollution is observed by high-relief minerals associated with plagioclase, quartz, and opaque minerals; all of these minerals are derived from protolithic (rhyolite) that presents three clays types: acicular, not acicular, and granular with a high hardness from silica. The clays also present coloration due to a chlorite presence, as shown in Figure 4(b).

The POM confirms the high content of SiO2, FeS, and FeO in certain samples from Alumbres and Huayacocotla, and the different minerals are associated with the kaolin. Employing this technique is important for the determination and contrasting of the interferences present in each area.

3.3. Scanning Electron Microscopy (SEM). The kaolin morphologies are similar in both areas, but different mineral species are observed in the Alumbres samples. In Figure 5, there are particles with crystalline habit of the subhedral to euhedral type with well-defined plans; these could correspond to kaolinite (Figure 5(a)), dickite (Figure 5(b)), and illite (Figure 5(c)) in aggregate form.

Morphological analysis by SEM was performed on the Huayacocotla samples, allowing the identification of kaolinite in H-RMC-2 and H-RMC-3 and dickite in H-RMC-6, H-RMC-12, and H-RMC-16. In the Alumbres samples, illite in A-RMC-25, kaolinite in A-RMC-31, A-RMC-38, and A-RMC-39, and dickite in A-RMC-45 were identified. The elemental semiquantitative analyses performed on these samples are shown in Table 3; they were also identified by SEM-EDS and showed the same characteristics that were mentioned previously.

TABLE 1: Oxide percentage of kaolin majority elements of Huayacocotla.

	Huayacocotla zone, % oxide (SD)							
Sample	SiO_2	Al_2O_3	FeO (T)	TiO_2	CaO	K_2O	Na_2O	SO_2
H-RMC-1	36.1 (1.5)	19.2 (2.5)	0.2 (0.1)	0.6 (0.1)	0.2 (0.1)	1.8 (0.3)	1.4 (0.3)	0.6 (0.1)
H-RMC-2	76.8 (2.3)	15.3 (1.3)	0.4 (0.1)	0.2 (0.1)	0.6 (0.0)	0.5 (0.2)	1.5 (0.6)	0.3 (0.0)
H-RMC-3	66.9 (2.2)	19.5 (1.9)	0.1 (0.1)	0.2 (0.0)	0.4 (0.0)	1.1 (0.2)	1.5 (0.4)	0.6 (0.1)
H-RMC-4	47.3 (2.0)	29.4 (1.4)	6.2 (0.4)	0.9 (0.1)	0.6 (0.0)	0.9 (0.1)	1.6 (0.4)	0.7 (0.2)
H-RMC-5	51.4 (2.2)	30.7 (0.9)	0.2 (0.1)	1.7 (0.3)	0.2 (0.0)	0.1 (0.1)	0.4 (0.2)	0.6 (0.1)
H-RMC-6	72.1 (2.1)	17.7 (1.0)	0.3 (0.1)	0.3 (0.0)	0.1 (0.1)	0.2 (0.1)	0.7 (0.4)	0.2 (0.0)
H-RMC-7	77.1 (1.2)	6.3 (1.7)	0.1 (0.1)	0.3 (0.1)	0.1 (0.0)	1.3 (0.2)	0.4 (0.1)	0.3 (0.1)
H-RMC-8	73.5 (1.3)	15.1 (1.6)	0.1 (0.1)	0.5 (0.0)	0.2 (0.0)	0.4 (0.4)	0.4 (0.3)	0.2 (0.0)
H-RMC-9	66.7 (1.4)	19.6 (1.2)	0.4 (0.2)	0.9 (0.1)	0.1 (0.1)	0.1 (0.1)	0.3 (0.1)	0.4 (0.1)
H-RMC-11	44.3 (1.2)	29.9 (1.3)	0.5 (0.1)	1.3 (0.3)	0.2 (0.0)	1.2 (0.2)	0.6 (0.1)	0.3 (0.1)
H-RMC-12	36.6 (1.4)	24.7 (0.9)	0.3 (0.2)	0.3 (0.1)	0.2 (0.0)	1.5 (0.1)	1.6 (0.4)	1.0 (0.4)
H-RMC-14	72.0 (1.4)	16.3 (0.9)	0.1 (0.0)	0.2 (0.1)	0.6 (0.1)	0.4 (0.2)	1.3 (0.2)	0.6 (0.2)
H-RMC-15	74.3 (1.7)	16.5 (1.2)	0.2 (0.0)	0.3 (0.2)	0.3 (0.1)	0.3 (0.3)	0.3 (0.2)	0.4 (0.1)
H-RMC-16	84.1 (1.6)	12.3 (1.0)	0.1 (0.1)	0.1 (0.0)	0.1 (0.0)	0.1 (0.1)	0.2 (0.1)	1.1 (0.4)
H-RMC-17	60.6 (1.7)	22.1 (1.2)	0.1 (0.0)	0.1 (0.0)	0.1 (0.0)	1.3 (0.1)	0.6 (0.2)	2.1 (0.6)
H-RMC-18	73.6 (1.1)	19.0 (1.4)	0.1 (0.0)	0.3 (0.1)	0.1 (0.0)	0.5 (0.2)	0.4 (0.2)	1.9 (0.5)
H-RMC-19	82.6 (1.7)	11.7 (1.3)	0.1 (0.0)	0.1 (0.0)	0.4 (0.0)	0.2 (0.1)	0.1 (0.1)	0.1 (0.0)
H-RMC-20	78.3 (1.9)	16.7 (1.1)	0.1 (0.0)	0.2 (0.1)	0.1 (0.0)	0.6 (0.2)	0.2 (0.1)	0.2 (0.0)
H-RMC-21	80.0 (1.6)	16.7 (1.9)	0.1 (0.1)	0.2 (0.0)	0.1 (0.1)	0.2 (0.1)	0.2 (0.1)	0.3 (0.1)
H-RMC-22	79.8 (1.2)	14.7 (1.1)	0.2 (0.1)	0.3 (0.2)	0.1 (0.0)	0.1 (0.1)	0.1 (0.1)	0.7 (0.2)
H-RMC-23	78.4 (1.8)	16.8 (1.4)	0.2 (0.0)	0.2 (0.1)	<0.1	0.2 (0.1)	0.1 (0.1)	0.9 (0.3)
H-RMC-32	21.7 (2.5)	27.9 (2.1)	0.1 (0.1)	0.2 (0.1)	<0.1	2.1 (0.1)	0.6 (0.1)	0.3 (0.0)
H-RMC-33	71.3 (1.7)	11.0 (1.3)	0.1 (0.0)	0.2 (0.0)	0.1 (0.1)	1.5 (0.1)	0.3 (0.1)	0.3 (0.0)
H-RMC-35	65.0 (1.4)	26.9 (1.4)	0.1 (0.1)	0.7 (0.2)	0.1 (0.1)	0.5 (0.2)	0.4 (0.3)	0.3 (0.0)
H-RMC-36	60.5 (1.7)	30.7 (1.3)	0.2 (0.1)	0.9 (0.1)	0.1 (0.1)	0.1 (0.1)	0.5 (0.3)	0.4 (0.1)
H-RMC-37	66.3 (1.9)	15.7 (1.7)	0.1 (0.0)	0.3 (0.1)	<0.1	1.4 (0.2)	0.5 (0.1)	0.4 (0.1)
Mean	65.3 (16.3)	19.3 (6.6)	0.4 (1.2)	0.4 (0.4)	0.2 (0.2)	0.7 (0.6)	0.6 (0.5)	0.6 (0.5)

FeO (T) was determined as total iron.

TABLE 2: Oxide percentage of kaolin majority elements of Alumbres.

	Alumbres zone, oxides % (SD)							
Sample	SiO_2	Al_2O_3	FeO (T)	TiO_2	CaO	K_2O	Na_2O	SO_2
A-RMC-25	60.9 (1.5)	13.6 (1.4)	14.0 (0.7)	0.4 (0.1)	0.1 (0.1)	1.6 (0.29)	0.1 (0.1)	0.5 (0.2)
A-RMC-26	41.6 (1.9)	16.3 (1.3)	19.0 (0.4)	1.2 (0.1)	0.2 (0.3)	1.5 (0.2)	0.1 (0.0)	0.4 (0.2)
A-RMC-27	68.9 (1.8)	16.6 (1.5)	0.5 (0.5)	0.2 (0.0)	0.1 (0.1)	1.5 (0.1)	0.4 (0.1)	0.3 (0.1)
A-RMC-28	87.7 (2.5)	11.5 (1.6)	0.2 (0.1)	0.3 (0.1)	0.1 (0.1)	0.7 (0.1)	0.4 (0.2)	0.3 (0.1)
A-RMC-29	60.6 (1.8)	36.0 (1.3)	0.3 (0.0)	1.3 (0.3)	0.1 (0.1)	0.1 (0.2)	0.4 (0.1)	0.3 (0.1)
A-RMC-30	73.6 (1.9)	21.4 (1.5)	0.2 (0.1)	0.3 (0.0)	0.2 (0.2)	0.6 (0.1)	0.2 (0.0)	0.6 (0.2)
A-RMC-31	78.5 (1.5)	21.4 (1.8)	0.1 (0.0)	0.2 (0.0)	0.2 (0.2)	0.1 (0.1)	0.2 (0.0)	0.1 (0.0)
A-RMC-38	87.8 (2.1)	7.8 (1.9)	1.5 (0.2)	0.4 (0.1)	0.1 (0.1)	0.2 (0.2)	0.1 (0.0)	0.4 (0.1)
A-RMC-39	51.2 (1.8)	31.3 (1.3)	6.3 (0.2)	1.3 (0.3)	0.4 (0.6)	0.1 (0.0)	0.3 (0.0)	0.4 (0.1)
A-RMC-40	78.7 (2.1)	18.6 (1.5)	0.2 (0.2)	0.3 (0.2)	0.4 (0.6)	0.7 (0.1)	0.1 (0.0)	0.4 (0.2)
A-RMC-41	58.6 (1.6)	22.6 (1.2)	0.1 (0.1)	0.2 (0.0)	0.4 (0.5)	1.7 (0.2)	0.4 (0.0)	0.4 (0.1)
A-RMC-42	83.4 (2.1)	13.4 (1.7)	0.1 (0.0)	0.4 (0.2)	0.1 (0.1)	1.2 (0.4)	0.2 (0.0)	0.4 (0.2)
A-RMC-43	84.5 (1.8)	10.7 (1.9)	0.1 (0.1)	0.1 (0.0)	0.3 (0.4)	0.8 (0.2)	0.2 (0.0)	0.4 (0.2)
A-RMC-44	51.3 (1.8)	26.4 (1.3)	0.1 (0.1)	0.2 (0.1)	0.4 (0.6)	1.5 (0.2)	0.1 (0.0)	0.4 (0.2)
A-RMC-45	72.5 (1.7)	22.4 (1.5)	0.1 (0.1)	0.2 (0.2)	0.4 (0.5)	0.4 (0.6)	0.1 (0.0)	0.4 (0.1)
Mean	69.3 (14.6)	19.3 (7.8)	2.9 (5.8)	0.5 (0.4)	0.2 (0.1)	0.8 (0.6)	0.2 (0.1)	0.4 (0.1)

FeO (T) was determined as total iron.

3.4. X-Ray Diffraction (XRD). The representative samples of kaolin in the two zones were selected for their high and low content of SiO2 to be analyzed by XRD, and they were evaluated with the Software Match V.3 and cards PDF-02 of the diffraction patterns. The major crystalline phases observed in Huayacocotla samples were andalusite (Al_2SiO_5), kaolinite ($Al_2Si_2O_5(OH)_4$), and cristobalite (SiO_2), and the major crystalline phases in Alumbres were quartz (SiO_2) and andalusite. The minority crystalline phases were quartz and kaolinite in Huayacocotla and pyrite (FeS_2) and dickite ($Al_2Si_2O_5(OH)_4$) in Alumbres. Finally, quartz, illite, sillimanite (Al_2SiO_5), and magnetite (Fe_3O_4) were identified as vestige mineral phases in both zones, as shown in Table 4.

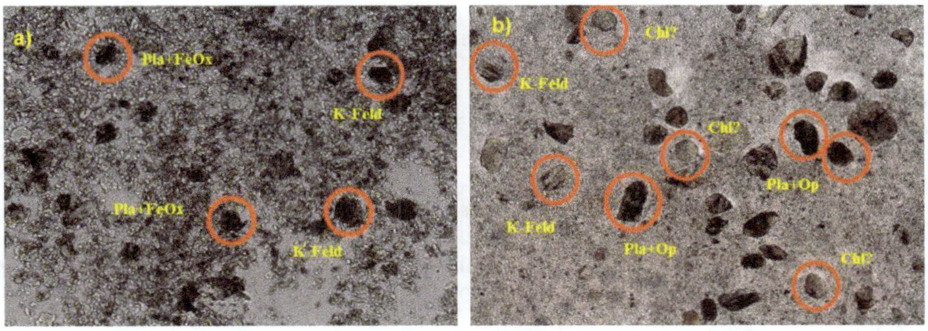

FIGURE 3: Polarization optical microscopy of Huayacocotla kaolin. (a) K-Feld and plagioclase with iron oxides (Pla + FeOx). (b) Spheroidal minerals, possible chlorite (Chl), plagioclase with opaque minerals (Pla + Op), K-Feld. Conditions: parallel Nichols, objective 10X, and 0.98 mm photography horizontal field.

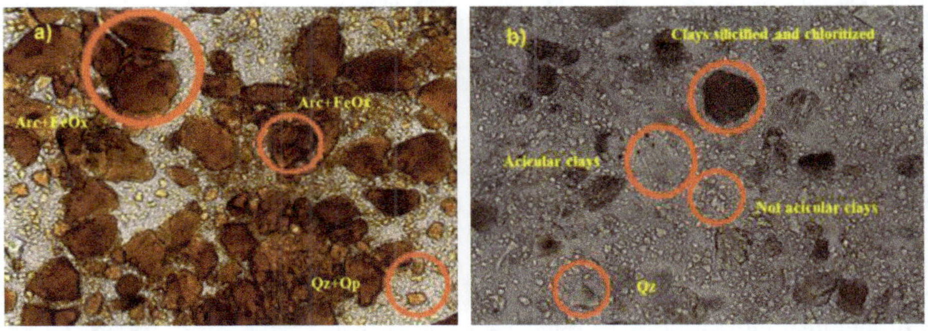

FIGURE 4: Minerals present in kaolin. (a) Minerals with high oxidation, clay with iron oxide (Arc + FeOx). (b) Minerals acicular, not acicular, and granular with possible chlorite (Chl) content. POM with parallel Nichols, objective 10X, and 0.98 mm photography horizontal field.

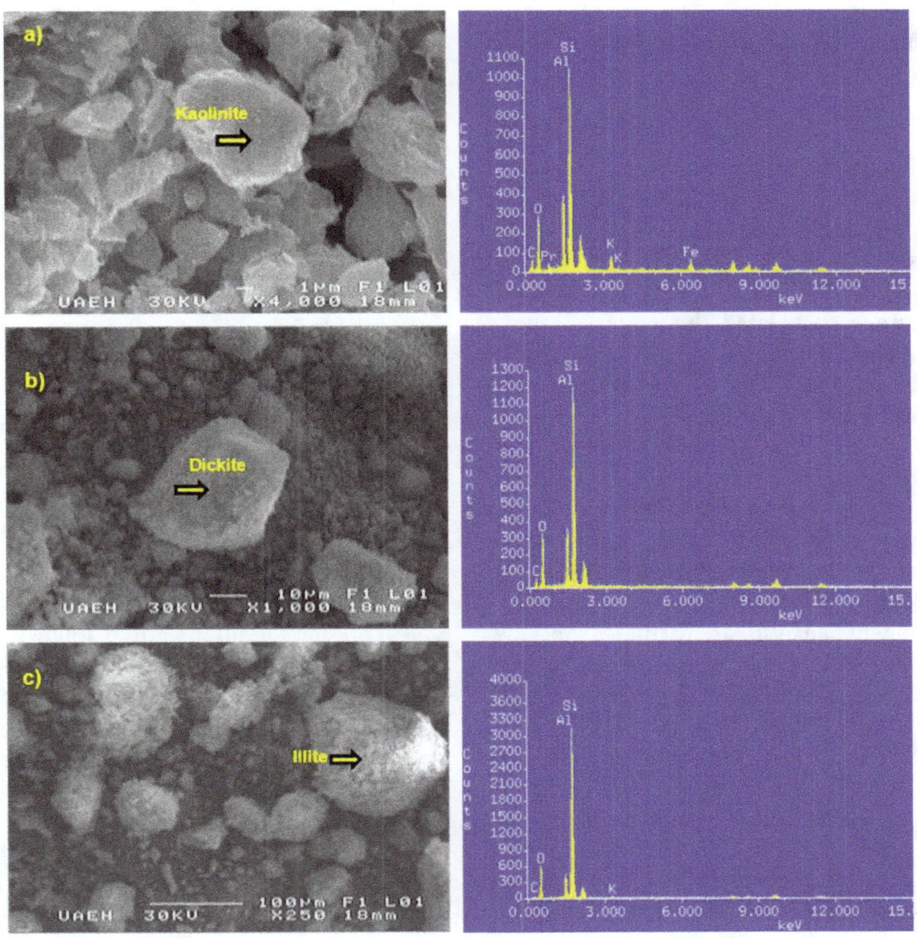

FIGURE 5: Kaolin minerals determined by SEM and EDS. (a) Kaolinite in H-RMC-2. (b) Dickite in H-RMC-16. (c) Illite in A-RMC-25.

TABLE 3: Semiquantitative analysis of the kaolin samples by SEM in percentage weight (SD).

Compounds	H-RMC-2	H-RMC-3	H-RMC-6	H-RMC-16	A-RMC-25	A-RMC-31	A-RMC-38	A-RMC-39
SiO_2	69.8 (4.6)	66.0 (5.2)	76.4 (5.0)	80.8 (5.4)	88.7 (7.4)	81.1 (6.8)	76.9 (4.6)	54.9 (3.6)
Al_2O_3	21.2 (1.4)	31.7 (2.8)	21.7 (2.4)	18.7 (2.4)	10.0 (2.4)	18.0 (3.6)	22.1 (3.4)	41.6 (4.6)
Na_2O	0.1 (0.1)	0.1 (0.1)	0.1 (0.1)	0.1 (0.1)	0.1 (0.1)	0.4 (0.1)	0.1 (0.1)	0.1 (0.1)
MgO	1.0 (0.4)	0.2 (0.1)	0.1 (0.1)	0.1 (0.1)	0.3 (0.1)	0.1 (0.1)	0.2 (0.1)	0.2 (0.1)
K_2O	2.7 (0.8)	1.1 (0.6)	0.1 (0.1)	0.1 (0.1)	0.2 (0.1)	0.2 (0.1)	0.1 (0.1)	0.1 (0.1)
CaO	0.1 (0.1)	0.1 (0.1)	0.1 (0.1)	0.1 (0.1)	0.1 (0.1)	0.1 (0.1)	0.1 (0.1)	0.1 (0.1)
FeO	4.5 (1.4)	0.4 (0.1)	0.1 (0.1)	0.1 (0.1)	0.4 (0.1)	0.1 (0.1)	0.4 (0.1)	1.3 (0.4)
TiO_2	0.7 (0.2)	0.5 (0.1)	1.6 (0.4)	0.2 (0.1)	0.2 (0.1)	0.2 (0.1)	0.2 (0.1)	1.7 (0.4)

The lower diffractogram of Figure 6 shows the main phases of the kaolin sample H-RMC-21 without heat treatment, which shows the kaolinite ($Al_2Si_2O_5(OH)_4$ with PDF 00-003-0052) and andalusite (Al_2SiO_5, PDF 00-039-0376). The XRD analysis was performed on kaolin H-RMC-21 that was sintered at 1450°C, and the phases that are observed are β-cristobalite (SiO_2, PDF 01-082-1407) and pseudomullite ($3(Al_2O_3.SiO_2)$, PDF 00-002-1160), as seen in the upper part of Figure 7. The transformation of the phases occurs as the temperature increases, and the silica is transformed into its different polymorphic forms, such as low quartz (573°C), high quartz (867°C), tridymite (>1400°C), and cristobalite (>1450°C). Simultaneously, there is a reaction of alumina with silica that originates presence of pseudomullite. It is not possible to observe the mullite phase formation due to the higher alumina content and the formation of eutectic, which favors the dissolution of alumina in a transitional phase of liquid silica above 1500°C [22, 23]. Diffractograms of kaolin treated at temperatures up to 1600°C have the same phases.

TABLE 4: Analysis of principal and secondary phases of the kaolin samples at Huayacocotla and Alumbres by XRD.

Sample	Principal phase (>50%)	Secondary phase (10–40%)	Trace (<5%)
Huayacocotla			
H-RMC-4	An	Kln	—
H-RMC-9	Cr	Kln	Mag
H-RMC-14	Cr	Kln	Qz
H-RMC-21	An	Kln	Mag
H-RMC-32	An	Qz	—
H-RMC-35	Kln	Qz	—
Alumbres			
A-RMC-25	Qz	Py	Ill
A-RMC-26	Qz	Py	Kln, Ill
A-RMC-40	An	Dck	Cr, Ill, Sil
A-RMC-45	An	Dck	Qz,

An = andalusite, Cr = cristobalite, Qz = quartz, Kln = kaolinite, Dck = dickite, Py = pyrite, Mag = magnetite, Ill = illite, Sil = sillimanite.

3.5. Refractoriness Test (PCE). The pyrometric cone equivalent (PCE) test for the kaolin samples from Huayacocotla and Alumbres can help us to determine the stability temperature (T_{St}), the softening (T_{So}), and/or vitrification (T_V) of the kaolin.

As the temperature was increased, the samples acquired a reddish tone that varied according to the iron content of the samples; however, after 1200°C, the test cones recovered their white color, except those whose iron content was higher than 5%.

According to the temperature employed in the pyrometric tests, it was possible to observe stability temperatures below 1300°C and up to 1600°C and softening temperatures and vitrification from approximately 1600°C and above, as is shown in Figure 7.

Principal component analysis is useful in the determination of the parameters that influence the thermal stability of kaolin for which the PCE values (Figure 2) and oxides composition data (Tables 1 and 2) are used to standardize each variable to the mean zero, and the unit variance is used to make all variables have the same weight in the analysis to obtain the Pearson correlation (Table 5).

It is observed in Table 5 that the stability temperature of kaolin varies directly with the content of SiO_2 and inversely with the content of K_2O and that the contents of Al_2O_3, FeO, and CaO correlate directly with contents of TiO_2, K_2O, and Na_2O, respectively.

The principal component analyses of kaolin from the Huayacocotla-Alumbres region are shown in Figure 8, where components F1 and F2 are the main weights. In this figure, the stability temperature correlates positively with F1 and F2 components, negatively with K_2O and FeO; the stability temperature does not correlate with Al_2O_3 or SiO_2 but correlates negatively with the content of Na_2O, CaO, and SO_2.

The softening temperature of kaolin is due to presence of K_2O, Na_2O, CaO, FeO, and SO_2. The vitrification temperature depends on the content of SiO_2, whereas high stability depends on the Al_2O_3 content.

3.6. Thermogravimetric Analysis (TGA). The TGA obtained for the kaolin samples is shown in Figure 9; the samples present characteristic sigmoidal curves and significant loss of mass, which starts at 400°C and ends near 850°C.

Working conditions are as follows: thermogravimetric analyzer, Mettler Toledo, Mod. TGA/SDTA851^e, with a range from 50 to 1100°C with a heating rate of 10°C min^{-1}, a N_2 atmosphere, and a flow rate of 100 $ml{\cdot}min^{-1}$.

It was possible to observe desorption of surface water in the samples H-RMC-4, A-RMC-25, and A-RMC-26 from 50 to 100°C.

The samples started their decomposition between 456°C and 576°C. The hydroxyl groups were removed from the crystal network of ore, and kaolinite was transformed to

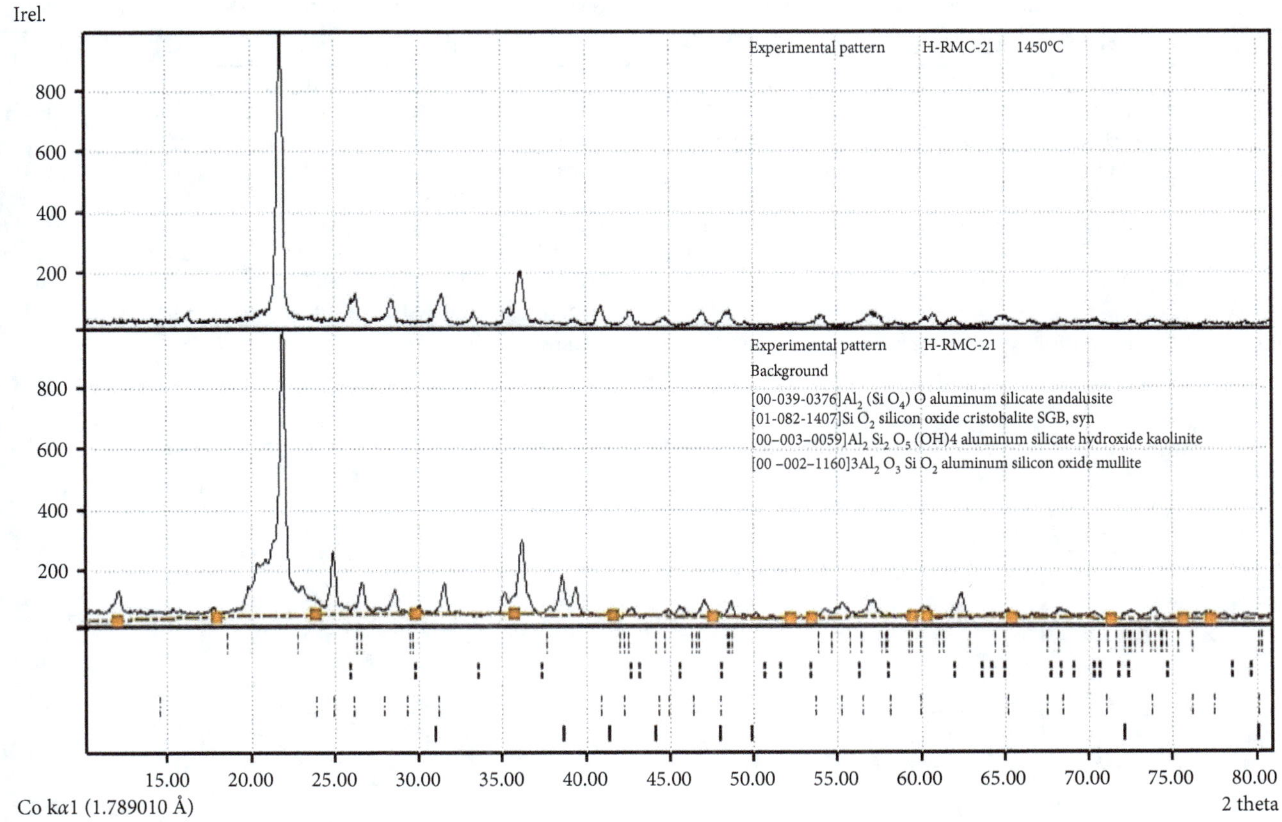

FIGURE 6: Diffractogram of the sample H-RMC-21 with and without thermal treatment.

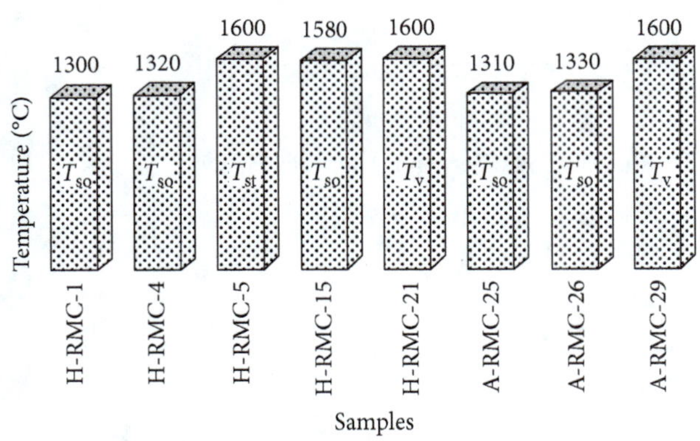

FIGURE 7: Kaolin thermal treatment of Huayacocotla, Ver., and Alumbres, Hgo., samples. Stability temperature (T_{St}), softening (T_{So}), and/or vitrification (T_V) of the kaolin.

metakaolinite with the water loss of all samples; this derivative is low quartz according to the SDTA data of A-RMC-44 of Figure 9 by equilibrium 2 [6, 24–27]:

$$\underset{\text{Kaolinite}}{Si_2Al_2O_5(OH)_4} \xrightarrow{456-576^\circ C} \underset{\text{Metakaolinite}}{Al_2O_32SiO_2} + 2H_2O \quad (2)$$

The samples H-RMC-1, H-RMC-12, H-RMC-21, A-RMC-29, and A-RMC-38 present low mass loss, which is characteristic of well-crystallized kaolinite. The H-RMC-4 samples, A-RMC-25, A-RMC-26, and A-RMC-44 exhibit a high mass loss. This behavior can be attributed to two factors. The first is water elimination retained between blades of ore, and the second is determined by reorganization of other present minerals and silica polymorphs, such as low and high quartz and tridymite.

In the heating stage, which starts at 739°C and continues up to 850°C, there is mass loss in samples that is mainly attributable to the impurities present in kaolin that contains carbonates, etc. Once these dehydration processes and decomposition are completed, the samples behave stably up to approximately 950°C.

The kaolin samples having low losses of mass at temperatures under 950°C is indicative of high purity. So as seen in Figure 10, the samples H-RMC-21, A-RMC-29, and A-RMC-38 have a high purity, while the samples H-RMC-4, A-RMC-26, and A-RMC-44 are less pure.

The TGA data of A-RMC sample were converted to SDTA, and at approximately 1000°C, the formation of mullite can also be observed, as is shown in Figure 9, according to the equilibrium 3:

$$\underset{\text{Metakaolinite}}{2(Al_2O_32SiO_2)} \xrightarrow{989-1009^\circ C} \underset{\text{Mullite}}{Al_4Si_2O_{10}} + \underset{\beta-\text{cristobalite}}{2SiO_2} \quad (3)$$

4. Conclusions

The principal mineral phases identified in the Huayacocotla samples are cristobalite, andalusite, and kaolinite; and the

TABLE 5: Pearson correlation of kaolin from Huayacocotla Ver. and Alumbres Hgo.

Variables	SiO_2	Al_2O_3	FeO	TiO_2	CaO	K_2O	Na_2O	SO_2	T_{St}
SiO_2	**1**	−0.191	−0.344	−0.519	−0.284	−0.643	−0.546	−0.637	0.620
Al_2O_3	−0.191	**1**	−0.427	**0.727**	0.236	−0.530	0.331	0.172	0.421
FeO	−0.344	−0.427	**1**	0.062	0.006	0.634	−0.284	0.036	−0.650
TiO_2	−0.519	**0.727**	0.062	**1**	0.056	−0.219	0.026	0.223	0.235
CaO	−0.284	0.236	0.006	0.056	**1**	0.044	0.704	0.661	−0.339
K_2O	−0.643	−0.530	0.634	−0.219	0.044	**1**	0.272	0.385	**−0.940**
Na_2O	−0.546	0.331	−0.284	0.026	0.704	0.272	**1**	**0.733**	−0.430
SO_2	−0.637	0.172	0.036	0.223	0.661	0.385	**0.733**	**1**	−0.487
St_T (°C)	0.620	0.421	−0.650	0.235	−0.339	**−0.940**	−0.430	−0.487	**1**

Values in bold are different from 0 with a significance level $a = 0.05$; T_{St} is the stability temperature.

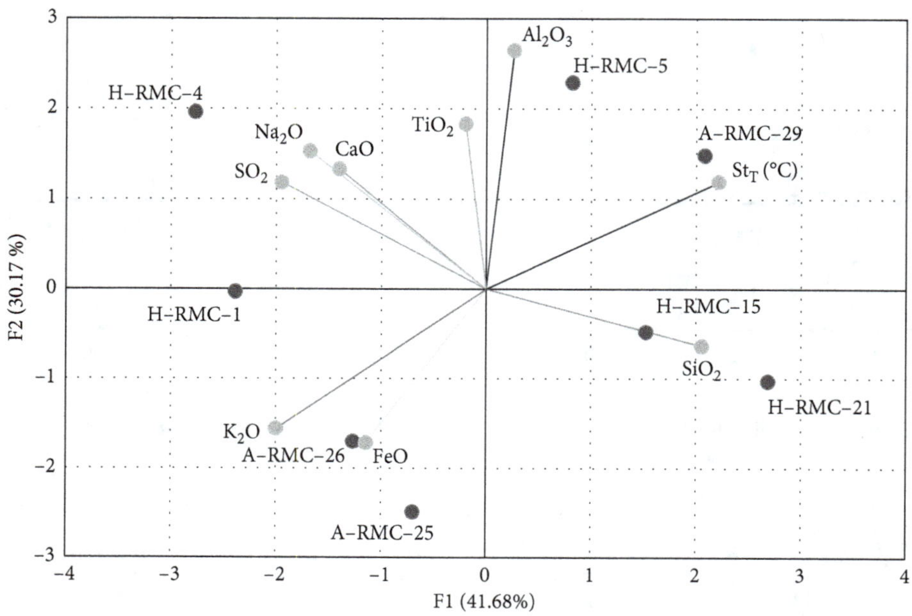

FIGURE 8: Principal component analyses of kaolin from Huayacocotla Ver. and Alumbres Hgo.

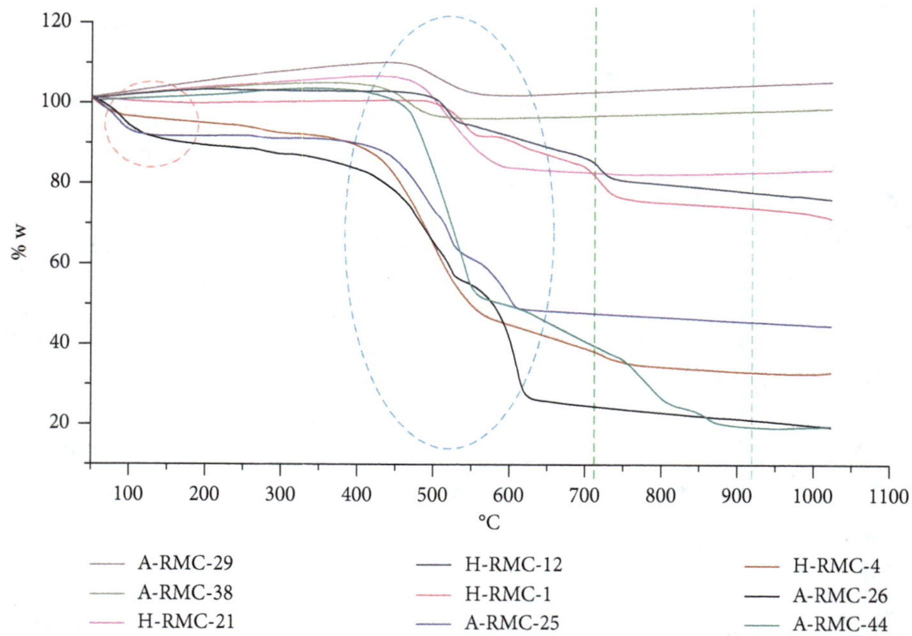

FIGURE 9: Thermograms of the kaolin samples from Huayacocotla, Ver., and Alumbres, Hgo.

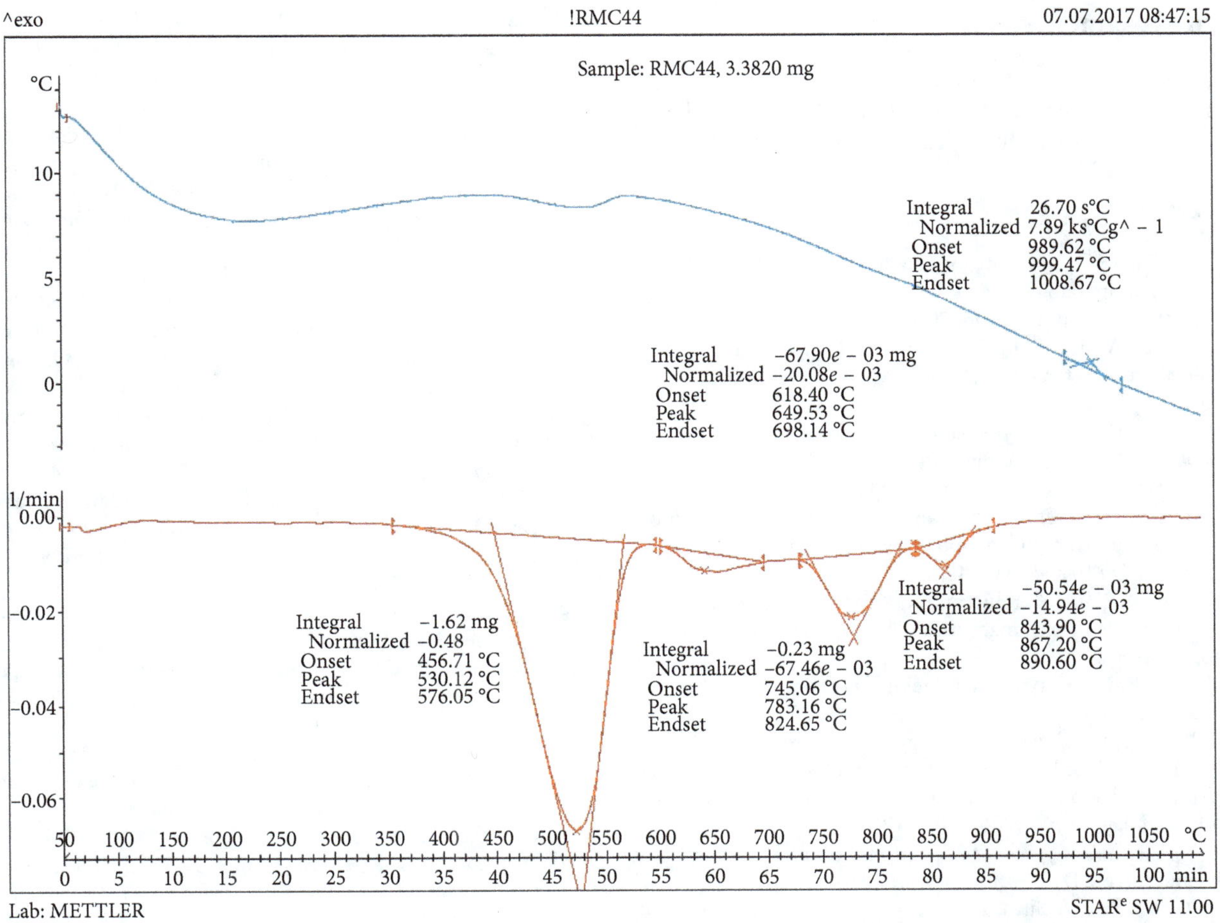

FIGURE 10: Thermogram for the A-RMC 44 sample from Alumbres, Hgo. Working condition: thermogravimetric analyzer, Mettler Toledo, Mod. TGA/SDTA851^{e}, range from 50°C up to 1100°C with a heating rate of 10°C min^{-1}, N_2 atmosphere, flow rate of 100 ml·min^{-1}.

principal mineral phases identified in the Alumbres samples are quartz and cristobalite. The secondary mineral phases detected in the Huayacocotla samples were kaolinite and quartz, and the secondary mineral phase detected in the Alumbres samples was dickite.

The kaolin alteration grade is different in the two areas; in Huayacocotla, there is hydrothermal alteration with the presence of primary minerals, such as plagioclase, feldspar, and quartz. In Alumbres, there is variation of clays and the presence of pyrite ores.

Finally, the trace mineral phases that have been determined in some samples are pyrite and sphalerite due to the calco-alkaline character of the argillic rock (kaolin), which has strong alteration accompanied by silicification.

The presence of mineral phases based on silica is due to the nature of the hosting rock, which comprises massive rhyolite and ignimbrite with some horizons of the rhyolitic domain. Therefore, the oxides and hydroxides with a greater presence in the kaolin samples are hematite (Fe_2O_3), limonite ($FeO(OH){\cdot}nH_2O$), and goethite ($FeO(OH)$) due to the late hydrothermal formation associated with meteoric leaching.

The crystalline morphology is similar for the subhedral type in both zones; this allowed us to identify the presence of kaolinite and dickite in Huayacocotla and illite and kaolinite in Alumbres.

The kaolin thermal stability is approximately 1300°C, and it rises up to 1600°C with increases of the Al_2O_3 content. When the SiO_2 content increases, the thermal stability of kaolin decreases, and its vitrifying temperature is lower than the stability temperature. The presence of K_2O, Na_2O, CaO, and SO_2 reduces kaolin's stability temperature to the softening temperature of the material.

Kaolin that tolerated temperatures between 1300°C and 1600°C can be used as a coating for low-temperature furnaces, while that which presented a refractory character of the silico-aluminous type due to the content of SiO_2 (70–90%) and Al_2O_3 (10–30%) can be used as a high-temperature refractory material since it shows a softening temperature and stability up to 1600°C.

A structural formula estimation of the minerals was not possible due to the high content of secondary and tertiary minerals.

Conflicts of Interest

The authors declare that they have no conflicts of interest.

Acknowledgments

The authors are grateful to CONACYT, PRODEP, and Autonomous University of the State of Hidalgo for the financial support and facilities provided to support this work.

References

[1] G.-I. Ekosse, "Kaolin deposits and occurrences in Africa: geology, mineralogy and utilization," *Applied Clay Science*, vol. 50, no. 2, pp. 212–236, 2010.

[2] M. Garcia-Valles, T. Pi, P. Alfonso et al., "Mineralogical and thermal characterization," *Clay Minerals*, vol. 50, no. 3, pp. 405–416, 2015.

[3] D. Gardner, *A Study of Mineral Impurities within the Georgia Kaolins*, Georgia State University, Atlanta, Georgia, USA, 2016.

[4] M. Jordan, M. Montero, S. Meseguer, and T. Sanfeliu, "Influence of firing temperature and mineralogical composition on bending strength and porosity of ceramic tile bodies," *Applied Clay Science*, vol. 42, no. 1-2, pp. 259–265, 2008.

[5] K. Singh, S. Kaur, H. Kaur, and K. Kaur, "Multifaceted role of clay minerals in pharmaceuticals," *Future Science OA*, vol. 1, no. 3, pp. 1–9, 2015.

[6] M. Valásková, "Clays, clay minerals and cordierite ceramics: a review," *Ceramics-Silikáty*, pp. 331–340, 2015.

[7] H. P. Alves, J. B. Silva, L. F. Campos, S. M. Torres, R. P. Dutra, and D. A. Macedo, "Preparation of mullite based ceramics from clay–kaolin waste mixtures," *Ceramics International*, vol. 42, no. 16, pp. 19086–19090, 2016.

[8] A. K. Chakraborty, *Phase Transformation of Kaolinite Clay*, Springer, New Delhi, India, 2014.

[9] W. Gonçalves, V. Silva, R. Menezes, G. Neves, H. Lira, and L. Santana, "Microstructural, physical and mechanical behavior of pastes containing clays and alumina waste," *Applied Clay Science*, vol. 137, pp. 259–265, 2016.

[10] P. Schroeder and G. Erickson, "Kaolin: from anciente porcelains to nanocomposites," *Elements*, vol. 10, no. 3, pp. 177–182, 2014.

[11] J. Srodon, V. Drits, D. McCarty, J. Hsieh, and D. Eberl, "Quantitative X-ray diffraction analysis of clay-bearing rocks from random preparations," *Clays and Clay Minerals*, vol. 49, no. 6, pp. 514–528, 2001.

[12] C. L. Fernández, *NORMA Oficial Mexicana NOM-021-RECNAT-2000, Que establece las especificaciones de fertilidad, salinidad y clasificación de suelos. Estudios, muestreo y análisis. Diario Oficial*, 2002.

[13] EPA, *Method 3052 Microwave Assisted Acid Digestion of Silicious and Organically Based Matrices, EPA SW-846, Revision 3*, Environmental Protection Agency, Washington, DC, USA, 2004.

[14] E. Muller, M. Mesko, D. Moraes, M. Korn, and E. Flores, "Wet digestion using microwave heating," *Microwave-Assisted Sample Preparation for Trace Element Determination*, pp. 99–142, Elsevier, Rio de Janeiro, Brazil, 2014.

[15] J. Walsh and R. Howie, "An evaluation of the performance of an inductively coupled plasma source spectrometer from the determination of the major and trace constituents of silicate rocks and minerals," *Mineralogical Magazine*, vol. 43, no. 332, pp. 967–974, 1980.

[16] D. Murphy, *Fundamentals of Light Microscopy and Electronic Imaging*, Wiley, London, UK, 2001.

[17] E. Pirard, "Multispectral imaging of ore minerals in optical microscopy," *Mineral Magazine*, vol. 68, no. 2, pp. 323–333, 2004.

[18] ASTM C24-01, *Standardd Test Method for Pyrometric Cone Equivalent (PCE) of Fireclay and High Alumina Refractory Materials C24-01*, ASTM International, West Conshohocken, PA, USA, 2001.

[19] Microsoft, Recuperado, 2016, website@xlstat.com, https://www.xlstat.com/es/descargar.

[20] R. Gaines, H. Skinner, E. Foord, B. Mason, and A. Rosenzweig, *Dana's New Mineraalogy: The System of Mineralogy of James Dwight and Edward Salisbury*, John Wiley & Sons Inc, Somerset, NJ, USA, 1997.

[21] M. Albert, L. Ortega, R. Lunar, F. Martínez, and R. Piña, "Mineralogy of the hydrothermal alteration in the Námafjall geothermal field (Iceland)," *Revista de la Sociedad Española de Mineralogía*, vol. 1, no. 15, pp. 25-26, 2011.

[22] K. Liu and G. Thomas, "Time-temperature-transformation curves for kaolinite-alfa-alumina," *Journal of the American Ceramic Society*, vol. 77, no. 6, pp. 545–552, 1994.

[23] H. Rezaire, W. Rainforth, and W. Lee, "Mullite evolution in ceramics derived from kaolinite, kaolinite with added alfa-alumina and Sol-Gel precursors," *British Ceramic Transactions*, vol. 96, no. 5, pp. 181–187, 1997.

[24] J. Shackelford and R. Doremus, *Ceramic and Glasss Materials*, Springer, New York, NY, USA, 2008.

[25] E. Jhonson and S. Arshad, "Hydrothermally synthesized zeolites based on kaolinite: a review," *Applied Clay Science*, vol. 97-98, pp. 215–221, 2014.

[26] B. Sonuparlak, M. Sarikaya, and I. Askay, "Spinel phase formation during the 980°C, exothermic reaction in the Kaolinite-to Mullite reaction series," *Journal of the American Ceramic Society*, vol. 70, no. 11, pp. 837–842, 1987.

[27] D. J. Duval, S. H. Risbud, and J. F. Shackelford, "Mullite," *Ceramic and Glass Materials: Structure, Properties and Processing*, p. 202, Berlin, Germany, 2008.

18

Experimental Research on Microstructure and Physical-Mechanical Properties of Expansive Soil Stabilized with Fly Ash, Sand, and Basalt Fiber

Qin-yong Ma ,[1,2] Zi-ming Cao ,[1,2] and Pu Yuan [1,2]

[1]*Engineering Research Center of Underground Mine Construction, Ministry of Education, Anhui University of Science and Technology, Huainan, Anhui 232001, China*
[2]*School of Civil Engineering and Architecture, Anhui University of Science and Technology, Huainan, Anhui 232001, China*

Correspondence should be addressed to Zi-ming Cao; ziming_cao@foxmail.com

Academic Editor: Carlo Santulli

Expansive soil is a source of risk to the foundations or subgrade engineering. Stabilization of expansive soil is imperative for practical engineering. A series of laboratory experiments were performed to analyze the physical-mechanical properties and microstructures of stabilized soil. Three stabilizers used in this study are fly ash, sand, and basalt fiber. Different percentages of fly ash (0, 5, 10, 15, and 20%), sand (0, 8, 16, and 24%), and basalt fiber (0 and 0.4%) were added by weight into natural soil. Experimental results indicate that the optimum moisture content of stabilized soil increases with the increase of fly ash content for a given sand content, whereas the maximum dry density shows a decreasing trend. The variation trend of optimum moisture content and maximum dry density turns reverse with the increase of sand content for a given fly ash content. Plasticity index is decreased by both increasing fly ash content and sand content. It is found that the maximum unconfined compressive strength and optimum growth rate of strength are obtained by selected mixtures of 10% fly ash, 8% sand, and 0.4% basalt fiber contents. As the analysis of complementary effect suggests, most of the mixt treatments applied in this study have produced good results associated with the strength enhancement of expansive soil. In line with the results of SEM tests, the connection among clay particles has been enhanced through the generation of hydration products (C-S-H and AFt) of fly ash. The filling effect of sand has increased the integrality and compactness of stabilized soil. Moreover, the gripping effect between fibers and soil particles notably improves the strength of stabilized soil. The effect of sand on reinforced soil with 0.4% basalt fiber increases the interfacial force between fibers and soil particles.

1. Introduction

Expansive soil is a kind of high-plasticity clay with strong hydrophilic mineral components, such as montmorillonite and illite. It presents the swelling property when absorbing water and shrinking property when losing water [1, 2]. Due to the unstable behavior of expansive soil, it is easy to cause a series of engineering disasters, inclusive of differential settlement of subgrade, incline and subsidence of buildings, and collapse of slope occurrence. Besides, the particular features of expansive soil will bring severe safety hazards and economic losses to expansive soil areas. Given these situations, improving the physical and mechanical properties of expansive soil is indispensable for stability of foundation and subgrade engineering.

In recent decades, how to find a feasible and economical method to improve engineering properties of expansive soil has posed a huge challenge to geotechnical engineers. In the meantime, some researchers have investigated various methods for expansive soil stabilization. The effect of lime, cement, and fly ash on the physical-mechanical behaviors of soil samples is also confirmed by the research of modified methods on expansive soils. As the test results of Yong and Ouhadi [3] suggested, lime or cement stabilization is practical to abate the deterioration of bearing capacity of expansive soil. As Khemissa and Mahamedi [4] reported, the cement-lime mixture amendment has improved the strength and durability of expansive clay remarkably. Furthermore, the effect of cement and lime on the plasticity, swelling behaviors, and strength properties of expansive soils is also confirmed by Al-Mukhtar et al. [5], Estabragh et al. [6], Sharma et al. [7], and Thyagaraj et al. [8] in their studies of improvement on expansive soils. Phanikumar and Sharma [9], Rao and Rao [10], and Feng et al. [11] found that fly ash is a suitable stabilizer to reduce the swelling property of expansive soil and improve its mechanical performance. Accordingly, chemical stabilization has been considered a practical and effective solution to reduce swell-shrink characteristics and improve physical-mechanical properties of expansive soil. From the perspective of economy and environmental protection, the industrial wastes, such as fly ash, must be treated in an appropriate way. Fly ash can be added scientifically in expansive soils to modify their plasticity index, free swelling ratio, and mechanical properties [12]. When fly ash is added into expansive soils, the reaction with clay particles will improve the strength, consolidation, and drainage characteristics [13]. In aspect of physical stabilization, sand is a nonexpansive soil with minimum plasticity and has no type of charge on its surface. The addition of sand into expansive soil helps to balance the electrical charge in expansive soil and then mitigates the expansion [14]. In addition, sand is broadly applied in civil engineering materials by its abundant source and low price. Another physical stabilization of expansive soil is fiber reinforcement. The basic principle of fiber-reinforced expansive soil is that it utilizes the frictional force between fibers and soil particles to constrain the deformation of soil, and the physical-mechanical properties of expansive soil improve significantly [15–18]. Given that basalt fiber is characterized by environmental protection, good compatibility, and superior mechanical properties [19, 20], it is also selected as an additive into the soil specimens during the experiment. The previous investigations primarily concentrated on the macro-characteristics of expansive soil, and limited studies are available on microstructural analysis of stabilized expansive soil. The experiment of microstructure not only reflects the internal structure changes before and after the improvement of soil specimens but also explains the intrinsic mechanism of soil stabilization [21, 22].

In the following chapters of this paper, an experimental study has been conducted for expansive soil stabilized with fly ash, sand, and basalt fiber. This paper aims to study the effect of the noted three additives on physical-mechanical and microstructural characteristics of expansive soil which can be satisfactorily applied for construction of foundations, subgrade, road, and even driveways after reasonable and effective treatments.

2. Experimental Materials and Methods

2.1. Experimental Materials. The test expansive soil specimens were procured from a subway construction site in Hefei, Anhui Province. Hefei is known as a typical expansive soil area. The test results of physical properties of natural soil are shown in Table 1 and Figure 1.

The fly ash was collected from the thermal power plant in Huainan, Anhui Province. The primary chemical compositions of fly ash are SiO_2, Al_2O_3, Fe_2O_3, and CaO. The percentage of SiO_2, Al_2O_3, Fe_2O_3, and CaO is 40.12, 23.58, 6.39, and 15.65, respectively. Other chemical compositions have taken up 14.26%. Based on the chemical composition of fly ash, it can be classified as class-C fly ash according to ASTM C618 Standard Specification [23].

The sand applied in this study is commercially available Xiamen sand. The fineness modulus of sand reaches 2.31.

Basalt fiber is a material made from extremely fine fibers of basalt, encompassed by the minerals plagioclase, pyroxene, and olivine. Basalt fiber has been broadly used in scientific experiments by its excellent mechanical characteristics and green features. The major performance indices of the basalt fiber used in this study are listed in Table 2.

2.2. Specimens Preparation and Experimental Methods. In this study, the soil specimens blended with various additives were tested to determine their physical-mechanical and microstructural properties. Fly ash and sand were added into the soil samples in different percentages on dry weight basis of expansive soil. The content of fly ash falls into five grades, i.e., 0%, 5%, 10%, 15%, and 20%. Besides, the content of sand falls into four grades, i.e., 0%, 8%, 16%, and 24%. The basalt fibers were dispersed into soil samples at 0% and 0.4%.

The physical performance index of expansive soil specimens was determined in line with the pertinent tests specified in GB/T 50123-1999 [24]. All the soil specimens of the unconfined compression test keep the same moisture content of 18.9% and the same dry density of $1.76\,g/cm^3$. The loading rate of unconfined compression test was fixed at 1 mm/min. The preparation procedures of stabilized soil specimen are as follows: (1) the dry soil samples were crushed and passed through a 2 mm sieve; (2) the sifted soil particles were put in a drying oven at a constant temperature of 105°C for more than 24 h; (3) the dry soil, fly ash, sand, basalt fiber, and distilled water were fully mixed before the preparation of the soil specimens, and then the mixtures were put in an airtight container for 24 h to make soil samples moisture equilibrium; (4) the prepared mixtures were placed into a compacting mold with a height of 100 mm and a diameter of 50 mm at least five times; (5) the prepared samples were kept in a standard curing chamber at a constant temperature of 20 ± 2°C and 95% relative humidity for 28 days. The soil specimens were investigated through the scanning electron microscopy (SEM) method to observe the changes in the microstructure. The SEM observations were conducted on blocks taken from a cylindrical sample with a height of 100 mm and a diameter of 50 mm. All the SEM samples were kept in a dryer for 24 h before the microstructural experiment. In addition, the SEM samples were gold coated by a sputtering technique to make them conductive after drying process. The SEM examinations were performed using a Quanta FEG 650 scanning electron microscope manufactured by FEI Corporation.

TABLE 1: Physical properties of expansive soil.

Specific gravity	Optimum moisture content (%)	Maximum dry density (g/cm^3)	Liquid limit (%)	Plastic limit (%)	Plasticity index (%)	Free swell ratio (%)
2.71	18.9	1.76	42.8	22.0	20.8	46

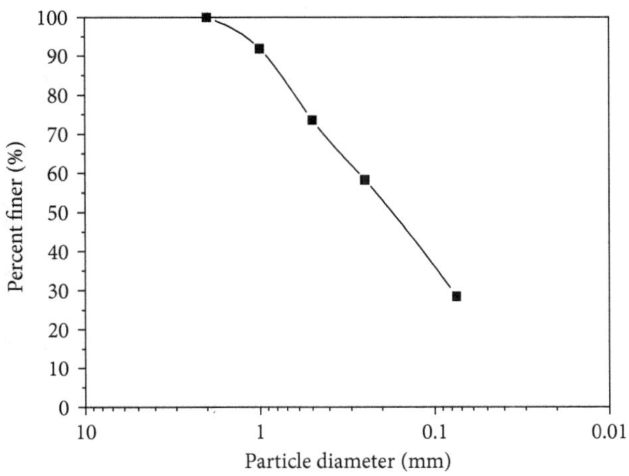

FIGURE 1: Grain-size distribution curve.

TABLE 2: Physical and mechanical properties of basalt fiber.

Density (g/cm^3)	Average length (mm)	Tensile strength (GPa)	Elasticity modulus (GPa)	Elongation at break (%)
2.8	9	3.5~4.0	90~110	3.2

3. Physical-Mechanical Test Results and Discussions

3.1. Specific Gravity. Specific gravity of natural expansive soil and stabilized soil samples obtained from the pycnometer is shown in Figures 2(a) and 2(b). The specific gravity value for expansive soil is 2.71. It is observed from the fitting results shown in Figure 2(a) that the specific gravity of stabilized soil samples decreases almost linearly with increasing of fly ash content. In contrast, the specific gravity of stabilized soil specimens increases almost linearly with the increase of sand content. The reason is that the virgin soil samples are heavier particles compared with fly ash and lighter particles compared with sand.

3.2. Compaction Characteristics. To study the water stability of stabilized soil, compaction tests were performed on stabilized expansive soil with different percentages of fly ash and sand. The curves of the optimum moisture content and maximum dry density can be acquired by performing the compaction tests. This compaction test adopts a light compaction test, and each group of test has prepared five soil specimens with different moisture contents. Each soil specimen was compacted in three layers and compacted 25 times per layer [24]. The results of compaction tests are shown in Figure 3. When the dosage of sand is 0%, the optimum moisture content gradually increases from 18.9% to 21.6% with the increase of fly ash content, as presented in Figure 3(a). However, the value of optimum moisture content changes from 15.3% to 18.9% with the addition of 0% fly ash to the virgin soil specimens. With the constant dosage of fly ash, the value of optimum moisture content decreases with increasing percentage of sand. It is clear as shown in Figure 3(a) that the value of optimum moisture content has the fastest reduction rate with 24% sand content.

With increasing of fly ash content, Figure 3(b) presents a reducing trend of maximum dry density of stabilized soil. In contrast, the maximum dry density shows an increasing trend with the increase of sand content. When the sand content is 0%, the maximum dry density gradually decreases from 1.76 g/cm^3 to 1.43 g/cm^3 with increasing of fly ash content. This is because the relative volumetric mass of fly ash is smaller than that of expansive soil. With the fly ash content of 0%, the maximum dry density of stabilized soil samples increases from 1.76 g/cm^3 to 2.05 g/cm^3 as the sand content increases from 0% to 24%. This is probably because the relative volumetric mass of sand is higher than that of soil particles. With the sand dosage of 24%, the maximum dry density of stabilized soil has the fastest growth rate except for the soil samples with 10% fly ash and 24% sand.

3.3. Plastic Limit, Liquid Limit, and Plasticity Index. Plastic limit, liquid limit, and plasticity index of expansive soil specimens stabilized with fly ash or sand are shown in Figures 4–6, respectively.

From Figures 5 and 6, it can be seen that the liquid limit and plasticity index decrease with the increase of fly ash content. However, the plastic limit increases with the increase of fly ash content in Figure 4. Under the sand dosage of 0%, the value of the plastic limit increases from 22.0% to 27.5%, the value of the liquid limit decreases from 42.8% to 34.9%, and the value of the plasticity index notably decreases from 20.8% to 7.4% with increasing of fly ash dosage. The nonplastic characteristics of stabilized soil improve with the increase of fly ash dosage in soil specimens. With the increase of fly ash content, the high cations (Al^{3+}, Fe^{3+}, and Ca^{2+}) of fly ash components exchange with the low cations (K^+ and Na^+) on the clay particle surface [11]. There is an ion exchange interaction between fly ash and soil particles with the presence of fly ash, which improves the flocculation of clay particles [9,11]. Therefore, the plasticity index decreases significantly with the reduction of the amount of clay particles and hydrophilic minerals. The results obtained in this study are similar to Feng et al. [11] and Kolay and Ramesh [14]. It is found that the plasticity index is a good indicator of swelling behaviors, namely, a lower plasticity index indicates a weaker expansibility [21]. In addition, the liquid limit decreases slightly with the increase of sand content. The plastic limit generally shows an increasing

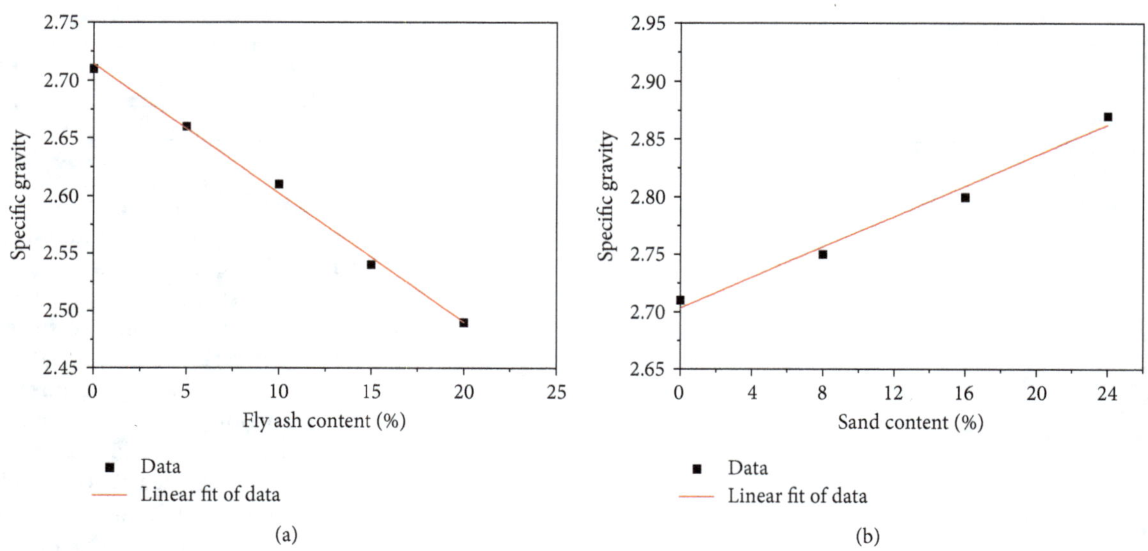

FIGURE 2: Specific gravity of stabilized expansive soil. (a) Expansive soil stabilized with fly ash. (b) Expansive soil stabilized with sand.

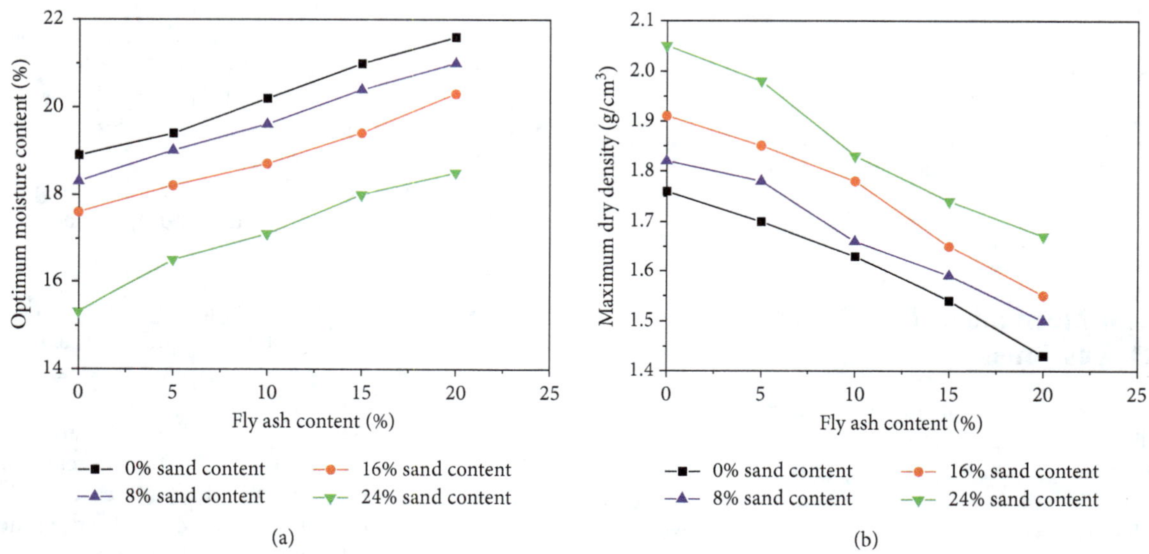

FIGURE 3: The compaction characteristics of stabilized expansive soil. (a) Optimum moisture content. (b) Maximum dry density.

trend with the increase of sand content. When the sand dosage is 24%, the incremental gradient of the plastic limit value is the largest because sand is virtually noncohesive.

3.4. Unconfined Compressive Tests

3.4.1. Axial Stress-Strain Curves. Figures 7 and 8 present the relationships between axial stress and strain, with regard to unreinforced and fiber-reinforced soil samples with different fly ash and sand contents. All the stress-strain curves in Figures 7 and 8 show a similar variation tendency, and they fall into three stages, i.e., elastic stage, plastic stage, and failure stage. In the elastic stage, the stress increases linearly with the increase of the strain. In the plastic stage, the incremental gradient of stress slows down, yet the stress continues to increase progressively. The increase of the incremental gradient of stress is approximately zero, and no conspicuous trend of further growth is observed. In the last stage, the axial stress decreases as the axial strain increases. This stage indicates the deterioration of soil samples. The soil samples have the same conditions in Figures 7(a) and 7(b), which suggest that the dosage of sand is 0%, and the fly ash contents are 0%, 5%, 10%, 15%, and 20%, respectively. The only difference is that all the soil samples in Figure 7(b) added 0.4% basalt fiber. It can be observed that the axial stress of the stabilized soil samples increases when the fly ash content is lower than 10% and decreases when the fly ash content exceeds 10%. All the values of maximum axial stress in Figure 7(b) are notably larger than that in Figure 7(a). From the beginning of the plastic stage to the end of the failure stage, all stress-strain curves in Figure 7(b) have

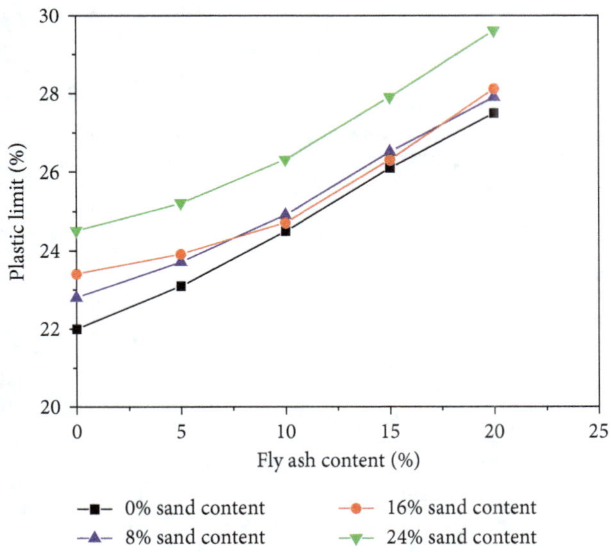

FIGURE 4: Plastic limit of stabilized expansive soil.

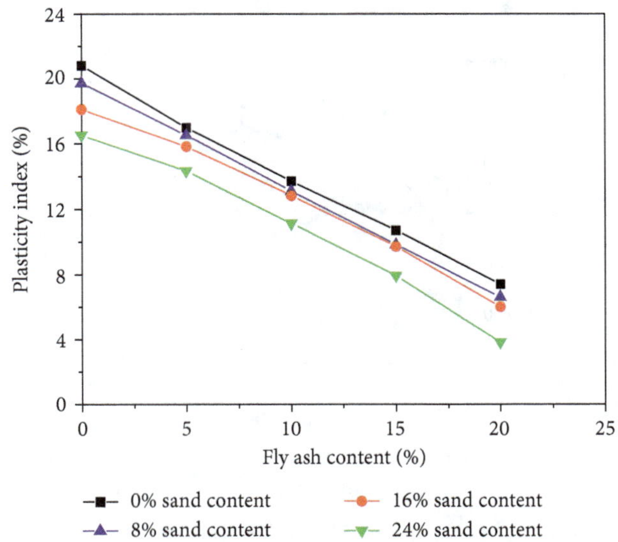

FIGURE 6: Plasticity index of stabilized expansive soil.

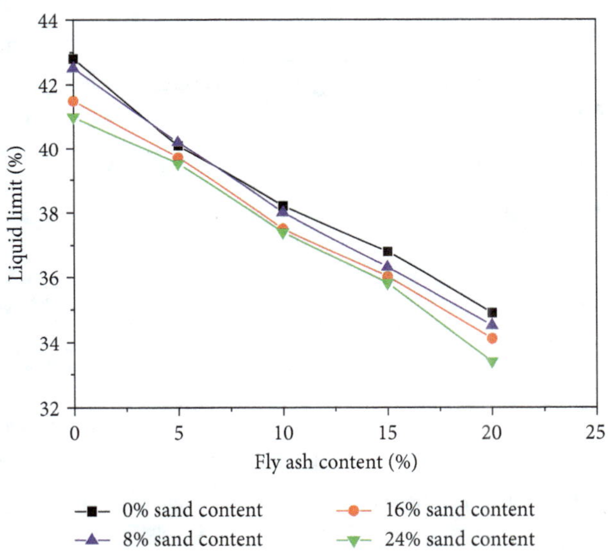

FIGURE 5: Liquid limit of stabilized expansive soil.

changed gently and maintained a relatively flat variation trend for a long time. Conversely, the stress-strain curves in Figure 7(a) have changed rapidly. Given a reasonable interpretation, a stable three-dimensional space structure system is formed inside soil with the addition of basalt fiber, which enhances the interfacial force between fibers and soil particles [20]. With the interaction of the bonding force and frictional resistance between fibers and soil particles, the basalt fiber sustains the tensile stress caused by the further deformation of the soil samples under loading. Accordingly, fiber-reinforced soil effectively constrains the continuous development of soil deformation, which shows a significant increase in maximum axial stress as compared with the stabilized soil samples with no basalt fiber.

The effect of sand dosage on the strength enhancement can be found in Figure 8(a). The effect of sand dosage on the maximum axial stress of sand-stabilized soil shows a similar feature with the effect of fly ash dosage on maximum axial stress. It is found that the optimal content of sand for the maximum axial stress of sand-stabilized soil is 8%. When the dosage of sand is rather more, the maximum axial stress of sand-stabilized soil decreases obviously. The reinforcement effect of basalt fiber on sand-stabilized soil is similar to the effect of fiber on fly ash-stabilized soil. The occurrence of maximum axial stress of fiber-sand-stabilized soil is corresponding to 8% sand content and 0.4% basalt fiber content. The maximum axial stress has a higher enhancement of fiber-sand-stabilized soil compared with sand-stabilized soil, as illustrated in Figure 8(b).

3.4.2. Unconfined Compressive Strength. Unconfined compressive strength is determined by the ultimate strength of soil that resists axial pressure under no lateral pressure. And the unconfined compressive strength is one of the important indices to reflect the mechanical properties of stabilized soil, and it often serves as a crucial basis for the design of road, subgrade, and foundation. The unconfined compression tests were performed with natural expansive soil and stabilized soil with fly ash, sand, and basalt fiber. The results of these tests are illustrated in Figure 9. The unconfined compressive strength of stabilized soil specimens increases progressively with both increasing fly ash content and sand content. The optimum contents of fly ash and sand for maximum unconfined compressive strength of stabilized soil were found to be 10% and 8%, respectively. For fly ash-stabilized soil, the maximum value of unconfined compressive strength suggested that with the increase of fly ash dosage, the occurrence of flocculation between fly ash particles and soil particles is improved until the fly ash dosage reaches its optimum value. When the fly ash content exceeds its optimum value, the excessive fly ash particles remaining in soil specimens are equal to the increase of silt particles in total soil particles. As a result, the strength of soil specimens decreases gradually with the addition of excessive

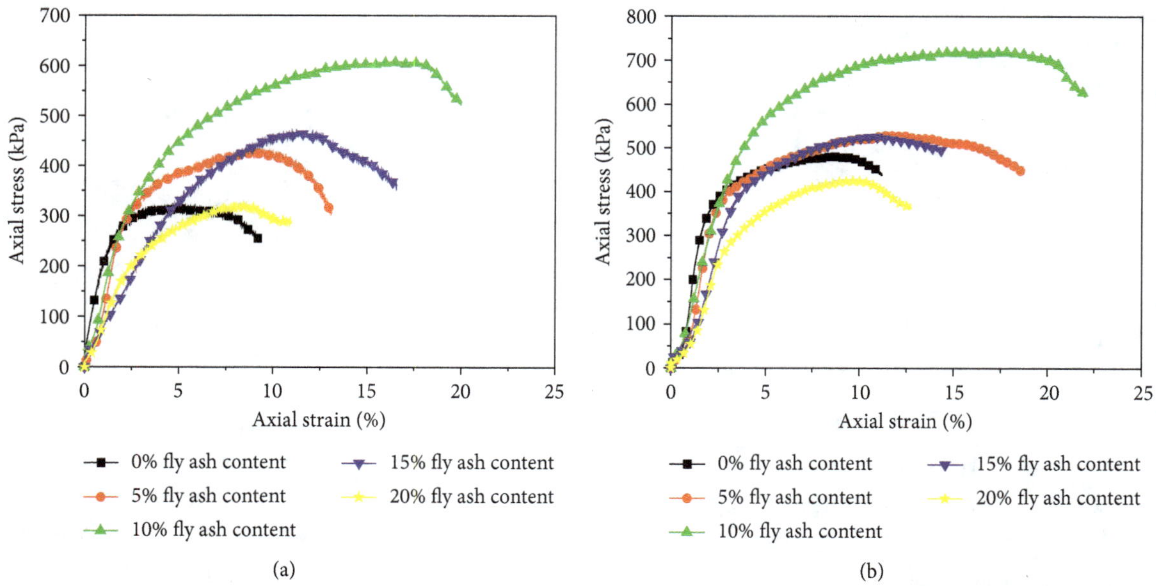

FIGURE 7: Stress-strain curves of stabilized expansive soil under different fly ash contents. (a) Without basalt fiber. (b) With 0.4% basalt fiber.

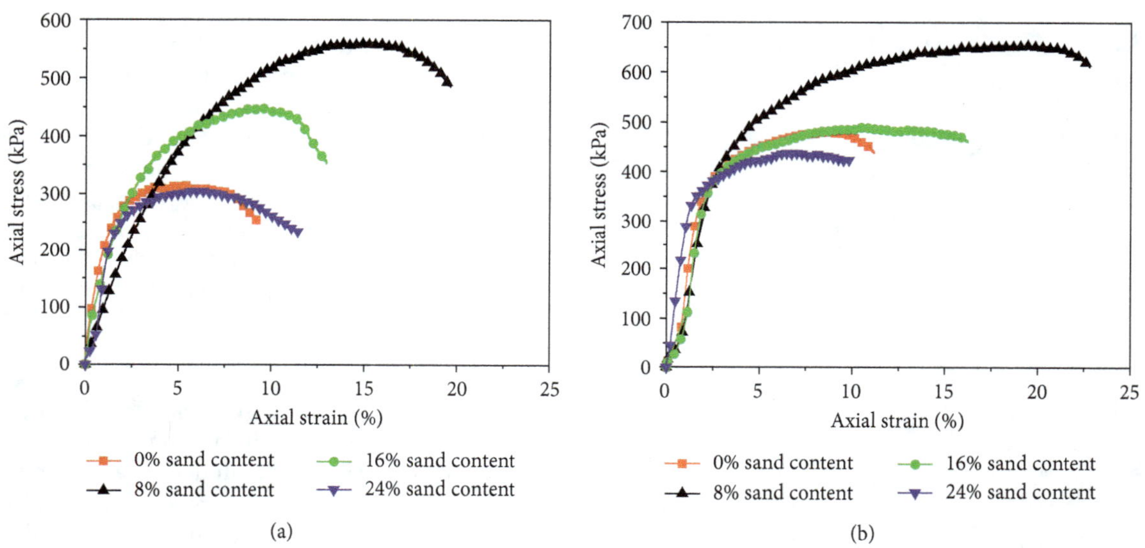

FIGURE 8: Stress-strain curves of stabilized expansive soil under different sand contents. (a) Without basalt fiber. (b) With 0.4% basalt fiber.

fly ash. Furthermore, the fly ash itself has no cohesion, which directly leads to the reduction of the strength of stabilized soil. For the soil stabilized with sand, the filling effect of sand increases the compactness and integrality of internal structure of stabilized soil, and the strength of stabilized soil also presents an evident uptrend. Moreover, the sand, considered as a hard particle, can effectively prevent the relative sliding of crack appeared inside the soil structures [25]. With the further increase of sand content, there are more sand-sand joint surfaces occurring inside the soil samples. Therefore, the unconfined compressive strength increases firstly and then decreases with the increase of sand content. As the results in Figures 9(a) and 9(b) show, the interaction between fly ash and sand enables the maximum strength of stabilized soil specimens under the optimum values for fly ash and sand content. It can be considered that the binding force among the fly ash, sand, and soil particles reaches the biggest value with the mixt addition of 10% fly ash and 8% sand into expansive soil. Besides, all the maximum strength values in Figure 9(b) are larger than those shown in Figure 9(a) because of the fiber reinforcement to stabilized soil.

The growth rate of unconfined compressive strength is an important index to measure the strength enhancement of stabilized soil. The growth rate of unconfined compressive strength can be defined by the following equation:

$$\eta = \frac{\sigma_x - \sigma_0}{\sigma_0} \times 100\%, \tag{1}$$

where σ_0 denotes the unconfined compressive strength of untreated soil samples and σ_x refers to the unconfined compressive strength of stabilized soil with different additives.

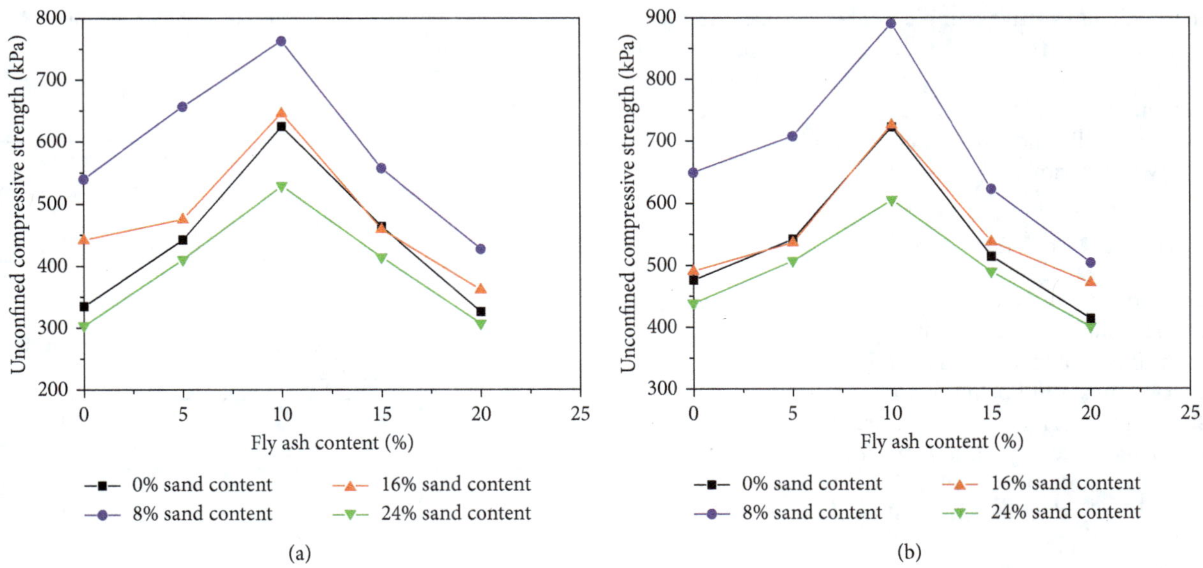

FIGURE 9: Effect of different additives on unconfined compressive strength of soil specimens. (a) Without basalt fiber. (b) With 0.4% basalt fiber.

The calculation results of the growth rate are listed in Tables 3 and 4. It is found in Table 3 that the growth rate of unconfined compressive strength of stabilized soil samples increases with increasing of fly ash content and sand content. When the fly ash content is 10% and sand content is 8%, the growth rate is peaked at 127.71% for stabilized soil with 0% basalt fiber. Moreover, it is observed that the growth rate of unconfined compressive strength of stabilized soil specimens increases notably with the addition of 0.4% basalt fiber. The maximum strength growth rate of soil specimens in Figure 9(b) reaches 165.76%, which shows an increase of 29.79% compared with the maximum strength growth rate in Figure 9(a). Accordingly, the role of fiber in strength improvement of stabilized soil is not negligible.

TABLE 3: Growth rate of unconfined compressive strength of stabilized soil with 0% basalt fiber.

Sand content (%)	Fly ash content (%)				
	0	5	10	15	20
0	0	31.97	86.30	38.36	−2.71
8	61.13	96.08	127.71	66.31	27.46
16	31.89	41.83	92.64	36.95	7.83
24	−9.20	22.59	57.84	23.73	−8.28

TABLE 4: Growth rate of unconfined compressive strength of stabilized soil with 0.4% basalt fiber.

Sand content (%)	Fly ash content (%)				
	0	5	10	15	20
0	42.11	61.76	115.60	53.45	23.38
8	32.58	111.31	165.76	85.81	50.46
16	46.41	60.23	116.54	60.52	40.77
24	31.01	51.43	80.69	46.13	19.41

3.4.3. Complementary Effect. To verify whether there is a complementary effect among various additives on enhancing the strength of stabilized soil. The definition of β_i and β_j are the ratio of strength growth coefficient for the soil specimens stabilized with one additive (inclusive of fly ash, sand, or basalt fiber) compared with the untreated soil specimens. The definition of β_{ij} is the ratio of strength growth coefficient for the soil specimens stabilized with mixtures of two additives compared with untreated soil specimens. The equations are as follows:

$$\beta_i = \frac{\sigma_i}{\sigma_0},\quad \beta_j = \frac{\sigma_j}{\sigma_0},\quad \beta_{ij} = \frac{\sigma_{ij}}{\sigma_0}, \tag{2}$$

where σ_0 denotes the unconfined compressive strength of untreated expansive soil samples; σ_i and σ_j are the unconfined compressive strength of the soil samples stabilized with one additive, i.e., fly ash or sand or basalt fiber; and σ_{ij} is the unconfined compressive strength of the soil samples stabilized with two additives.

The complementary coefficient θ is defined by the following equation:

$$\theta = \frac{\beta_{ij}^2}{\beta_i \cdot \beta_j}. \tag{3}$$

When $\theta \geq 1$, it is positive complementary effect for the strength enhancement of stabilized soil; when $\theta < 1$, it is negative complementary effect for the strength enhancement of stabilized soil. The calculation results of the complementary coefficient are listed in Tables 5–7. It is concluded that most of the stabilized soil specimens have a characteristic of positive complementary effect. The above results demonstrate that the soil specimens stabilized with mixtures of two additives have a more obvious strength enhancement than that stabilized with only one additive. Besides, the results indicated that there is a good combination between various additives used in this study.

4. Microstructural Analysis

To better understand the observed macroscopic behaviors of stabilized soil such as plasticity index, compaction strength, and unconfined compressive strength, the soil microstructure observation should be performed. On the one hand, the microstructure of soil reveals the internal structures of soil and the formation of hydration products and aggregations inside soil. On the other hand, it is an important factor to determine the physical, mechanical, and other engineering properties of soil [26].

4.1. Effect of Fly Ash Content. The SEM images of expansive soil samples at different fly ash contents after 28 days of curing period are given in Figure 10. It is presented that the microstructure of virgin expansive soil samples and fly ash-stabilized soil samples is completely different. At low magnification, Figures 10(a), 10(b), and 10(c) suggest that the pores inside soil samples are gradually filled with the hydration products with increasing of fly ash dosage. For the soil specimens stabilized with 0% fly ash, it can be observed that the structure of soil aggregates is composed of soil particles and inter-aggregates pores. The effect of fly ash on expansive soil samples is shown in Figures 10(b) and 10(c), and a granular structure with hydration products is clearly observed. This implies that the addition of fly ash promotes the occurrence of pozzolanic reaction and helps in binding the unbonded soil particles.

At 1500 times magnification, it is found that hydration products generated in the inter-aggregates pores, as shown in Figures 10(e) and 10(f). There is a cation exchange reaction (substitution of the Na^+ or K^+ ions by the Al^{3+} or Ca^{2+} ions) between fly ash and soil particles with the increase of fly ash content. The results of cation exchange reaction presented a trend of flocculation of particles [21], as shown in Figures 10(e) and 10(f). The gels of calcium silicate hydrates are the products of SiO_2 in fly ash reacted with Ca^{2+} ions and OH^- ions (formed by ionization of Ca $(OH)_2$) [9], which can effectively increase the strength and integrality of stabilized soil. However, the fly ash cannot react with soil particles completely if the amount of fly ash is excessive. It is clear as shown in Figure 10(f) that when the addition of fly ash is in a proportion of 20%, the spherical fly ash particles are still present among soil particles without reaction. Since fly ash is a nonstick material, the unreacted fly ash particle remained inside soil samples reduces the interparticle connection and the bonding force between soil particles. The unconfined compressive strength of stabilized soil with excessive fly ash (such as 20% fly ash dosage) decreases accordingly in previous tests. At 5000 times magnification, it is clear

Table 5: Complementary coefficient of soil specimens stabilized with fly ash and sand.

Sand content (%)	Fly ash content (%)				
	0	5	10	15	20
0	1	1.32	1.86	1.38	0.97
8	1.61	1.81	1.73	1.24	1.04
16	1.32	1.16	1.51	1.03	0.91
24	0.91	1.25	1.47	1.22	0.95

Table 6: Complementary coefficient of soil specimens stabilized with fly ash and basalt fiber.

Basalt fiber content (%)	Fly ash content (%)				
	0	5	10	15	20
0	1	1.32	1.86	1.38	0.97
0.4	1.42	1.40	1.76	1.20	1.10

Table 7: Complementary coefficient of soil specimens stabilized with sand and basalt fiber.

Basalt fiber content (%)	Sand content (%)			
	0	8	16	24
0	1	1.61	1.32	0.91
0.4	1.42	1.64	1.14	1.33

that the calcium silicate hydrates (C-S-H) and calcium sulfoaluminate hydrates (AFt) were filled in the inter-aggregates pores as shown in Figures 10(h) and 10(i). Yet, there are many inter-aggregates pores as shown in Figure 10(g). Moreover, the microstructure observed in the fly ash-stabilized soil specimens by SEM can be considered as compact, which indicates an intact structure. The micrograph of soil samples treated with fly ash shows a dense fabric with cementitious products or aggregations. The formation of aggregations is responsible for the mitigation of the swelling potential of expansive soils, so the plasticity index presents a decreasing trend in Figure 6.

4.2. Effect of Sand Content. Figure 11 presents the SEM images of expansive soil samples stabilized with 10% fly ash at different sand contents after 28 days of curing period. It is directly shown that all the stabilized soil samples have the same fly ash content of 10% and the interparticle pores inside soil samples gradually decrease with the increase of sand content. With the addition of sand, the pores between soil particles are filled with sand and the inter-aggregates pores are also filled with sand simultaneously. Thus, the integrality and compactness of stabilized soil improve remarkably in the presence of sand.

4.3. Effect of Fiber Reinforcement. Figure 12 gives the SEM images of the effect of fiber reinforcement on microstructure of stabilized expansive soil. Figure 12(a) presents a SEM image of the untreated soil specimen. It presents that the internal structure of expansive soil is a turbulent

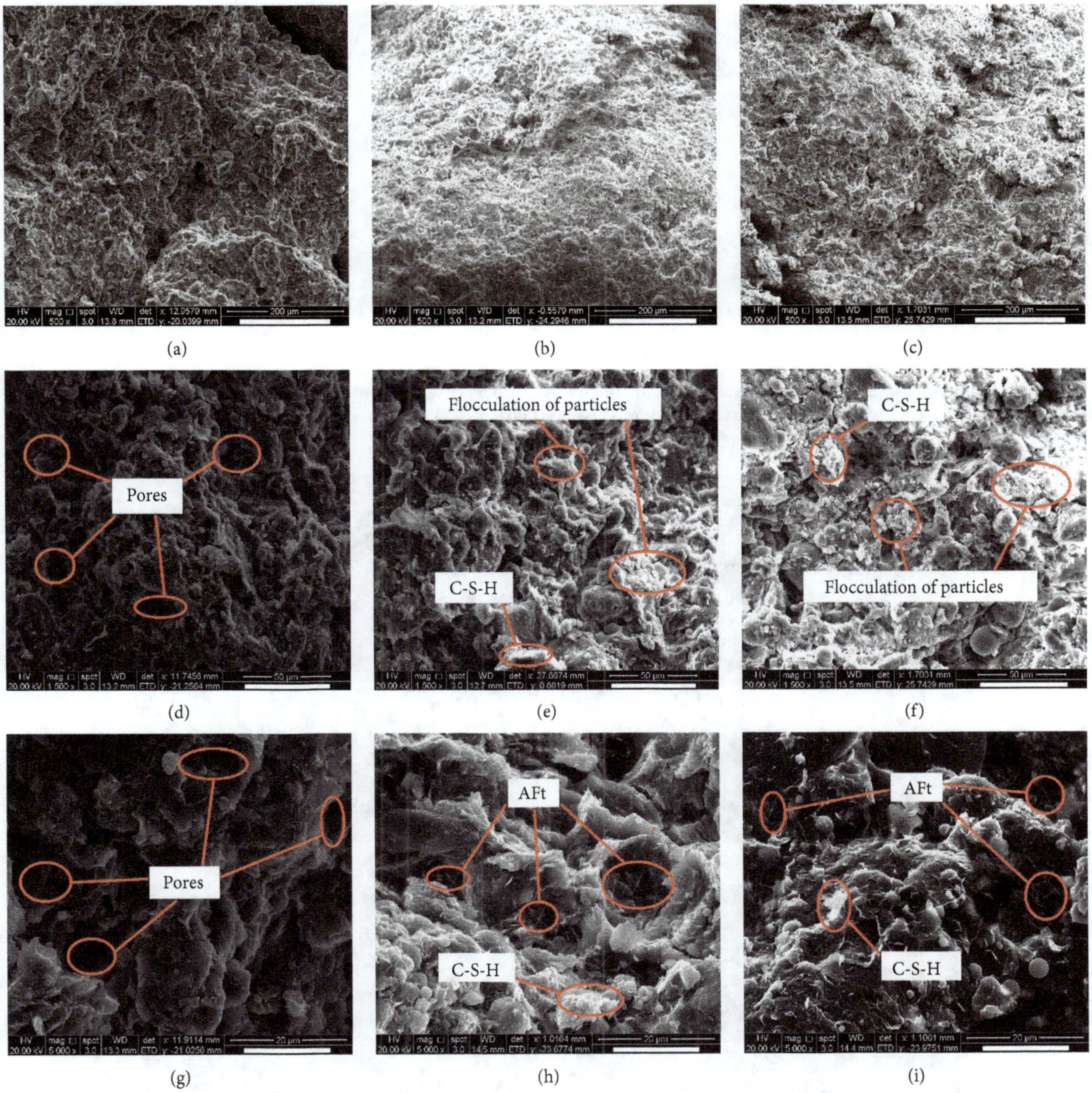

FIGURE 10: Microstructure of stabilized expansive soil under different fly ash contents. (a) 0% fly ash (×500). (b) 10% fly ash (×500). (c) 20% fly ash (×500). (d) 0% fly ash (×1500). (e) 10% fly ash (×1500). (f) 20% fly ash (×1500). (g) 0% fly ash (×5000). (h) 10% fly ash (×5000). (i) 20% fly ash (×5000).

structure, and the result complies with the experimental result of Mao and Xia [26]. Figure 12(b) shows the shape and location of basalt fibers appeared inside stabilized soil specimens. It can be easily seen that the basalt fibers are tightly wrapped by the ambient clay particles. Due to the gripping interactions between fibers and soil particles, they jointly bear the external loading [27, 28]. When stabilized soil is subjected to external loading, the interfacial force between clay particles and fibers increases gradually to resist external loading [29]. As a result, the strength of stabilized soil evidently increases with the addition of basalt fiber, and this point has been confirmed in Section 3.4.2 from a macroscopic perspective. At higher magnification, Figures 12(c) and 12(d) show this gripping effect between fibers and soil particles more clearly.

4.4. Effect of Sand Content on Fiber Reinforcement. The SEM images of the effect of sand on microstructure of expansive soil stabilized with 0.4% basalt fiber are given in Figure 13. With the certain percentage of fly ash (10%), the superficial roughness of fiber increases with the increase of sand content. It is clearly observed that the basalt fiber presented in Figures 13(b) and 13(c) has a rugged surface with the addition of sand to stabilized soil. Figures 13(e) and 13(f) show more clearly. The surface of the basalt fiber has

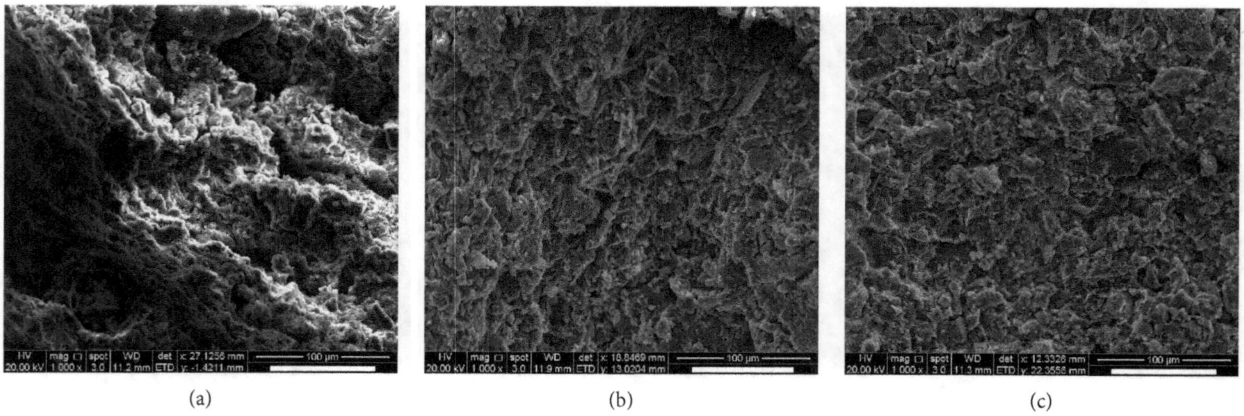

Figure 11: Microstructure of expansive soil stabilized with 10% fly ash at different sand contents. (a) 0% sand and 10% fly ash (×1000). (b) 8% sand and 10% fly ash (×1000). (c) 24% sand and 10% fly ash (×1000).

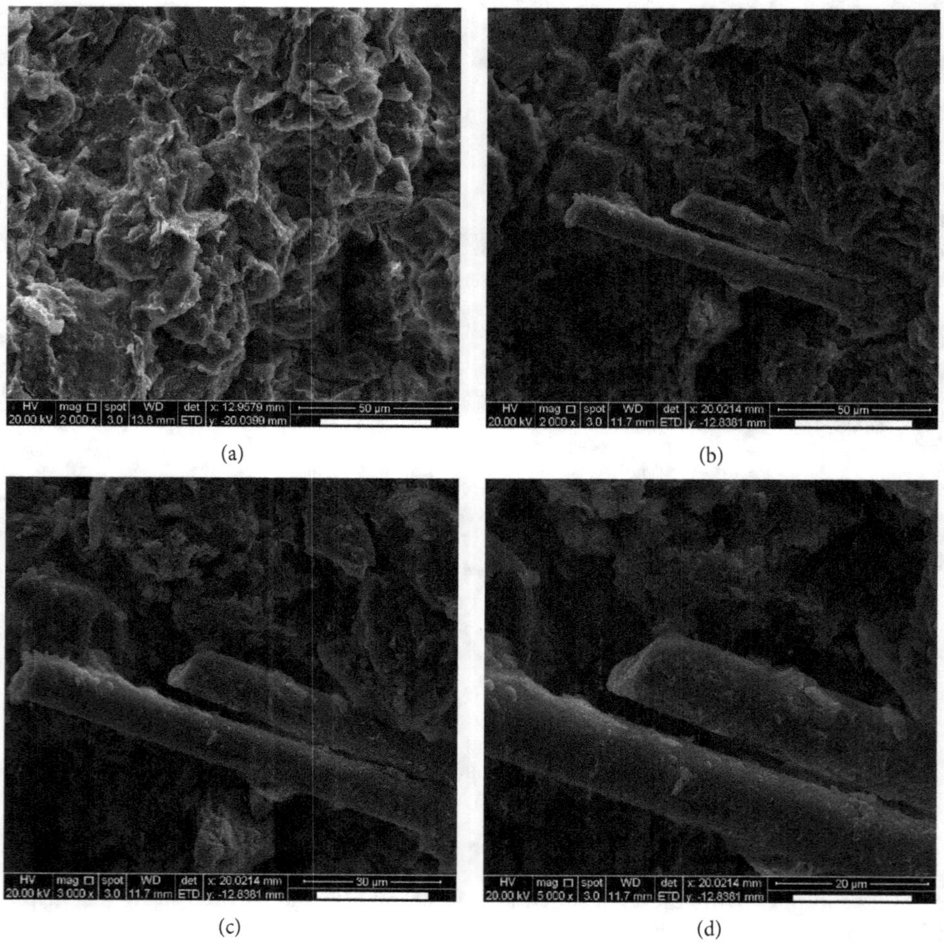

Figure 12: Effect of fiber reinforcement on microstructure of stabilized expansive soil. (a) Without basalt fiber (×2000). (b) With 0.4% basalt fiber (×2000). (c) With 0.4% basalt fiber (×3000). (d) With 0.4% basalt fiber (×5000).

mechanically damaged because of the impact between sand grains and fibers during mixing or compressing [25]. The hardness of the sand is higher than that of the basalt fiber. Under the action of the external loading, the fiber surface undergoes extrusion deformation and becomes rugged. Undoubtedly, the friction force between fibers and soil particles increases with the increase of the surface area between fibers and soil particles. As a result, the presence of sand contributes to the connection between soil particles and the enhancement of interfacial force between fibers and soil particles. From a macro point of view, the strength of stabilized soil increases significantly and the fiber-stabilized soil presents a positive complementary effect with the addition of sand.

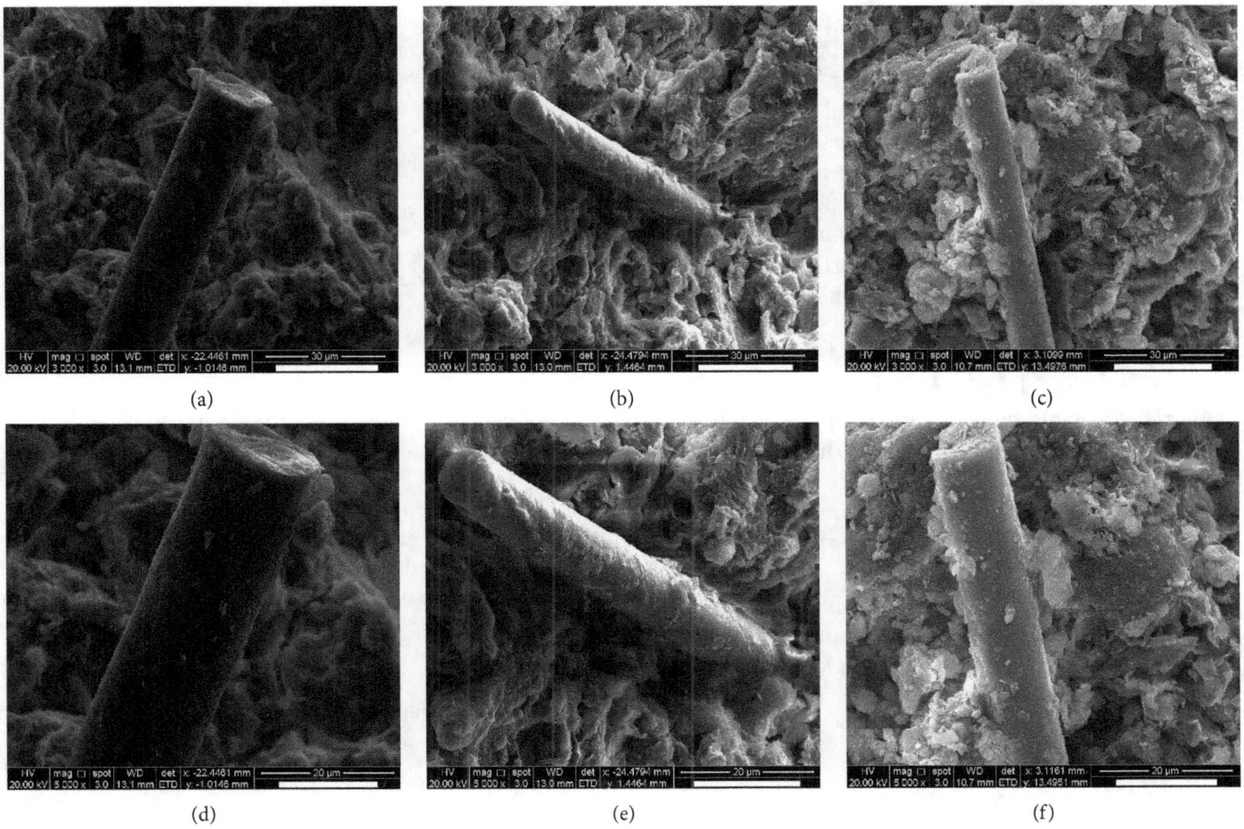

FIGURE 13: Effect of sand content on microstructure of fiber-stabilized soil. (a) 0% sand and 10% fly ash (×3000). (b) 8% sand and 10% fly ash (×3000). (c) 24% sand and 10% fly ash (×3000). (d) 0% sand and 10% fly ash (×5000). (e) 8% sand and 10% fly ash (×5000). (f) 24% sand and 10% fly ash (×5000).

5. Conclusions

This paper analyzes the physical-mechanical properties and microstructures of stabilized expansive soil by performing a series of laboratory experiments. Then, the effect of various additives on physical performances, unconfined compressive strength, and microstructure is highlighted in this study. Conclusions are drawn as follows:

(1) The physical properties of stabilized soil are different with the original expansive soil. With the addition of fly ash, specific gravity of stabilized soil goes on decreasing as the percentage of fly ash increases from 0% to 20%. In contrast, specific gravity of stabilized soil is increasing with the increase of sand content. The optimum moisture content of stabilized soil shows an increasing trend with the addition of fly ash, and the optimum moisture content shows a reducing trend with the increase of sand content. Yet, the maximum dry density of stabilized soil decreases with the increase of fly ash content and increases with the increase of sand content. The plastic limit and liquid limit of original expansive soil are 22% and 42.8%, respectively. With the addition of fly ash or sand, there is reduction in the liquid limit value, whereas the plastic limit value presents an increasing trend. The plasticity index of stabilized soil decreases with the increase of fly ash or sand content.

(2) Unconfined compression test is performed to clarify the strength improvement of stabilized soil with various additives. It is found that the failure mode of stabilized soil with fiber can be transformed from brittle failure to plastic failure. The maximum unconfined compressive strength and the best growth rate of strength are acquired for mixt treatments with 10% fly ash and 8% sand contents. All the performances of stabilized soil have further upgraded with the addition of 0.4% basalt fiber. Therefore, it is concluded that the addition of fly ash, sand, and basalt fiber mixture can be considered as an efficient stabilization method for strength enhancement of expansive soil. As the analysis of complementary effect suggests, most of the mixt treatments applied in this study have produced good results concerned with the enhancement of strength of expansive soil. Moreover, the analysis of complementary effect indicates that there is a good combination between different stabilizers used in this study.

(3) Owing to the generation of hydration products of fly ash, the connection among clay particles has been enhanced. At 5000 times magnification, it is easily found that the calcium silicate hydrates (C-S-H) and

calcium sulfoaluminate hydrates (AFt) are filled in the inter-aggregates pores in SEM images. Besides, the flocculation of dispersed clay particles occurs under the ionization of divalent and trivalent cations with the addition of fly ash. For the content of fly ash is 20%, the spherical fly ash particles still exist among soil particles without reaction in SEM images. The unreacted fly ash particles remained in the soil particles will reduce the interparticle connection. The suitable proportion of sand has a good filling effect on inter-aggregates pores; hence, the integrality and compactness of stabilized soil have increased. The gripping effect between fibers and soil particles notably increases the strength of stabilized soil. Furthermore, the surface of the fiber becomes rugged due to the impact between sand grains and fibers in the process of mixing or compressing, and the friction force between fibers and soil particles increases directly. As a result, the effect of sand on reinforced soil with 0.4% basalt fiber can improve the connection between soil particles and the interfacial force between fibers and soil particles, which suggests that the use of mixt additives shows a positive effect on the stabilization of expansive soil.

Conflicts of Interest

The authors declare that they have no conflicts of interest regarding the publication of this paper.

Acknowledgments

The authors sincerely thank the Engineering Research Center of Underground Mine Construction, Ministry of Education, and the Anhui University of Science and Technology for providing the experiment conditions. And the authors would like to acknowledge the College of Civil Engineering and Architecture, Zhejiang University, for their support during completion of SEM tests.

References

[1] A. Seco, F. Ramírez, L. Miqueleiz, and B. García, "Stabilization of expansive soils for use in construction," *Applied Clay Science*, vol. 51, no. 3, pp. 348–352, 2011.

[2] S. B. Ikizler, M. Vekli, E. Dogan, M. Aytekin, and F. Kocabas, "Prediction of swelling pressures of expansive soils using soft computing methods," *Neural Computing and Applications*, vol. 24, no. 2, pp. 473–485, 2014.

[3] R. N. Yong and V. R. Ouhadi, "Experimental study on instability of bases on natural and lime/cement-stabilized clayey soils," *Applied Clay Science*, vol. 35, no. 3-4, pp. 238–249, 2007.

[4] M. Khemissa and A. Mahamedi, "Cement and lime mixture stabilization of an expansive overconsolidated clay," *Applied Clay Science*, vol. 95, pp. 104–110, 2014.

[5] M. Al-Mukhtar, A. Lasledj, and J. F. Alcover, "Behaviour and mineralogy changes in lime-treated expansive soil at 20°C," *Applied Clay Science*, vol. 50, no. 2, pp. 191–198, 2010.

[6] A. R. Estabragh, M. R. S. Pereshkafti, B. Parsaei, and A. A. Javadi, "Stabilised expansive soil behaviour during wetting and drying," *International Journal of Pavement Engineering*, vol. 14, no. 4, pp. 418–427, 2013.

[7] R. S. Sharma, B. R. Phanikumar, and B. V. Rao, "Engineering behavior of a remolded expansive clay blended with lime, calcium chloride, and rice-husk ash," *Journal of Materials in Civil Engineering*, vol. 20, no. 8, pp. 509–515, 2008.

[8] T. Thyagaraj, S. M. Rao, P. S. Suresh, and U. Salini, "Laboratory studies on stabilization of an expansive soil by lime precipitation technique," *Journal of Materials in Civil Engineering*, vol. 24, no. 8, pp. 1067–1075, 2012.

[9] B. R. Phanikumar and R. S. Sharma, "Volume change behavior of fly ash-stabilized clays," *Journal of Materials in Civil Engineering*, vol. 19, no. 1, pp. 67–74, 2007.

[10] A. S. Rao and M. R. Rao, "Swell-shrink behaviour of expansive soils under stabilized fly ash cushions," in *Proceedings of the 12th International Conference on International Association for Computer Methods and Advances in Geomechanics*, pp. 561–564, Goa, India, December 2008.

[11] M. G. Feng, S. X. Chen, S. Yu, and J. Ma, "Laboratory study on water stability of fly ash-treated expansive soil," *Rock and Soil Mechanics*, vol. 28, no. 9, pp. 1889–1893, 2007, in Chinese.

[12] X. Kang, L. Ge, G. C. Kang, and C. Mathews, "Laboratory investigation of the strength, stiffness, and thermal conductivity of fly ash and lime kiln dust stabilised clay subgrade materials," *Road Materials and Pavement Design*, vol. 16, no. 4, pp. 928–945, 2015.

[13] S. K. Mohanty, P. K. Pradhan, and C. R. Mohanty, "Consolidation and drainage characteristics of expansive soil stabilized with fly ash and dolochar," *Geotechnical and Geological Engineering*, vol. 34, no. 5, pp. 1435–1451, 2016.

[14] P. K. Kolay and K. C. Ramesh, "Reduction of expansive index, swelling and compression behavior of kaolinite and bentonite clay with sand and class C fly ash," *Geotechnical and Geological Engineering*, vol. 34, no. 1, pp. 87–101, 2016.

[15] S. Y. Lei and W. T. Ding, "Experimental investigation on restraining the swell of expansive soil with fibre-reinforcement," *Chinese Journal of Geotechnical Engineering*, vol. 27, no. 4, pp. 482–485, 2005, in Chinese.

[16] Y. S. Deng, P. Wu, M. H. Zhao, and B. Z. Duan, "Strength of expansive soil reinforced by polypropylene fiber under optimal water content," *Rock and Soil Mechanics*, vol. 38, no. 2, pp. 349–360, 2017, in Chinese.

[17] J. Prabakar and R. S. Sridhar, "Effect of random inclusion of sisal fibre on strength behaviour of soil," *Construction and Building Materials*, vol. 16, no. 2, pp. 123–131, 2002.

[18] T. Yetimoglu and O. Salbas, "A study on shear strength of sands reinforced with randomly distributed discrete fibers," *Geotextiles and Geomembranes*, vol. 21, no. 2, pp. 103–110, 2003.

[19] X. Qin, A. Q. Shen, Y. C. Guo, Z. N. Li, and Z. H. Lv, "Characterization of asphalt mastics reinforced with basalt fibers," *Construction and Building Materials*, vol. 159, pp. 508–516, 2018.

[20] L. Gao, G. H. Hu, N. Xu, J. Y. Fu, C. Xiang, and C. Yang, "Experimental study on unconfined compressive strength of basalt fiber reinforced clay soil," *Advances in Materials Science and Engineering*, vol. 2015, Article ID 561293, 8 pages, 2015.

[21] Y. L. Chen, "Microstructure of expansive soil from Yunnan Province," *Chinese Journal of Geotechnical Engineering*, vol. 35, no. S1, pp. 334–339, 2013, in Chinese.

[22] S. Horpibulsuk, R. Rachan, A. Chinkulkijniwat, and Y. Raksachon, "Analysis of strength development in cement-stabilized silty clay from microstructural considerations," *Construction and Building Materials*, vol. 24, no. 10, pp. 2011–2021, 2010.

[23] American Society for Testing and Materials (ASTM), *ASTM C 618–03, Standard Specification for Coal Fly Ash and Raw or Calcined Natural Pozzolan for Use as a Mineral Admixture in Concrete*, Annual Book of ASTM Standards, Philadelphia, USA, 2003.

[24] Ministry of Construction of the People's Republic of China, *GB/T 50123–1999 Standard for Soil Test Method*, China Planning Press, Beijing, China, 1999, in Chinese.

[25] C. S. Tang, B. Shi, W. Gao, Y. Cai, and J. Liu, "Study on effects of sand content on strength of polypropylene fiber reinforced clay soil," *Chinese Journal of Rock Mechanics and Engineering*, vol. 26, no. S1, pp. 2968–2973, 2007, in Chinese.

[26] T. E. Mao and L. Xia, "Experimental research on microstructure of expansive soil in north of Hubei Province," *Journal of Huazhong University of Science and Technology (Urban Science Edition)*, vol. 27, no. 2, pp. 48–52, 2010, in Chinese.

[27] C. S. Tang, B. Shi, and K. Gu, "Microstructural study on interfacial interaction between fiber reinforcement and soil," *Journal of Engineering Geology*, vol. 19, no. 4, pp. 610–614, 2011, in Chinese.

[28] C. S. Tang, B. Shi, and L. Z. Zhao, "Interfacial shear strength of fiber reinforced soil," *Geotextiles and Geomembranes*, vol. 28, no. 1, pp. 54–62, 2010.

[29] N. Cristelo, V. M. C. F. Cunha, M. Dias, A. T. Gomes, T. Miranda, and N. Araújo, "Influence of discrete fibre reinforcement on the uniaxial compression response and seismic wave velocity of a cement-stabilised sandy-clay," *Geotextiles and Geomembranes*, vol. 43, no. 1, pp. 1–13, 2015.

19

Aggregate Evolution Mechanism during Ion-Adsorption Rare Earth Ore Leaching

Kui Zhao,[1,2] Yulong Zhuo ,[1,2] Xiaojun Wang ,[1,2] and Wen Zhong[1,2]

[1]*Jiangxi Key Laboratory of Mining Engineering, Jiangxi University of Science and Technology, Jiangxi 341000, China*
[2]*School of Resources and Environment Engineering, Jiangxi University of Science and Technology, Jiangxi 341000, China*

Correspondence should be addressed to Yulong Zhuo; 879408217@qq.com and Xiaojun Wang; xiaojun7903@126.com

Academic Editor: Ali Nazari

The phenomenon of particle aggregation occurs when ammonium chloride is used as a leaching reagent to infiltrate rare earth samples. To reveal the formation and evolution mechanisms of aggregates, a self-developed column leaching experimental device was employed in conjunction with nuclear magnetic resonance technology. The relationships among the amount of rare earth leaching, the evolution of the microscopic pore structure, the porosity, and the leaching time were obtained. A comparative analysis of pure water and ammonium chloride test groups revealed that aggregates were present only in the latter. Consequently, the results of comprehensive analyses indicate that the formation of aggregates is a temporary particle deposition phenomenon caused by the settling of fine soil particles migrating from the top to the bottom of a sample. Furthermore, chemical exchanges constitute the main cause of aggregate formation.

1. Introduction

Ion-adsorption rare earth ore, also known as weathering shell leaching rare earth ore, was first discovered in a foot cave in Longnan County, Jiangxi Province, in the late 1960s. Ion-adsorption rare earth ore is found in shallow, weathered rock formations [1–5]. Currently, mining is conducted through a relatively mature process known as in situ leaching and mining, which uses mainly chemical exchange reactions composed of a leaching solution during the seepage process of an ore body to recover rare earth cations, thereby extracting rare earth elements [6–9]. The rare earth leaching rate depends on the degree of the chemical exchange and the seepage of the rare earth mother liquor in the ore body. This ion-adsorption rare earth leaching process includes both analytical and antiadsorption processes. The exchange of rare earth cations is performed under dynamic equilibrium conditions. Furthermore, various parameters of the leaching ore body, including the concentration of the leaching solution, the cation activity, and the temperature, are positively related to the intensity of the chemical exchange. A previous experimental study found that the rare earth leaching rate is affected by changes in the microscopic pore structure during the seepage process [10]. Therefore, research on the relationship between the evolution of the pore structure and the leaching rate during the rare earth leaching process is highly significant for the development of green extraction technologies and for the efficient recovery of ion-adsorption rare earth ores.

In recent years, with the extensive use of microcomputed tomography (CT) technology [11–13], scholars worldwide have begun to study the evolution of the microstructures of rock and soil media and their influencing mechanisms. Among those researchers, Kodali et al. used micro-CT to study the distribution of microcracks inside copper ore particles [14]. Similarly, Nosrati et al. used micro-CT to scan the internal structure of granulated nickel ore and reported the distribution of the microscopic structure [15], while Yan et al. introduced microscopic parameters to describe the pore characteristics of soil [16]. Zhuo et al. used nuclear magnetic resonance (NMR) scanning to study the relationship between the pore size distribution and the weakening of the intensity of ion-adsorption rare earth leaching processes, thereby revealing the mechanism through which the ore body strength becomes weakened during leaching [17, 18]. Wang et al. studied the variations in the permeability characteristics of rare earth ore bodies under seepage-chemical coupling using an indoor column leaching test [10].

The mineral leaching rate, which indicates the extent to which a metal is leached, represents one of the criteria for the leaching processes of various metals. With the continuous expansion of the scale of the production of mineral resources, the grade of infiltration has gradually decreased, and increasing numbers of scholars have initiated research on the mechanisms behind the leaching rate. Xiao et al. proposed the concept of ion-adsorption rare earth ore leaching using magnesium sulfate as a leaching agent from an environmental protection perspective; with a magnesium sulfate concentration of 0.2 mol/L, the rare earth leaching rate can exceed 95% [19]. In addition, Li et al. used citrate as a leaching agent to leach the weathering crust of rare earth ore; they postulated that the coordination of citrate ions with rare earth ions can promote rare earth ion exchanges among ores to increase the rare earth ore recovery rate [20].

During the in situ leaching of ion-adsorption rare earth ores, the pore structure of the ore body changes dynamically with the injection strength and the duration of liquid injection. This series of changes to the pore structure directly affects the permeability characteristics of the ore body. Moreover, the seepage of the leaching solution affects the stability of the slope. Existing research shows that particles experience both migration and blockage during the soil seepage process. However, few reports have been published regarding the disappearance of particles, the relationship between the blockage and disappearance of particles, and the chemical exchanges that occur during the leaching of rare earth ores. Based on NMR technology, this paper studies the evolutionary mechanism of the pore structure during the leaching of rare earth ore bodies and finds the relationship between internal structural changes and the leaching rate of ore bodies. This study also provides an important research basis for the leaching and extraction of ion-adsorption rare earth ores.

2. Experiments

2.1. Materials and Characterization. Rare earth ore samples were taken from a rare earth mine in Longnan County of southern China. The grade of Longnan ion-adsorption rare earth ore is low, mainly in the range of a few thousands. However, rare earth elements, especially heavy rare earth elements, with complete distributions are abundant. Moreover, the content of radioactive elements is low; as a result, most ore deposits are nonradioactive. The rare earth elements are mainly adsorbed onto the surfaces of ionic clay minerals in fully weathered layers.

For the following procedure, we followed the approach of Zhuo et al. [21]. The samples were acquired by collecting rare earth ores from the mine, and the sample sizes coincided with the effective NMR detection area. The diameter-to-height ratio ranged from 40 mm to 60 mm. The soil samples were air-dried before the tests; hence, the water content was 0%. During the remodelling process, the density and moisture content of the original soil samples were maintained to the maximum extent possible, as shown in Figure 1. Each sample was pretested by comparing the densities to determine the appropriate number of compaction events; this number was determined to be 3, and the compaction height was determined to be 30 cm. A standard compaction device with a compaction hammer diameter of 40 mm was employed. After phase detection, the rare earth content of the original sample was determined to be 0.070 ± 0.003%, which conforms to the experimental requirements. The physical parameters of the remodelled rare earth sample are provided in Table 1.

2.2. Chemical Replacement. Rare earth elements are found in strong weathered granites in the form of ions. An NH_4Cl solution is typically used during the leaching process [22, 23]. The rare earth cations adsorbed onto the ore body are replaced by NH_4^+, which has relatively active chemical properties, as shown in Formula (1) [24]. This strong chemical reaction occurs throughout the leaching process. The ion exchange that occurs during leaching involves the adjustment and reconstruction of the mineral microstructure, leading to changes in the crystal structure chain. The seepage effect of the leaching solution also impacts the microstructure of the ore body.

$$\begin{aligned}&[Al_4(Si_4O_{10})(OH)_8]_m \cdot nRE^{3+}{}_{(s)} + 3nNH_4^+{}_{(aq)}\\&\quad = [Al_4(Si_4O_{10})(OH)_8]_m \cdot (NH_4^+)_{3n(s)} + nRE^{3+}{}_{(aq)}.\end{aligned} \tag{1}$$

2.3. Column Leaching Test. The column leaching test device designed and developed in-house is adopted in this study. As shown in Figure 2, the column leaching test device consists of (from top to bottom) the leaching zone, filter paper, remodelled rare earth samples, filter paper, leaking liquid, liquid collection funnel, and liquid collection apparatus. The specific test steps are as follows:

(1) Put the remodelled rare earth sample into the column leaching device. Connect the various parts of the device as shown in Figure 2.

(2) First, saturate the remodelled rare earth sample by injecting 15 mL of pure water into the top of the sample; according to a preliminary test, when the injection volume is 15 mL, a magnetic resonance imaging (MRI) scan can obtain four images encompassing an even distribution of aggregates, which is helpful for constraining the evolution mechanism of aggregates. Based on a preliminary exploratory test, the designed injection rate is 1 mL/min. The process is considered complete if no liquid leakage is observed in the collection funnel within a span of 3 min; subsequently, the rare earth sample is taken for NMR scanning. Record the volume of the recovered liquid and the porosity test results at this time. After the NMR scanning test, repeat the previous steps until the differences between the two recovered liquids and the

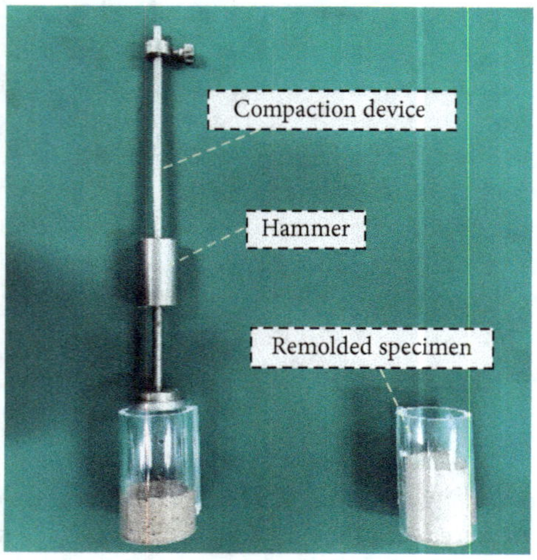

FIGURE 1: Manufacturing process of the remodelled soil specimens.

TABLE 1: Physical parameters of the remodelled rare earth ore sample.

Parameter type	Geometric dimensions (mm)		Density (g/cm^3)	Moisture content (%)	Quality (g)
	Diameter	Height			
Value	40	60	1.75	15	113

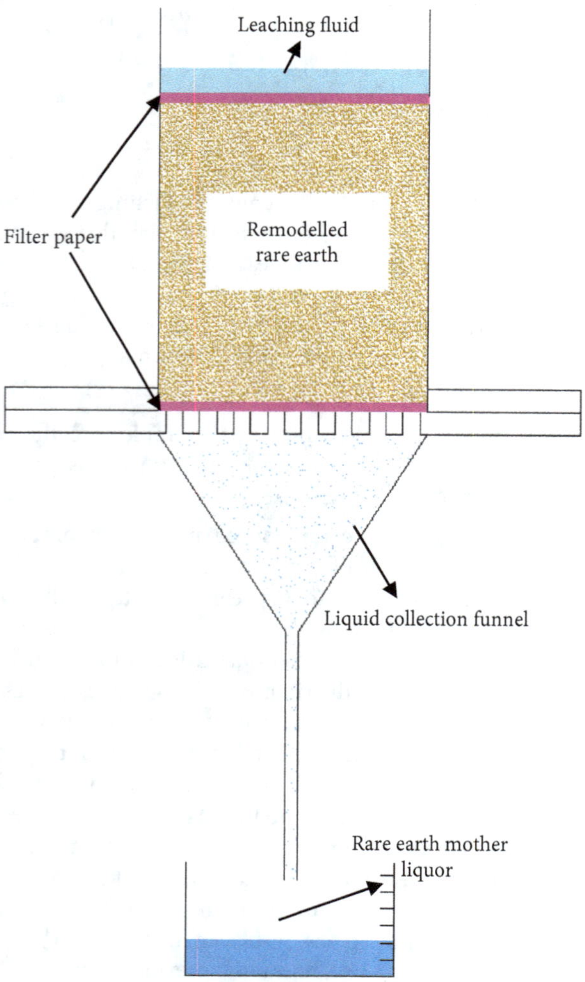

FIGURE 2: Column leaching test device.

porosity test results are less than 0.1, at which the sample is considered to be saturated.

(3) Begin the leaching test. The samples are divided into two groups: one group has a leaching solution composed of pure water and the other group contains a leaching solution of ammonium chloride. The pH value of the solution is maintained at 6.5. The test procedure is the same as that in step (2), but the criteria marking the end of the experiment are different.

(4) Terminate the experiment. The test is concluded when the leaching of the rare earth ion content cannot be determined by EDTA titration.

2.4. Microstructure Tests. NMR, which constitutes a new type of detection technology, has been widely applied in the field of geotechnical engineering [25]. This technology is most importantly characterized by its ability to rapidly, accurately, and quantitatively measure the microstructure porosity and pore size distribution while keeping the structures of the rock and soil intact (i.e., nondestructive testing). Therefore, NMR represents the most suitable detection technology for an analysis of the microstructural evolution of a rare earth ore body during the leaching process.

The micropore structure of the samples was measured using a PQ-OO1-type Mini-NMR (Suzhou Niumag Analytical Instrument Corporation, Suzhou, China), as shown in Figure 3, in which the permanent magnets possess a magnetic field intensity of 0.52 T. During the experiments, the valid experimental area of the sample was 60 mm × 60 mm, and the temperature of the permanent magnets was kept at 32 ± 0.01°C to ensure the stability and uniformity of the experimental magnetic field. A test sample was obtained after each leaching test. After detection, leaching was performed on the same sample; when detecting the microstructure of a sample, the homogeneity of the sample must be ensured to analyse the microstructural evolution of the rare earth ore body during the leaching process.

2.5. Titration Test. The rare earth ion content in the rare earth mother liquor during the test was determined by titration with disodium EDTA, which constituted the titration reagent. The buffer solution was dissolved in pure water with hexamethylenetetramine, and 10 mL of concentrated hydrochloric acid was added. The masking reagent was composed of 1% sulfosalicylic acid, 5% ascorbic acid, and 5% acetylacetone. Xylene orange was used as the agent. According to national standard GB/T 14635-2008 "Rare earth metals and their compounds: determination of total rare earth contents" in combination with the actual conditions of the mother liquor collected during the test, 5 mL of the rare earth mother liquor was titrated for each experiment. The rare earth ion leaching concentration in the secondary rare earth mother liquor is calculated according to the following formula:

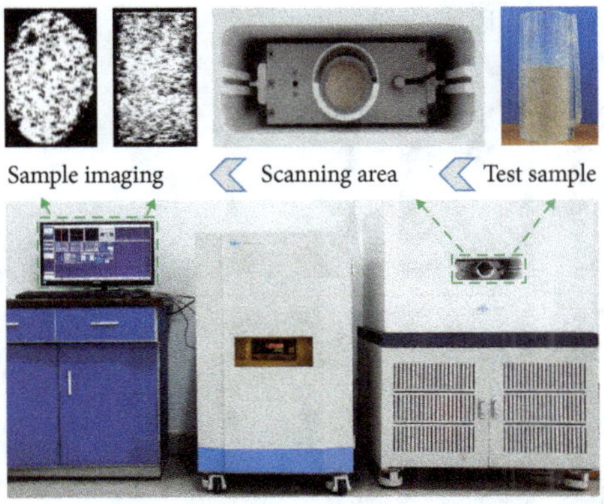

FIGURE 3: PQ-OO1-type Mini-NMR.

$$C_{RE} = \frac{(C_{EDTA} * (B - A))}{5}, \quad (2)$$

where C_{RE} is the rare earth ion concentration, mL/mg; C_{EDTA} is the titrant standard concentration, mL/mg; B is the burette reading after titration, mL; A is the burette reading before titration, mL; and the denominator of 5 indicates that 5 mL of the rare earth mother liquor is taken.

3. Results and Discussion

3.1. Rare Earth Ion Leaching Rate. Disodium EDTA titration recovery of the rare earth mother liquor was employed to determine the rare earth ion leaching concentration. The leaching quality of rare earth ions was calculated according to the collected volume of rare earth mother liquor. Considering the slight differences in the recovered volume of the mother liquor, the rare earth leaching effect on secondary rare earth ions was characterized by the leaching quality of rare earth ions. According to the results of previous research, pure water does not chemically interact with rare earth ions, indicating that the content of rare earth ions in the mother liquor recovered from a pure water leaching test group should be almost zero [10, 17, 18]. The recovered distribution of the rare earth ion mass from the column leaching test on the samples from the ammonium chloride group is shown in Figure 4, where V_{RE} represents the volume of the recycled rare earth mother liquid.

According to the titration results and the recovered volume of rare earth mother liquor, the quality of each leached rare earth sample was determined. Figure 4 shows the rare earth mass leaching results from the 1st test to the 9th test. The ore leaching process subsequent to the "tail" phenomenon in the rare earth column leaching test (i.e., the leaching qualities following the peak value of the rare earth leaching quality) presents a long duration, and the leaching qualities of the rare earth samples are low. The 8th and 9th leaching rare earth masses shown in Figure 4 are below 1% of the peak mass (the 3rd leaching amount); these leaching times indicate a tail phenomenon [17, 18], and this is not the focus of this study. Overall, the leaching quality of rare earth ions showed either a sharp increase or a sharp drop, and no rare earth leaching occurred during the initial leaching stage. The peak rare earth leaching quality was observed after the third and fourth leaching iterations, and the combined rare earth quality of these two leaching processes was 49.71 mg, reaching 71% of the total rare earth content of the sample (a single rare earth sample with a rare earth content of approximately 70 mg was measured in the prior period). After the fourth test, the amount of rare earth leaching decreased significantly, and a tail phenomenon occurred after the eighth leaching iteration.

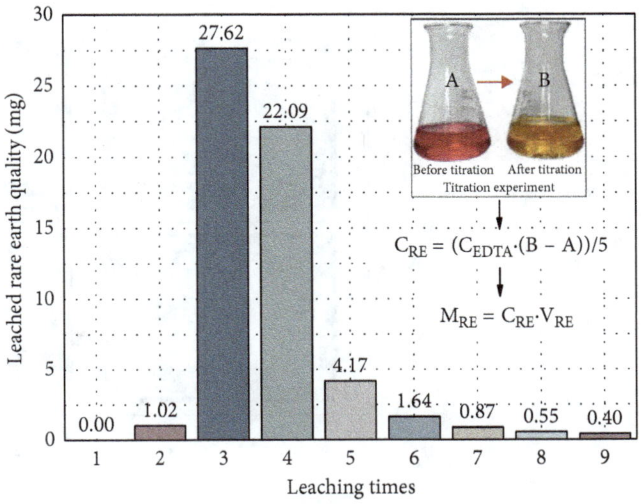

FIGURE 4: Recovery of the rare earth mass ion distribution.

3.2. Aggregate Phenomenon and Its Components. By analysing the NMR inversion imaging results from a large number of column leaching test samples, the inversion imaging pattern of the sample showed no obvious change with an increase in the infiltration time when the infiltration liquid was composed of pure water. In contrast, when the infiltration liquid was ammonium chloride, sample particles (called aggregates in this article) appeared and migrated from the top to the bottom (the dark areas in Figure 5), and they disappeared from the top down with an increase in the leaching time. All samples for this test were saturated prior to testing to consider differences in the seepage velocities of the samples. The test was considered complete when 15 mL of the leaching solution thoroughly flowed through the soil sample into the mother liquor collection container. Inverted images of the sample obtained when leaching pure water with ammonium chloride are shown in Figure 5. Because this article mainly addresses the presence of aggregates, only the leaching iterations with aggregates up to their disappearance are displayed in Figure 5, and two images after the aggregates disappeared are also given to demonstrate that no aggregates appeared in follow-up leaching iterations.

As shown in Figure 5, aggregates exist only in the sample group in which the leaching solution is ammonium chloride, and they appear only in the first four leachings. After the fourth leaching, the aggregates almost disappear and do not reappear. Moreover, aggregates are not observed in the pure water leaching test group. The red arrows in Figure 5 indicate the movement of the aggregates throughout the leaching process. During the initial stage of infiltration of the

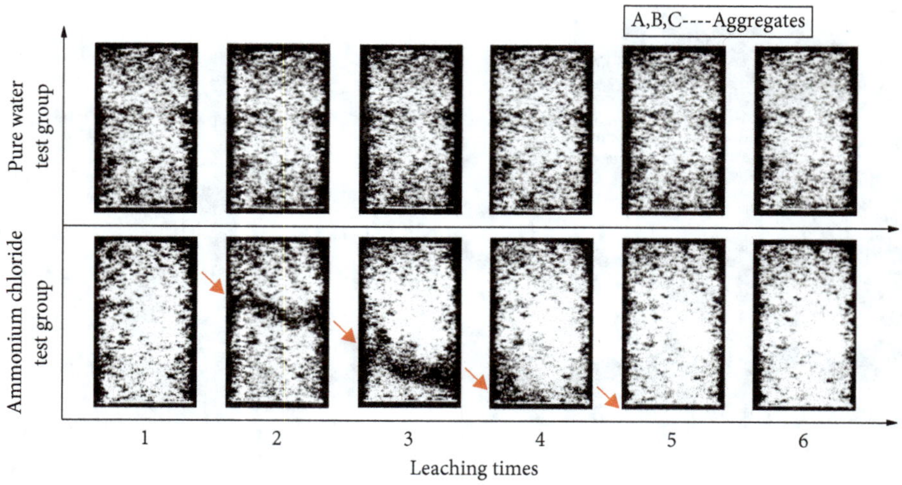

FIGURE 5: Inversion images of the sample during the column leaching tests.

leaching solution, the aggregates are concentrated mainly near the surface of the liquid leaching solution in the upper part of the sample. With the continuous infiltration of the leaching solution, the aggregates move to the middle of the sample and become readily visible following the completion of the second leaching test. After the third leaching, the aggregates move to the bottom of the adjacent specimen; the amount of aggregates has significantly increased since the previous leaching. After the fourth leaching, the aggregates mostly disappear and remain only in the peripheral part of the sample. In subsequent experiments, no aggregates are discernible, and the image formation of the ammonium chloride group is whiter than that of the pure water group. Furthermore, the pure water group specimens show no change from the beginning to the end of the image, and the black and white areas are staggered.

According to the MRI scanning results, the aggregates within the leaching sample were selected for electron microscopy and energy spectrum analysis. The test results from a comparative analysis and random selection of the original soil samples are shown in Table 2 and Figure 6. Figure 6(a) shows that the rare earth particles are distributed in layers and have a greater number of pores, thereby increasing the specific surface area and improving the clay adsorption capacity. Figures 6(b)–6(d) show that the aggregates include a large number of solid particles. Brighter regions indicate that the particles are densely packed, while darker regions indicate that the degree of particle aggregation is relatively low, indicating that the aggregate region is not characterized by a discernible distribution.

Based on the degree of aggregation of solid particles, the energy spectrum was randomly selected, and the mean elemental mass percent distributions of the original soil samples containing common clay constituent elements, such as carbon, oxygen, aluminium, and silicon, were obtained, and the results are shown in Table 2 and Figure 7. Among the constituent elements, oxygen, aluminium, and silicon account for relatively large proportions. The main components of the aggregates in areas A, B, and C are carbon, oxygen, aluminium, and silicon; these elements represent substantial portions of the clay particles. In other words, the aggregates are not a new material, rather, they constitute large agglomerations of local silting phenomena caused by the obstructed migration of clay particles.

By comparing the original soil samples with the aggregate element distributions from areas A, B, and C, it is relatively evident that the scandium, yttrium, and lanthanum contents in the solid particles within the aggregate region are almost zero, while the contents of the other elements are essentially equivalent. The cations among the rare earth elements can be exchanged with ammonia ions, pulling them from the solid particles into the solution. Therefore, the agglomerated microparticles first chemically exchange cations with the ammonia ions in the solution and then migrate downward under the effect of percolation as the solution moves from the top down.

According to other research findings [26], soil samples experience the migration of particles during water infiltration. However, no obvious aggregates were observed in the pure water infiltration test sample, whereas the aggregates were obvious in the ammonium chloride test group and were present in all of the samples, indicating that simple seepage was not the main cause of aggregate formation.

3.3. Microscopic Pore Characteristics of the Samples. The ratio of pores within a certain radius of a sample to the total number of pores is called the ratio of the pore radius. The distributions of pore radii of different sizes can reflect changes in the microscopic pore structure of a sample during the infiltration of the leaching solution. Accordingly, the T_2 spectrum is used to detect the proportion of each site within the sample through a predetermined spot. The size of the T_2 spectrum is positively related to the pore size; that is, the distribution of the T_2 spectrum can be used to characterize the pore distribution of the sample. To better reveal the aggregate formation mechanism, the T_2 spectrum distributions with pure water and ammonium chloride as the leaching solutions are shown in Figure 8, in which the 0th time indicates the initial pore distribution when the sample is saturated.

TABLE 2: Elemental mass percent distribution at each measuring point.

Spectrum	C	O	Al	Si	Fe	La	Ce	Nd
(a) Mass percent distribution of the original soil sample								
360	5.19	51.13	9.14	20.08	—	1.23	2.89	0.17
361	6.80	36.48	13.67	29.99	—	—	—	0.17
362	9.80	56.99	10.75	15.84	—	—	—	—
363	4.71	63.45	16.22	14.43	0.44	—	—	—
364	5.27	59.17	15.98	14.80	0.61	—	1.36	0.06
365	—	56.17	21.60	20.06	1.18	—	—	—
366	7.09	59.27	18.36	14.90	—	—	—	—
Mean value:	6.48	54.66	15.10	18.58	0.74	1.23	2.12	0.13
Sigma:	1.89	8.84	4.31	5.59	0.38	0.00	1.08	0.07
Sigma mean:	0.71	3.34	1.63	2.11	0.14	0.00	0.41	0.02
	C	O	Al	Si	Fe			
(b) Mass percent distribution of spot soil sample A								
353	7.61	59.00	17.38	15.74	0.26			
354	7.13	56.94	18.87	17.05	—			
355	5.09	56.26	20.20	18.44	—			
356	5.52	60.24	17.75	16.49	—			
357	6.30	55.94	19.71	18.04	—			
358	6.79	58.46	18.50	16.01	0.25			
359	5.71	52.87	21.68	19.73	—			
Mean value:	6.31	57.10	19.16	17.36	0.25			
Sigma:	0.92	2.43	1.50	1.45	0.01			
Sigma mean:	0.35	0.92	0.57	0.55	0.00			
(c) Mass percent distribution of spot soil sample B								
346	5.92	59.36	15.38	19.25	0.08			
347	6.34	60.93	13.04	18.66	1.02			
348	5.08	56.91	22.14	15.84	0.03			
349	5.61	60.20	16.07	16.86	1.25			
350	5.59	58.79	14.89	20.66	0.06			
351	6.84	55.61	16.47	17.21	3.86			
352	5.68	58.59	16.33	19.28	0.15			
Mean value:	5.87	58.63	13.32	18.25	0.92			
Sigma:	0.57	1.84	4.30	1.68	1.39			
Sigma mean:	0.22	0.69	1.62	0.63	0.52			
(d) Mass percent distribution of spot soil sample C								
339	6.27	52.49	16.68	24.55	—			
340	8.12	54.26	18.28	19.33	—			
341	6.57	56.3	15.44	21.68	—			
342	7.35	55.26	16.99	20.39	—			
343	5.84	63.19	11.24	19.39	0.33			
344	6.38	56.57	18.65	18.39	—			
345	7.47	63.96	11.97	16.59	—			
Mean value:	6.86	57.43	15.61	20.04	0.33			
Sigma:	0.81	4.42	2.94	2.54	0.00			
Sigma mean:	0.30	1.67	1.11	0.96	0.00			

To facilitate this analysis, the area with T_2 spectrum values of less than 1 ms is defined as pore A, the area with T_2 spectrum values between 1 ms and 10 ms is defined as pore B, the area with T_2 spectrum values between 10 ms and 100 ms is defined as pore C, the area with T_2 spectrum values between 100 ms and 1000 ms is defined as pore D, and the area with T_2 spectrum values greater than 1000 ms is defined as pore E; the pore areas increase sequentially from A to E. Figures 8(a) and 8(b) show that the initial pore distributions of the sample are approximately the same; in addition, a trend is observed where two small central areas become one large area. At the end of the first leaching, the overall pore size distributions of the two groups of specimens did not change significantly. The changes were mainly concentrated in regions D and E, and there was a shift from region E to region D. Specifically, the diameters of some large-diameter pores decreased. At the end of the second and third leachings, the two groups of samples continued to follow the trend of a reduction in the diameters of large-diameter pores, and the trend of the change in the pore size of the ammonium chloride group sample was significantly stronger than that of the pure water group sample. After the fourth leaching of the pure water group sample, the pore size of each area fluctuated in a small range; that is, the pore size tended to be roughly constant. In contrast, after the fourth leaching of the ammonium chloride group sample, the numbers of small-diameter pores in areas B and C greatly diminished, while the numbers of large-diameter pores in those areas greatly increased; subsequently, the pore diameters in the various leaching regions also tended to remain approximately constant. The pore size distribution for the overall leaching process can be divided into two stages: a dynamic change period and a quiet period. Studies [25] have shown that changes in the sample skeleton, particle generation, and migration are all important causes of variations in the pore diameter.

The micropore structures of the samples were measured using a PQ-OO1-type Mini-NMR. After the test, various pore radius distributions were determined. The sum of the ratios of the pore radius for each size is called the porosity of the sample. The error of the NMR test results is ±1%; therefore, to reduce the influence of instrumental errors on the test results, each test was performed thrice (i.e., 3 measurements were obtained) to provide an arithmetic average for the porosity of the test sample. Considering the influence of the expansion of the soil samples on the test analysis, the height of each sample was measured before each test, and the height of the sample was measured four times before the test along the wall surface of the device. The average value was taken as the average height of the test sample. During the test, the soil sample did not slide along the inner tube wall. Furthermore, there was no gap between the soil sample and the inner wall of the device. In addition, because the column leaching test device is composed of an acrylic sheet, the diameter of the device did not change during the test, which means that the soil sample was fixed in the radial direction with a constrained size of 50 mm. The specific physical parameters and porosities of the samples are shown in Table 3.

Table 3 shows that the volumes of the two groups of samples remained unchanged after the onset of leaching (i.e., subsequent to the initiation of the first leaching). The initial porosities of the samples (referred to as the 0th time before the first experiment) were basically equivalent. With an increasing number of leachings, the porosities of both samples gradually increased, but the increase in the porosity of the pure water sample was small, and no obvious growth in the porosity was observed. The overall increase in the porosity of the ammonium chloride sample was significantly higher than that of the pure water sample, and the porosities after the first, second, third, and fourth leachings exhibited large increases; after subsequent leachings, however, the

Figure 6: Layout of spectral scanning measurement points of the soil samples. (a) Original. (b) Spot A. (c) Spot B. (d) Spot C.

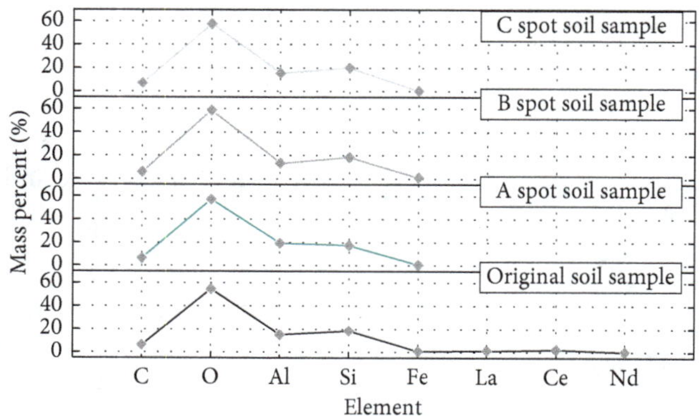

Figure 7: Elemental mass percent distribution.

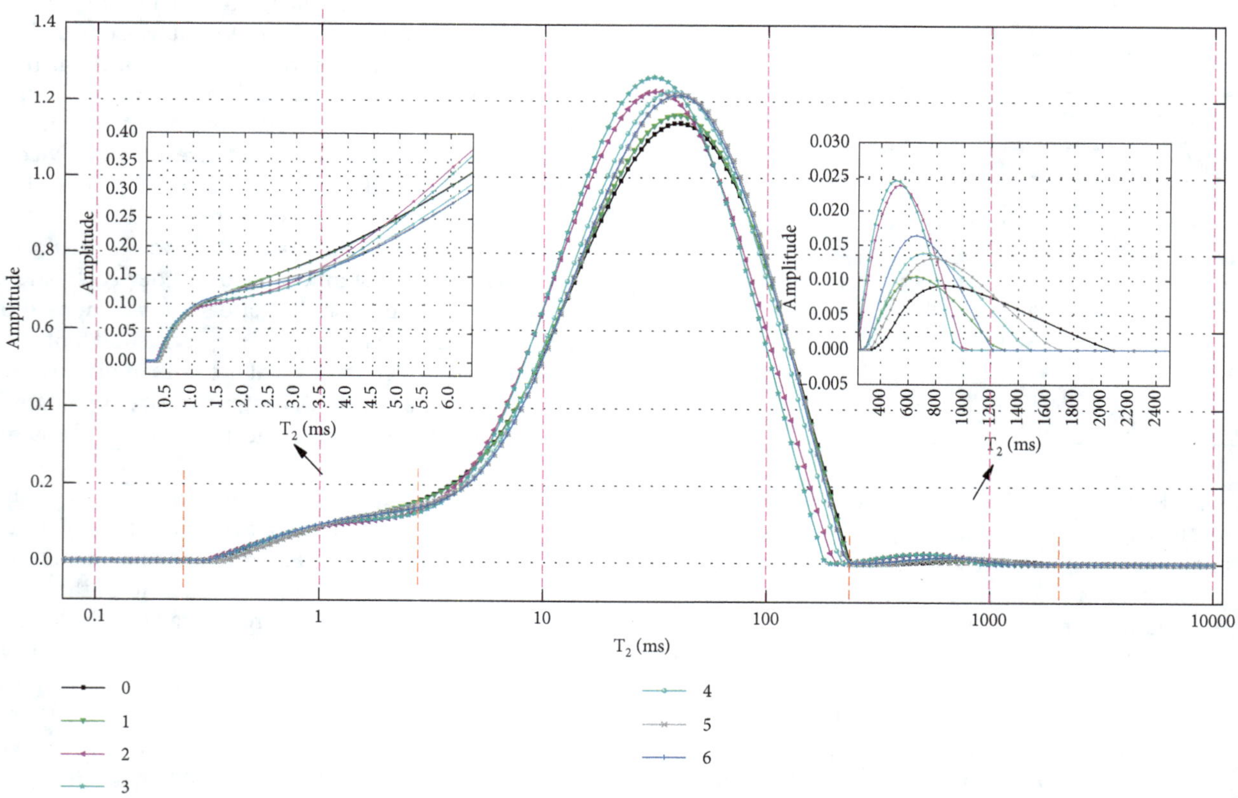

Figure 8: Continued.

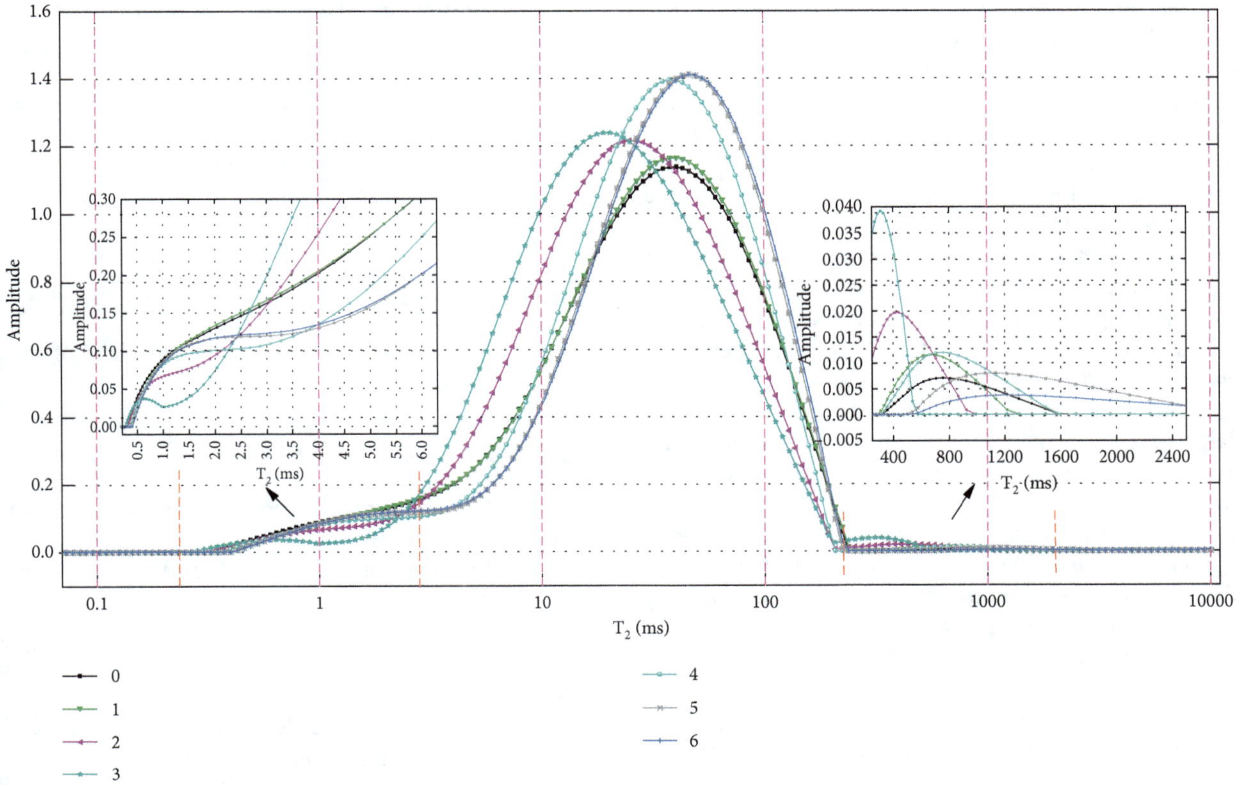

(b)

FIGURE 8: Sample leaching T_2 spectrum distributions. (a) Pure water leaching T_2 spectrum distribution. (b) Ammonium chloride leaching T_2 spectrum distribution.

TABLE 3: Physical parameters and porosity distributions of the samples.

Frequency	Sample height (mm)					Volume (mm^3)	Porosity (%)			
	Test 1	Test 2	Test 3	Test 4	Average value		Test 1	Test 2	Test 3	Average value
(a) Pure water leaching test group										
0	59	59	60	60	59.5	74.732	45.116	45.443	45.437	45.332
1	59	60	59	60	59.5	74.732	45.683	45.653	45.663	45.666
2	59	59	60	60	59.5	74.732	45.880	45.783	45.764	45.809
3	59	59	60	60	59.5	74.732	45.653	45.862	45.443	45.652
4	59	59	60	60	59.5	74.732	45.880	46.081	46.153	46.038
5	59	59	60	60	59.5	74.732	46.235	45.879	46.545	46.230
6	59	59	60	60	59.5	74.732	46.503	45.667	46.259	46.143
(b) Ammonium chloride leaching test group										
0	58	59	59	58	58.5	73.476	45.258	45.392	45.267	45.305
1	59	60	58	59	59	74.104	45.919	45.880	46.004	45.934
2	58	60	59	59	59	74.104	47.458	47.562	47.498	47.506
3	59	59	59	59	59	74.104	48.767	48.971	48.908	48.882
4	58	60	59	59	59	74.104	49.711	49.906	49.011	49.542
5	59	60	59	58	59	74.104	49.968	49.083	49.784	49.612
6	59	60	59	58	59	74.104	49.544	49.899	49.568	49.670

porosity fluctuated only within a certain range. The constant volume of the ammonium chloride sample indicates that there was no significant change in the overall support skeleton within the sample during the leaching process, and the increase in the porosity was a result of the combined effect of seepage and chemical exchanges in addition to the ionic strength and viscosity. In contrast, the change in the porosity during pure water leaching was not significant, indicating that seepage was not the main cause of the increase in the porosity.

3.4. Aggregate Formation Mechanism and Its Effect on the Leaching Rate. Aggregates existed only in the ammonium chloride leaching group, and the aggregates began to germinate at the end of the first leaching. The presence of aggregates was the most significant after the second, third, and fourth leachings, and the ion leaching amount peaked during those times. However, after the fourth leaching, aggregates no longer appeared, and the ion leaching rate decreased sharply, indicating that the strong ion exchange effect is closely related to the production of aggregates. In combination with the pore distribution and the changes in the porosity of the ammonium chloride sample, the strong chemical exchanges did not affect the main support skeleton of the soil sample. However, under the combined effect of seepage and chemical exchanges in addition to the ionic strength and viscosity, a large number of fine clay particles and large particles were removed and transferred into the solution. With the sharp increase in the fine particle content, the original seepage flow channels became congested. Over time, the particles gradually aggregated to form the black areas of aggregation shown in Figure 5. The decrease in ion leaching observed upon comparing the ammonium leaching before and after the leaching of the ammonium chloride group sample indicates that the exchange effect was weakened, the changes in the pore distribution and porosity tended to become stable, and the aggregates disappeared. This strong degree of chemical exchange is an indispensable factor in the formation of aggregates, while the ionic strength and viscosity are not the main causes of aggregate formation.

During the initial stage of aggregate formation, the interior of the sample was saturated with water. At this time, the ammonium chloride solution squeezed pure water out from inside the sample, and no rare earth ions were leached. At the end of the second leaching, the aggregates had migrated to the middle of the sample, and the amount of rare earth leaching was reduced. After the third and fourth leachings, the aggregates had reached the bottom of the sample, and the aggregates were leached with large amounts of rare earth ions. At this time, the amount of rare earth leaching peaked, accounting for 71% of the total amount of leaching. This demonstrates that the upper and lower adjacent areas of the aggregates contained large amounts of rare earth ions.

4. Conclusions

(1) No exchange reactions occurred, and no aggregates were produced between the pure water and rare earth ions. The porosity did not change significantly during the process of pure water seepage.

(2) When ammonium chloride was leached, fine soil particles were removed from the large granular soil by chemical exchanges and were transferred into the solution. This process was partially blocked by seepage during migration from the top to the bottom of the sample, resulting in the formation of aggregates. As the number of leachings increased, the aggregates continued to migrate downward.

(3) In the third and fourth test leachings, the amounts of leached rare earth ions were significantly greater than those at other times. The amount of rare earth leaching peaked at the third leaching iteration. This is because the aggregates had reached nearly the bottom of the sample during the third leaching, and thus, the chemical reaction is more complete. After the third leaching, the amount of leaching continuously decreased.

Conflicts of Interest

The authors declare that they have no conflicts of interest.

Authors' Contributions

K.Z and Y.L.Z. conceived and designed the experiments, analysed the data, and wrote the paper. X.J.W. and W.Z. performed most of the experiments and analysed the data.

Acknowledgments

This research was financially supported by the National Natural Science Foundation of China (nos. 51564012 and 51504102), the Science and Technology Project funded by the Education Department of Jiangxi Province (GJJ150653), the Qingjiang Excellent Young Talents of Jiangxi University of Science and Technology, and the Jiangxi University of Science and Technology Excellent Doctoral Thesis Training Project (YB2017001).

References

[1] T. S. Qiu, D. M. Zhu, X. H. Fang, Q. H. Zeng, G. K. Gao, and H. L. Zhu, "Leaching kinetics of ionic rare-earth in ammonia-nitrogen wastewater system added with impurity inhibitors," *Journal of Rare Earths*, vol. 32, no. 12, pp. 1175–1183, 2014.

[2] K. Sanematsu, Y. Kon, A. Imai et al., "Geochemical and mineralogical characteristic of ion-adsorption type REE mineralization in Phuket, Thailand," *Mineralium Deposita*, vol. 48, no. 4, pp. 437–451, 2013.

[3] L. Zhang, K. X. Wu, L. K. Chen, P. Zhu, and Y. H. Ou, "Over view of metallogenic features of ion-adsorption type REE deposits in southern Jiangxi Province," *Journal of the Chinese society of rare earths*, vol. 1, p. 10, 2015.

[4] J. Li, C. Y. Li, B. W. Li, S. B. Wang, J. C. Han, and J. L. Wang, "Acid leaching of low grade rare earth concentrate by microwave heating," *Chinese Journal of Rare Metals*, vol. 38, no. 5, p. 839, 2014.

[5] R. A. Chi, Z. J. Li, C. Peng, G. C. Zhu, and S. M. Xu, "Partitioning properties of rare earth ores in China," *Rare Metals*, vol. 24, no. 3, p. 205, 2005.

[6] Q. B. Zhang, Y. X. Hua, C. Y. Xu et al., "Non-haloaluminate ionic liquids for low-temperature electrodeposition of rare-earth metals-A review," *Journal of Rare Earths*, vol. 33, no. 10, pp. 1017–1025, 2015.

[7] L. Zhao, Z. Dong, G. L. Ma, and W. J. Yuan, "Solution extraction of several lanthanides from nitric acid with is ohexyl-BTP

in [Cnmim][NTf_2] ionic liquid," *Journal of Rare Earths*, vol. 33, no. 11, pp. 1182–1188, 2015.

[8] X. Q. Sun, Y. Ji, J. Chen, and J. T. Ma, "Solvent impregnated resin prepared using task-specific ionic liquids for rare earth separation," *Journal of Rare Earths*, vol. 27, no. 6, pp. 932–936, 2009.

[9] R. X. Wang, B. Y. Xie, P. Yu et al., "Selection of leaching agent and optimization of column leaching process of ion-absorbed rare earth deposits," *Chinese Journal of Rare Metals*, vol. 39, no. 11, pp. 1060–1064, 2015.

[10] X. J. Wang, Y. L. Zhuo, S. Q. Deng et al., "Experimental research on the impact of ion exchange and infiltration on the microstructure of rare earth orebody," *Advances in Materials Science and Engineering*, vol. 2017, Article ID 4762858, 8 pages, 2017.

[11] J. Dewanckele, T. De Kock, M. A. Boone et al., "4D imaging and quantification of pore structure modifications inside natural building stones by means of high resolution X-ray CT," *Science of The Total Environment*, vol. 416, no. 2, pp. 436–448, 2012.

[12] J. R. Kyle and R. A. Ketcham, "Application of high resolution X-ray computed tomography to miner-al deposit origin, evaluation, and processing," *Ore Geology Reviews*, vol. 65, no. 4, pp. 821–839, 2015.

[13] L. J. Munkholm, R. J. Heck, and B. Deen, "Soil pore characteristics assessed from X-ray micro-CT derived images and correlations to soil friability," *Geoderma*, vol. 181-182, no. 7, pp. 22–29, 2012.

[14] P. Kodali, N. Dhawan, T. Depci et al., "Particle damage and exposure analysis in HPGR crushing of selected copper ores for column leaching," *Minerals Engineering*, vol. 24, no. 13, pp. 1478–1487, 2011.

[15] A. Nosrati, K. Quast, D. F. Xu et al., "Agglomeration and column leaching behaviour of nickel laterite ores: effect of ore mineralogy and particle size distribution," *Hydrometallurgy*, vol. 146, no. 5, pp. 29–39, 2014.

[16] X. Q. Yan, Y. G. Fang, and P. Zhang, "Experiment study on the effects of bentonite on the micropore structure characteristics of soil," *Chinese Journal of Geotechnical Engineering*, vol. 33, no. 8, pp. 1302–1307, 2011.

[17] Y. L. Zhuo, X. J. Wang, S. R. Cao et al., "Study on relationship between pore structure and strength weakening of rare earth ore under seepage," *Gold Science and Technology*, vol. 25, no. 5, pp. 101–106, 2017.

[18] Y. L. Zhuo, X. J. Wang, K. Zhao et al., "Study on strength weakening mechanism of iron-adsorbed rare earth ore during displacement reaction," *Chinese Rare Earths*, vol. 38, no. 6, pp. 57–63, 2017.

[19] Y. F. Xiao, Y. Y. Chen, Z. Y. Feng et al., "Leaching characteristics of ion-adsorption type rare earths ore with magnesium sulfate," *Transactions of Nonferrous Metals Society of China*, vol. 25, no. 11, pp. 3784–3790, 2015.

[20] Q. Li, Z. Y. He, Z. Y. Zhang et al., "Studies on coordination leaching of weathered crust elution-deposited rare earth ore with citrate," *Chinese Rare Earths*, vol. 36, no. 1, pp. 18–22, 2015.

[21] X. J. Wang, Y. L. Zhuo, K. Zhao et al., "Experimental measurements of the permeability characteristics of rare earth ore under the hydro-chemical coupling effect," *RSC Advances*, vol. 8, no. 21, pp. 11652–11660, 2018.

[22] G. Q. Deng and Y. M. Yang, "A review of the mining technologies of ion-absorbed rare earth mineral," *Chinese Rare Earths*, vol. 37, no. 3, p. 129, 2016.

[23] R. A. Chi, J. Tian, X. P. Luo, Z. G. Xu et al., "The basic research on the weathered crust elution-deposited rare earth ores," *Nonferrous Metals Science and Engineering*, vol. 3, no. 4, p. 1, 2012.

[24] R. A. Chi and J. Tian, *Weathered Crust Elution-Deposited Rare Earth Ores*, Nova Science Publishers, New York, NY, USA, 2008.

[25] K. P. Zhuo, J. L. Li, Y. J. Xu et al., "Experimental study of NMR characteristics in rock under freezing and thawing cycles," *Chinese Journal of Rock Mechanics and Engineering*, vol. 31, no. 4, p. 731, 2012.

[26] Q. Q. Huang, *Experimental study on permeability change rule of ion-absorbed rare-earth in ore leaching process*, Jiangxi University of Science and Technology, Ganzhou, China, 2014.

20

A Nonlinear Strength Criterion for Frozen Sulfate Saline Silty Clay with Different Salt Contents

Yanhu Zhao,[1,2] Yuanming Lai ,[1,2] Jing Zhang,[1,2] and Chong Wang[3]

[1]*State Key Laboratory of Frozen Soil Engineering, Northwest Institute of Eco-Environment and Resources, Chinese Academy of Sciences, Lanzhou 730000, China*
[2]*University of Chinese Academy of Sciences, Beijing 100049, China*
[3]*Key Laboratory of Mechanics on Disaster and Environment in Western China, The Ministry of Education of China and School of Civil Engineering and Mechanics, Lanzhou University, Lanzhou 730000, China*

Correspondence should be addressed to Yuanming Lai; ymlai@lzb.ac.cn

Academic Editor: Jun Liu

It has been proven that the mechanical properties of frozen saline soils are different from frozen soils and unfrozen saline soils. In this paper, in order to study the effects of the salt contents on the strength characteristics of frozen soils, a series of conventional triaxial compression tests are carried out for frozen saline silty clay with Na_2SO_4 contents 0.0, 0.5, 1.5, and 2.5% under confining pressures from 0 MPa to 18 MPa at −6°C, respectively. The experimental results show that the strength of frozen saline silty clay presents obvious nonlinearity, the strength of frozen saline silty clay increases with increasing confining pressures at first, but with a further increase in confining pressures, the strength decreases because of pressure melting and crushing phenomena under high confining pressures, and salt contents have an important influence on strength of frozen saline silty clay. A strength criterion is proposed on the basis of the experimental results. The strength criterion could well reflect the nonlinear strength characteristic of frozen saline silty clay and the influence of salt contents on frozen saline silty clay.

1. Introduction

The essential difference between frozen saline soils and unfrozen soils is the ice and salt existing in frozen saline soils, which has an important influence on cohesive strength of the soil particles. The effect of salt content on the mechanical properties of frozen saline silty clay is considerable. Research results suggest that sulfate soil is widely distributed in the western China [1]. With the implementation of the Belt and Road Initiative, infrastructure constructions in the western China, such as metro engineering, tunnel construction, bridges, and other engineering activities, will be of top priority. The strength of frozen soils has great design guidance for construction in the western China. So far, research on strength characteristics of frozen saline silty clay is rarely reported. The strength characteristics of frozen saline soils are urgently needed.

The Mohr–Coulomb strength criterion, which has linear strength envelopes, is one of the widely used strength criteria nowadays. However, a majority of experimental results show that the strength envelope for most of the geomaterials is nonlinear. The linear Mohr–Coulomb strength criterion is not suitable for most of the geomaterials, such as frozen soils and sand. Using the nonlinear strength criterion to reflect strength characteristics of most of the geomaterials is more reasonable. There are some nonlinear strength criteria for geomaterials. So far, for sandy soils, there are two popular strength criteria, MN strength criterion [2] and LD strength criterion [3], respectively. Later, many researchers [4, 5] proposed new strength criteria on the basis of them. Yao et al. [6] proposed a unified hardening (UH) model for overconsolidated clays based on the MN criterion and the corresponding transforming stress method. Frozen soils are made up of mineral particles, ice inclusions, liquid water, and gaseous inclusions [7]. Compared with unfrozen soils, the strength and deformation characteristics of frozen soils are more complex owing to the existence of ice, so these strength criteria are unreasonable for frozen soils. A series of studies [8–23] on the strength and deformation characteristics of frozen soils were carried out. The experimental results showed that the strength of frozen soils increases with increasing confining pressures at first, but with a further increase in confining pressures, the strength decreases for pressure melting and crushing phenomena under high confining pressures [21, 24, 25]. And many strength criteria were presented for frozen soils. Fish [26] and Ma et al. [27] presented parabolic yield criteria for frozen soils, respectively. Lai et al. [21] presented a nonlinear Mohr criterion which can describe the pressure melting and crushing phenomena of frozen soils. Yang et al. [28, 29] established a new strength criterion for frozen soils based on MN and LD criteria, and then, Yang et al. [30] presented a better one based on the Hoek–Brown criterion.

Compared with general frozen soils, the strength criterion of frozen saline silty clay is more complex than that of general frozen soils for the existence of salt crystals. Liao et al. [31] found that the existence of salt lowers the freezing temperature and salt content can influence the strength of frozen soils. It was proven that the content of sodium sulfate has an important influence on the mechanical properties of frozen soils [32], so strength criteria for general frozen soils and unfrozen soils are not available for frozen saline silty clay. Liao et al. [31] established a strength criterion for frozen sodium sulfate saline soils, which can reflect the influence of salt content on strength. The studies mentioned above are the latest developments in strength criterion and represent a great advance in the investigation on this domain.

In this paper, a series of conventional triaxial compression tests were carried out for frozen saline silty clay with Na_2SO_4 contents 0.0, 0.5, 1.5, and 2.5% under confining pressure from 0 MPa to 18 MPa at −6°C, respectively. Based on the experimental results, a new strength criterion was proposed to describe strength characteristics of frozen saline silty clay, which can well reflect the nonlinear strength characteristic of frozen saline silty clay and the influence of salt content on strength.

2. Test Conditions and Results

2.1. Description of the Test and Test Conditions. The soil studied was taken from Qinghai–Tibet plateau, and its grain size composition is given in Table 1. The soil studied contains lots of salinity, so we must desalinate the soil firstly. In order to desalinate the soil, salt in saline silty clay was removed with distilled water until the total amount of soluble salts is approximately zero. An electronic conductivity pen was used to measure the electronic conductivity of the soil during desalinating. When the electronic conductivity is close to 0, it is considered that the total amount of soluble salts is approximately zero. Secondly, the desalinated soil sample was dried at 105°C for 24 h and then pulverized and filtered with a 2 mm sieve. Finally, the sieved soils were selected to make up the sodium sulfate saline silty clay samples whose salt contents were 0.0, 0.5, 1.5 and 2.5%, respectively. The process of making samples is as follows: (1) mix the soil with the salt; (2) mix the mixture of soil and salt with water; (3) samples were made by testing mold. The experimental apparatus is the MTS-810 material testing system. The cylindrical soil samples with 61.8 mm in diameter and 125 mm in height were prepared. The test procedure was as follows: (1) the soil samples were saturated with the vacuum method so that the saturated water content of the soil sample is approximately 22%; (2) the soil samples were frozen at −6°C for 24 h using the cooling equipment; (3) the soil samples were put into a pressure cell and were kept for 2 h at −6°C using the incubator; (4) the soil samples were loaded under the confining pressures of 0–18 MPa; and (5) triaxial stress was applied. The loading rate during the testing process is 1.25 mm/min according to [33].

Table 1: Grain size composition of saline silty clay (unit: %).

>0.50 mm	0.50–0.25 mm	0.25–0.075 mm	0.075–0.005 mm	<0.005 mm
0.10	0.53	7.8	70.34	21.23

2.2. Test Results. Figure 1 shows the typical stress-strain curves of frozen saline silty clay with Na_2SO_4 content 1.5% under confining pressures from 0 MPa to 18 MPa at −6°C, respectively.

From Figure 1, it is found that the confining pressures have a large influence on the deviator stress-axial strain relationship of frozen saline silty clay. The frozen saline silty clay presents strain hardening, and the stress-strain curves have no peak values for $\varepsilon_a \le 15\%$, when confining pressures range from 0 MPa to 18 MPa at −6°C. The frozen saline silty clay produces volume contraction at first and then produces volume expansion when confining pressure <5 MPa. But with a further increase in confining pressures, the frozen saline silty clay only produces volume contraction. The strength of frozen saline silty clay increases with increasing confining pressures, when confining pressure <10 MPa. The strength of frozen saline silty clay decreases with increasing confining pressure, when confining pressure >10 MPa. According to the research of Lai et al. [22], in this paper, the values of $\sigma_1 - \sigma_3$ at $\varepsilon_a = 15\%$ are taken as the strength of frozen saline silty clay when confining pressures range from 0 MPa to 18 MPa. Figure 2 shows the stress-strain curves of frozen saline silty clay with Na_2SO_4 contents 0, 0.5, 1.5, and 2.5% under the confining pressure of 8 MPa at −6°C, respectively. From Figure 2, it can be seen that the salt contents have an important influence on the strain-strain relationship.

3. Strength Theory of Frozen Saline Silty Clay

Generally, the failure condition can be given by the general state of stress:

$$F(\sigma_{ij}, k_i) = 0, \tag{1}$$

where F is the strength function, σ_{ij} is the second-rank symmetric stress tensor, and k_i is the strength parameter.

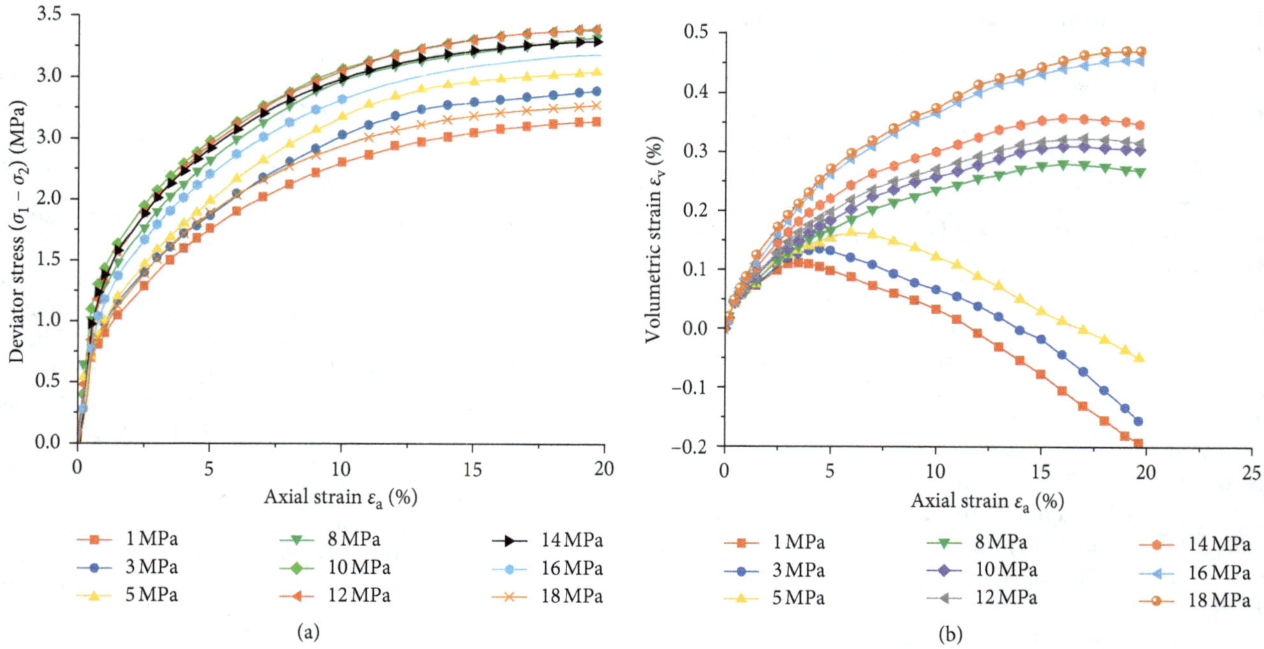

FIGURE 1: The typical stress-strain curves of frozen saline silty clay with salt content 1.5%.

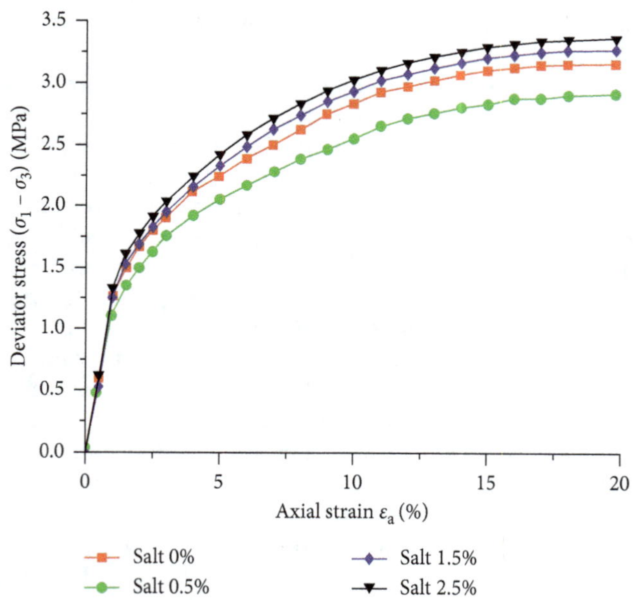

FIGURE 2: The typical stress-strain curves of frozen saline silty clay with different salt contents under a confining pressure of 8 MPa.

Equation (1) can also be written by principal stress invariants I_1, I_2, and I_3 in the principal stress space and p, q, and θ in the $p-q-\theta$ stress space:

$$F(I_1, I_2, I_3, k_j) = 0, \quad (2)$$

$$F(p, q, \theta, k_j) = 0, \quad (3)$$

where I_1, I_2, and I_3 are the first, second, and third principal stress invariants, respectively. p is the hydrostatic pressure and q is the deviatoric stress. According to the research of Liao et al. [31], Equation (3) can be expressed as

$$q = f_{p-q} g(\theta), \quad (4)$$

where f_{p-q} stands for the critical state curve on the $p-q$ plane, $g(\theta)$ represents the shape function on the deviatoric plane, and θ is Lode's angle.

3.1. Critical Strength Curves of Frozen Saline Silty Clay in $p-q$ Plane. Referring to the research of Chu [34], the critical strength curves of frozen saline silty clay with different salt contents are shown in Figure 3. From Figure 3, the following can be observed. (1) The CSL of frozen saline silty clay with different salt contents passes through the different isotropic tensile points. The reason is that the existence of salt has an important influence on the cohesive strength of frozen soils. (2) The CSL of frozen saline silty clay with different salt contents is nonlinear, and the slope of the CSLs is positive when hydrostatic pressure is relatively small. When hydrostatic pressure reaches a certain value, the CSL is gradually bended downward with the increase of hydrostatic pressure, and the slope of the CSL is negative accordingly. (3) The slope k of the CSL of frozen saline silty clay under different salt contents is different when the hydrostatic pressure $p < 8$ MPa, in detail $k_{0.5} < k_0 < k_{1.5} < k_{2.5}$. And the relationship between shear strength and hydrostatic pressure can be expressed by

$$f(p, q) = \frac{a(s)(p/p_a)^2 + b(s)(p/p_a) + c(s)}{(p/9.8687 p_a) + d(s)} p_a - q = 0, \quad (5)$$

where p_a is the standard atmospheric pressure, p_a = 0.10133 MPa. The parameters $a(s)$, $b(s)$, $c(s)$, and $d(s)$ were gotten by the fitting curves. $a(s)$, $b(s)$, $c(s)$, and $d(s)$ are the material parameters related to salt contents at −6°C.

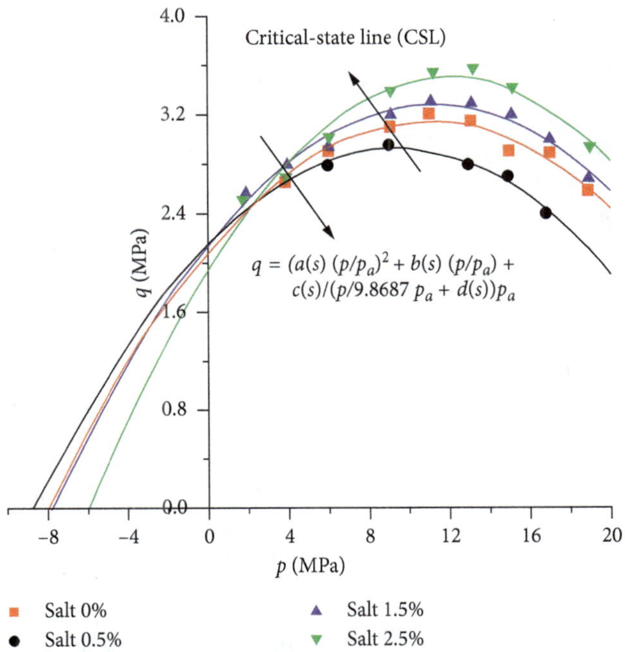

FIGURE 3: Test values and fitting curves of CSL with different salt contents.

$$\begin{aligned} a(s) &= -2.815\times10^5 s^3 + 1.07\times10^6 s^2 - 9.754\times10^5 s \\ &\quad - 2.009\times10^3, \\ b(s) &= -5.455\times10^5 s^3 - 2.044\times10^6 s^2 + 1.848\times10^6 s \\ &\quad - 4.487\times10^3, \\ c(s) &= 6.729\times10^5 s^3 - 2.591\times10^6 s^2 + 2.381\times10^6 s \\ &\quad + 4.983\times10^3, \\ d(s) &= 2.127\times10^5 s^3 - 3.318\times10^5 s^2 + 6.573\times10^4 s \\ &\quad + 2.351\times10^4. \end{aligned} \tag{6}$$

Based on the Mohr strength theorem, the failure envelope can be written as

$$F(\sigma,\tau,\sigma_1,\sigma_3) = \left(\sigma - \frac{\sigma_1+\sigma_3}{2}\right)^2 + \tau^2 - \left(\frac{\sigma_1-\sigma_3}{2}\right)^2 = 0. \tag{7}$$

For the conventional triaxial tests, $p = (1/3)(\sigma_1 + 2\sigma_3)$, $q = \sigma_1 - \sigma_3$. Substituting p, q into Equation (7), Equation (7) can be written as

$$F(p,q) = \left(\sigma - p - \frac{q}{6}\right)^2 + \tau^2 - \left(\frac{q}{2}\right)^2 = 0. \tag{8}$$

Based on the envelope theory, we can obtain the following formula:

$$\frac{\partial F}{\partial p}\frac{\partial f}{\partial q} - \frac{\partial F}{\partial q}\frac{\partial f}{\partial p} = 0. \tag{9}$$

Differentiating Equations (5) and (8), we can get

$$\frac{\partial f}{\partial q} = -1, \tag{10}$$

$$\frac{\partial f}{\partial p} = \frac{[(p/9.8687p_a) + d(s)]\left[a(s)\left(p/(p_a)^2\right) + b(s)(1/p_a)\right] - (1/9.8687p_a)\left[a(s)(p/p_a)^2 + b(s)(p/p_a) + c(s)\right]}{[(p/9.8687p_a) + d(s)]^2} p_a, \tag{11}$$

$$\frac{\partial F}{\partial q} = -\frac{1}{3}\left(\sigma - p - \frac{q}{6}\right) - \frac{q}{2}, \tag{12}$$

$$\frac{\partial F}{\partial p} = -2\left(\sigma - p - \frac{q}{6}\right). \tag{13}$$

By substituting Equations (10), (11), (12), and (13) into Equation (9), we can get the following relationship

$$\sigma = \frac{[(1/3)(p+(q/6))-(q/2)]\{[[(p/9.8687p_a) + d(s)][a(s)(p/(p_a)^2) + b(s)(1/p_a)] - (1/9.8687p_a)[a(s)(p/p_a)^2 + b(s)(p/p_a) + c(s)]]/[[(p/(9.8687p_a)) + d(s)]^2]p_a\} + 2(p+(q/6))}{2 + (1/3)\{[[(p/9.8687p_a) + d(s)][a(s)(p/(p_a)^2) + b(s)(1/p_a)] - (1/9.8687p_a)[a(s)(p/p_a)^2 + b(s)(p/p_a) + c(s)]]/[[(p/(9.8687p_a)) + d(s)]^2]p_a\}}. \tag{14}$$

By substituting Equation (14) into Equation (8), we can get the shear tress τ as follows:

$$\tau = \sqrt{\left(\frac{q}{2}\right)^2 - \left\{\frac{[(1/3)(p+(q/6))-(q/2)]\{[[(p/9.8687p_a) + d(s)][a(s)(p/(p_a)^2) + b(s)(1/p_a)] - (1/9.8687p_a)[a(s)(p/p_a)^2 + b(s)(p/p_a) + c(s)]]/[[(p/9.8687p_a) + d(s)]^2]p_a\} + 2(p+(q/6))}{2 + (1/3)\{[(p/9.8687p_a) + d(s)][[a(s)(p/(p_a)^2) + b(s)(1/p_a)] - (1/9.8687p_a)[a(s)(p/p_a)^2 + b(s)(p/p_a) + c(s)]]/[[(p/9.8687p_a) + d(s)]^2]p_a\}} - p - \frac{q}{6}\right\}^2}. \tag{15}$$

The friction angle φ of frozen saline silty clay can be calculated by

$$\varphi = \arctan\left(\frac{d\tau}{d\sigma}\right) = \arctan\left(-\frac{\partial f/\partial\sigma}{\partial f/\partial\tau}\right) = \arctan\left(-\frac{\sigma - p - (q/6)}{\tau}\right). \tag{16}$$

From Equations (14), (15), and (16), it can be seen that the friction angle φ of frozen saline silty clay is related to salt content. Using Equations (14), (15), and (16), we can get the curves between friction angle φ and hydrostatic pressure with different salt contents 0.0, 0.5, 1.5, and 2.5%, shown in Figure 4.

From Figure 4, the following can be observed. (1) Under any same hydrostatic pressure, the friction angles reached a minimum value when salt content is equal to 0.5%. (2) Under each kind of salt content, the friction angles decrease with the increase of hydrostatic pressure. (3) Salt content has an important influence on the friction angles when the hydrostatic pressure $p < 18$ MPa. The influence of salt content on the friction angles is less obvious when $p > 18$ MPa. This phenomenon is contrary to the finding of Liao et al. [31]. (4) Under any salt content, when the hydrostatic pressure p is beyond a certain value, the friction angles obtained by Equation (16) will become negative. The main reason to result in this phenomenon is the pressure melting phenomenon under high confining pressures. The greater the confining pressure, the more obvious the phenomenon so that its strength is gradually reduced, and the strength envelope bends downwards gradually, so the calculated friction angles is negative.

Using Equations (14) and (15), we can get the shear strength envelope of frozen saline silty clay in the $\sigma - \tau$ space, shown in Figure 5. From Figure 5, it can be seen that the strength envelope is able to reflect the strength of frozen saline silty clay considering the influence of confining pressures well.

3.2. Strength Formulation in π Plane. The experiments show that failure curves of some geomaterials in π plane are close to the SMP criterion [35], so we take the shape function of the SMP criterion in π plane as the shape function of frozen saline silty clay in the π plane. Matsuoka et al. [36] gave the shape function for the SMP criterion in π plane as follows:

$$g(\theta) = \frac{\sqrt{3}\left(\sqrt{8 + \sin^2\varphi_0} - \sin\varphi_0\right)}{4\sqrt{2 + \sin^2\varphi_0}\cos\vartheta}, \tag{17}$$

$$\vartheta = \frac{1}{3}\cos^{-1}\left[-\left(\frac{3}{2 + \sin^2\varphi_0}\right)^{3/2}\sin\varphi_0\cos 3\theta\right] \quad (0^\circ \le \theta \le 60^\circ), \tag{18}$$

$$\tan\varphi_0 = \frac{2\sqrt{2}}{3}\tan\varphi, \tag{19}$$

where φ_0 is the friction angle in the SMP.

In this paper, the shape function for frozen saline silty clay in π plane can be written as

$$g(\theta) = \frac{\sqrt{3}\left(\sqrt{8 + \sin^2\varphi_0} - \sin\varphi_0\right)}{4\sqrt{2 + \sin^2\varphi_0}\cos\left\{(1/3)\cos^{-1}\left[-(3/(2 + \sin^2\varphi_0))^{3/2}\sin\varphi_0\cos 3\theta\right]\right\}} \quad (0^\circ \le \theta \le 60^\circ). \tag{20}$$

Finally, using Equations (4), (5), and (20), the entire strength criterion for frozen saline silty clay can be expressed as

$$F = \frac{a(s)(p/p_a)^2 + b(s)(p/p_a) + c(s)}{(p/9.8687p_a) + d(s)}p_a \times \left[\frac{\sqrt{3}\left(\sqrt{8 + \sin^2\varphi_0} - \sin\varphi_0\right)}{4\sqrt{2 + \sin^2\varphi_0}\cos\vartheta}\right] - q = 0. \tag{21}$$

We take the failure curve when salt content is 1.5% as an example to study failure curves of frozen saline silty clay. The failure surface and the shape function curves of this strength criterion with salt content 1.5% are shown in Figure 6.

From Figures 6(a) and 6(b), it can be seen that the failure surface for frozen saline silty clay in principal stress space with salt content 1.5% is a curved surface, whose opening increases first and then decreases. The shape function in π

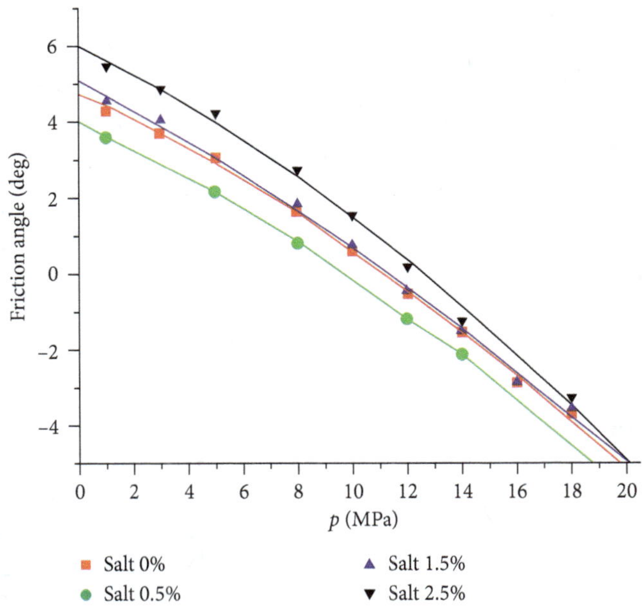

FIGURE 4: Relationship between hydrostatic pressure and friction angle of frozen saline silty clay with different salt contents.

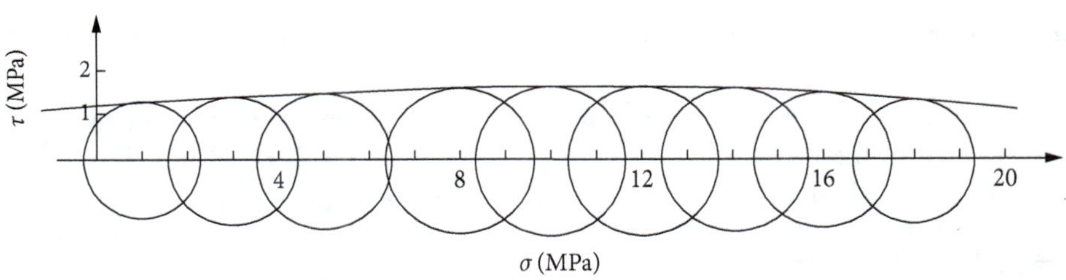

FIGURE 5: The shear strength envelope of frozen saline silty clay.

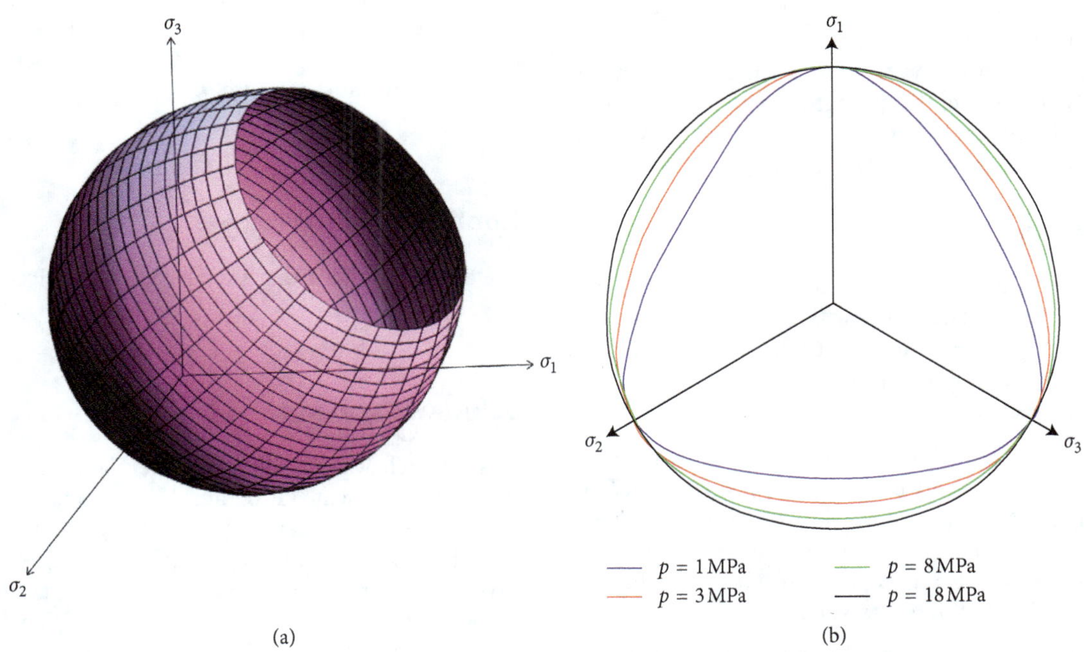

FIGURE 6: The failure surface and the shape function curves for frozen saline silty clay with salt content 1.5%: (a) principal stress space; (b) π plane.

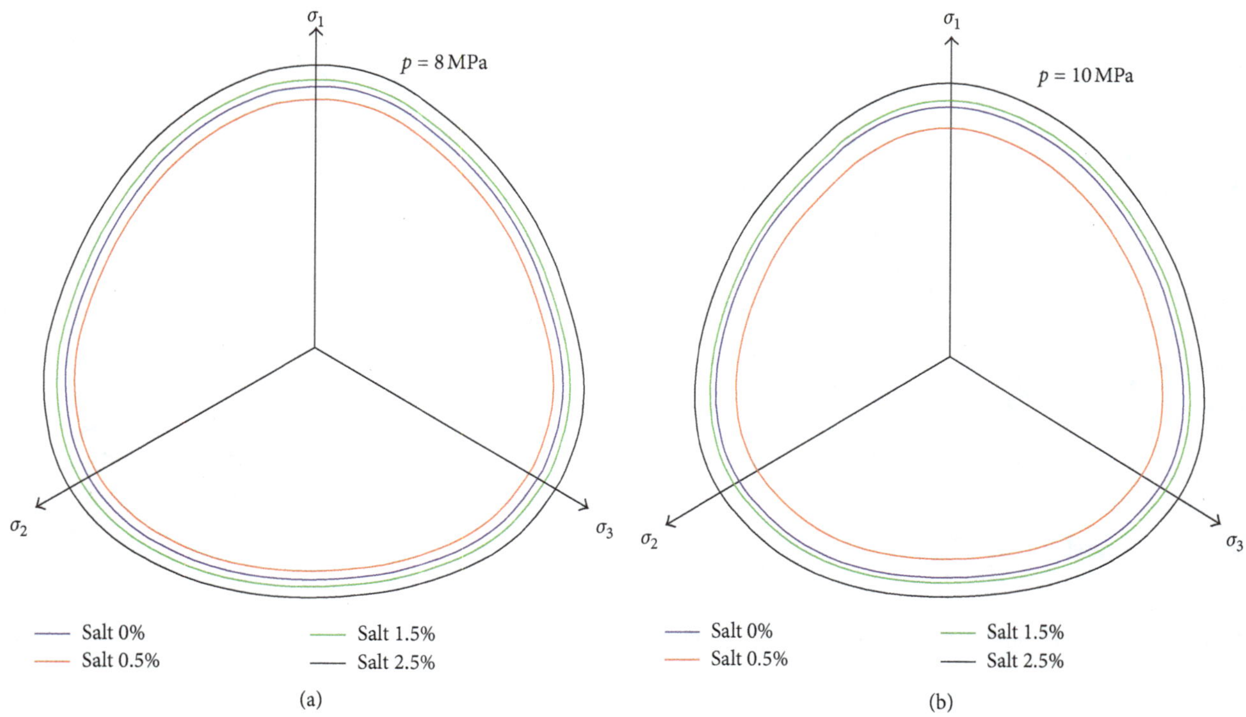

FIGURE 7: The shape function curves in the π plane with different salt contents: (a) $p = 8$ MPa; (b) $p = 10$ MPa.

plane is a curved triangle when the hydrostatic pressure $p = 1$ MPa; with the increase of the hydrostatic pressure, the shape function changes from curved triangle to circle, and the shape function is approximately a circle when the hydrostatic pressure $p = 18$ MPa. The shape function curves with different salt contents in the π plane are shown in Figure 7. From Figures 7(a) and 7(b), it can be seen that the shape function changes versus salt content variation from 0.0% to 2.5% when the hydrostatic pressure p is a constant. The innermost curve is the shape function curve of frozen saline silty with salt content 0.5%, while the outermost curve is the shape function curve of frozen saline silty with salt content 2.5%. Therefore, the strength reached a minimum value when salt content is equal to 0.5%, while it reached a maximum value when salt content is equal to 2.5%.

4. Conclusions

A series of conventional triaxial compression tests were carried out for frozen saline silty with Na_2SO_4 contents 0.0, 0.5, 1.5, and 2.5% under confining pressures ranging from 0 MPa to 18 MPa at −6°C, respectively. From the test results, the following conclusions can be made:

(1) The strength of frozen saline silty clay presents obvious nonlinearity under each salt content, and the strength of frozen saline silty clay increases with increasing confining pressures under low confining pressures, but the strength decreases with a further increase in confining pressures.

(2) The salt contents have an important influence on strength of frozen saline silty clay. The strength reaches a minimum value when salt content is equal to 0.5%, while it reaches a maximum value when salt content is equal to 2.5%.

(3) The slope k of the CSL of frozen saline silty clay under different salt contents is different when the hydrostatic pressure $p < 8$ MPa. In this study, it is $k_{0.5} < k_0 < k_{1.5} < k_{2.5}$.

(4) The strength criterion proposed in this paper could well reflect the effects of confining pressures and salt contents on strength properties of frozen saline silty clay. The shape function in the π plane changes from curved triangle to circle with the increases of the hydrostatic pressure.

Conflicts of Interest

The authors declare that they have no conflicts of interest.

Acknowledgments

This research was supported by the National Key Research and Development Program of China (Grant No. 2018YFC0809605), Key Research Program of Frontier Sciences of Chinese Academy of Sciences (Grant No. QYZDY-SSW-DQC015), National Key Basic Research Program of China (973 Program No. 2012CB026102), National Natural Science Foundation of China (Grant Nos. 41230630 and 41701068), and Program of the State Key Laboratory of Frozen Soil Engineering (Grant No. SKLFSE-ZT-23).

References

[1] X. Z. Xu, J. C. Wang, L. X. Zhang, and Z. X. Tao, *Mechanisms of Frost Heave and Soil Expansion of Soils*, Science Press, Beijing, China, 1995.

[2] H. Matsuoka and T. Nakai, "Stress-deformation and strength characteristics of soil under three different principal stresses," *Proceedings of the Japanese Society of Civil Engineers*, vol. 1974, no. 232, pp. 59–70, 1974.

[3] P. V. Lade and J. M. Duncan, "Elasto-plastic stress-strain theory for cohesionless soil," *Journal of the Geotechnical Engineering Division*, vol. 101, no. 10, pp. 1037–1053, 1975.

[4] A. Gajo and D. Muir Wood, "A kinematic hardening constitutive model for sands: the multiaxial formulation," *International Journal Numerical and Analytical Methods in Geomechanics*, vol. 23, no. 9, pp. 925–965, 1999.

[5] J. M. Pestana, A. J. Whittle, and L. A. Salvati, "Evaluation of a constitutive model for clays and sands: Part I-sand behavior," *International Journal Numerical and Analytical Methods in Geomechanics*, vol. 26, no. 11, pp. 1097–1121, 2002.

[6] Y. P. Yao, W. Hou, and A. N. Zhou, "UH model: three-dimensional unified hardening model for overconsolidated clays," *Geotechnique*, vol. 59, no. 5, pp. 451–469, 2009.

[7] N. A. Tsytovich, *The Mechanics of Frozen Ground*, Science Press, Beijing, China, 1985.

[8] O. B. Andersland and I. Ainouri, "Time-dependent strength behavior of frozen soils," *Journal of the Soil Mechanics and Foundation Division*, vol. 96, no. 4, pp. 1249–1265, 1970.

[9] B. D. Alkire and O. B. Andersland, "The effect of confining pressure on the mechanical properties of sand-ice materials," *Journal of Glaciology*, vol. 12, no. 66, pp. 469–481, 1973.

[10] F. D. Haynes and J. A. Karalius, *Effect of Temperature on the Strength of Frozen Silt*, Cold Regions Research and Engineering Laboratory, Hanover, NH, USA, 1977.

[11] S. S. Vialov, *Determination of Strength and Creep for Artificially Frozen Soils*, Stroyizdat, Leningr.otdelenie, Leningrad, Russia, 1981.

[12] V. R. Parameswaran and S. J. Jones, "Triaxial testing of frozen sand," *Journal of Glaciology*, vol. 27, no. 95, pp. 147–155, 1981.

[13] T. H. W. Baker, S. J. Jones, and V. R. Parameswaran, "Confined and unconfined compression tests of frozen sand," in *Proceedings of 4th Canadian Permafrost Conference National Research Council of Canada*, pp. 387–392, Canada, 1982.

[14] J. M. Ting, R. T. Martin, and C. C. Ladd, "Mechanisms of strength for frozen sand," *Journal of Geotechnical Engineering*, vol. 109, no. 10, pp. 1286–1302, 1983.

[15] Y. L. Zhu and D. L. Carbee, "Uniaxial compressive strength of frozen silt under constant deformation rates," *Cold Regions Science and Technology*, vol. 9, no. 1, pp. 3–5, 1987.

[16] Y. L. Zhu and D. L. Carbee, *Creep and Strength Behaviour of Frozen Silt in Uniaxial Compression*, Cold Regions Research and Engineering Laboratory, Hanover, NH, USA, 1987.

[17] I. E. Guryanov and W. Ma, "Strength characteristics of frozen soil under loading and unloading," *Journal of Glaciology and Geocryology*, vol. 18, no. 1, pp. 53–57, 1995, in Chinese.

[18] W. Ma, Z. W. Wu, L. X. Zhang, and X. X. Chang, "Analyses of process on the strength decrease in frozen soils under high confining pressure," *Cold Regions Science and Technology*, vol. 29, no. 1, pp. 1–7, 1999.

[19] W. Ma and X. X. Chang, "Analyses of strength and deformation of an artificially frozen soil wall in underground engineering," *Cold Regions Science and Technology*, vol. 34, no. 1, pp. 11–17, 2002.

[20] D. Y. Wang, W. Ma, X. X. Chang, and A. G. Wang, "Study on the resistance to deformation of artificially frozen soil in deep alluvium," *Cold Regions Science and Technology*, vol. 42, no. 3, pp. 194–200, 2005.

[21] Y. M. Lai, H. B. Cheng, Z. H. Gao, and X. X. Chang, "Stress–strain relationships and nonlinear Mohr strength criterion of frozen sand clay," *Chinese Journal of Rock Mechanics and Engineering*, vol. 26, no. 8, pp. 1612–1617, 2007.

[22] Y. M. Lai, L. Jin, and X. X. Chang, "Yield criterion and elasto-plastic damage constitutive model for frozen sandy soil," *International Journal of Plasticity*, vol. 25, no. 6, pp. 1177–1205, 2009.

[23] Y. M. Lai, Y. G. Yang, X. X. Chang, and S. Y. Li, "Strength criterion and elastoplastic constitutive model of frozen silt in generalized plastic mechanics," *International Journal of Plasticity*, vol. 26, no. 10, pp. 1461–1484, 2010.

[24] E. Chamberlain, C. Groves, and R. Perham, "The mechanical behaviour of frozen earth materials under high pressure triaxial test conditions," *Géotechnique*, vol. 22, no. 3, pp. 469–483, 1972.

[25] J. L. Qi and W. Ma, "A new criterion for strength of frozen sand under quick triaxial compression considering effect of confining pressure," *Acta Geotechnica*, vol. 2, no. 3, pp. 221–226, 2007.

[26] A. M. Fish, "Strength of frozen soil under a combined stress state," in *Proceedings of 6th International Symposium on Ground Freezing*, vol. 1, pp. 135–145, A.A. Balkema, Beijing, China, September 1991.

[27] W. Ma, Z. Wu, and X. X. Chang, "Effect of confining pressure on strength behaviour of frozen soil," *Chinese Journal of Geotechnical Engineering*, vol. 17, no. 5, pp. 7–11, 1995.

[28] Y. G. Yang, Y. M. Lai, and J. B. Li, "Laboratory investigation on the strength characteristic of frozen sand considering effect of confining pressure," *Cold Regions Science and Technology*, vol. 60, no. 3, pp. 245–250, 2010.

[29] Y. G. Yang, Y. M. Lai, Y. H. Dong, and S. Y. Li, "The strength criterion and elastoplastic constitutive model of frozen soil under high confining pressures," *Cold Regions Science and Technology*, vol. 60, no. 2, pp. 154–160, 2010.

[30] Y. G. Yang, F. Gao, and Y. M. Lai, "Modified Hoek-Brown criterion for nonlinear strength of frozen soil," *Cold Regions Science and Technology*, vol. 86, pp. 98–103, 2013.

[31] M. K. Liao, Y. M. Lai, and C. Wang, "A strength criterion for frozen sodium sulfate saline soil," *Canadian Geotechnical Journal*, vol. 53, no. 7, pp. 1176–1185, 2016.

[32] X. S. Wan, *Laboratory investigation on salt crystals precipitation of sulfate saline soil and salt transfer and salt-heaving mitigation of embankment in cold regions*, Ph.D. thesis, University of Chinese Academy of Sciences, Huairou, China, 2015.

[33] People's Republic of China National Standard GB/T 50123-1999, *Standard for Soil Test Method*, China Planning Press, Beijing, China, 1999.

[34] J. Chu, "An experimental examination of the critical state and other similar concepts for granular soils," *Canadian Geotechnical Journal*, vol. 32, no. 6, pp. 1065–1075, 1995.

[35] Y. P. Yao, J. Hu, A. N. Zhou, T. Luo, and N. D. Wang, "Unified strength criterion for soils, gravels, rocks, and concretes," *Acta Geotechnica*, vol. 10, no. 6, pp. 749–759, 2015.

[36] H. Matsuoka, Y. P. Yao, and D. Sun, "The cam-clay models revised by the SMP criterion," *Soils and Foundations*, vol. 39, no. 1, pp. 81–95, 1999.

21

A New Method to Simulate Free Surface Flows for Viscoelastic Fluid

Yu Cao,[1,2,3] Xiao-Guang Ren,[1,2] Xiao-Wei Guo,[1,2] Miao Wang,[1,2] Qian Wang,[1,2] Xin-Hai Xu,[1,2] and Xue-Jun Yang[1,2]

[1] *College of Computer, National University of Defense Technology, Changsha, Hunan 410073, China*
[2] *State Key Laboratory of High Performance Computing, National University of Defense Technology, Changsha, Hunan 410073, China*
[3] *School of Chemical Engineering and Analytical Science, Manchester Institute of Biotechnology, University of Manchester, Manchester M13 9PL, UK*

Correspondence should be addressed to Xiao-Guang Ren; renxiaoguang@nudt.edu.cn

Academic Editor: Olanrewaju Ojo

Free surface flows arise in a variety of engineering applications. To predict the dynamic characteristics of such problems, specific numerical methods are required to accurately capture the shape of free surface. This paper proposed a new method which combined the Arbitrary Lagrangian-Eulerian (ALE) technique with the Finite Volume Method (FVM) to simulate the time-dependent viscoelastic free surface flows. Based on an open source CFD toolbox called OpenFOAM, we designed an ALE-FVM free surface simulation platform. In the meantime, the die-swell flow had been investigated with our proposed platform to make a further analysis of free surface phenomenon. The results validated the correctness and effectiveness of the proposed method for free surface simulation in both Newtonian fluid and viscoelastic fluid.

1. Introduction

In recent years, there is a growing interest in free surface simulations (e.g., die-swell and bubble growing). Die-swell, as a typical phenomenon seen in viscoelastic fluids, is widely investigated. Numerical simulations of extrusion flow need to address free surface problems. Numerical schemes for free surface treatment can be classified into the fixed grid (interface capturing) and the moving grid (interface tracking) methods. Fixed grid methods capture the interface in a Eulerian description, including Volume of Fluid (VOF) method [1], Level Set method [2], and Marker and Cell (MAC) method [3], in which the interface is reconstructed through numerical schemes in these methods. Though these methods could address sharp interface and large deformation issues well, the position of the free surface is usually not precise enough. On the contrary, moving grid methods explicitly track the position of interface by computational grids which guarantees the precision of the interface position, but the quality of the mesh might be reduced during mesh moving in large deformation cases. Arbitrary Lagrangian-Eulerian (ALE) Method [4] is one of the moving grid methods. It combines the advantage of the Lagrangian method in tracking interface and the efficiency of Eulerian method in solving governing equations. Therefore, it has been widely used in wave-like free surface simulations.

ALE method is usually combined with Finite Element Method (FEM) [5], in which the traction boundary condition on the interface could be treated in a neat and exact way by integrating the force in the momentum equation by parts. However, with the development of Finite Volume Method (FVM), FVM system is more attractive in fluid simulations for its good property in local conservation. Therefore, combining ALE method with FVM is capable of representing the position of free surface by the computational mesh directly, while benefitting from a good conservation property with FVM. Tuković and Jasak [6] developed a platform to simulate Newtonian fluid free surface problems by using the ALE method in FVM system. In order to get the accurate velocity on the surface, a new surface mesh is coupled with the traditional mesh, but it increases the computational cost and complexity. Moreover, it could not tackle with viscoelastic fluids. The greatest challenge of doing this is setting proper boundary conditions on the free surface.

In this paper, a novel numerical method is proposed for 2D viscoelastic extrusion flow simulations. To the best of the authors' knowledge, this paper is the first implementation for combining ALE technique and FVM in studying time-dependent viscoelastic extrusion simulations, which extends the application fields of ALE method. The main contribution of this paper may be summarized as follows:

(i) We addressed the challenges in coupling the ALE with FVM method to simulate time-dependent viscoelastic extrusion flows. The obstacle of setting proper free surface boundary conditions is well addressed by deriving the Dirichlet pressure boundary condition and introducing a balance force imposed on the free surface. The proposed method is capable of dealing with both Newtonian fluids and viscoelastic fluids.

(ii) We implemented a coupled ALE-FVM simulation platform based on OpenFOAM (Open Source Field Operation and Manipulation) [7]. The novel numerical method is validated by comprehensively comparing with previous literature. Compared to Tuković and Jasak's work [6], our approach significantly reduced the storage and computation cost by abandoning the additional surface mesh. Moreover, based on OpenFOAM, our free surface numerical solver could be easily scaled to massive processor cores for large-scale parallel simulations.

The rest of the paper is organized as follows. Details of the proposed method are presented in Section 2. The numerical results and discussion are reported in Section 3. Conclusions are drawn in Section 4.

2. Methodology

In this section, the mathematical formulation of Oldroyd-B model and numerical algorithms used in the paper will be presented.

2.1. Mathematical Model. The governing equations for isothermal and incompressible viscoelastic fluid flows include the mass (continuity) conservation equation, momentum equation, and constitutive equation. The forms of those equations for the Oldroyd-B model are given by

$$\begin{gathered} \rho\left(\frac{\partial \mathbf{u}}{\partial t} + (\mathbf{u}-\mathbf{w})\cdot\nabla\mathbf{u}\right) = -\nabla p + \nabla\cdot\boldsymbol{\tau} + \eta_s\Delta\mathbf{u}, \\ \nabla\cdot\mathbf{u} = 0, \\ \boldsymbol{\tau} + \lambda\overset{\nabla}{\boldsymbol{\tau}} = 2\eta_p\mathbf{D}, \end{gathered} \tag{1}$$

where ρ is the fluid density, $\mathbf{u}$ is the fluid velocity, $\mathbf{w}$ is the velocity of the cell, p is the pressure, and $\boldsymbol{\tau}$ is the polymeric contribution for stress tensor. η_s and η_p are the solvent viscosity and polymer viscosity, respectively. λ is the characteristic relaxation time, and $\mathbf{D}$ is the rate of deformation tensor, which can be expressed as

$$\mathbf{D} = \frac{1}{2}\left(\nabla\mathbf{u} + (\nabla\mathbf{u})^T\right). \tag{2}$$

The upper-convected time derivative of the polymer stress tensor is defined as

$$\overset{\nabla}{\boldsymbol{\tau}} = \frac{\partial\boldsymbol{\tau}}{\partial t} + (\mathbf{u}-\mathbf{w})\cdot\nabla\boldsymbol{\tau} - (\nabla\mathbf{u})^T\cdot\boldsymbol{\tau} - \boldsymbol{\tau}\cdot\nabla\mathbf{u}. \tag{3}$$

For the convenience of comparison with others' works, variables above are usually converted into their dimensionless form. New dimensionless variable Reynolds number Re and Weissenberg number Wi are introduced by scaling the following variables as

$$\begin{aligned} x' &= \frac{x}{L}, \\ y' &= \frac{y}{L}, \\ \mathbf{u}' &= \frac{\mathbf{u}}{U}, \\ p' &= \frac{pL}{\eta U}, \\ \boldsymbol{\tau}' &= \frac{\boldsymbol{\tau}L}{\eta U}, \\ t' &= \frac{L}{U}t, \\ \mathrm{Re} &= \frac{\rho UL}{\eta}, \\ \mathrm{Wi} &= \lambda\frac{U}{L}, \\ \beta &= \frac{\eta_s}{\eta} \end{aligned} \tag{4}$$

in which L is the characteristic length (the half width of the die in this work), U is the characteristic velocity (the average velocity in this work), and η is the sum of the η_s and η_p; namely, total viscosity $\eta = \eta_s + \eta_p$. β is the viscosity ratio parameter.

2.2. Numerical Algorithm

2.2.1. Implementation of Free Surface Boundary Condition. The most difficult problems in die-swell simulation are the implementation of free surface boundary condition. Boundary conditions imposed on the free surface could be divided into two categories, the kinematic boundary condition and the dynamic boundary condition. The kinematic boundary condition requires that the normal component of the fluid velocity equals the normal component of the interfacial grid velocity, as denoted by

$$\mathbf{u}\cdot\mathbf{n} = \mathbf{w}\cdot\mathbf{n}. \tag{5}$$

The equation above means the relative speed between the fluid and the grid perpendicular to the interface is zero. In OpenFOAM, the velocity of fluid is stored at the cell center, which means it would be hard to get the velocity of interfacial grid directly. However, the flux of the fluid through each cell face could be obtained easily. Therefore, the velocity of interfacial face could be calculated according to the net mass flux on the surface. A detailed description about the implementation of the kinematic boundary condition can be found in [6, 8]. After moving the interfacial nodes at the interface, the internal nodes need to be moved according to the solution of solving a Laplace equation:

$$\nabla \cdot (\Gamma \nabla \mathbf{u}) = 0, \tag{6}$$

where Γ is the diffusion coefficient. The solver chosen to move nodes in this work is Laplace Face Decomposition [9], which could maintain a good quality of the mesh in large deformation cases. And the diffusion coefficient chosen is inversely proportional to the square of distance from the free surface. Regenerating the internal grids in this way could form lines of equal potential and decrease the nonorthogonality which in turn increases the quality of the mesh.

The dynamic boundary condition is also called the traction boundary condition. Ignoring the surface tension, it means the traction exerted from the fluid to the atmosphere should be equal to the atmospheric forces acting on the fluid. Without loss of generality, the atmosphere pressure is set as $p_0 = 0$. So the boundary condition is given by

$$\boldsymbol{\sigma} \cdot \mathbf{n} = (-p\mathbf{I} + 2\beta\mathbf{D} + \boldsymbol{\tau}) \cdot \mathbf{n} = (-p\mathbf{I} + \mathbf{T}) \cdot \mathbf{n} = 0, \tag{7}$$

where $\boldsymbol{\sigma}$ is the total stress tensor, namely, the Cauchy stress tensor across the free surface, and $\mathbf{T} = 2\beta\mathbf{D} + \boldsymbol{\tau}$ is the extra stress tensor. When solving equations in OpenFOAM, PISO algorithm requires the term $-\nabla p$ and the term $\eta_s \Delta \mathbf{u}$ exists separately in the equation. However, the term $\boldsymbol{\sigma}$ is a combination of p, $\boldsymbol{\tau}$, and $\mathbf{D}$. Therefore, this boundary condition is hard to satisfy directly. It has been known that $-\nabla p + \nabla \cdot \boldsymbol{\tau} + \eta_s \Delta u = \nabla \cdot \boldsymbol{\sigma}$, and in FVM systems, every term in the equation should be integrated over the cell volume V. Therefore, the right side of (1) is actually solved as $\int_V (-\nabla p + \nabla \cdot \boldsymbol{\tau} + \eta_s \Delta u) dV = \int_V (\nabla \cdot \boldsymbol{\sigma}) dV$. According to Gauss's theorem, the spatial derivative term could be converted to integrals over the cell surface S bounding the volume. Then, we have

$$\begin{aligned}\int_V (\nabla \cdot \boldsymbol{\sigma})\, dV &= \int_S d\mathbf{S} \cdot \boldsymbol{\sigma} = \sum_a \boldsymbol{\sigma}_a \cdot \mathbf{n}_a S_a \\ &= \sum_{a \neq b} \boldsymbol{\sigma}_a \cdot \mathbf{n}_a S_a + \boldsymbol{\sigma}_b \cdot \mathbf{n}_b S_b,\end{aligned} \tag{8}$$

where $\mathbf{S}$ is the face area vector and $\mathbf{n}$ is the unit normal vector of face area. a and b represent faces around a cell and free surface faces, respectively.

The dynamic boundary condition requires $\boldsymbol{\sigma}_b \cdot \mathbf{n}_b = 0$; however, it would not be the case if any of the boundary conditions is improper. Noting that the boundary condition would only influence the calculation of the cell that owns the boundary patch, we introduce a balance force $\boldsymbol{\sigma}^* = -\boldsymbol{\sigma}$ on the free surface (it is zero elsewhere including the internal field and other boundary fields) to compensate the influence of nonzero boundary force. By adding the term $\nabla \cdot \boldsymbol{\sigma}^*$ into the original momentum equation, the right side of (1) can be solved as

$$\begin{aligned}&\int_V (-\nabla p + \nabla \cdot \boldsymbol{\tau} + \eta_s \Delta u + \nabla \cdot \boldsymbol{\sigma}^*)\, dV \\ &= \int_V (\nabla \cdot \boldsymbol{\sigma} + \nabla \cdot \boldsymbol{\sigma}^*)\, dV = \int_S d\mathbf{S} \cdot (\boldsymbol{\sigma} + \boldsymbol{\sigma}^*) \\ &= \sum_a \boldsymbol{\sigma}_a \cdot \mathbf{n}_a S_a - \boldsymbol{\sigma}_b \cdot \mathbf{n}_b S_b = \sum_{a \neq b} \boldsymbol{\sigma}_a \cdot \mathbf{n}_a S_a.\end{aligned} \tag{9}$$

It can be seen that the influence of the total stress imposed on the free surface is offset in this way, which means the traction boundary condition is satisfied equivalently. The momentum equation is actually solved in the form:

$$\begin{aligned}&\rho \left(\frac{\partial \mathbf{u}}{\partial t} + (\mathbf{u} - \mathbf{w}) \cdot \nabla \mathbf{u} \right) \\ &= -\nabla p + \nabla \cdot \boldsymbol{\tau} + \eta_s \Delta \mathbf{u} + \langle \nabla \cdot \boldsymbol{\sigma}^* \rangle_{\Gamma_f},\end{aligned} \tag{10}$$

where Γ_f indicates the term is only applied on the free surface boundary. Since the SIMPLE algorithm is used to iteratively solve the linear momentum equation and the continuity equation, a proper pressure boundary condition is required during the step of solving the Poisson equation. According to the force balance in the normal direction on the interface, Dirichlet boundary condition for the pressure is given by

$$p = \mathbf{n} \cdot \mathbf{T} \cdot \mathbf{n}. \tag{11}$$

It is important to note that the pressure condition here is just a derived quantity, which has arisen only because of the need to solve the Poisson equation for p in isolation. While the traction boundary condition is the real physical quantity, it must be satisfied on the fluid. Provided that p and $\mathbf{u}$ are solved jointly, no pressure boundary condition is needed anymore [10].

2.2.2. DEVSS Method for Numerical Stability. In order to enhance the stability of the linear momentum equation, especially for high Wi number flow, the discrete elastic split stress (DEVSS) numerical strategy proposed by Matallah et al. [11] is used. The momentum equation can be rewritten as

$$\begin{aligned}&\rho \left(\frac{\partial \mathbf{u}}{\partial t} + (\mathbf{u} - \mathbf{w}) \cdot \nabla \mathbf{u} \right) \\ &= -\nabla p + \nabla \cdot \Sigma_p + \eta_t \Delta \mathbf{u} + \langle \nabla \cdot \boldsymbol{\sigma}^* \rangle_{\Gamma_f},\end{aligned} \tag{12}$$

where $\eta_t = \eta_s + \alpha$ and $\nabla \cdot \Sigma_p$ is the added artificial viscosity ratio. The term $\nabla \cdot \Sigma_p = \nabla \cdot \boldsymbol{\tau} - \alpha \Delta \mathbf{u}$ is treated as a source term. A typical choice of α is η_p. This strategy greatly increases numerical stability and can be applied in modelling various constitutive models.

2.2.3. The Simulation Procedure. Apart from the treatment of the free surface boundary conditions, the simulation procedure is similar to [6] as shown below:

(1) For a new time step t_n, the velocity field $\mathbf{u}^*$, the pressure field p^*, the polymer stress field $\boldsymbol{\tau}^*$, and the fluid mass fluxes $\dot{m}_f$ are obtained from the previous time step. The initial net mass fluxes through the interface faces φ_n are calculated according to the previous interface position by subtracting the fluid mass fluxes by the volumetric face fluxes $\dot{V}_f$.

(2) Move the interface mesh points so that the net mass flux through the interface could be compensated and the mass conservation law is satisfied.

(3) According to the displacement of the interface mesh points, the position of internal mesh points is obtained by solving the Laplacian equation.

(4) Beginning of the outer iteration loop:

   (a) getting the new estimated polymer stress tensor $\boldsymbol{\tau}^{**}$ by solving the specified constitutive equation with previous velocity field $\mathbf{u}^*$;

   (b) setting up the pressure boundary condition and estimating the balance force $\boldsymbol{\sigma}^*$;

   (c) calculating the latest velocity field $\mathbf{u}^{**}$ and the pressure filed p^{**} with SIMPLE algorithm;

   (d) moving the interface mesh points similar to step (2) until the prescribed accuracy is reached.

(5) If the programme has not reached the final time, return to step (1).

Step (4)(d) is performed again in order to avoid the operation of step (3), which is a time-consuming step. Such procedure is reasonable provided that the displacement of the interface in one time step is smaller than the thickness of the nearest interface cell. And this condition could be easily satisfied by decreasing the time interval.

2.2.4. Other General Numerical Schemes Used in the Simulation. In addition to the approaches described above, under-relaxation algorithm is also used to enhance the stability of the numerical simulation, where the relaxation parameters for p and U are 0.3 and 0.5, respectively.

Backward Euler method with second order is used for time derivatives. In terms of the linear system solver, PCG (Preconditioned Conjugate Gradient method) with AMG (Algebraic Multi-Grid method) preconditioning is used for pressure, while BiCGstab (Bi-Conjugate Gradient Stabilized method) with Diagonal Incomplete Lower-Upper (DILU) preconditioning is used for velocity and stress [12, 13]. Absolute tolerances are 1.0×10^{-6} for the velocity and 1.0×10^{-7} for pressure and stress, respectively.

3. Results and Discussion

3.1. Basic Geometric Shape and Boundary Conditions. A sketch of the geometry and the boundary conditions for

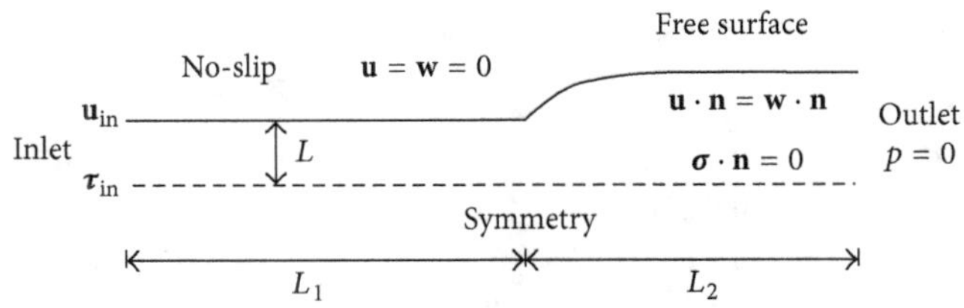

FIGURE 1: Die-swell flow configurations.

modelling the Newtonian and Oldroyd-B fluid is shown in Figure 1. The half width of the die L is chosen as the characteristic length. The lengths of the upstream and the downstream are set as L_1 and L_2, respectively. They should be long enough to ensure the flow before entry of the die exit and further downstream at the outlet is fully developed. In this paper, length settings are $L = 1$ and $L_1 = L_2 = 8$.

As the pressure, the velocity, and the polymer stress are solved in a coupled way in our system, we usually give either all the correct boundary conditions or any one of them on the boundary. In order to avoid unnecessary numerical failure near the inlet, the conditions of fully developed flow are imposed on the inlet boundary. The velocity along x direction at the inlet is $\mathbf{u}_{\text{in}} = 1.5(1 - y^2)$, while the velocity along y direction is zero. Therefore, the average velocity of the inflow is $\bar{\mathbf{u}} = 1$, and the shear rate near the wall is $\dot{\gamma}_w = 3$. The polymer stress tensor at the inlet for the Oldroyd-B model is set according to the analytical solution. No-slip boundary condition is set at the wall. The symmetry boundary condition is set at the centreline for all fields. Symmetry boundary condition is provided by OpenFoam itself. It means there is neither convection flux nor diffusion flux across the plane; that is, the vertical velocity and the shear stress component are zero ($\mathbf{u}_x = 0$ and $\boldsymbol{\sigma}_{xy} = 0$). We assumed that the flow at the outlet has been already fully developed; therefore, the pressure is set as zero at the outlet. Note that the pressure here is just a reference pressure which is used to solve the governing equations. In terms of the free surface, the pressure is initially set as fixed value zero, and it would change dynamically during the simulation (see (11)). Without special declaration, zero gradient boundary condition is set for the remaining boundaries.

3.2. Validation of Computational Model with Newtonian Fluid. By setting Re = 0.01, Wi = 0, $\beta = 1$, and $\varepsilon = 0$, Oldroyd-B model degrades to Newtonian fluid. Under isothermal condition and neglecting the effects of gravity and surface tension, the swelling ratio, defined as the maximum diameter of the extrudate divided by the diameter of the die, is 1.193 from our simulation. It is in an excellent agreement with the results of Mitsoulis et al. [14] and Georgiou and Boudouvis [15], which is around 1.19.

3.3. Validation of Computational Model with Viscoelastic Fluid Using Oldroyd-B Model. Based on K-BKZ models, Tanner

Table 1: Mesh parameters for convergence validation with Wi = 0.1.

Name	Size	$\Delta x_{\min}/L$	$\Delta y_{\min}/L$	S_R
M1	160 * 30	0.1	0.02	1.1677
M2	320 * 30	0.05	0.02	1.1727
M3	160 * 60	0.1	0.01	1.1691
M4	640 * 30	0.025	0.02	1.1789
M5	160 * 120	0.1	0.005	1.1697
M6	360 * 60	0.025	0.01	1.1770
M7	400 * 60	0.0125	0.01	1.1787

[16] derived an analytical formula to account for the viscous and elastic contributions to the swell ratio as

$$S_R = 0.19 + \left[1 + \frac{1}{12}\left(\frac{N_1}{\boldsymbol{\sigma}_{xy}}\right)_w^2\right]^{1/4}, \quad (13)$$

where $N_1 = \boldsymbol{\sigma}_{xx} - \boldsymbol{\sigma}_{yy}$ is the first normal stress difference and $\boldsymbol{\sigma}_{xy}$ is the shear stress. Both of them are evaluated at the wall. Introducing the parameter of recoverable shear $R_s = (N_1/2\boldsymbol{\sigma}_{xy})_w$, the swelling ratio can be estimated as

$$S_R = 0.19 + \left[1 + \frac{1}{3}R_s^2\right]^{1/4}. \quad (14)$$

For the Oldroyd-B model, the recoverable shear of a fully developed Poiseuille flow is equal to

$$R_s = \left[\frac{2\mathrm{Wi}\,(1-\beta)\,\dot{\gamma}^2}{2\dot{\gamma}}\right]_w = (1-\beta)\,\mathrm{Wi}\dot{\gamma}_w. \quad (15)$$

Since in our case $\dot{\gamma}_w = 3$ and $\beta = 0.1$, the recoverable shear is $R_s = 2.7\mathrm{Wi}$.

Mesh convergence was tested at $\mathrm{Wi} = 0.1$ with seven meshes, which are listed in Table 1. $\Delta x_{\min}/L$ and $\Delta y_{\min}/L$ are the relative minimum length along x-axis and y-axis, respectively. These cells appear near the exit of the die. Through the comparison of all these seven meshes, it is found that the value of $\Delta x_{\min}/L$ plays a dominant role when determining the welling ratio. Furthermore, if $\Delta y_{\min}/L \le 0.01$ and $\Delta x_{\min}/L \le 0.025$, the relative error of the swelling ratio is below 0.2%. The solution convergence is reached by using M6. The time interval is $\Delta t = 10^{-6}$ for M7 and $\Delta t = 10^{-5}$ for other cases (except for M1).

The value of pressure and polymer stress near the singularity is illustrated in Figure 2, where x is in the range of $[-0.4, 0.4]$, along with different mesh densities (M4, M6) at $\mathrm{Wi} = 0.1$. It is shown that the two curves are very close to each other, which indicates that the given mesh density is sufficient. In terms of the tendency, the pressure gradually declines at first and then experiences a steep fall followed by a mild elevation before the exit. Subsequently, it dramatically drops from positive to negative and approaches zero afterwards. As shown in Figures 2(b) and 2(c), there is a highest point for $\boldsymbol{\tau}_{xx}$ and lowest point for $\boldsymbol{\tau}_{yy}$, where both points are around the exit. Furthermore, the curve of $\boldsymbol{\tau}_{yy}$ is sharper than the one of $\boldsymbol{\tau}_{xx}$. As x increases, the value of $\boldsymbol{\tau}_{xy}$ decreases at first, reaches the bottom, and sharply grows to an apogee, before it declines to zero gradually. In terms of the trend of the profiles as well as the maximum and the minimum values near the singularity, the results of the proposed method are close to those of the spectral element method with 3rd order polynomial adopted in Claus work [17], which again demonstrates that the current mesh is capable of providing reliable results. The small deviation in terms of the positions of the maximum value and the minimum value are attributed to the fact that the velocity is stored on the cell center in this paper and on the cell vertex in [17], respectively.

In general, with the increase of Wi, the profile of components of the polymer stress should be similar but with a greater value. However, the figure for $\boldsymbol{\tau}_{yy}$ is much different from the intuition. Figure 3 shows that the nadir for Wi = 0.1 and Wi = 0.2 which is negative becomes the peak for Wi = 0.5 which is positive. The force along y direction caused by the polymer stress components $\boldsymbol{\tau}_{yy}$ is $n_y\boldsymbol{\tau}_{yy}$, where n_y is vertical component of the standard interfacial normal. Thus, the more likely n_y is approaching to one (corresponding to a small swelling ratio with a gentle curve), the greater the influence of $\boldsymbol{\tau}_{yy}$ would be. As the force increases, the extrusion is growing larger.

The swelling ratio is given in Figure 4 with a comparison with others' works. Note that, in their works, the definition of relative parameters could be different. For example, Russo and Phillips [18] and Tomé et al. [19] used the full width as the characteristic length. We just altered them to be the same one in our work. Moreover, Crochet and Keunings [20] and Tomé et al. [19] took the Reynolds number as 0 and 0.5, respectively. Russo and Phillips [18] tested two situations with Re = 0.001 and Re = 0.5 and the results shed light on the fact that no significant change was achieved by varying the Reynolds number when Re is small enough. All of these three papers took $\beta = 1/9$ in their similar simulations. Our simulation is kind of similar to Crochet and Keunings [20] and Tomé et al. [19] who varied the relaxation time λ in different cases; therefore, the trend of the curves is the same. However, Russo and Phillips [18] changed the input velocity, that is, both shear rate near the wall $\dot{\gamma}_w$ and the Weissenberg number Wi are changed. In his case, the recoverable shear rate is far different from the one in our simulation even with the same Wi. That is the reason the trend of his curve seems different from others. The swelling ratio of all these curves shown increases with the increase of the Weissenberg number. From Figure 4, our results are above Crochet and Keunings [20] and Tomé et al. [19] but below Russo and Phillips [18]. With the increase of Wi number, the elastic effect plays a dominant role in determining the die-swell. In addition, slight drop of swelling ratio at low Wi number illustrated in Figure 4 is due to the negative value of $\boldsymbol{\tau}_{yy}$ and was also observed in [20].

The steady state profiles of the free surface at various Wi numbers are shown in Figure 5. It can be seen that a larger Wi value would result in an increased slope of the curve. This is because the first normal stress at the exit of the channel for bigger Wi number is larger than the smaller ones. Such phenomenon also indicates the main factor of the swelling is the value of the first normal stress near the exit of the die. Under higher Wi number, the hexahedron cell is deformed

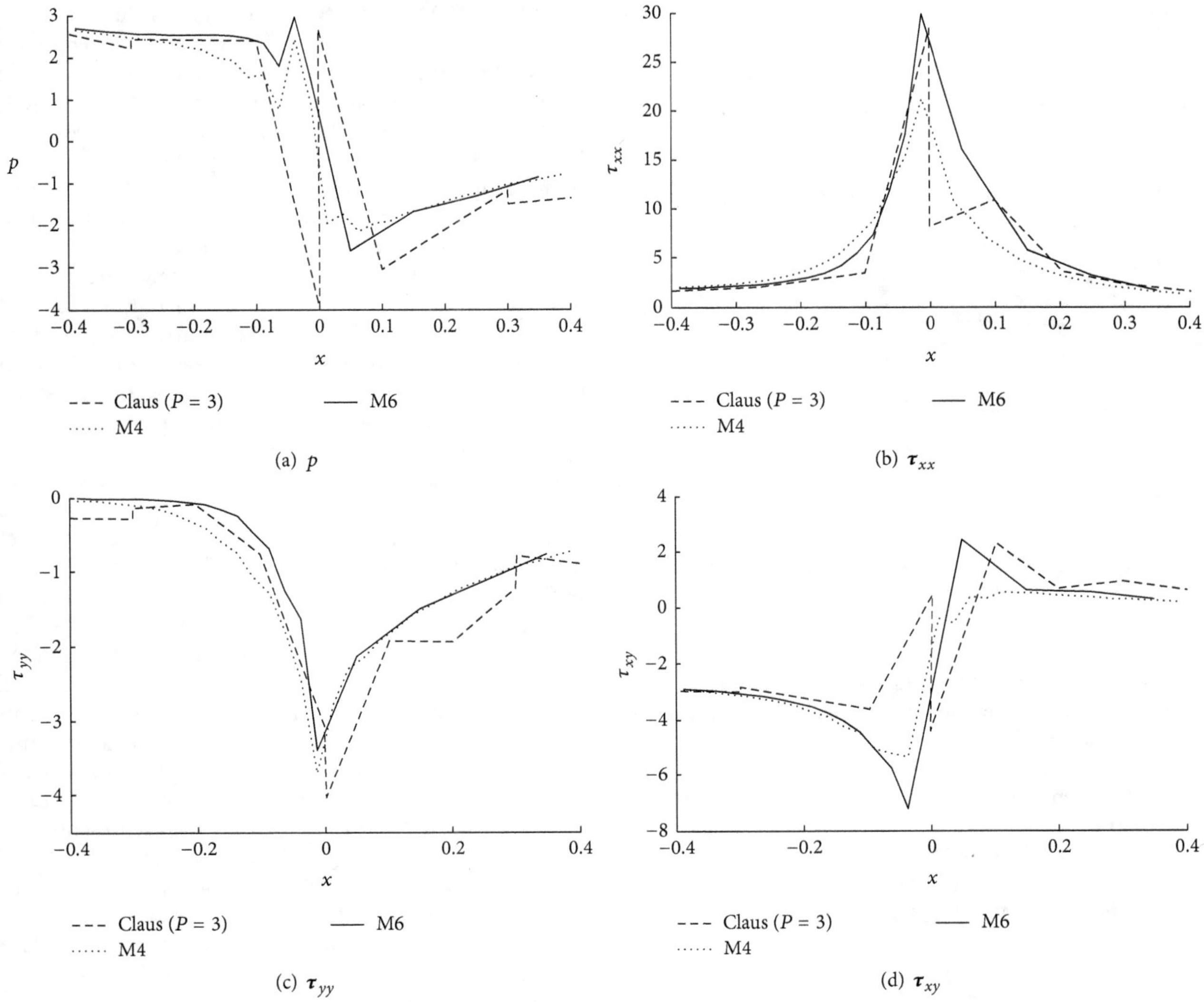

FIGURE 2: Pressure and components of polymer stress near the singularity at Wi = 0.1.

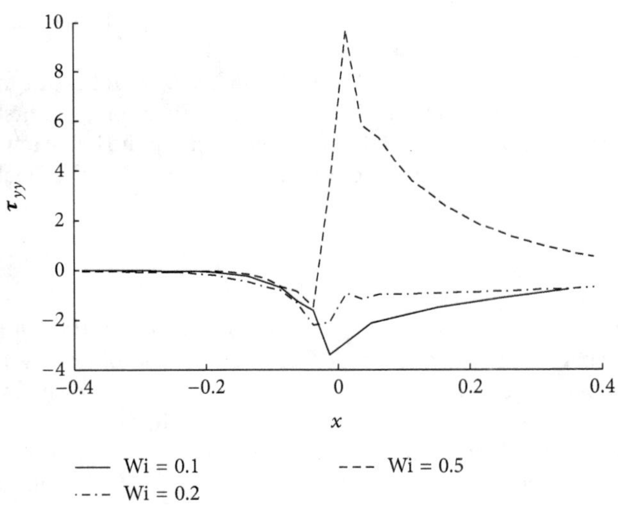

FIGURE 3: τ_{yy} near the singularity for various Wi.

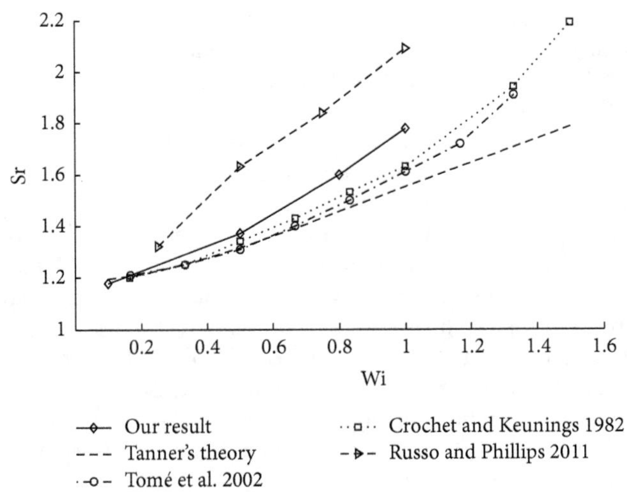

FIGURE 4: Swelling ratio with the increase of Wi.

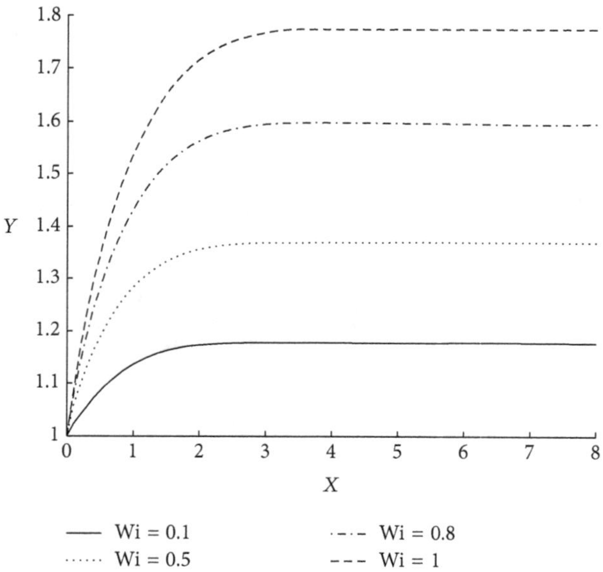

FIGURE 5: Free surface profile for various Wi for $\beta = 0.1$.

greatly which increases the mesh nonorthogonality. The maximum mesh nonorthogonality has reached as high as 40 at Wi = 1.0 and the threshold in OpenFOAM is usually set as 70. Though the nonorthogonality value could be reduced by mesh refinement, the computational time would increase. The maximum swelling ratio obtained from our simulation is about 2.26 at Wi = 2.5 using Mesh M1.

Contour plots of velocity components U_x for Wi = 0.1, 0.5, 0.8, 1.0 and U_y for Wi = 0.5, 1.0 are displayed in Figure 6. By increasing Wi number, the horizontal components of the velocity U_x increase along y directions. The contour line of $U_x = 1.4$ expands upward and forward. The vertical components of the velocity U_y also increase due to larger elastic effects which results in a larger swelling ratio.

The profile of the first normal stress difference N_1 just before the exit (at $x = -0.04$, and the exit is at $x = 0$) of the die for a large range of Wi is shown in Figure 7. These lines are sampled along y-axis at fixed x. Apparently, the maximum value of each line increases as Wi becomes larger, which is consistent with (15). All of these four lines begin with zero at the middle of the die ($y = 0$). As for Wi = 1.0, it goes up gradually until it approaches the wall, where it grows instantly to the peak. Subsequently, it drops dramatically. The profiles of other Wi values smaller than 1.0 are something like shifting the curve of Wi = 1.0 downwards to the right. In detail, there is no peak in the case of Wi = 0.1. Generally speaking, a larger Wi results in a higher peak, which indicates the first normal stress dominates the extrusion process.

Decreasing the viscosity ratio β from 1 to 0 means gradually increasing the contribution of elastic stress and decreasing the effects of purely viscous stress. When β reaches zero, the model degrades to the upper-convected Maxwell (UCM) model. In order to investigate the role of β in die-swell simulations, simulations under $\beta = 0.5$ and $\beta = 0.9$ are carried out. Figure 8 illustrates the trend that the swelling ratio would increase as Wi number increases, regardless of the value of β. Moreover, the result departs from Tanners theory obviously. Considering an extreme case, as β approaches one, the swelling ratio almost remains a constant even for a large value of Wi according to Tanners theory. However, this does not match the experimental result. For the case $\beta = 0.9$, Wi = 1 and the case $\beta = 0.1$, Wi = 1, the recoverable shear R_s are 3 and 2.7, respectively. According to (15), the swelling ratio of the former one should be larger than the later one, while the fact is contrary. Russo and Phillips [18] attribute the discrepancy to the fact that Tanners formula derived from a KBK-Z integral constitutive equation only considered a single relaxation time; however, the Oldroyd-B model owns a retardation time in addition to the relaxation time. This result might be related to the fact that the elongational response for $\beta = 0.1$ is much stronger than the one for $\beta = 0.9$ as illustrated in Figure 9. Thus, the extensional response also plays an important role in swelling process and should be taken into account when predicting the final swelling ratio. The simulation results in [21] demonstrate that flows with a sufficiently high elongational viscosity are able to swell to an apparently large state even at a relatively low flow rate, ignoring their performance in simple shear flows. Nevertheless, decreasing the viscosity ratio would lead to an increase in the recoverable shear and, in turn, increases the value of swelling ration according to (15). By comparing the value in Figures 4 and 8, it is reasonable to say that Tanners theory still makes some sense. Apart from that, with a careful comparison, it is found that when the viscosity ratio β is very large (for the case of $\beta = 0.9$), which represents a very dilute fluid, the swelling ratio suffers a significant drop. In comparison, when β is relatively small (for the case of $\beta = 0.5$), the swelling ratio is much less sensitive. This phenomena was also observed in [22]. Note that because of different definitions of the viscosity ratio, β in this paper is equivalent to $1 - \beta$ in their works.

As can be seen from (13), the value of the shear stress would not change once the shear rate is determined. Consequently, the first normal stress difference is the only variable which determines the swelling ratio. Figures 7 and 10 demonstrate that a high proportion of polymeric stress contribution would result in a larger value of N_1, which would provide greater power to swell. According to the analytical solution of the Oldroyd-B model, smaller β would result in a higher value of the first normal stress. Therefore, the first normal stress plays a vital role in determining the behaviour of extrusion flows and could be used to predict the final results of swelling ratio.

4. Conclusion

In this work, a new method to track the free surface for viscoelastic fluids was proposed. The obstacle of setting a proper free surface boundary condition was well resolved by deriving a Dirichlet pressure boundary condition and introducing a balance force. The proposed numerical approach was implemented opon OpenFOAM, which is widely used in CFD simulations [23, 24]. Comparing with previous literature, the results showed that our solutions were close to those with other methods (e.g., FEM, Spectral Element

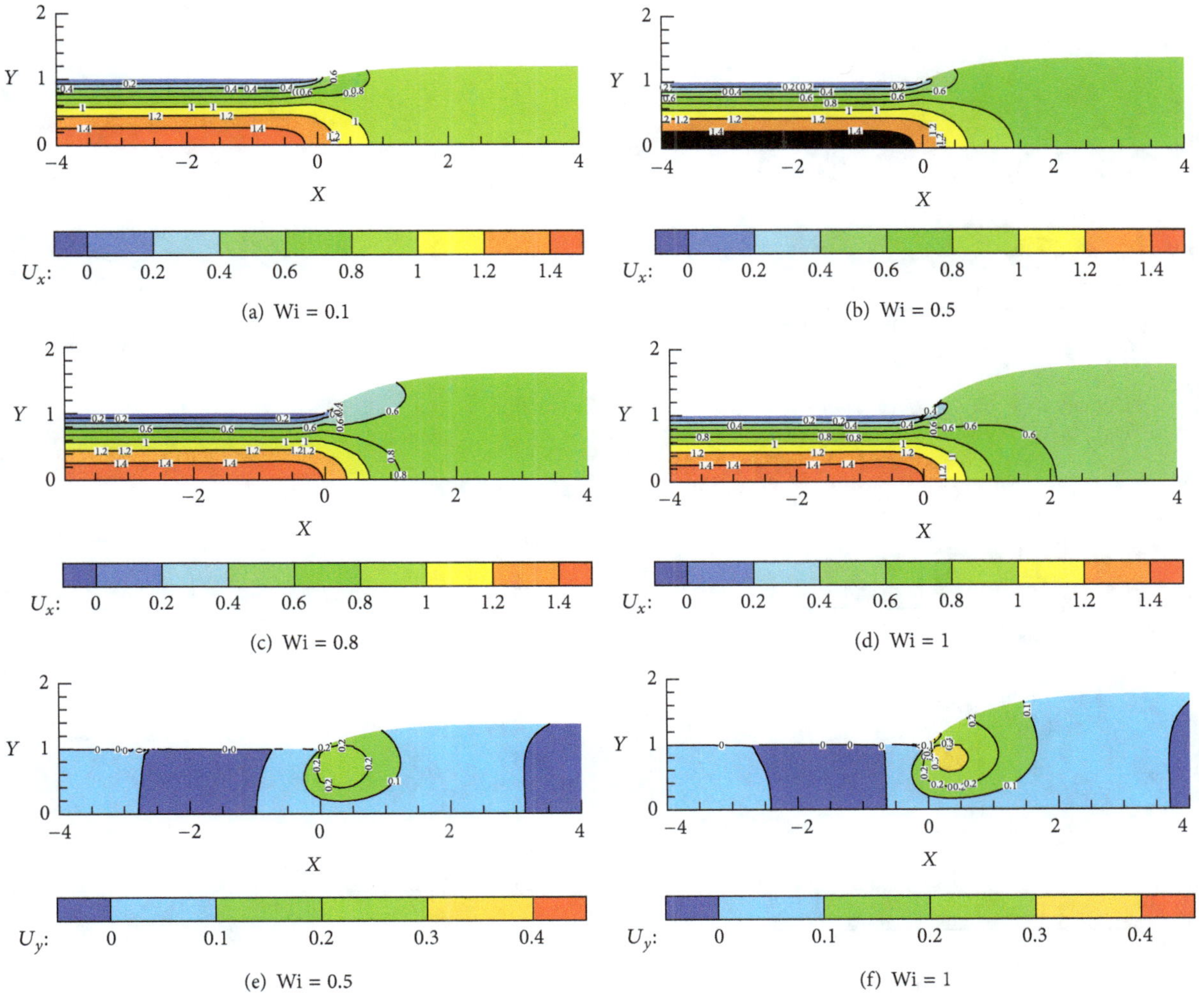

FIGURE 6: Contour plots of velocity components U_x (a)–(d) and U_y (e)-(f) for a range of Wi.

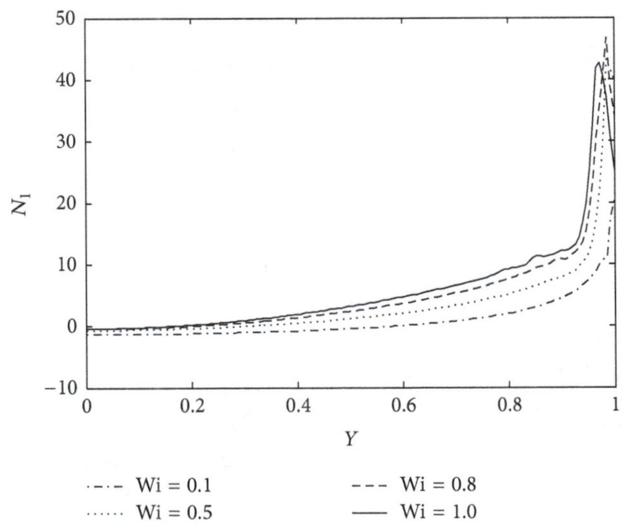

FIGURE 7: First normal stress difference at $x = -0.04$ for various Wi with $\beta = 0.1$.

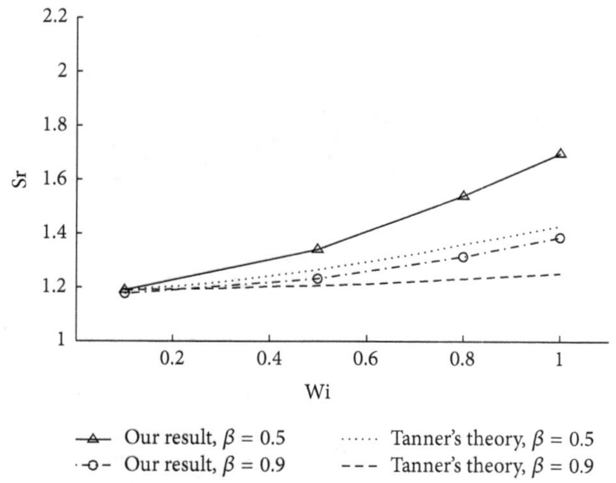

FIGURE 8: Swelling ratio with the increase of Wi for $\beta = 0.5$ and $\beta = 0.9$.

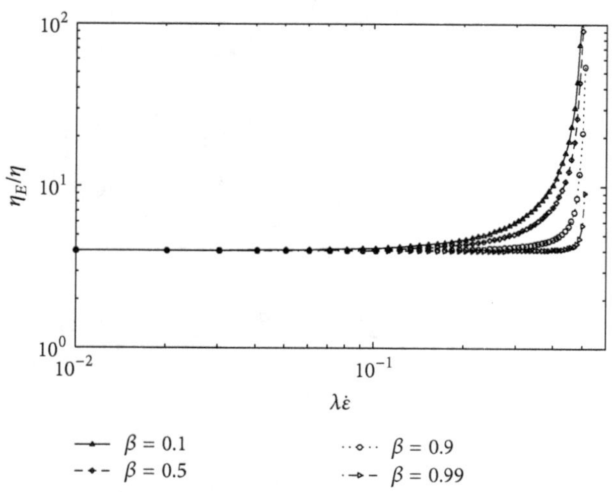

FIGURE 9: Planar extensional viscosity for Oldroyd-B model with various β.

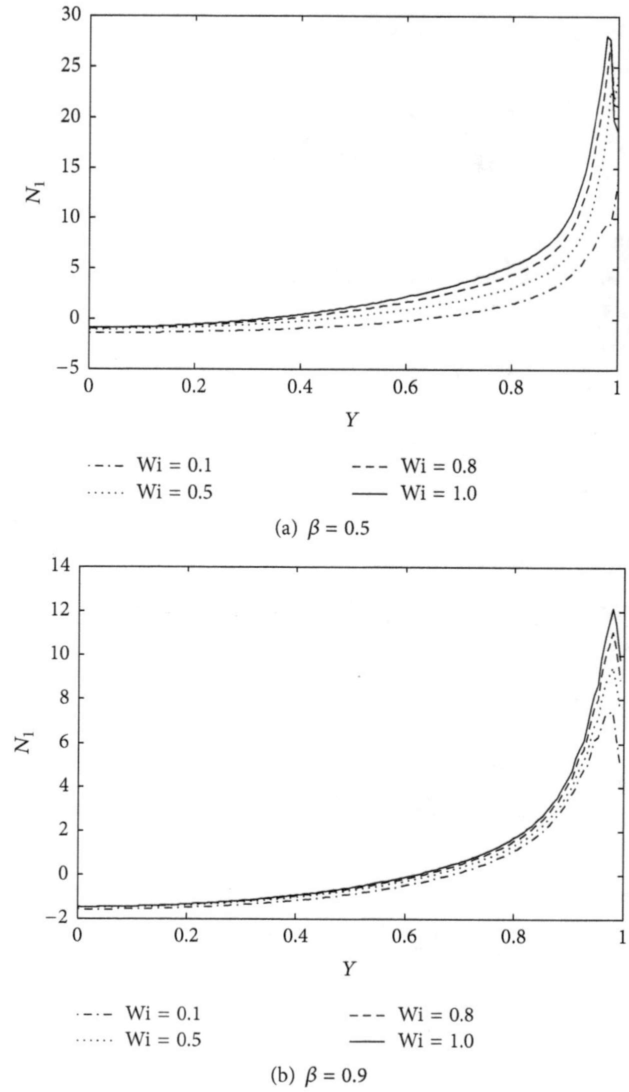

FIGURE 10: First normal stress difference at $x = -0.04$ for various Wi with $\beta = 0.5$ and $\beta = 0.9$.

Method, and MAC), which validated the correctness and effectiveness of the proposed numerical method. Moreover, the results showed that the swelling ratio would decline gradually with the increase of the viscosity ratio. In the meantime, brief analysis of the influence of the viscosity ratio was performed. Moreover, providing appropriate constitutive equations, investigation of die-swell behaviour for new polymeric materials can be performed based on our new platform. In addition, since the FVM systems are suitable for parallelization, the free surface simulations in our platform would be easily scaled to massive processors for parallel computing.

Conflict of Interests

The authors declare that there is no conflict of interests regarding the publication of this paper.

Acknowledgments

Thanks are due to the computational resources provided by the University of Manchester, partial funding from the National Natural Science Foundation of China under Grants no. 61221491, no. 61303071, and no. 61120106005, and Open fund (no. 201503-01 and no. 201503-02) from State Key Laboratory of High Performance Computing.

References

[1] J. L. Favero, A. R. Secchi, N. S. M. Cardozo, and H. Jasak, "Viscoelastic fluid analysis in internal and in free surface flows using the software OpenFOAM," *Computers and Chemical Engineering*, vol. 34, no. 12, pp. 1984–1993, 2010.

[2] S. Osher and J. A. Sethian, "Fronts propagating with curvature-dependent speed: algorithms based on Hamilton-Jacobi formulations," *Journal of Computational Physics*, vol. 79, no. 1, pp. 12–49, 1988.

[3] S. McKee, M. F. Tomé, V. G. Ferreira et al., "The MAC method," *Computers & Fluids*, vol. 37, no. 8, pp. 907–930, 2008.

[4] C. W. Hirt, A. A. Amsden, and J. L. Cook, "An arbitrary lagrangianeulerian computing method for all flow speeds (reprinted from the Journal of Computational Physics, vol 14, pg 227–253, 1974)," *Journal of Computational Physics*, vol. 135, no. 2, pp. 203–216, 1997.

[5] M. Fortin and D. Esselaoui, "A finite-element procedure for viscoelastic flows," *International Journal for Numerical Methods in Fluids*, vol. 7, no. 10, pp. 1035–1052, 1987.

[6] Z. Tuković and H. Jasak, "A moving mesh finite volume interface tracking method for surface tension dominated interfacial fluid flow," *Computers & Fluids*, vol. 55, pp. 70–84, 2012.

[7] H. Jasak, "OpenFOAM: open source CFD in research and industry," *International Journal of Naval Architecture and Ocean Engineering*, vol. 1, no. 2, pp. 89–94, 2009.

[8] S. Muzaferija and M. Perić, "Computation of free-surface flows using the finite-volume method and moving grids," *Numerical Heat Transfer, Part B: Fundamentals*, vol. 32, no. 4, pp. 369–384, 1997.

[9] H. Jasak and Ž. Tuković, "Automatic mesh motion for the unstructured finite volume method," *Transactions of Famena*, vol. 30, no. 2, pp. 1–20, 2004.

[10] N. Phanthien and R. Tanner, "Viscoelastic finite volume method," *Rheology Series*, vol. 8, pp. 331–359, 1999.

[11] H. Matallah, P. Townsend, and M. F. Webster, "Recovery and stress-splitting schemes for viscoelastic flows," *Journal of Non-Newtonian Fluid Mechanics*, vol. 75, no. 2-3, pp. 139–166, 1998.

[12] M. A. Ajiz and A. Jennings, "A robust incomplete Choleski-conjugate gradient algorithm," *International Journal for Numerical Methods in Engineering*, vol. 20, no. 5, pp. 949–966, 1984.

[13] J. Lee, J. Zhang, and C.-C. Lu, "Incomplete LU preconditioning for large scale dense complex linear systems from electromagnetic wave scattering problems," *Journal of Computational Physics*, vol. 185, no. 1, pp. 158–175, 2003.

[14] E. Mitsoulis, G. C. Georgiou, and Z. Kountouriotis, "A study of various factors affecting Newtonian extrudate swell," *Computers & Fluids*, vol. 57, pp. 195–207, 2012.

[15] G. C. Georgiou and A. G. Boudouvis, "Converged solutions of the Newtonian extrudate-swell problem," *International Journal for Numerical Methods in Fluids*, vol. 29, no. 3, pp. 363–371, 1999.

[16] R. I. Tanner, *Engineering Rheology*, Oxford University Press, Oxford, UK, 2002.

[17] S. Claus, *Numerical simulation of complex viscoelastic flows using discontinuous galerkin spectral/hp element methods [Ph.D. thesis]*, 2013.

[18] G. Russo and T. N. Phillips, "Spectral element predictions of die-swell for Oldroyd-B fluids," *Computers & Fluids*, vol. 43, no. 1, pp. 107–118, 2011.

[19] M. F. Tomé, N. Mangiavacchi, J. A. Cuminato, A. Castelo, and S. McKee, "A finite difference technique for simulating unsteady viscoelastic free surface flows," *Journal of Non-Newtonian Fluid Mechanics*, vol. 106, no. 2-3, pp. 61–106, 2002.

[20] M. J. Crochet and R. Keunings, "Finite element analysis of die swell of a highly elastic fluid," *Journal of Non-Newtonian Fluid Mechanics*, vol. 10, no. 3-4, pp. 339–356, 1982.

[21] A. Al-Muslimawi, H. R. Tamaddon-Jahromi, and M. F. Webster, "Simulation of viscoelastic and viscoelastoplastic die-swell flows," *Journal of Non-Newtonian Fluid Mechanics*, vol. 191, pp. 45–56, 2013.

[22] J.-R. Clermont and M. Normandin, "Numerical simulation of extrudate swell for Oldroyd-B fluids using the stream-tube analysis and a streamline approximation," *Journal of Non-Newtonian Fluid Mechanics*, vol. 50, no. 2-3, pp. 193–215, 1993.

[23] X.-W. Guo, S. Zou, X. Yang, X.-F. Yuan, and M. Wang, "Interface instabilities and chaotic rheological responses in binary polymer mixtures under shear flow," *RSC Advances*, vol. 4, no. 105, pp. 61167–61177, 2014.

[24] X.-W. Guo, W.-G. Yang, X.-H. Xu, Y. Cao, and X.-J. Yang, "Non-equilibrium steady states of entangled polymer mixtures under shear flow," *Advances in Mechanical Engineering*, vol. 7, no. 6, pp. 1–10, 2015.

Permissions

All chapters in this book were first published in AMSE, by Hindawi Publishing Corporation; hereby published with permission under the Creative Commons Attribution License or equivalent. Every chapter published in this book has been scrutinized by our experts. Their significance has been extensively debated. The topics covered herein carry significant findings which will fuel the growth of the discipline. They may even be implemented as practical applications or may be referred to as a beginning point for another development.

The contributors of this book come from diverse backgrounds, making this book a truly international effort. This book will bring forth new frontiers with its revolutionizing research information and detailed analysis of the nascent developments around the world.

We would like to thank all the contributing authors for lending their expertise to make the book truly unique. They have played a crucial role in the development of this book. Without their invaluable contributions this book wouldn't have been possible. They have made vital efforts to compile up to date information on the varied aspects of this subject to make this book a valuable addition to the collection of many professionals and students.

This book was conceptualized with the vision of imparting up-to-date information and advanced data in this field. To ensure the same, a matchless editorial board was set up. Every individual on the board went through rigorous rounds of assessment to prove their worth. After which they invested a large part of their time researching and compiling the most relevant data for our readers.

The editorial board has been involved in producing this book since its inception. They have spent rigorous hours researching and exploring the diverse topics which have resulted in the successful publishing of this book. They have passed on their knowledge of decades through this book. To expedite this challenging task, the publisher supported the team at every step. A small team of assistant editors was also appointed to further simplify the editing procedure and attain best results for the readers.

Apart from the editorial board, the designing team has also invested a significant amount of their time in understanding the subject and creating the most relevant covers. They scrutinized every image to scout for the most suitable representation of the subject and create an appropriate cover for the book.

The publishing team has been an ardent support to the editorial, designing and production team. Their endless efforts to recruit the best for this project, has resulted in the accomplishment of this book. They are a veteran in the field of academics and their pool of knowledge is as vast as their experience in printing. Their expertise and guidance has proved useful at every step. Their uncompromising quality standards have made this book an exceptional effort. Their encouragement from time to time has been an inspiration for everyone.

The publisher and the editorial board hope that this book will prove to be a valuable piece of knowledge for researchers, students, practitioners and scholars across the globe.

List of Contributors

ShanShan Li
College of Civil Engineering and Architecture, Shandong University of Science and Technology, Qingdao 266590, China

Dayong Li
College of Civil Engineering and Architecture, Shandong University of Science and Technology, Qingdao 266590, China
College of Civil Engineering, Fuzhou University, Fuzhou 350116, China

Jin Liu, Xiaohui Qi, Da Zhang, Qiao Feng and Yong Wang
School of Earth Sciences and Engineering, Hohai University, Nanjing 210098, China

Debi Prasanna Kanungo
CSIR-Central Building Research Institute (CBRI), Roorkee 247667, India

Muawia A. Dafalla
Bugshan Research Chair in Expansive Soils, Department of Civil Engineering, King Saud University, Riyadh 11421, Saudi Arabia

Wei Zhang and Xiaohong Bai
Department of Civil Engineering, Taiyuan University of Technology, Taiyuan, Shanxi 030024, China

David Du
Department of Civil Engineering, Taiyuan University of Technology, Taiyuan, Shanxi 030024, China
Subsea 7 (US) LLC, 17220 Katy Freeway, Suite 100, Houston, TX 77094, USA

Son Tung Pham
Faculty of Geology & Petroleum Engineering, Department of Drilling & Production Engineering, Ho Chi Minh City University of Technology-Vietnam National University, Ho Chi Minh City, Vietnam

Zhaolin Li, Lianguo Wang, Yinlong Lu, Wenshuai Li and Kai Wang
State Key Laboratory for Geomechanics and Deep Underground Engineering, China University of Mining and Technology, Xuzhou, Jiangsu 221116, China

Zheng Lu, Shaohua Xian and Ran Fang
State Key Laboratory of Geomechanics and Geotechnical Engineering, Institute of Rock and Soil Mechanics, Chinese Academy of Sciences, Wuhan 430071, China

Xiaowen Wu
State Key Laboratory of Geomechanics and Geotechnical Engineering, Institute of Rock and Soil Mechanics, Chinese Academy of Sciences, Wuhan 430071, China
School of Civil Engineering, Architecture and Environment, Hubei University of Technology, Wuhan 430068, China

Zhi Hu
School of Civil Engineering, Architecture and Environment, Hubei University of Technology, Wuhan 430068, China

Shahid Ikramullah Butt, Umer Asgher, Umar Mushtaq, Yasar Ayaz, Mohsin Jamil and Muhammad Kamal Amjad
School of Mechanical & Manufacturing Engineering, National University of Sciences and Technology (NUST), Islamabad, Pakistan

Riaz Ahmed
Directorate of Quality Assurance, National University of Sciences and Technology (NUST), Islamabad, Pakistan

Faping Zhang
School of Mechanical Engineering, Beijing Institution of Technology, Beijing, China

Federico Aguayo, Anthony Torres, Tate Talamini and Kevin Whaley
Department of Engineering Technology, Texas State University, San Marcos, TX, USA

Haibo Wang and Ping Yang
Department of Civil Engineering, Nanjing Forestry University, Nanjing, Jiangsu 210037, China

An Deng
School of Civil, Environmental and Mining Engineering, University of Adelaide, Adelaide, SA, Australia

Ying Lai
Postdoctoral Fellow, MOE Key Laboratory of Soft Soils and Geoenvironmental Engineering, College of Civil Engineering and Architecture, Zhejiang University, 866 Yuhangtang Road, Hangzhou, Zhejiang, China

Bin Zhu
MOE Key Laboratory of Soft Soils and Geoenvironmental Engineering, College of Civil Engineering and Architecture, Zhejiang University, 866 Yuhangtang Road, Hangzhou, Zhejiang, China

Xiangtian Xu
Institute of Transportation, Inner Mongolia University, Hohhot 010070, China

Jin Liu, Qiao Feng, Yong Wang, Yuxia Bai, Jihong Wei and Zezhuo Song
School of Earth Sciences and Engineering, Hohai University, Nanjing 210098, China

Walid Oueslati and Mahdi Meftah
UR 05/13-01 Physique des Matériaux Lamellaires et Nano-Matériaux Hybrides (PMLNMH), Faculté des Sciences de Bizerte, Université de Carthage, 7021 Zarzouna, Tunisia

Kang Fei and Jinxin Xu
Institute of Geotechnical Engineering, Yangzhou University, Yangzhou 225127, China

Wanpeng Huang and Shaojie Chen
State Key Laboratory of Mining Disaster Prevention and Control, Shandong University of Science and Technology, Qingdao 266590, China
College of Mining and Safety Engineering, Shandong University of Science and Technology, Qingdao 266590, China

Wenbin Xing and Kai Wu
College of Mining and Safety Engineering, Shandong University of Science and Technology, Qingdao 266590, China

Yang Liu
College of Mechanical and Electronic Engineering, Shandong University of Science and Technology, Qingdao 266590, China

Huizhi Zhang and Jifeng Liu
Key Laboratory of Engineering Materials and Structural Reinforcement, Sanming University, Fujian Province University, Fujian, China
Department of Architecture and Civil Engineering, Sanming University, Fujian, China

L. M. Romero-Guerrero, R. Moreno-Tovar, A. Arenas-Flores and F. Pérez-Moreno
Área Académica de Ciencias de la Tierra y Materiales, Universidad Autónoma del Estado de Hidalgo, Carr. Pachuca-Tulancingo Km. 4.5 Col. Carboneras, 42184 Mineral de la Reforma, Hidalgo, Mexico

Y. Marmolejo Santillán
Área Académica de Química, Universidad Autónoma del Estado de Hidalgo, Carr. Pachuca-Tulancingo Km. 4.5 Col. Carboneras, 42184 Mineral de la Reforma, Hidalgo, Mexico

Qin-yong Ma, Zi-ming Cao and Pu Yuan
Engineering Research Center of Underground Mine Construction, Ministry of Education, Anhui University of Science and Technology, Huainan, Anhui 232001, China
School of Civil Engineering and Architecture, Anhui University of Science and Technology, Huainan, Anhui 232001, China

Kui Zhao, Yulong Zhuo, Xiaojun Wang and Wen Zhong
Jiangxi Key Laboratory of Mining Engineering, Jiangxi University of Science and Technology, Jiangxi 341000, China
School of Resources and Environment Engineering, Jiangxi University of Science and Technology, Jiangxi 341000, China

Yanhu Zhao, Yuanming Lai and Jing Zhang
State Key Laboratory of Frozen Soil Engineering, Northwest Institute of Eco-Environment and Resources, Chinese Academy of Sciences, Lanzhou 730000, China
University of Chinese Academy of Sciences, Beijing 100049, China

Chong Wang
Key Laboratory of Mechanics on Disaster and Environment in Western China, The Ministry of Education of China and School of Civil Engineering and Mechanics, Lanzhou University, Lanzhou 730000, China

Miao Wang
College of Computer, National University of Defense Technology, Changsha, Hunan 410073, China
State Key Laboratory of High Performance Computing, National University of Defense Technology, Changsha, Hunan 410073, China

M. Fattahi
Department of Mechanical Engineering, Tabriz Branch, Islamic Azad University, Tabriz, Iran

Yidong Wang and Yue Xi
Department of Civil Engineering, Shanghai Jiao Tong University, Shanghai 200240, China

Index

A
Acoustic Emission, 56-58, 64, 66, 70-71
Alkali Silica Reactivity, 89-90, 93, 95, 97
Alluvial Silt, 100-101, 103, 107, 110-111
Atmospheric Stress, 137, 149, 151

B
Basalt Fiber, 136, 197-199, 204, 206-208

C
Clay Material, 26, 72-73, 76
Clay-aggregate Mixtures, 154-157, 161
Clayey Soil Mixtures, 1-2, 10
Creep State, 164-165, 167-168, 170, 174-175
Crumbs Content, 1-2
Csd, 139
Csl, 222-223, 226

D
Dickite, 186, 189-192, 195
Different Salt Contents, 220, 222-224, 226
Dry-wet Cycling, 72-73, 75-76
Dynamic Behavior, 154-156, 160-161

E
Electric Resistance, 72-73
Electromagnetic Field, 113-114, 116, 118, 121, 124-125
Expansive Soil, 25, 27, 32, 72, 112, 197-209

F
Fly Ash, 10, 89-90, 100-107, 109-112, 176, 197-204, 206-209
Foundry Sand, 89-95, 97-98
Frost Actions, 100, 103, 105-111
Frozen Saline Silty Clay, 220-222, 224-226

G
Geobelt-reinforced Cushion, 41-42
Geobelt-reinforced Gravel, 34-35
Geobelt-reinforced Thin, 42
Geomechanical Models, 47
Glass Beads, 155, 245
Granulated Ground Blast Furnace Slag, 90, 97

H
Harris Corner Detector Technique, 78, 81
Hydromechanical Models, 46-47, 53-54
Hydrothermal Alteration, 186, 189, 195-196

I
Interlamellar Space, 137-138, 150
Ion-adsorption, 210-211, 218-219
Iron Oxide, 186, 191
Iron Powder, 113-116, 118, 121, 124-126

K
Kaolin, 154, 186-196
Kaolinite, 138, 155, 179, 181, 186-187, 189-193, 195-196, 208

M
Machine Vision, 78-80, 88
Mdds, 103
Mineral Compositions, 177-179, 183-184
Mixed-layer Structure, 137, 142
Mohr-coulomb Strength Criterion, 220

N
Na-montmorillonite, 137-138, 152-153
Nonlinear Strength Criterion, 220
Normalized Convolution, 88
Nuclear Magnetic Resonance, 210

O
Oil Field, 47, 53, 55
Omcs, 103

P
Physical-mechanical Properties, 197-198, 207
Polarization Optical Microscopy, 186, 191
Polymer-fiber, 128-130, 132-135
Polyurethane Soil Stabilizer, 11-12, 21, 23
Positioning System, 78-79, 84, 86, 88

R
Rare Earth Ore, 210-212, 219
Red-layers, 177-180, 183-185
Reinforced Sand, 11-13, 22-23, 44-45, 113, 121, 128-129, 131, 134-136
Rock Specimen, 64, 164-170, 173-175
Rock Strength, 46-47, 57, 60-61, 64, 70, 165, 184
Rock's Dynamic Characteristics, 164, 175

S
Sand Casting Process, 78, 88
Sand Production Rate, 46-47, 50, 52, 54
Sand-clay Liners, 25-27

Scrap Tire Crumbs, 1-3, 5, 8-9
Sem, 12, 21, 23, 128-129, 135, 177-181, 183, 186-187, 189, 191-192, 197-198, 204-205, 208
Shear Strength of Sand, 128, 135, 162
Shield Cutters, 177-178
Siltstone Specimen, 164
Soft Silt, 34-35, 39
Stress Load, 166-167
Supplementary Cementitious Material, 89, 98
Swell of Mixtures, 25, 30
Swelling Property, 197-198

T
Testing Equipment, 115, 164-166
Thermogravimetric Analysis, 186, 188
Thin Cushion, 34, 36
True Triaxial Compression, 56-59, 61-67, 69-70

U
Ucs, 47, 49-51, 53-54, 130, 133-135
Ultrahigh Strength Concrete, 89, 98
Unconfined Compressive, 47, 53-54, 72, 100-101, 111-112, 128, 130, 132-133, 135-136, 204, 207-208
Unified Hardening, 220, 227

X
Xrd, 137-140, 142-153, 177-179, 181-183, 186-187, 190, 192

www.ingramcontent.com/pod-product-compliance
Lightning Source LLC
LaVergne TN
LVHW081018210926
848485LV00004B/895